Algebraic Curves and Their Applications

CONTEMPORARY MATHEMATICS

724

Algebraic Curves and Their Applications

Lubjana Beshaj
Tony Shaska
Editors

AMS
AMERICAN MATHEMATICAL SOCIETY

2010 *Mathematics Subject Classification.* Primary 11G30, 11G50, 11G32, 11T71, 11T06, 14H37, 14H40, 14H45, 14H52, 14H55.

Library of Congress Cataloging-in-Publication Data

Names: Beshaj, Lubjana, 1986- editor. | Shaska, Tony, 1967- editor.
Title: Algebraic curves and their applications / Lubjana Beshaj, Tony Shaska Editors.
Description: Providence, Rhode Island : American Mathematical Society, [2019] | Series: Contemporary mathematics ; volume 724 | Includes bibliographical references.
Identifiers: LCCN 2018040058 | ISBN 9781470442477 (alk. paper)
Subjects: LCSH: Curves, Algebraic.
Classification: LCC QA565 .A447 2019 | DDC 516.3/52–dc23
LC record available at https://lccn.loc.gov/2018040058

DOI: https://doi.org/10.1090/conm/724

∞ The paper used in this book is acid-free and falls within the guidelines
established to ensure permanence and durability.
Visit the AMS home page at `https://www.ams.org/`

10 9 8 7 6 5 4 3 2 1   24 23 22 21 20 19

Contents

Preface

Algebraic curves are one of the most classical objects in mathematics. Their study led to the development of many branches of mathematics, such as invariant theory, Riemann surfaces, and algebraic geometry, with some of the biggest names in mathematics, such as Abel, Jacobi, Riemann, Weierstrass, and Noether, contributing to a beautiful theory and laying the foundation for what is known today as modern algebraic geometry.

In the last few decades, with the development of computational techniques and the significant growth in computer power, algebraic curves have found many applications, most notably in algebraic coding theory, cryptography, and dynamical systems. The goal of this book is to focus on such applications.

This volume consists of a variety of papers that combine classical questions on algebraic curves with more recent developments and applications. While some of the papers are long surveys covering such topics as automorphisms of algebraic curves, Abelian and Jacobian varieties, and cryptography, other papers are focused on new trends, such as tropical geometry, arithmetic statistics of the moduli space of curves, and applications of curves to differential equations. Below we briefly describe each contribution.

Bekker and Zarhin study families of elliptic curves with rational torsion points of even order. They construct explicitly versal families of elliptic curves with rational points of order 4, 6, 8, 10, and 12, respectively. They also construct versal families of elliptic curves in characteristic 2 that admit a rational point of order 4 or 8.

Beshaj and Guest study the weighted moduli spaces of sextics. This is a continuation of a series of papers on the distribution of points in the moduli space of genus two for which the field of moduli is not a field of definition. In this paper, they use the weighted moduli height to investigate the distribution of fine moduli points in the moduli space of genus two curves. They show that for any genus two curve with equation $y^2 = f(x)$, its weighted moduli height is $\mathfrak{h}(\mathfrak{p}) \leq 2^3\sqrt{3 \cdot 5 \cdot 7} \cdot H(f)$, where $H(f)$ is the minimal naive height of the curve. Based on the weighted moduli height $\mathfrak{h}$, they create a database of genus two curves defined over $\mathbb{Q}$ with small $\mathfrak{h}$ and show that for small such height ($\mathfrak{h} < 5$) about 30% of points are fine moduli points.

Ford studies a family of non-normal double planes associated to hyperelliptic curves. He generalizes the construction of a non-normal rational affine double plane $X \to \mathbb{A}^2_k$ together with a one-to-one homomorphism from the subgroup of torsion elements in the Picard group of C to the Brauer group of X to the situation where C is an arbitrary affine variety.

Otake and Shaska study the discriminant of certain quadrinomials. They give an explicit formula for the discriminant $\Delta_f(x)$ of the quadrinomials of the form $f(x) = x^n + t(x^2 + ax + b)$. The proof uses Bezoutians of polynomials. This extends the work of Selmer, Mori, and others.

Dokchitser, Dokchitser, Maistret, and Morgan explore three combinatorial descriptions of semistable types of hyperelliptic curves over local fields: dual graphs, their quotient trees by the hyperelliptic involution, and configurations of the roots of the defining equation ("cluster pictures"). They construct explicit combinatorial one-to-one correspondences between the three, which furthermore respect automorphisms and allow one to keep track of the monodromy pairing and the Tamagawa group of the Jacobian. They introduce a classification scheme and a naming convention for semistable types of hyperelliptic curves and types with a Frobenius action. This is the higher genus analogue of the distinction between good, split, and non-split multiplicative reduction for elliptic curves.

Stout studies formal deformations of algebraic spaces and generalizations of the motivic Igusa-zeta function. He generalizes the notion of the auto-Igusa-zeta function to formal deformations of algebraic spaces. By incorporating data from all algebraic transformations of local coordinates, this function can be viewed as a generalization of the traditional motivic Igusa-zeta function. Furthermore, the author introduces a new series, which he calls the canonical auto-Igusa-zeta function, whose coefficients are given by the quotient stacks formed from the coefficients of the auto-Igusa-zeta function modulo change of coordinates.

Mandili and Shaska study computing heights on weighted projective spaces. They extend the concept of height on projective spaces to that of weighted height on weighted projective spaces and show the basic properties of this height and how it can be used to study hyperelliptic/superelliptic curves. Some examples are provided from the weighted moduli space of binary sextics and octavics.

Beshaj and Polak study the moduli space of genus three hyperelliptic curves via the weighted projective space of binary octavics. This enables them to create a database of genus three hyperelliptic curves of weighted moduli height $\mathfrak{h} = 1$. Genus three hyperelliptic curves are some of the most interesting curves in cryptography due to the many interesting properties of their Jacobians. It is the first time that a database on such curves is compiled and organized.

Broughton, Shaska, and Wootton give a complete survey of automorphism groups of algebraic curves defined over a field F of characteristic $p \geq 0$. This is a classical problem with a rich and beautiful history with many notable mathematicians involved, such as Klein, Friecke, Wimann, Hurwitz, MacBeath, Accola, Singerman, and Stichtenoth. The paper gives a comprehensive review of the main results in zero characteristic and in positive characteristic. They display complete tables of groups for all characteristics for genus $g = 3$ and $g = 4$ and give algorithms for how this can be done for any genus. This paper will be valuable for many mathematicians who work in the area and want a single source for automorphisms of algebraic curves. Several open problems are suggested throughout the paper.

Cortázar and Costa study the algebraic classification of subgroups of hyperbolic planar crystallographic groups. A planar Euclidean or hyperbolic crystallographic group Δ is a subgroup of the group of isometries of the Euclidean plane $\mathbb{E}^2$, respectively, the hyperbolic plane $\mathbb{H}^2$, with compact orbit space. These groups are classified algebraically by a symbol called signature and an equivalence relation defined on the set of signatures. In 1990, A. H. M. Hoare gave an algorithm to obtain the signature of a finite index subgroup of a planar crystallographic group. Recently, authors Cortázar and Costa completed the algorithm of Hoare and implemented it in GAP.

Bujalance, Conder, Costa, and Izquierdo study regular dessins d'enfants with $4g$ automorphisms and Wiman's curve. They show that with a few exceptions, every regular dessin d'enfant with genus g having exactly $4g$ automorphisms is embedded in Wiman's curve of type II.

Hidalgo studies an explicit descent of real algebraic varieties. In this paper, he gives an explicit method on how to compute the equations for real algebraic varieties. The main tools used by the author are provided by the theory of fields of moduli and fields of definition. The method used is a consequence of Weil's descent theorem.

Previato focuses on curves on isomonodromy and isospectral deformations: Painlevé VI as a case study. Certain integrable dynamical problems can be described by isospectral deformations; under certain restrictions, the spectrum is an algebraic curve. On the other hand, certain isomonodromy deformations may have an associated algebraic curve, under certain restrictions on the monodromy group. This manuscript is a survey of the study on the relation between isomonodromy and isospectral deformations. The author shows a non-trivial result of the problem in Benes and Previato [J. Phys. A 43 (2010), 434006, 14 pp.] and, based on the result, shows the current state of the investigations of the relation.

Boston and Hao introduce a family of codes called *quasi-quadratic residuecodes* in algebraic coding theory. These codes are interesting because their weight distributions play an important role in a famous conjecture in coding theory known as *Goppa's conjecture*. Furthermore, the weight of their codewords has a close relation with the number of points on corresponding hyperelliptic curves. In the second part, they implement a heuristic model to estimate the limiting behavior of the number of points on related hyperelliptic curves over $\mathbb{F}_p$ as p goes to infinity. For primes $p < 50$, explicit calculations are given. These results provide evidence that Goppa's conjecture is dubious.

Frey and Shaska give the latest developments in the theory of curves, Jacobians, and cryptography. In the first part, they provide the necessary mathematical background on Abelian varieties, their torsion points, Honda-Tate theory, and Galois representations, with emphasis on Jacobian varieties and hyperelliptic Jacobians. In the second part, they focus on applications of Abelian varieties in cryptography and treat separately elliptic curve cryptography and genus two and three cryptography, including Diffie-Hellman key exchange, index calculus in Picard groups, isogenies of Jacobians via correspondences, and applications to discrete logarithms. Several open problems and new directions are suggested.

We hope this book will be helpful to all mathematicians working with algebraic curves, especially those who are crossing over from other areas of mathematics. Special thanks to all of the authors who contributed papers, the referees for all of their work and effort, and the AMS production staff for their help in preparing this book.

Lubjana Beshaj and Tony Shaska

Contemporary Mathematics
Volume **724**, 2019
https://doi.org/10.1090/conm/724/14582

Families of elliptic curves with rational torsion points of even order

Boris M. Bekker and Yuri G. Zarhin

ABSTRACT. We construct explicitly versal families of elliptic curves with rational points of order 4, 6, 8, 10, and 12, respectively. In the last section, we construct versal families of elliptic curves in characteristic 2 that admit a rational point of order 4 or 8.

1. Introduction

This paper is a follow up of [**1**]. Our aim is to give an explicit construction of versal families of elliptic curves with a torsion point of a certain (small) order. The problem of constructing such families goes back to Beppo Levi [**7**, **8**] and is closely related to certain modular curves of genus zero. However, our approach based on the explicit formulas for "halves" of points on elliptic curves [**1**, Sect. 2] is quite elementary and does not use the involved theory of modular curves. It seems to us that our families of elliptic curves are defined by formulas that are slightly easier to handle than the formulas that define Kubert's families [**5**].

Here are our main results.

THEOREM 1.1. *Let K be a field with* $\mathrm{char}(K) \neq 2$. *Let E be an elliptic curve over K.*

(1) *Let a be a nonzero element of K. The following conditions are equivalent.*
   (4i) *$E(K)$ contains exactly one point of order 2 and a point of order 4.*
   (4ii) *There exists a nonzero element $b \in K$ such that $a^2 + 4b$ is not a square in K and E is K-isomorphic to the elliptic curve*

$$\mathcal{E}_{a,b}^{(4)} : y^2 = \left(x^2 + (a^2 + 2b)x + b^2\right) x.$$

(2) *The following conditions are equivalent.*
   (8i) *$E(K)$ contains exactly one point of order 2 and a point of order 8.*
   (8ii) *There exists $t \in K \setminus \{0, \pm 1\}$ such that $2t^2 - 1$ is not a square in K and E is K-isomorphic to the elliptic curve*

$$\mathcal{E}_t^{(8)} : y^2 = \left(x^2 + 2\frac{t^4 + 2t^2 - 1}{(t^2 - 1)^2}x + 1\right) x.$$

2010 *Mathematics Subject Classification.* Primary 14H52, 11G05.

The second author was partially supported by Simons Foundation Collaboration grant #585711.

(3) *The following conditions are equivalent.*
 (6i) $E(K)$ *contains exactly one point of order* 2 *and a point of order* 6.
 (6ii) *There exists* $t \in K \setminus \{0, -4, 1/2\}$ *such that* $t^2 + 4t$ *is not a square in* K *and* E *is* K*-isomorphic to the elliptic curve*

$$\mathcal{E}_t^{(6)} : y^2 = (x+1)(x^2 + (t^2+2t)x + t^2).$$

(4) *The following conditions are equivalent.*
 (12i) $E(K)$ *contains exactly one point of order* 2 *and a point of order* 12.
 (12ii) *There exists* $t \in K \setminus \{0, \pm 1, \pm\sqrt{-1}\}$ *such that* $(t^2+1)(3t^2-1)$ *is not a square in* K, $3t^2+1 \neq 0$, *and* E *is* K*-isomorphic to the elliptic curve*

$$\mathcal{E}_t^{(12)} : y^2 = (x+1)\left(x^2 + \frac{8t^2(t^2+1)(t^4+4t^2-1)}{(t^2-1)^4}x + \frac{16t^4(t^2+1)^2}{(t^2-1)^4}\right).$$

(5) *The following conditions are equivalent.*
 (10i) $E(K)$ *contains exactly one point of order* 2 *and a point of order* 10.
 (10ii) *There exists* $t \in K \setminus \{0, \pm 1, (-1\pm\sqrt{5})/2, 2\pm\sqrt{5}\}$ *such that* $t(t^2+t-1)$ *is not a square in* K *and* E *is* K*-isomorphic to the elliptic curve*

$$\mathcal{E}_t^{(10)} : y^2 = (x+1)\left(x^2 + \frac{8t^2(t^3+t^2-t+1)}{(t-1)^2(t+1)^4}x + \frac{16t^4}{(t-1)^2(t+1)^4}\right).$$

REMARKS 1.2. (1) In each of (8ii,6ii,12ii,10ii) of Theorem 1.1 we assume that $t \in K$ does *not* take certain finitely many exceptional values and the value of a certain polynomial at t is *not* a square in K. These restrictions on t mean that the corresponding cubic polynomial in x is defined, has *no repeated roots* and has exactly one root in K, i.e., $\mathcal{E}_t^{(n)}$ is an *elliptic curve* over K that has exactly one K-point of order 2 for all such t with $n = 8$, 6, 12, 10 respectively. It follows from Theorem 1.1 that $\mathcal{E}_t^{(n)}(K)$ contains a torsion point of order n.

(2) The conditions $a \neq 0$, $b \neq 0$ and $a^2 + 4b$ is not a square in K mean that the cubic polynomial in x of Theorem 1.1(4ii) has *no repeated roots* and has exactly one root in K. This means that $\mathcal{E}_{a,b}^{(4)}$ is an *elliptic curve* over K that has exactly one K-point of order 2 for all such $a, b \in K$. $\mathcal{E}_{a,b}^{(4)}(K)$ contains exactly two points of order 4, namely, $(-b, ab)$ and $(-b, -ab)$. On the other hand, dividing both sides of the equation for $\mathcal{E}_{a,b}^{(4)}$ by a^6 and introducing

$$\tilde{x} = \frac{x}{a^2},\ \tilde{y} = \frac{y}{a^3},\ \tilde{b} = \frac{b}{a^2},$$

we obtain that $\mathcal{E}_{a,b}^{(4)}$ is K-isomorphic to

$$\mathcal{E}_{1,\tilde{b}}^{(4)} : \tilde{y}^2 = \left(\tilde{x}^2 + (1+2\tilde{b})\tilde{x} + \tilde{b}^2\right)\tilde{x}.$$

(3) $\mathcal{E}_t^{(8)}(K)$ contains exactly two points of order 4, namely,

$$\left(1, \frac{2t}{1-t^2}\right),\ \left(1, -\frac{2t}{1-t^2}\right)$$

and exactly four points of order 8, namely,

$$\left(\frac{1+t}{1-t}, \frac{-2t}{(1-t)^2}\right),\ \left(\frac{1-t}{1+t}, \frac{2t}{(1+t)^2}\right)$$

and their negatives

$$\left(\frac{1+t}{1-t}, \frac{2t}{(1-t)^2}\right), \left(\frac{1-t}{1+t}, \frac{-2t}{(1+t)^2}\right).$$

(4) $\mathcal{E}_t^{(6)}(K)$ contains the points $(0,t)$ and $(0,-t)$ of order 3 and the points $(-2t, t-2t^2)$ and $(-2t, 2t^2-t)$ of order 6. If $\mathrm{char}(K)=3$, then there are no other points of order 3 or 6 in $\mathcal{E}_t^{(6)}(K)$.

(5) $\mathcal{E}_t^{(12)}(K)$ contains the points

$$\left(0, \frac{4t^2(t^2+1)}{(t^2-1)^2}\right), \left(0, -\frac{4t^2(t^2+1)}{(t^2-1)^2}\right)$$

of order 3, and exactly two points of order 4, namely,

$$\left(-\frac{3t^2+1}{t^2-1}, -\frac{8t^3(3t^2+1)}{(t^2-1)^3}\right), \left(-\frac{3t^2+1}{t^2-1}, \frac{8t^3(3t^2+1)}{(t^2-1)^3}\right).$$

(6) $\mathcal{E}_t^{(10)}(K)$ contains the point $\left(0, 4t^2/(t-1)(t+1)^2\right)$ of order 5 and exactly one point $(-1,0)$ of order 2.

The paper is organized as follows. First three sections deal with division by 2 in $E(K)$ under various assumptions about the existence of K-points of order 2 on the elliptic curve E. Our goal is to obtain explicit formulas that will be used in the next five sections containing a construction of versal families of elliptic curves with rational points of order 4,8, 6, 12, and 10, respectively (and with exactly one rational point of order 2). The last section deals with versal families of (ordinary) elliptic curves in characteristic 2 that admit a rational point of order 4 or 8.

2. Review of [1, section 2]

Let K be a field with $\mathrm{char}(K) \ne 2$ and $\bar{K}$ an algebraic closure of K. Let $f(x) \in K[x]$ be a monic cubic polynomial without repeated roots and $\{\alpha_1, \alpha_2, \alpha_3\} \subset \bar{K}$ the set of roots of $f(x)$. Clearly

$$f(x) = (x-\alpha_1)(x-\alpha_2)(x-\alpha_3) \in \bar{K}[x].$$

Let

$$E = E_f : y^2 = f(x)$$

be an elliptic curve over K and ∞ its only infinite point (the zero of the group law). We have

$$E(K) = \{(x_0, y_0) \in K^2 \mid y_0^2 = f(x_0)\} \amalg \{\infty\}$$
$$\subset \{(x_0, y_0) \in \bar{K}^2 \mid y_0^2 = f(x_0)\} \amalg \{\infty\} = E(\bar{K}).$$

The points

$$W_i = (\alpha_i, 0) \in E(\bar{K}) \ (1 \le i \le 3)$$

are the only points of order 2 on E.

REMARK 2.1. A root α_i lies in K if and only if $W_i = (\alpha_i, 0) \in E(K)$. This implies that $E(K)$ contains exactly one point of order 2 if and only if $f(x)$ has exactly one root in K.

The following assertion is pretty well-known [**2**, **3**] (see also [**1**]).

THEOREM 2.2. *Suppose that all the roots α_i of $f(x)$ lie in K, i.e., all three points of order 2 on E lie in $E(K)$. Let $P = (x_0, y_0) \in E(K)$. Then $P \in 2E(K)$ if and only if all $x_0 - \alpha_i$ are squares in K. In addition, each point $Q \in E(\bar{K})$ with $2Q = P$ actually lies in $E(K)$.*

In what follows we discuss the divisibility by 2 of points in $E(K)$ when not necessarily all roots of $f(x)$ lie in K.

Let $P = (x_0, y_0) \in E(\bar{K})$. There are precisely four points $Q \in E(\bar{K})$ such that $2Q = P$. The following explicit construction of all *halves* Q's was described in [**1**, Sect. 2]. Let us choose square roots

$$r_i = \sqrt{x_0 - \alpha_i} \in \bar{K} \ (1 \le i \le 3). \tag{1}$$

(There are exactly eight choices of such triples (r_1, r_2, r_3).) We have

$$r_1 r_2 r_3 = \sqrt{(x_0 - \alpha_1)(x_0 - \alpha_2)(x_0 - \alpha_3)} = \pm y_0. \tag{2}$$

Since all α_i's are *distinct*, $r_i \ne \pm r_j$ if $i \ne j$. This implies that

$$r_i \pm r_j \ne 0 \tag{3}$$

for all $i \ne j$. Let us consider only the triples that satisfy

$$r_1 r_2 r_3 = -y_0. \tag{4}$$

(There are exactly four choices of such triples (r_1, r_2, r_3).) Let us put

$$s_1 = s_1(r_1, r_2, r_3) = r_1 + r_2 + r_3, \ s_2 = s_2(r_1, r_2, r_3) = r_1 r_2 + r_2 r_3 + r_3 r_1. \tag{5}$$

Then the point

$$Q = Q_{r_1, r_2, r_3} = (x_1, y_1) := (x_0 + s_2, -y_0 - s_1 s_2) = \tag{6}$$

$$(x_0 + r_1 r_2 + r_2 r_3 + r_3 r_1, -(r_1 + r_2)(r_2 + r_3)(r_3 + r_1))$$

lies in $E(\bar{K})$ and satisfies

$$2Q = P \in E(\bar{K}).$$

In addition,

$$l := -s_1 = -(r_1 + r_2 + r_3) \tag{7}$$

is the slope of the tangent line $\mathcal{L}_Q$ to E at Q [**1**, Proof of Th. 2.1]. More precisely, $\mathcal{L}_Q$ passes through the point $-P = (x_0, -y_0)$ and is defined by the equation

$$y = l(x - x_0) - y_0.$$

Distinct choices of triples (r_1, r_2, r_3) give rise to distinct halves Q_{r_1, r_2, r_3}'s. More precisely [**1**, Th. 2.3],

(8)

$$Q_{r_1, -r_2, -r_3} = Q_{r_1, r_2, r_3} + W_1, Q_{-r_1, r_2, -r_3} = Q_{r_1, r_2, r_3} + W_2, Q_{-r_1, -r_2, r_3} = Q_{r_1, r_2, r_3} + W_3.$$

(That is how we get all four $Q \in E(\bar{K})$ with $2Q = P$.) Conversely, if we are given $Q = (x_1, y_1) \in E(\bar{K})$, then $Q = Q_{r_1, r_2, r_3}$ with

$$r_i = -\frac{y_1}{2} \cdot \left(-\frac{1}{x_1 - \alpha_i} + \frac{1}{x_1 - \alpha_j} + \frac{1}{x_1 - \alpha_k} \right) \tag{9}$$

for each permutation i, j, k of $1, 2, 3$.

The formulas discussed above almost immediately lead to the following statement.

THEOREM 2.3. *Suppose that* $P = (x_0, y_0) \in E(K)$. *Suppose that the square roots* $r_i = \sqrt{x_0 - \alpha_i} \in \bar{K}$ $(i = 1, 2, 3)$ *satisfy*

$$r_1 r_2 r_3 = -y_0; \ s_1 = r_1 + r_2 + r_3 \in K, \ s_2 = r_1 r_2 + r_2 r_3 + r_3 r_1 \in K. \tag{10}$$

Then the point $Q_{r_1,r_2,r_3} \in E(\bar{K})$ *defined by formula* (6) *enjoys the following properties:*

$$Q_{r_1,r_2,r_3} \in E(K), \ 2Q_{r_1,r_2,r_3} = P.$$

Conversely, suppose that $Q \in E(K)$ *satisfies* $2Q = P$. *Then there exists precisely one triple* $\{r_1, r_2, r_3\}$ *of square roots* $r_i = \sqrt{x_0 - \alpha_i} \in \bar{K}$ $(i = 1, 2, 3)$ *that satisfy* (10) *and such that* $Q = Q_{r_1,r_2,r_3}$.

PROOF. Suppose that the square roots r_i $(i = 1, 2, 3)$ satisfy (10). This implies that both s_1 and s_2 defined in (5) lie in K. The point $Q_{r_1,r_2,r_3} = (x_1, y_1)$ defined by formula (6) lies in $E(\bar{K})$ and $2Q = P$. Since, $x_0, y_0, s_1, s_2 \in K$, formulas (6) imply that $x_1, y_1 \in K$, i.e., $Q \in E(K)$.

Conversely, suppose that $P = 2Q$ with $Q = (x_1, y_1) \in E(K)$. Then

$$Q \in E(K) \subset E(\bar{K}); \ x_1, y_1 \in K.$$

There exists exactly one triple $\{r_1, r_2, r_3\}$ of square roots $r_i = \sqrt{x_0 - \alpha_i} \in \bar{K}$ satisfying (1) such that $r_1 r_2 r_3 = -y_0$ and $Q = Q_{r_1,r_2,r_3}$. In light of (6), $s_2 = x_1 - x_0$. Since both $x_0, x_1 \in K$, we obtain that $s_2 \in K$. Since $Q \in E(K)$, the slope l of the tangent line to E at Q lies in K. In light of (7), $l = -s_1$ and therefore $s_1 \in K$. To summarize: $r_1 r_2 r_3 = -y_0$, $r_1 + r_2 + r_3 = s_1 \in K$, and $s_2 \in K$. In other words, the triple $\{r_1, r_2, r_3\}$ satisfies conditions (10). □

COROLLARY 2.4. *Suppose that* $P = (x_0, y_0) \in E(K)$. *Then the following conditions are equivalent.*

(i) $P \in 2E(K)$, *i.e., there exists* $Q \in E(K)$ *such that* $2Q = P$.

(ii) *One may choose square roots* $r_i = \sqrt{x_0 - \alpha_i} \in \bar{K}$ *in such a way that*

$$s_1 = r_1 + r_2 + r_3 \in K, \ s_2 = r_1 r_2 + r_2 r_3 + r_3 r_1 \in K.$$

(iii) *One may choose square roots* $r_i = \sqrt{x_0 - \alpha_i} \in \bar{K}$ *in such a way that*

$$r_1 r_2 r_3 = -y_0; \ s_1 = r_1 + r_2 + r_3 \in K, \ s_2 = r_1 r_2 + r_2 r_3 + r_3 r_1 \in K.$$

PROOF. By Theorem 2.3, one has only to check that (ii) implies (iii). Indeed, suppose we are given a triple (r_1, r_2, r_3) of square roots $r_i = \sqrt{x_0 - \alpha_i}$ such that

$$s_1 = r_1 + r_2 + r_3 \in K, s_2 = r_1 r_2 + r_2 r_3 + r_3 r_1 \in K.$$

By equality (2), $r_1 r_2 r_3 = \pm y_0$. Replacing if necessary (r_1, r_2, r_3) by $(-r_1, -r_2, -r_3)$ (and s_1 by $-s_1$), we may and will assume that $r_1 r_2 r_3 = -y_0$ and still $s_1, s_2 \in K$. This proves that (ii) implies (is actually equivalent to) (iii). □

3. Division by 2

In this section we discuss division by 2 in $E(K)$ when $E(K)$ contains exactly one point of order 2.

Let K be a field with $\mathrm{char}(K) \neq 2$. Let α be an element of K,

$$g(x) = x^2 + px + q \in K[x]$$

a monic irreducible quadratic polynomial (in particular, α is *not* a root of $g(x)$), and $K_g = K[x]/g(x)K[x]$ the corresponding quadratic field extension of K. We write $\mathbf{x}$ for the image of x in K_g with respect to the natural surjective homomorphism

$$K[x] \twoheadrightarrow K[x]/g(x)K[x] = K_g.$$

Clearly, $\mathbf{x}^2 + p\mathbf{x} + q = 0$. Let $\iota : K_g \to K_g$ be the only nontrivial K-linear automorphism (involution) of K_g. We write

$$\mathrm{Tr} : K_g \to K,\ z \mapsto z + \iota(z),\ \mathrm{Norm} : K_g \to K,\ z \mapsto z \cdot \iota(z)$$

for the trace and norm maps attached to K_g/K. We have

$$z^2 - \mathrm{Tr}(z) + \mathrm{Norm}(z) = (z - z)(z - \iota(z)) = 0\ \forall z \in K_g.$$

For each $a \in K$ we have

$$g(a) = \mathrm{Norm}(a - \mathbf{x}).$$

We also have

$$\iota(\mathbf{x}) = -p - \mathbf{x},\ \mathrm{Norm}(\mathbf{x}) = q,\ \mathrm{Tr}(\mathbf{x}) = -p.$$

In addition, we have the following equality in $K_g[x]$:

$$g(x) = x^2 + px + q = (x - \mathbf{x})(x - \iota\mathbf{x}) = (x - \mathbf{x})(x + p + \mathbf{x}).$$

Let us consider the elliptic curve

$$E = \mathcal{E}_{\alpha,p,q} : y^2 = f(x) := (x - \alpha)g(x). \tag{11}$$

Clearly, $E(K)$ contains exactly one point of order 2, namely, $(\alpha, 0)$. On the other hand, $E(K_g)$ contains two remaining points of order 2, namely, $(\mathbf{x}, 0)$ and $(\iota\mathbf{x}, 0) = (-p - \mathbf{x}, 0)$.

We start with the following result that was inspired by a paper of Schaefer [**6**].

THEOREM 3.1. *A point $P = (x_0, y_0) \in E(K)$ lies in $2E(K)$ if and only if $x_0 - \mathbf{x}$ is a square in K_g. If this is the case, then $x_0 - \alpha$ is a square in K.*

The proof of Theorem3.1 is coming later in the text.

REMARK 3.2. (i) If $x_0 \in K$, then $x_0 - \mathbf{x} \neq 0$, because $\mathbf{x} \notin K$.

(ii) Suppose that $\mathbf{r} \in K_g$ and $r \in K$ satisfy

$$r^2 = x_0 - \alpha,\ \mathbf{r}^2 = x_0 - \mathbf{x}.$$

Then

$$(r \cdot \mathrm{Norm}(\mathbf{r}))^2 = r^2 \cdot \mathrm{Norm}(x_0 - \mathbf{x}) = (x_0 - \alpha)g(x_0) = y_0^2.$$

This means that

$$r \cdot \mathrm{Norm}(\mathbf{r}) = \pm y_0. \tag{12}$$

In addition,

$$\begin{aligned}(\mathrm{Tr}(\mathbf{r}))^2 = (\mathbf{r} + \iota(\mathbf{r}))^2 = \mathbf{r}^2 + 2\mathbf{r}\iota(\mathbf{r}) + \iota(\mathbf{r})^2 = \left(\mathbf{r}^2 + \iota(\mathbf{r}^2)\right) + 2\mathrm{Norm}(\mathbf{r}) \\ = \mathrm{Tr}(x_0 - \mathbf{x}) + 2\mathrm{Norm}(\mathbf{r}) = 2x_0 + p + 2\mathrm{Norm}(\mathbf{r}),\end{aligned}$$

i.e.,

$$(\mathrm{Tr}(\mathbf{r}))^2 = 2x_0 + p + 2\mathrm{Norm}(\mathbf{r}). \tag{13}$$

Notice also that

$$\operatorname{Norm}(r+\mathbf{r}) = (r+\mathbf{r})\iota(r+\mathbf{r}) = (r+\mathbf{r})(r+\iota(\mathbf{r}))$$
$$= r^2 + r(\mathbf{r}+\iota(\mathbf{r})) + \mathbf{r}\iota(\mathbf{r}) = (x_0-\alpha) + r\operatorname{Tr}(\mathbf{r}) + \operatorname{Norm}(\mathbf{r}),$$

i.e.,

$$\operatorname{Norm}(r+\mathbf{r}) = (x_0-\alpha) + r\operatorname{Tr}(\mathbf{r}) + \operatorname{Norm}(\mathbf{r}). \tag{14}$$

The next assertion describes explicitly both $\frac{1}{2}P$'s in $E(K)$.

THEOREM 3.3. *Suppose that $P=(x_0,y_0)\in E(K)$. Suppose that $r\in K$ and $\mathbf{r}\in K_g$ satisfy*

$$r^2 = x_0-\alpha,\ \mathbf{r}^2 = x_0-\mathbf{x},\ r\operatorname{Norm}(\mathbf{r}) = -y_0.$$

Then:

(i) *the points*

$$Q_{r,\mathbf{r}} := Q_{\mathbf{r},\iota(\mathbf{r}),r}(x_1,y_1)$$
$$:= (x_0+\operatorname{Tr}(\mathbf{r})r+\operatorname{Norm}(\mathbf{r}), -y_0-(r+\operatorname{Tr}(\mathbf{r}))\left(\operatorname{Tr}(\mathbf{r})r+\operatorname{Norm}(\mathbf{r})\right))$$
$$= (\alpha+\operatorname{Norm}(r+\mathbf{r}), -\operatorname{Tr}(\mathbf{r})\operatorname{Norm}(r+\mathbf{r}))$$

and

$$Q_{r,-\mathbf{r}} := Q_{-\mathbf{r},-\iota(\mathbf{r}),r} = (x_{-1},y_{-1})$$
$$:= (x_0-\operatorname{Tr}(\mathbf{r})r+\operatorname{Norm}(\mathbf{r}), -y_0-(r-\operatorname{Tr}(\mathbf{r}))\left(-\operatorname{Tr}(\mathbf{r})r+\operatorname{Norm}(\mathbf{r})\right))$$
$$= (\alpha+\operatorname{Norm}(r-\mathbf{r}), \operatorname{Tr}(\mathbf{r})\operatorname{Norm}(r-\mathbf{r}))$$

are distinct points of $E(K)$, which satisfy

$$2Q_{r,\mathbf{r}} = 2Q_{r,-\mathbf{r}} = P\in E(K).$$

(ii) *The tangent lines to E at $Q_{r,\mathbf{r}}$ and $Q_{r,-\mathbf{r}}$ are defined by the equations*

$$\mathcal{L}_{r,\mathbf{r}}: y = l_{r,\mathbf{r}}(x-x_0)-y_0 \quad \text{and } \mathcal{L}_{r,-\mathbf{r}}: y = l_{r,-\mathbf{r}}(x-x_0)-y_0,$$

respectively, where

$$l_{r,\mathbf{r}} = -\left(r+\operatorname{Tr}(\mathbf{r})\right),\ l_{r,-\mathbf{r}} = -\left(r-\operatorname{Tr}(\mathbf{r})\right).$$

REMARK 3.4. Suppose we are given $\mathbf{r}\in K_g$ with $\mathbf{r}^2 = x_0-\mathbf{x}$. It follows from Theorem 3.1 and (12) that there exists precisely one $r\in K$ such that $r^2 = x_0-\alpha$ and $r\operatorname{Norm}(\mathbf{r}) = -y_0$.

PROOF OF THEOREMS 3.1 AND 3.3. We start with the proof of Theorem 3.1. Suppose that $P=(x_0,y_0)\in 2E(K)$. Since $E(K_g)$ contains all three points of order 2 on E and $(\mathbf{x},0)$ is one of them, it follows from Theorem 2.2 that $x_0-\mathbf{x}$ is a square in K_g.

Conversely, suppose that $x_0-\mathbf{x}$ is a square in K_g. This means that there exists $\mathbf{r}\in K_g$ such that

$$x_0-\mathbf{x} = \mathbf{r}^2 = (-\mathbf{r})^2.$$

By Remark 3.2(i), $\mathbf{r}$ does not lie in K. In particular, $\mathbf{r}\neq 0$. We have

$$x_0-\iota(\mathbf{x}) = \iota(x_0-\mathbf{x}) = \iota(\mathbf{r}^2) = \iota(\mathbf{r})^2.$$

This implies that

$$g(x_0) = (x_0 - \mathbf{x})(x_0 - \iota\mathbf{x}) = (\mathbf{r}\iota(\mathbf{r}))^2 = \mathrm{Norm}(\mathbf{r})^2$$

is a square in K, because $\mathrm{Norm}(\mathbf{r}) \in K$. Since $(x_0, y_0) \in E(K)$,

$$y_0^2 = f(x_0) = (x_0 - \alpha)g(x_0) = (x_0 - \alpha)\mathrm{Norm}(\mathbf{r})^2$$

is a square in K. This implies that $x_0 - \alpha$ is also a square in K, because the norm of nonzero $\mathbf{r}$ does *not* vanish. It follows that $x_0 - \alpha$ is a square in K_g as well. By Remark 3.2(ii), we may choose

$$r = \sqrt{x_0 - \alpha} \in K \subset K_g$$

in such a way that

$$r\mathrm{Norm}(\mathbf{r}) = r\mathrm{Norm}(-\mathbf{r}) = r \cdot \mathbf{r} \cdot \iota(\mathbf{r}) = -y_0.$$

Since all three $x_0 - \mathbf{x}$, $x_0 - \iota(\mathbf{x})$, and $x_0 - \alpha$ are squares in K_g, Theorem 2.2 implies that

$$P = (x_0, y_0) \in 2E(K_g);$$

by formula (6) combined with the condition $r\mathrm{Norm}(\mathbf{r}) = -y_0$ and equality (14), the two different choices $(\mathbf{r}, \iota(\mathbf{r}), r)$ and $(-\mathbf{r}, -\iota(\mathbf{r}), r)$ of the corresponding square roots give us that

$$\begin{aligned} Q_{r,\mathbf{r}} &= (x_0 + (\mathbf{r} + \iota(\mathbf{r}))r + \mathbf{r}\iota(\mathbf{r}), -y_0 - (r + \mathbf{r} + \iota(\mathbf{r}))\,((\mathbf{r} + \iota(\mathbf{r}))r + \mathbf{r}\iota(\mathbf{r}))) \\ &= (x_0 + \mathrm{Tr}(\mathbf{r})r + \mathrm{Norm}(\mathbf{r}), -y_0 - (r + \mathrm{Tr}(\mathbf{r}))\,(\mathrm{Tr}(\mathbf{r})r + \mathrm{Norm}(\mathbf{r}))) \\ &= (\alpha + \mathrm{Norm}(r + \mathbf{r}), -\mathrm{Tr}(\mathbf{r})\mathrm{Norm}(r + \mathbf{r})) \end{aligned}$$

and

$$\begin{aligned} Q_{r,-\mathbf{r}} &= (x_0 - (\mathbf{r} + \iota(\mathbf{r}))r + \mathbf{r}\iota(\mathbf{r}), -y_0 - (r - \mathbf{r} - \iota(\mathbf{r}))\,((-\mathbf{r} - \iota(\mathbf{r}))r + \mathbf{r}\iota(\mathbf{r}))) \\ &= (x_0 - \mathrm{Tr}(\mathbf{r})r + \mathrm{Norm}(\mathbf{r}), -y_0 - (r - \mathrm{Tr}(\mathbf{r}))\,(-\mathrm{Tr}(\mathbf{r})r + \mathrm{Norm}(\mathbf{r}))) \\ &= (\alpha + \mathrm{Norm}(-r + \mathbf{r}), \mathrm{Tr}(\mathbf{r})\mathrm{Norm}(r - \mathbf{r})) \end{aligned}$$

are distinct points of $E(K_g)$ such that

$$2Q_{r,\mathbf{r}} = 2Q_{r,-\mathbf{r}} = P \in E(K).$$

However, it is clear that they both lie in $E(K)$, since all three r, $\mathrm{Tr}(\mathbf{r})$, and $\mathrm{Norm}(\mathbf{r})$ lie in K. This ends the proof of Theorem 3.1. On the other hand, $Q_{r,\mathbf{r}}$ and $Q_{r,-\mathbf{r}}$ are exactly the points that appear in the statement of Theorem 3.3; this proves Theorem 3.3(i) as well. Theorem 3.3(ii) follows from the explicit formula (7) for the slope l of the tangent line at $\frac{1}{2}P$. □

REMARK 3.5. If $P = (x_0, y_0) \in 2E(K)$, then $g(x_0)$ is a square in K. Indeed, we have

$$y_0^2 = f(x_0) = g(x_0)(x_0 - \alpha), \; g(x_0) = \mathrm{Norm}(x_0 - \mathbf{x}).$$

Since $x_0 - \mathbf{x}$ is a square in K_g, its norm $g(x_0)$ is a square in K.

EXAMPLE 3.6. Suppose $g(x) = x^2 + px + q \in K[x]$ is an irreducible quadratic polynomial over K and $-\mathbf{x}$ is a square in K_g, i.e., there exist $u, v \in K$ such that

$$-\mathbf{x} = (u\mathbf{x} + v)^2$$

in K_g. This means that the polynomial $(ux + v)^2 + x$ is divisible by $x^2 + px + q$ in $K[x]$, i.e., the polynomial

$$(2uv + 1 - u^2p)x + (v^2 - u^2q) = (ux + v)^2 + x - u^2(x^2 + px + q)$$

of degree < 2 is divisible by quadratic x^2+px+q in $K[x]$ and therefore is identically zero, i.e.,

$$2uv+1=u^2\cdot p,\ v^2=u^2q.$$

In addition, $u\neq 0$, since $-\mathbf{x}\in K_g\setminus K$. This implies that there exists $b\in K$ such that

$$q=b^2,\ v=ub.$$

It follows that

$$u^2\cdot p=2u^2b+1,\ (p-2b)u^2=1.$$

This means that

$$p=2b+\frac{1}{u^2}, q=b^2,\ v=ub.$$

Let us put

$$a:=\frac{1}{u}\in K\setminus\{0\}.$$

Then we obtain that

$$p=a^2+2b,\ q=b^2,$$

$$g(x)=x^2+(a^2+2b)x+b^2;\ u=\frac{1}{a},\ v=\frac{b}{a},$$

and

$$\left(\frac{1}{a}\mathbf{x}+\frac{b}{a}\right)^2=-\mathbf{x}$$

in K_g. Since $g(x)$ is irreducible over K, its discriminant

$$(a^2+2b)^2-4b^2=a^4+4a^2b=a^2(a^2+4b)$$

is *not* a square in K, i.e, a^2+4b is *not* a square in K. In addition, the constant term b^2 of $g(x)$ does *not* vanish, i.e., $b\neq 0$. Now let us consider the K-point $W=(0,0)$ of the elliptic curve

$$\mathcal{E}_{a,b}^{(4)}=\mathcal{E}_{0,a^2+2b,b^2}:y^2=\left(x^2+(a^2+2b)x+b^2\right)x.$$

Clearly, W has order 2. It follows from Theorem 3.1 that W is divisible by 2 in $E(K)$, i.e., there are two points Q_+ and Q_- of order 4 in $\mathcal{E}_{a,b}^{(4)}(K)$ such that

$$2Q_+=2Q_-=W.$$

In order to find Q_+ and Q_- explicitly, let us apply Theorem 3.3. We have

$$r=0,\ \mathbf{r}=\frac{1}{a}(\mathbf{x}+b);$$

$$\mathrm{Tr}(\mathbf{r})=\frac{1}{a}(-p)+2\cdot\frac{b}{a}=-\frac{(a^2+2b)}{a}+\frac{2b}{a}=-a,$$

$$\mathrm{Norm}(\mathbf{r})=\frac{\mathrm{Norm}(\mathbf{x}+b)}{a^2}=\frac{g(-b)}{a^2}=-b.$$

Now the explicit formulas of Theorem 3.3 give us

$$Q_+=(-b,ab),\ Q_-=(-b,-ab).$$

4. Another criterion of divisibility by 2

We keep the notation of Section 3. However, we drop the assumption that $g(x) = x^2 + px + q \in K[x]$ is irreducible over K and assume only that it has *no* multiple roots. Let α_1, α_2 be (distinct) roots of $g(x)$ in $\bar{K}$. We have

$$\alpha_1 + \alpha_2 = -p, \; \alpha_1\alpha_2 = q. \tag{15}$$

THEOREM 4.1. *Let α be an element of K that is not a root of $g(x)$. Let us consider the elliptic curve*

$$E : y^2 = (x - \alpha)g(x)$$

over K. Let

$$W_3 = (\alpha_3, 0) = (\alpha, 0) \in E(K)$$

be a point of order 2.

Let $P = (x_0, y_0)$ be a K-point of E. Then P is divisible by 2 in $E(K)$ if and only if the following two conditions hold.

(i) *The difference $x_0 - \alpha$ is a square in K.*
(ii) *There exist square roots $r = \sqrt{x_0 - \alpha} \in K$ and*

$$r_i = \sqrt{x_0 - \alpha_i} \in \bar{K} \; (i = 1, 2)$$

such that

$$r_1 r_2 r = -y_0; \; r_1 + r_2 \in K. \tag{16}$$

In addition, if square roots r_1, r_2, r satisfy conditions (ii), then

$$Q_+ = Q_{r_1,r_2,r} = (x_0 + (r_1 + r_2)r + r_1 r_2, -y_0 - ((r_1 + r_2) + r)((r_1 + r_2)r + r_1 r_2)$$

and

$$Q_- = Q_{-r_1,-r_2,r} = (x_0 - (r_1 + r_2)r + r_1 r_2, -y_0 - (-(r_1 + r_2) + r)(-(r_1 + r_2)r + r_1 r_2)$$

are distinct points of $E(K)$, which satisfy

$$2Q_+ = 2Q_- = P, \; Q_- = Q_+ + W_3.$$

Conversely, if $Q \in E(K)$ satisfies $2Q = P$, then there exist square roots r, r_1, r_2 that satisfy conditions (ii) and $Q = Q_{r_1,r_2,r}$.

REMARK 4.2. If $x_0 \neq \alpha$, i.e., $P \neq (\alpha, 0)$, then $r = \sqrt{x_0 - \alpha} \neq 0$ and

$$r_1 r_2 = -\frac{y_0}{r}.$$

It follows from Theorem 4.1 that

$$P = (x_0, y_0) \in E(K)$$

is divisible by 2 in $E(K)$ if and only if there exist square roots

$$r = \sqrt{x_0 - \alpha} \in K, \; r_i = \sqrt{x_0 - \alpha_i} \in \bar{K} \; (i = 1, 2)$$

such that

$$r_1 r_2 = -\frac{y_0}{r}, \quad r_1 + r_2 \in K. \tag{17}$$

PROOF OF THEOREM 4.1. Let us put

$$\alpha_3 = \alpha, \ W_3 = (\alpha_3, 0) \in E(K).$$

Suppose that conditions (i)-(ii) hold. Let us put

$$r_3 = r \in K.$$

We have

$$(r_1 + r_2)^2 = r_1^2 + 2r_1r_2 + r_2^2 = (x_0 - \alpha_1) + 2r_1r_2 + (x_0 - \alpha_2)$$
$$= 2x_0 - (\alpha_1 + \alpha_2) + 2r_1r_2 = (2x_0 + p) + 2r_1r_2.$$

Since all three $r_1 + r_2, x_0$, and p lie in K, we obtain that

$$r_1r_2 \in K.$$

Notice that

(18) $s_1 = r_1 + r_2 + r_3 = (r_1 + r_2) + r, \ s_2 = r_1r_2 + r_2r_3 + r_3r_1 = (r_1 + r_2)r + r_1r_2.$

Clearly, both s_1 and s_2 lie in K. Let us put

$$x_1 = x_0 + r_1r_2 + r_2r_3 + r_3r_1 = x_0 + s_2, \tag{19}$$

$$y_1 = -y_0 - (r_1 + r_2 + r_3)(r_1r_2 + r_2r_3 + r_3r_1) = -y_0 - s_1s_2.$$

Since all x_0, y_0, s_1, s_2 lie in K, both x_1 and y_1 also lie in K. By formula (6), the point $Q_+ = Q_{r_1,r_2,r_3} = (x_1, y_1)$ lies in $E(\bar{K})$ and satisfies $2Q_+ = P$. Since x_1 and y_1 lie in K, the point Q_+ actually lies in $E(K)$, and therefore P is divisible by 2 in $E(K)$. Similarly, the triple of square roots $(-r_1, -r_2, r_3)$ satisfies (1) and (4); in addition,

$$(-r_1) + (-r_2) = -(r_1 + r_2) \in K, (-r_1)(-r_2) = r_1r_2 \in K.$$

This implies that

$$Q_- = Q_{-r_1,-r_2,r_3} = Q_{r_1,r_2,r_3} + W_3 = Q_+ + W_3$$

also lies in $E(K)$ and satisfies $2Q_- = P$.

Conversely, suppose that $P = 2Q$ with $Q = (x_1, y_1) \in E(K)$. We claim that $x_0 - \alpha_3 = x_0 - \alpha$ is a square in K. Indeed, if $g(x)$ is irreducible over K, then our claim follows from Theorem 3.1. If $g(x)$ is reducible, i.e., α_1, α_2 and α_3 lie in K, then our claim follows from Theorem 2.2.

It follows from Theorem 2.3 that there exist square roots $r_i = \sqrt{x_0 - \alpha_i}$ such that

$$r_1r_2r_3 = -y_0, r_1 + r_2 + r_3 \in K, r_1r_2 + r_2r_1 + r_3r_1 \in K$$

and

$$Q = (x_1, y_1) = Q_{r_1,r_2,r_3}.$$

Notice that

$$r := r_3 = \sqrt{x_0 - \alpha_3} = \sqrt{x_0 - \alpha} \in K$$

and therefore

$$r_1r_2r = r_1r_2r_3 = -y_0, \ Q = Q_{r_1,r_2,r_3} = Q_{r_1,r_2,r}.$$

Since $r_3 = r \in K$, we conclude that $r_1 + r_2 = (r_1 + r_2 + r_3) - r_3 \in K$. □

REMARK 4.3. Let $Q = (x_1, y_1) \in E(K)$ with $y_1 \neq 0$ (i.e., Q is none of W_j) and $P = 2Q \in E(K)$. Then $Q = Q_{r_1,r_2,r_3}$ for suitable $r_k = \sqrt{x(P) - \alpha_k}$, and formulas (9) (with $i = 3$) tell us that

$$r_3 = -\frac{y_1}{2} \cdot \left(-\frac{1}{x_1 - \alpha} + \frac{1}{x_1 - \alpha_1} + \frac{1}{x_1 - \alpha_2} \right)$$
$$= -\frac{y_1}{2} \cdot \left(-\frac{1}{x_1 - \alpha} + \frac{2x_1 + p}{g(x_1)} \right).$$

In particular, if $x_1 = 0$ and $\alpha = -1$, then $y_1^2 = q$, $g(x_1) = q$ and

$$r_3^2 = \frac{q}{4} \cdot \left(-1 + \frac{p}{q} \right)^2 = \frac{(p-q)^2}{4q}$$

and

$$x(P) = x(2Q) = \alpha + r_3^2 = \frac{(p-q)^2}{4q} - 1. \tag{20}$$

THEOREM 4.4. *We keep the notation and assumptions of Theorem 4.1. Assume additionally that $x_0 \neq \alpha$, i.e., $P \neq (\alpha, 0)$. Then $P = (x_0, y_0)$ is divisible by 2 in $E(K)$ if and only if there exists $r \in K$ such that $r^2 = x_0 - \alpha$ and $(2x_0 + p)(x_0 - \alpha) - 2y_0 r$ is a square in K.*

In addition, if these equivalent conditions hold and we choose

$$T = \frac{\sqrt{(2x_0 + p)(x_0 - \alpha) - 2y_0 r}}{r} \in K,$$

then

$$\mathcal{Q}_{r,T} = (x_1, y_1) := \left(x_0 + rT - \frac{y_0}{r}, -y_0 - (r+T)\left(rT + \frac{y_0}{r} \right) \right)$$

and

$$\mathcal{Q}_{r,-T} = (x_{-1}, y_{-1}) := \left(x_0 - rT - \frac{y_0}{r}, -y_0 - (r-T)\left(-rT + \frac{y_0}{r} \right) \right)$$

are distinct points of $E(K)$, which satisfy

$$2\mathcal{Q}_{r,T} = 2\mathcal{Q}_{r,-T} = P \in E(K), \ \mathcal{Q}_{r,-T} = \mathcal{Q}_{r,T} + W_3.$$

Conversely, if $Q \in E(Q)$ satisfies $2Q = P$, then there exist

$$r = \sqrt{x_0 - \alpha} \in K \quad \text{and} \quad T = \frac{\sqrt{(2x_0 + p)(x_0 - \alpha) - 2y_0 r}}{r} \in K$$

such that $Q = \mathcal{Q}_{r,T}$.

PROOF. In light of Remark 4.2, the divisibility of P by 2 in $E(K)$ is equivalent to the existence of square roots $r = \sqrt{x_0 - \alpha} \in K$ and $r_i = \sqrt{x_0 - \alpha_i} \in \bar{K}$ $(i = 1, 2)$ such that

$$r_1 r_2 = -\frac{y_0}{r}, \ r_1 + r_2 \in K.$$

Suppose such r_1, r_2, r exist. Then $(r_1 + r_2)^2$ is a square in K. We have

$$\begin{aligned}(r_1 + r_2)^2 &= r_1^2 + 2r_1 r_2 + r_2^2 = (x_0 - \alpha_1) - 2\frac{y_0}{r} + (x_0 - \alpha_2) \\ &= 2x_0 + p - 2\frac{y_0}{r} = \frac{(2x_0 + p)(x_0 - \alpha) - 2y_0 r}{r^2}.\end{aligned} \tag{21}$$

This implies that $(2x_0 + p)(x_0 - \alpha) - 2y_0 r$ is a square in K.

Conversely, suppose that there exists $r \in K$ such that $r^2 = x_0 - \alpha$ and $(2x_0 + p)(x_0 - \alpha) - 2y_0 r$ is a *square* in K. This implies that

$$\frac{(2x_0 + p)(x_0 - \alpha) - 2y_0 r}{r^2} = \frac{(2x_0 + p)(x_0 - \alpha) - 2y_0 r}{x_0 - \alpha} = 2x_0 + p - \frac{2y_0}{r}$$

is a square in K. Let us put $r_3 = r$ and choose square roots $r_i = \sqrt{x_0 - \alpha_i} \in \bar{K}$ $(i = 1, 2)$ in such a way that $r_1 r_2 r_3 = -y_0$. (Since $r_3 = r \neq 0$, the only other choice of such a pair of square roots is $(-r_1, -r_2)$.) Let

$$T := r_1 + r_2.$$

We have

$$r_1 r_2 = -\frac{y_0}{r} \in K$$

and

$$T^2 = (r_1 + r_2)^2 = r_1^2 + 2r_1 r_2 + r_2^2 = (x_0 - \alpha_1) - 2\frac{y_0}{r} + (x_0 - \alpha_2) =$$

$$2x_0 + p - 2\frac{y_0}{r} = \frac{(2x_0 + p)(x_0 - \alpha) - 2y_0 r}{r^2}$$

is a *square* in K. This implies that

$$T = r_1 + r_2 \in K.$$

It follows from Remark 4.2 that P is divisible by 2 in $E(K)$. By Theorem 4.1 we obtain that

$$Q_+ = Q_{r_1, r_2, r} = (x_0 + (r_1 + r_2)r + r_1 r_2, -y_0 - ((r_1 + r_2) + r)((r_1 + r_2)r + r_1 r_2)$$

and

$$Q_- = Q_{-r_1, -r_2, r} = (x_0 - (r_1 + r_2)r + r_1 r_2, -y_0 - (-(r_1 + r_2) + r)(-(r_1 + r_2)r + r_1 r_2)$$

are distinct points of $E(K)$, which satisfy

$$2Q_+ = 2Q_- = P,\ Q_- = Q_+ + W_3.$$

Taking into account that

$$r_1 + r_2 = T,\ r_1 r_2 = -\frac{y_0}{r},$$

we obtain that

$$\mathcal{Q}_{r,T} = Q_+,\ \mathcal{Q}_{r,-T} = Q_-,$$

and therefore

$$\mathcal{Q}_{r,T}, \mathcal{Q}_{r,-T} \in E(K),\ \mathcal{Q}_{r,-T} = \mathcal{Q}_{r,T} + W_3,$$
$$2\mathcal{Q}_{r,T} = 2\mathcal{Q}_{r,-T} = P.$$

On the other hand, if $P = 2Q$ with $Q \in E(K)$, then it follows from Theorem 4.1 that there exist square roots $r = \sqrt{x_0 - \alpha} \in K$ and

$$r_i = \sqrt{x_0 - \alpha_i} \in \bar{K}\ (i = 1, 2)$$

such that

$$r_1 r_2 r = -y_0,\ r_1 + r_2 \in K$$

and $Q = Q_{r_1, r_2, r}$. This implies that $r_1 r_2 = -y_0/r$. Now if we put $T = (r_1 + r_2) \in K$, then (as we have already seen)

$$T^2 = (r_1 + r_2)^2 = (2x_0 + p)(x_0 - \alpha) - 2y_0 r \in K.$$

Now one can easily check that

$Q_{r_1,r_2,r} = (x_0 + (r_1 + r_2)r + r_1r_2, -y_0 - ((r_1 + r_2) + r)((r_1 + r_2)r + r_1r_2) = \mathcal{Q}_{r,T}$,

and therefore $Q = \mathcal{Q}_{r,T}$.

□

REMARK 4.5. It is known (3) that $r_1 + r_2 \neq 0$. In light of (21),

$$(2x_0 + p)(x_0 - \alpha) - 2y_0 r \neq 0.$$

THEOREM 4.6. *Let α be an element of K that is not a root of $g(x)$. Let us consider the elliptic curve*

$$E : y^2 = g(x)(x - \alpha)$$

over K. Let y_0 be a nonzero element of K. Then

$$P = (0, y_0) \in 2E(K) \subset E(K)$$

if and only if there exists $r \in K$ and a nonzero $T \in K$ such that

$$r^2 = -\alpha,\ g(x) = x^2 + \left(T^2 + 2\frac{y_0}{r}\right)x + \left(\frac{y_0}{r}\right)^2. \tag{22}$$

If (22) holds, then the equation for E becomes

$$y^2 = (x + r^2)\left(x^2 + \left(T^2 + 2\frac{y_0}{r}\right)x + \left(\frac{y_0}{r}\right)^2\right). \tag{23}$$

In addition,

$$\mathcal{Q}_{r,T} = \left(rT - \frac{y_0}{r}, -y_0 - (r + T)\left(rT - \frac{y_0}{r}\right)\right)$$

and

$$\mathcal{Q}_{r,-T} = \left(-rT - \frac{y_0}{r}, -y_0 - (r - T)\left(-rT - \frac{y_0}{r}\right)\right)$$

are distinct points of $E(K)$, which satisfy

$$2\mathcal{Q}_{r,T} = 2\mathcal{Q}_{r,-T} = P,\ \mathcal{Q}_{r,-T} = \mathcal{Q}_{r,T} + W_3.$$

Moreover, if $Q \in E(K)$ satisfies $2Q = P$, then there exist

$$r \in K \text{ and } T \in K$$

that enjoy property (22) and $Q = \mathcal{Q}_{r,T}$.

PROOF. We have the equation for E

$$y^2 = (x^2 + px + q)(x - \alpha).$$

Since $P = (x_0, y_0) = (0, y_0) \in E(K)$, we have $y_0^2 = q \cdot (-\alpha)$, which means that

$$q = \frac{y_0^2}{-\alpha}.$$

In light of Theorem 4.4, the inclusion $P \in 2E(K)$ is equivalent to the existence of nonzero $r, T \in K$ such that

$$r^2 = -\alpha,\ T^2 = \frac{(2x_0 + p)(x_0 - \alpha) - 2y_0 r}{r^2} = \frac{pr^2 - 2y_0 r}{r^2} = p - 2\frac{y_0}{r}.$$

This means that

$$q = \left(\frac{y_0}{r}\right)^2,\ p = T^2 + 2\frac{y_0}{r},$$

i.e.,

$$g(x) = x^2 + px + q = x^2 + \left(T^2 + 2\frac{y_0}{r}\right)x + \left(\frac{y_0}{r}\right)^2.$$

The assertions about $\mathcal{Q}_{r,T}$, $\mathcal{Q}_{r,-T}$, and $Q = \frac{1}{2}P$ follow from the corresponding assertions of Theorem 4.4. $\square$

REMARK 4.7. Dividing both sides of (23) by r^6 and making the substitution

$$\tilde{x} = \frac{x}{r^2}, \tilde{y} = \frac{y}{r^3}, t = \frac{T}{r}, \tilde{y}_0 = \frac{y_0}{r^3},$$

we obtain that E is isomorphic to the elliptic curve

$$\tilde{E}_{t,\tilde{y}_0} : \tilde{y}^2 = (\tilde{x}+1)\left(\tilde{x}^2 + \left(t^2 + 2\tilde{y}_0\right)\tilde{x} + \tilde{y}_0^2\right). \tag{24}$$

In addition, the isomorphism

$$E \cong \tilde{E}_{t,\tilde{y}_0} : \ (x,y) \mapsto (\tilde{x},\tilde{y}) = \left(\frac{x}{r^2}, \frac{y}{r^3}\right)$$

sends $W_3 = (-r^2, 0) \in E(K)$ to $\tilde{W}_3 = (-1, 0) \in \tilde{E}_{t,\tilde{y}_0}(K)$, P to $\tilde{P} = (0, \tilde{y}_0) \in \tilde{E}_{t,\tilde{y}_0}(K)$, $\mathcal{Q}_{r,T}$ to $\mathcal{Q}_{1,t} = (t - \tilde{y}_0, -\tilde{y}_0 - (1+t)(t - \tilde{y}_0))$, and $\mathcal{Q}_{r,-T}$ to $\mathcal{Q}_{1,-t} = (-t - \tilde{y}_0, -\tilde{y}_0 - (1-t)(-t - \tilde{y}_0))$. We also have

$$2\mathcal{Q}_{1,t} = 2\mathcal{Q}_{1,-t} = \tilde{P}, \ \mathcal{Q}_{1,-t} = \mathcal{Q}_{1,t} + \tilde{W}_3.$$

5. Elliptic curves with points of order 4

Let E be an elliptic curve over a field K with $\mathrm{char}(K) \neq 2$. Suppose that $E(K)$ contains exactly one point of order 2. Then E is K-isomorphic to

$$\mathcal{E}_{0,p,q} : y^2 = x(x^2 + px + q),$$

where $x^2 + px + q \in K[x]$ is an irreducible quadratic polynomial.

THEOREM 5.1. *The following conditions are equivalent.*

(i) *$E(K)$ contains a cyclic subgroup of order 4.*

(ii) *There exist nonzero $a, b \in K$ such that $a^2 + 4b$ is not a square in K and E is K-isomorphic to the elliptic curve*

$$\mathcal{E}_{a,b}^{(4)} : y^2 = x\left(x^2 + (a^2 + 2b)x + b^2\right).$$

PROOF. Suppose (i) holds. We may assume that $E = \mathcal{E}_{0,p,q}$. It follows from Example 3.6 that there are nonzero $a, b \in K$ such that $a^2 + 4b$ is not a square in K, $p = a^2 + 2b$, and $q = b^2$. This means that $E = \mathcal{E}_{a,b}^{(4)}$, i.e., (ii) holds.

Suppose (ii) holds. We may assume that

$$E = \mathcal{E}_{a,b}^{(4)} : y^2 = x\left(x^2 + (a^2 + 2b)x + b^2\right)$$

with nonzero $a, b \in K$ and $a^2 + 4b$ is *not* a square in K. (In particular, $a^2 + 4b \neq 0$.) Clearly, $W = (0, 0)$ is a point of order 2 in $E(K)$ and $P = (-b, ab) \in E(K)$. If we put $g(x) = x^2 + (a^2 + 2b)x + b^2$, then arguments of Example 3.6 show that W is divisible by 2 in $\mathcal{E}_{a,b}^{(4)}(K)$, and therefore $\mathcal{E}_{a,b}^{(4)}(K)$ contains a point of order 4. In fact, it contains exactly two points of order 4, namely, $(-b, ab)$ and $(-b, -ab)$. $\square$

REMARK 5.2. D. Kubert [**5**] described another family of elliptic curves

$$\mathcal{E}_{4,t} : y^2 + xy - ty = x^3 - tx^2$$

with point $Q = (0, 0)$ of order 4. The equation for $\mathcal{E}_{4,t}$ is equivalent to

$$y^2 + 2y\frac{(x-t)}{2} + \frac{(x-t)^2}{4} = (x-t)x^2 + \frac{(x-t)^2}{4}.$$

Now the change of variables

$$\tilde{x} = x - t, \ \tilde{y} = y + \frac{(x-t)}{2}$$

allows us to rewrite the equation as follows:

$$\tilde{y}^2 = \tilde{x}\left((\tilde{x}+t)^2 + \frac{\tilde{x}}{4}\right) = \tilde{x}\left(\tilde{x}^2 + (2t + \frac{1}{4})\tilde{x} + t^2\right).$$

This implies that $\mathcal{E}_{4,t}$ is K-isomorphic to $\mathcal{E}_{a,b}$ with $b = t, a = 1/2$.

THEOREM 5.3. *Let K be a field with $\mathrm{char}(K) \neq 2$, and let a, b be nonzero elements of K such that $a^2 + 4b$ is not a square in K. (In particular, $a^2 + 4b \neq 0$.) Let c, d be nonzero elements of K such that $c^2 + 4d \neq 0$. Then the elliptic curves $\mathcal{E}_{a,b}$ and $\mathcal{E}_{c,d}$ are K-isomorphic if and only if there exists a nonzero $u \in K$ such that*

$$c^2 + 2d = u^2(a^2 + 2b), \ u^4 b^2 = d^2.$$

PROOF. Since $a^2 + 4b$ is not a square in K, the quadratic polynomial $x^2 + (a^2 + 2b)x + b^2$ has *no* roots in K. This implies that $\mathcal{E}_{a,b}$ has exactly one point of order 2, namely, $W_{a,b} = (0,0)$. Let $\phi : \mathcal{E}_{a,b} \cong \mathcal{E}_{c,d}$ be a K-isomorphism. Since $W_{c,d} = (0,0)$ is a point of order 2 in $\mathcal{E}_{c,d}(K)$, it is the only point of order 2 in $\mathcal{E}_{c,d}(K)$ and ϕ sends $W_{a,b} = (0,0) \in \mathcal{E}_{a,b}$ to $W_{c,d} = (0,0) \in \mathcal{E}_{c,d}$. Let

$$\tilde{y}^2 = \tilde{x}\left(\tilde{x}^2 + (c^2 + 2d)\tilde{x} + d^2\right)$$

be the equation of $\mathcal{E}_{c,d}$ in coordinates $\tilde{x}, \tilde{y}$. According to [**10**, Sect. 1, p. 301, formula (1.13)], there exist a nonzero $u \in K$ and $\alpha, \beta, \gamma \in K$ such that the following conditions hold:

$$\tilde{x} \circ \phi = u^2 x + \alpha, \ \tilde{y} \circ \phi = u^3 y + \beta u^2 x + \gamma.$$

Since ϕ sends $(0,0) \in \mathcal{E}_{a,b}(K)$ to $(0,0) \in \mathcal{E}_{c,d}(K)$, we have $\alpha = 0 = \gamma$. Since ϕ commutes with the multiplication by -1, we conclude that $\beta = 0$, i.e., ϕ is defined by the formulas

$$\tilde{x} \circ \phi = u^2 x, \ \tilde{y} \circ \phi = u^3 y. \tag{25}$$

Now [**10**, Sect. 1, p. 301, formula (1.14)] implies that

$$u^2(a^2 + 2b) = (c^2 + 2d), \ u^4 b^2 = d^2. \tag{26}$$

This proves our assertion in one direction. In order to prove it in the opposite direction, let us assume that (26) holds. Then the formulas (25) define an isomorphism $\phi : \mathcal{E}_{a,b} \cong \mathcal{E}_{c,d}$. □

REMARK 5.4. Let us look more closely at equalities (26). They may be rewritten as follows:

$$u^2(a^2 + 2b) = (c^2 + 2d), \ d = \pm u^2 b.$$

It follows that $u^2(a^2 + 2b) = (c^2 \pm 2u^2 b)$, i.e., either

$$d = u^2 b, \ u^2(a^2 + 2b) = (c^2 + 2u^2 b),$$

and so

$$u^2 a^2 = c^2, \ c = \pm ua,$$

or

$$d = -u^2 b, \ u^2(a^2 + 2b) = (c^2 - 2u^2 b),$$

and therefore

$$u^2(a^2+4b)=c^2;$$

in the latter case a^2+4b is a square in K, which is not the case. So, we have

$$d=u^2b,\ dc=\pm ua,$$

and therefore $u=\pm c/a$, which implies that

$$b=\frac{da^2}{c^2}.$$

We have

$$(c^2+2d)=\left(\pm\frac{c}{a}\right)^2\left(a^2+2\cdot\frac{da^2}{c^2}\right)u^2=u^2(a^2+2b).$$

It follows that if a is a fixed nonzero element of K and E is an elliptic curve over K such that $E(K)$ contains a point of order 4 and exactly one point of order 2, then there exists exactly one (nonzero) $b\in K$ such that E is K-isomorphic to $\mathcal{E}_{a,b}$. In particular, if we take $a=1/2$, then we get that there is exactly one $t\in K$ such that E is K-isomorphic to $\mathcal{E}_{1/2,t}$.

6. Elliptic curves with points of order 8

Let E be an elliptic curve over a field K with $\operatorname{char}(K)\ne 2$. Suppose that $E(K)$ contains exactly one point of order 2 and two points of order 4. Then there exist *nonzero* $a,b\in K$ such that a^2+4b is *not* a square in K and E is K-isomorphic to

$$\mathcal{E}^{(4)}_{a,b}: y^2=x\left(x^2+(a^2+2b)x+b^2\right).$$

Recall that $\mathcal{E}^{(4)}_{a,b}(K)$ contains exactly two points of order 4, namely, $Q_+=(-b,ab)$ and $Q_-=(-b,-ab)=-Q_+$. It follows that the existence of a point of order 8 in $\mathcal{E}^{(4)}_{a,b}(K)$ is equivalent to the divisibility by 2 of Q_+ in $\mathcal{E}^{(4)}_{a,b}(K)$. By Theorem 4.4 and Remark 4.5, Q_+ is divisible by 2 if and only if $(-b)-0$ is a square in K and there exists $r\in K$ such that $r^2=x_0-\alpha$ and *nonzero* $(2x_0+p)(x_0-\alpha)-2y_0r$ is a square in K, where

$$\alpha=0, p=a^2+2b;\ x_0=-b, y_0=ab.$$

This means that $r^2=-b\ne 0$ and

$$(-2b+a^2+2b)(-b)-2abr=a^2r^2+2ar^2r=r^2(a^2+2ar)=r^2\left((a+r)^2-r^2\right)$$

is a nonzero square in K. It follows that Q_+ is divisible by 2 in $\mathcal{E}^{(4)}_{a,b}(K)$ if and only if there exist nonzero $r,T\in K$ such that

$$b=-r^2,\ T^2=(a+r)^2-r^2. \tag{27}$$

If this is the case, then Theorem 4.4 gives us two distinct halves of Q_+ in $\mathcal{E}^{(4)}_{a,b}(K)$, namely,

$$\mathcal{Q}_{r,T}=(x_1,y_1):=\left(-b+rT-\frac{ab}{r}, -ab-(r+T)\left(rT-\frac{ab}{r}\right)\right) \tag{28}$$

$$\mathcal{Q}_{r,-T}=(x_{-1},y_{-1}):=\left(-b-rT-\frac{ab}{r}, -ab-(r-T)\left(-rT-\frac{ab}{r}\right)\right);$$

$$2\mathcal{Q}_{r,T}=2\mathcal{Q}_{r,-T}=Q_+,\ \mathcal{Q}_{r,-T}=\mathcal{Q}_{r,T}+W_3.$$

This implies that both $\mathcal{Q}_{r,T}$ and $\mathcal{Q}_{r,-T}$ have order 8.

PROPOSITION 6.1. *$\mathcal{E}_{a,b}^{(4)}(K)$ contains a point of order 8 if and only if there exist nonzero $r, t \in K$ such that $t \neq 0, \pm 1$ and*

$$b = -r^2, \ a = \frac{2t^2 r}{1 - t^2}.$$

PROOF. Dividing the second equation in (27) by r^2, we get

$$\left(\frac{T}{r}\right)^2 = \left(\frac{a}{r} + 1\right)^2 - 1.$$

Using the well-known rational parametrization of the hyperbola, we obtain that this equality is equivalent to the existence of a $t \in K, t \neq \pm 1$, such that

$$\frac{a}{r} + 1 = \frac{1 + t^2}{1 - t^2}, \frac{T}{r} = \frac{2t}{1 - t^2}.$$

This means that

$$a = \frac{2t^2 r}{1 - t^2}, \ T = \frac{2tr}{1 - t^2} \tag{29}$$

for $t \in K$, $t \neq \pm 1$. Since $T \neq 0$, we have $t \neq 0$. We still need to find what does it mean in terms of r, t that $a^2 + 4b$ is *not* a square in K. We have

$$a^2 + 4b = \left(\frac{2t^2 r}{1 - t^2}\right)^2 + 4(-r^2) = 4r^2 \frac{2t^2 - 1}{(1 - t^2)^2}.$$

This implies that $a^2 + 4b$ is not a square in K if and only if $t \neq \pm 1$ and $2t^2 - 1$ is not a square. □

THEOREM 6.2. *Let E be an elliptic curve over a field K with* $\operatorname{char}(K) \neq 2$. *Suppose that $E(K)$ contains exactly one point of order 2. Then the following conditions are equivalent.*

(i) *$E(K)$ contains a cyclic subgroup of order 8.*
(ii) *There exists $t \in K$ such that*
   (1) *$t \neq 0, \pm 1$,*
   (2) *$2t^2 - 1$ is not a square in K,*
   (3) *E is K-isomorphic to the elliptic curve*

$$\mathcal{E}_t^{(8)} : y^2 = x\left(x^2 + 2\frac{t^4 + 2t^2 - 1}{(t^2 - 1)^2} x + 1\right).$$

PROOF. Combining Theorem 5.1 with Proposition 6.1, we conclude that the condition (i) is equivalent to the existence of nonzero $r, t \in K$, $t \neq \pm 1$, such that

$$a = \frac{2t^2 r}{1 - t^2}, b = -r^2$$

enjoy the following properties.

(a) $a \neq 0$.
(b) $a^2 + 4b$ is *not* a square in K.
(c) E is isomorphic over K to the elliptic curve

$$\mathcal{E}_{a,b}^{(4)} : y^2 = x\left(x^2 + (a^2 + 2b)x + b^2\right).$$

We notice that $a = 0$ if and only if $t = 0$; moreover,

$$b^2 = (-r^2)^2 = r^4,$$

$$a^2 + 2b = 2r^2 \frac{(t^4 + 2t^2 - 1)}{(t^2 - 1)^2}.$$

Consequently, the curve E is K-isomorphic to the curve

$$\mathcal{E}_{r,t}^{(8)} : y^2 = x\left(x^2 + 2r^2 \frac{t^4 + 2t^2 - 1}{(t^2 - 1)^2} x + r^4\right).$$

Dividing both sides by r^6 and using the change of variables $\tilde{x} = x/r^2$, $\tilde{y} = y/r^3$, we get the elliptic curve

$$\mathcal{E}_t^{(8)} : \tilde{y}^2 = \tilde{x}\left(\tilde{x}^2 + 2\frac{t^4 + 2t^2 - 1}{(t^2 - 1)^2}\tilde{x} + 1\right),$$

(where $t \neq 0, \pm 1$) that is K-isomorphic to E. It remains to notice that $a^2 + 4b$ is *not* a square in K if and only if $2t^2 - 1$ is not a square. $\square$

EXAMPLE 6.3. Let us describe explicitly the points of order 8 in $\mathcal{E}_t^{(8)}(K)$. Notice that $\mathcal{E}_t^{(8)} = \mathcal{E}_{a,b}^{(4)}$ with

$$a = \frac{2t^2}{1 - t^2}, \; b = -1, r = 1, T = \frac{2t}{1 - t^2}.$$

In particular, $\mathcal{E}_t^{(8)}(K)$ contains a point $Q_+ = (-b, ab) = (1, -\frac{2t^2}{1-t^2})$ of order 4. Clearly, the conditions (27) hold. Then (28) give us two distinct order 8 points in $\mathcal{E}_t^{(8)}(K)$, namely,

$$\mathcal{Q}_{r,T} = \left(1 + \frac{2t}{1-t^2} + \frac{2t^2}{1-t^2}, \frac{2t^2}{1-t^2} - \left(1 + \frac{2t}{1-t^2}\right)\left(\frac{2t}{1-t^2} + \frac{2t^2}{1-t^2}\right)\right)$$
$$= \left(\frac{t+1}{1-t}, \frac{-2t}{(1-t)^2}\right),$$

$$\mathcal{Q}_{r,-T} = \left(1 - \frac{2t}{1-t^2} + \frac{2t^2}{1-t^2}, \frac{2t^2}{1-t^2} - \left(1 - \frac{2t}{1-t^2}\right)\left(-\frac{2t}{1-t^2} + \frac{2t^2}{1-t^2}\right)\right)$$
$$= \left(\frac{1-t}{1+t}, \frac{2t}{(1+t)^2}\right);$$

$$2\mathcal{Q}_{r,T} = 2\mathcal{Q}_{r,-T} = Q_+, \; \mathcal{Q}_{r,-T} = \mathcal{Q}_{r,T} + W_3.$$

So, $\mathcal{E}_t^{(8)}(K)$ contains exactly four points of order 8, namely,

$$\left(\frac{1+t}{1-t}, \frac{-2t}{(1-t)^2}\right), \; \left(\frac{1-t}{1+t}, \frac{2t}{(1+t)^2}\right)$$

and their negatives

$$\left(\frac{1+t}{1-t}, \frac{2t}{(1-t)^2}\right), \; \left(\frac{1-t}{1+t}, \frac{-2t}{(1+t)^2}\right).$$

REMARK 6.4. D. Kubert [5] described another family of elliptic curves

$$\mathcal{E}_{8,d} : y^2 + (1-c)xy - by = x^3 - bx^2 \tag{30}$$

with point $Q = (0,0)$ of order 8, where

$$c = \frac{(2d-1)(d-1)}{d}, \quad b = (2d-1)(d-1).$$

Completing the square on the left-hand side of the equation, we get

$$\left(y + \frac{(1-c)x - b}{2}\right)^2 = x^3 - bx^2 + \left(\frac{(1-c)x - b}{2}\right)^2.$$

The curve given by the latter equation is K-isomorphic to

$$y^2 = x^3 + \left(\left(\frac{1-c}{2}\right)^2 - b\right)x^2 - \frac{(1-c)b}{2}x + \frac{b^2}{4}.$$

Since

$$\frac{1-c}{2} = \frac{-2d^2+4d-1}{2d},$$

we get

$$y^2 = x^3 + \left(\frac{-4d^4 - 4d^3 + 16d^2 - 8d + 1}{4d^2}\right)x^2$$
$$- \frac{(-2d^2+4d-1)(2d-1)(d-1)}{2d}x + \frac{(2d-1)^2(d-1)^2}{4}.$$

Multiplying both sides of the above equation by d^6 and replacing yd^3 by y and xd^2 by x, we get the equation with polynomial coefficients

$$y^2 = x^3 + \left(\frac{-4d^4 - 4d^3 + 16d^2 - 8d + 1}{4}\right)x^2$$
$$- \frac{(-2d^2+4d-1)(2d-1)(d-1)d^3}{2}x + \frac{d^6(2d-1)^2(d-1)^2}{4}.$$

The polynomial on the right can be factored as follows:

$$(x - (d^4 - d^3))\left(x^2 + \frac{1 - 8d(d-1)^2}{4}x - \frac{d^3(d-1)(2d-1)^2)}{4}\right).$$

Now using the change of variables $x - (d^4 - d^3) \to x$, we obtain that

$$y^2 = x\left(x^2 + \frac{8d^4 - 16d^3 + 16d^2 - 8d + 1}{4}x + d^4(d-1)^4\right).$$

Dividing both sides by $d^6(d-1)^6$ and using the change of variables $\tilde{x} = x/d^2(d-1)^2$, $\tilde{y} = y/d^3(d-1)^3$, we get the elliptic curve

$$\tilde{y}^2 = \tilde{x}\left(\tilde{x}^2 + \frac{8d^4 - 16d^3 + 16d^2 - 8d + 1}{4d^2(d-1)^2}\tilde{x} + 1\right).$$

It is easy to verify that

$$\frac{8d^4 - 16d^3 + 16d^2 - 8d + 1}{4d^2(d-1)^2} = 2\frac{t^4 + 2t^2 - 1}{(t^2-1)^2},$$

where $t = 2d-1$. This implies that $\mathcal{E}_{8,d}$ is isomorphic over K to $\mathcal{E}_t^{(8)}$ with $t = 2d-1$.

THEOREM 6.5. *Let K be a field with* $\operatorname{char}(K) \neq 2$. *Let $s, t \in K$ be nonzero elements of K such that $s, t \neq \pm 1$ and $2t^2 - 1, 2s^2 - 1$ are not squares in K.*

If K does not contain a primitive 4th root of 1, then the elliptic curves $\mathcal{E}_s^{(8)}$ and $\mathcal{E}_t^{(8)}$ are K-isomorphic if and only if either $s = \pm t$ or $s^2 + t^2 = 2s^2t^2$.

If K contains a primitive 4th root of 1, then $\mathcal{E}_s^{(8)}$ and $\mathcal{E}_t^{(8)}$ are K-isomorphic if and only if one of the equalities $s = \pm t$, $s^2+t^2 = 2s^2t^2$, or $s^4t^4+2s^2+2t^2 = 4s^2t^2+1$ is valid.

PROOF. It follows from Theorem 5.3 that $\mathcal{E}_s^{(8)}$ and $\mathcal{E}_t^{(8)}$ are K-isomorphic if and only if there exists $u \in K$ such that $u^4 = 1$ and

$$\frac{s^4 + 2s^2 - 1}{(s^2 - 1)^2} = u^2 \frac{t^4 + 2t^2 - 1}{(t^2 - 1)^2}.$$

If K does not contain a primitive 4th root of 1, then the latter equality is possible only if

$$\frac{s^4 + 2s^2 - 1}{(s^2 - 1)^2} = \frac{t^4 + 2t^2 - 1}{(t^2 - 1)^2},$$

which takes place if and only if one of the equalities $s = \pm t$ or $s^2 + t^2 = 2s^2t^2$ is valid. If K contains a primitive 4th root of 1, then we have

$$\frac{s^4 + 2s^2 - 1}{(s^2 - 1)^2} = \pm \frac{t^4 + 2t^2 - 1}{(t^2 - 1)^2},$$

which holds if and only if one of the equalities $s = \pm t$, $s^2 + t^2 = 2s^2t^2$, or $s^4t^4 + 2s^2 + 2t^2 = 4s^2t^2 + 1$ is valid.

□

7. Elliptic curves with point of order 6

Let K be a field with $\operatorname{char}(K) \neq 0$. Let E be an elliptic curve over K defined by the equation

$$y^2 = f(x),$$

where $f(x) \in K[x]$ is a monic cubic polynomial without repeated roots.

EXAMPLE 7.1. Let $t \in K \setminus 0, -4, 1/2$. Let us consider the elliptic curve

$$\mathcal{E}_t^{(6)} : y^2 = (x^2 + (t^2 + 2t)x + t^2)(x + 1)$$

over K. (We assume that $t \notin \{0, -4, 1/2\}$ in order to exclude the case when the cubic polynomial has a repeated root.) The group $E(K)$ contains the point $W_3 = (-1, 0)$ of order 2 and the point $P = (0, t)$. Let us put

$$y_0 = t, r = 1, T = t.$$

Then the curve $\mathcal{E}_t^{(6)}$ coincides with the elliptic curve (23) from Theorem 4.6. According to this theorem,

$$\mathcal{Q}_{r,T} = \left(rT - \frac{y_0}{r}, -y_0 - (r + T)\left(rT - \frac{y_0}{r}\right)\right)$$

and

$$\mathcal{Q}_{r,-T} = \left(-rT - \frac{y_0}{r}, -y_0 - (r - T)\left(-rT - \frac{y_0}{r}\right)\right)$$

are distinct points of $E(K)$, which satisfy

$$2\mathcal{Q}_{r,T} = 2\mathcal{Q}_{r,-T} = P, \ \mathcal{Q}_{r,-T} = \mathcal{Q}_{r,T} + W_3.$$

In our case

$$rT = t, \frac{y_0}{r} = t, rT - \frac{y_0}{r} = 0, -rT - \frac{y_0}{r} = -2t.$$

This implies that

$$\begin{aligned}\mathcal{Q}_{r,T} = (0, -y_0) = (0, -t) = -P, \ \mathcal{Q}_{r,T} + W_3 = \mathcal{Q}_{r,-T} \\ = (-2t, -t - (1-t)(-2t)) = (-2t, t - 2t^2).\end{aligned}$$

Since

$$\mathcal{Q}_{r,T} = -P, \ 2\mathcal{Q}_{r,T} = -P,$$

P and $\mathcal{Q}_{r,T}$ have order 3 while $\mathcal{Q}_{r,-T} = \mathcal{Q}_{r,T} + W_3$ has order 6. So, the point $(0, t) \in \mathcal{E}_t^{(6)}(K)$ has order 3 and the point $(-2t, t - 2t^2) \in \mathcal{E}_t^{(6)}(K)$ has order 6.

Notice that $\mathcal{E}_t^{(6)}(K)$ contains exactly one point of order 2 if and only if the discriminant $(t^2 + 2t)^2 - 4t^2 = t^2(t^2 + 4t)$ of $x^2 + (t^2 + 2t)x + t^2$ is *not* a square, i.e., if and only if $t^2 + 4t$ is *not* a square.

THEOREM 7.2. *The following conditions on E are equivalent.*

(i) *$E(K)$ contains a point of order* 6.
(ii) *There exists $t \in K \setminus \{0, -4, 1/2\}$ such that E is isomorphic over K to the elliptic curve*

$$\mathcal{E}_t^{(6)} : y^2 = (x + 1)(x^2 + (t^2 + 2t)x + t^2).$$

PROOF. Suppose $E(K)$ contains a point of order 6. This means that $E(K)$ contains a point of order 2 and a point of order 3. The existence of a point of order 2 in $E(K)$ means that $f(x)$ has a root in K say, α and one may represent $f(x)$ as a product

$$f(x) = g(x)(x - \alpha) \in K[x],$$

where

$$g(x) = x^2 + px + q \in K[x]$$

is a monic quadratic polynomial without repeated roots such that our $\alpha \in K$ is *not* a root of $g(x)$. Then $W_3 = (\alpha, 0) \in E(K)$ is a point of order 2.

Let $P = (x_0, y_0) \in E(K)$ be a point of order 3. Since $3 \neq 2$, we have $y_0 \neq 0$. Using the change of variables $x \to x - x_0$ (and replacing α by $\alpha - x_0$ and $g(x)$ by $g(x - x_0)$), we may and will assume that $x_0 = 0$, i.e., $P = (0, y_0)$. Since P has order 3, it lies in $2E(K)$, because $2(-P) = P$. Let us apply the last assertion of Theorem 4.6 to $P = (0, y_0)$ and $Q = -P = (0, -y_0)$. We obtain that there exist $r \in K$ and a nonzero $T \in K$ such that

$$r^2 = -\alpha, \ g(x) = x^2 + px + q = x^2 + \left(T^2 + 2\frac{y_0}{r}\right)x + \left(\frac{y_0}{r}\right)^2,$$

$$(0, -y_0) = Q = \mathcal{Q}_{r,T} = \left(rT - \frac{y_0}{r}, -y_0 - (r + T)\left(rT - \frac{y_0}{r}\right)\right).$$

Looking at the x-coordinates, we see that

$$rT - \frac{y_0}{r} = 0, \ rT = \frac{y_0}{r}, \ y_0 = r^2 T,$$

and therefore

$$g(x) = x^2 + (T^2 + 2rT)x + (rT)^2, \ f(x) = (x^2 + (T^2 + 2rT)x + (rT)^2)(x + r^2),$$

and the equation for E is

$$E : y^2 = f(x) = (x^2 + (T^2 + 2rT)x + (rT)^2)(x + r^2).$$

Dividing both sides of this equation by r^6, and making the change of variables $\tilde{x} = x/r^2, \tilde{y} = y/r^3$, we obtain that E is isomorphic to the elliptic curve

$$\mathcal{E}_t^{(6)} : \tilde{y}^2 = (\tilde{x}+1)(\tilde{x}^2 + (t^2+2t)\tilde{x} + t^2)$$

with $t = T/r \neq 0$. Since the polynomial $(\tilde{x}+1)(\tilde{x}^2+(t^2+2t)\tilde{x}+t^2)$ has no multiple roots, we conclude that $t \notin \{-4, 0, 1/2\}$.

The converse assertion follows from Example 7.1. □

REMARK 7.3. In Theorem 7.2 we do *not* assume that $\operatorname{char}(K) \neq 3$!

EXAMPLE 7.4. Let $K = \mathbb{F}_3$ be the 3-element field. Then there is exactly one element t in $\mathbb{F}_3 \setminus \{0, -4\}$, namely $t = 1$. It follows from Theorem 7.2 that E is an elliptic curve over K such that $E(\mathbb{F}_3)$ contains a point of order 6 if and only if E is $\mathbb{F}_3$-isomorphic to

$$\mathcal{E}_1^{(6)} : y^2 = (x+1)(x^2+1).$$

Let us consider the curve $\mathcal{E}_1^{(6)}$. It contains an $\mathbb{F}_3$-point of order 6, namely, $Q = (-2t, t-2t^2) = (-2, -1)$. Hence the whole group $\mathcal{E}_1^{(6)}(\mathbb{F}_3)$ has order divisible by 6. On the other hand, by Hasse's bound, the order of $\mathcal{E}_1^{(6)}(\mathbb{F}_3)$ does not exceed $3 + 2\sqrt{3} + 1 < 6 \cdot 2$. This implies that $\mathcal{E}_1^{(6)}(\mathbb{F}_3)$ has order 6 and therefore coincides with the cyclic group of order 6 generated by Q.

REMARK 7.5. D. Kubert described another family of elliptic curves

$$\mathcal{E}_{6,c} : y^2 + (1-c)xy - (c+c^2)y = x^3 - (c+c^2)x^2$$

with point $Q = (0,0)$ of order 6. The equation for $\mathcal{E}_{6,c}$ is equivalent to

$$\begin{aligned} y^2 + 2y\frac{(1-c)x-(c+c^2)}{2} + \frac{((1-c)x-(c+c^2))^2}{4} \\ = x^3 - (c+c^2)x^2 + \frac{((1-c)x-(c+c^2))^2}{4}. \end{aligned}$$

The left-hand side is equal to $(y + ((1-c)x - (c+c^2))/2)^2$ while the right-hand side splits into the product

$$(x-c)\left(x^2 + \frac{-3c^2-2c+1}{4}x - \frac{c^3+2c^2+c}{4}\right).$$

After the change of variables $\tilde{x} = x - (c^2+c)$, $\tilde{y} = y + ((1-c)x - (c+c^2))/2$, we get the equation

$$\tilde{y}^2 = (\tilde{x}+c^2)\left(\tilde{x}^2 + \frac{5c^2+6c+1}{4}\tilde{x} + \frac{c^2(c+1)^2}{4}\right).$$

Dividing both sides by c^6, we get the equation

$$\left(\frac{\tilde{y}}{c^3}\right)^2 = \left(\frac{\tilde{x}}{c^2}+1\right)\left(\left(\frac{\tilde{x}}{c^2}\right)^2 + \frac{5c^2+6c+1}{4c^2}\left(\frac{\tilde{x}}{c^2}\right) + \frac{(c+1)^2}{4c^2}\right).$$

Now the change of variables $\bar{x} = \tilde{x}/c^2$, $\bar{y} = \tilde{y}/c^3$ gives us the equation

$$\bar{y}^2 = (\bar{x}+1)\left(\bar{x}^2 + \frac{5c^2+6c+1}{4c^2}\bar{x} + \left(\frac{c+1}{2c}\right)^2\right),$$

which is nothing else but the equation of $\mathcal{E}_t^{(6)}$ with $t = (c+1)/2c$.

8. Elliptic curves with point of order 12

Let K be a field with $\mathrm{char}(K) \neq 2$, and let $t \in K \setminus \{0, -4, 1/2\}$ be such that $t^2 + 4t$ is not a square in K. Let us consider the elliptic curve

$$E := \mathcal{E}_t^{(6)} : y^2 = g(x)(x+1),$$

where

$$g(x) = x^2 + (t^2 + 2t)x + t^2$$

is a quadratic irreducible polynomial over K. Then $W = W_3 = (-1, 0)$ is the only point of order 2 in $E(K)$. We know that $E(K)$ contains a point of order 3. Hence $E(K)$ contains a point of order 12 if and only if W is divisible by 2 in $E(K)$. This is equivalent to the condition that $-1 - \mathbf{x}$ is a square in the quadratic extension

$$K_g = K[x]/g(x)K[x]$$

of K, i.e., there exist $u, v \in K$ such that

$$-1 - \mathbf{x} = (u\mathbf{x} + v)^2 \in K_g.$$

Clearly, such $u \neq 0$. (Here $\mathbf{x}$ is the image of x in K_g.) In other words, W is divisible by 2 in $E(K)$ if and only if there exist $u, v \in K$ such that $(ux + v)^2 - (-1 - x)$ is divisible by $g(x)$. This means that there exist $u, v \in K$ such that

$$u^2x^2 + (2uv + 1)x + (v^2 + 1) = (ux + v)^2 + x + 1$$

coincides with

$$u^2 g(x) = u^2x^2 + u^2(t^2 + 2t)x + u^2t^2,$$

i.e.,

$$2uv + 1 = u^2(t^2 + 2t), \; v^2 + 1 = u^2t^2.$$

Subtracting one equation from the other, we obtain that

$$v^2 - 2uv = -2tu^2.$$

Dividing both sides by u^2 and putting $\lambda = v/u$, we get

$$\lambda^2 - 2\lambda = -2t,$$

which gives us

$$t = \frac{2\lambda - \lambda^2}{2}, \; v = \lambda u.$$

This implies that $\lambda \neq 0$, since $t \neq 0$ and therefore $v \neq 0$. We have

$$(\lambda u)^2 + 1 = u^2 \left(\frac{2\lambda - \lambda^2}{2}\right)^2,$$

which means that

$$1 = u^2 \left(\left(\frac{2\lambda - \lambda^2}{2}\right)^2 - \lambda^2 \right) = u^2\lambda^2 \left(\left(\frac{2 - \lambda}{2}\right)^2 - 1 \right).$$

Consequently,

$$1 = v^2 \left(\left(1 - \frac{\lambda}{2}\right)^2 - 1 \right).$$

Putting $s = 1 - \lambda/2$ and $\mu = 1/v$, we get the equation for a hyperbola in (s, μ)-coordinates.

$$\mu^2 = s^2 - 1.$$

Using the standard parametrization

$$s=\frac{1+T^2}{1-T^2},\quad \mu=\frac{2T}{1-T^2}$$

of this hyperbola, we obtain

$$\lambda=\frac{4T^2}{T^2-1},\quad t=\frac{2\lambda-\lambda^2}{2}=-\frac{4T^2(T^2+1)}{(T^2-1)^2},$$
$$t^2+2t=\frac{8T^2(T^2+1)(T^4+4T^2-1)}{(T^2-1)^4},$$

and the equation of E takes the form

$$y^2=(x+1)\left(x^2+\frac{8T^2(T^2+1)(T^4+4T^2-1)}{(T^2-1)^4}x+\frac{16T^4(T^2+1)^2}{(T^2-1)^4}\right).$$

The condition $t\in K\setminus\{0,-4,1/2\}$ is equivalent to $T\notin\{0,\pm1,\pm\sqrt{-1}\}$, $3T^2-1\neq 0$, and $3T^2+1\neq 0$. Since

$$t^2+4t=\frac{16T^2(T^2+1)(3T^2-1)}{(T^2-1)^4},$$

t^2+4t is not a square in K if and only if $(T^2+1)(3T^2-1)$ is not a square in K.

We have proved the following statement.

THEOREM 8.1. *The following conditions on E are equivalent.*

(i) *$E(K)$ contains exactly one point of order 2 and a point of order 12.*

(ii) *There exists $T\in K\setminus\{0,\pm1,\pm\sqrt{-1}\}$ such that $3T^2+1\neq0$, $(T^2+1)(3T^2-1)$ is not a square, and E is isomorphic over K to the elliptic curve*

$$\mathcal{E}_T^{(12)}: y^2=(x+1)\left(x^2+\frac{8T^2(T^2+1)(T^4+4T^2-1)}{(T^2-1)^4}x+\frac{16T^4(T^2+1)^2}{(T^2-1)^4}\right).$$

REMARK 8.2. $\mathcal{E}_T^{(12)}(K)$ contains points $(0,\pm4T^2(T^2+1)/(T^2-1)^2)$ of order 3 and exactly two points of order 4. Applying the explicit formulas from Theorem 3.3 to $(x_0,y_0)=(-1,0)$ and taking into account that $r=x_0-\alpha=0$, $\mathbf{r}=u\mathbf{x}+v$,

$$\operatorname{Norm}(\mathbf{r})=u^2\operatorname{Norm}(\mathbf{x})+uv\operatorname{Tr}(\mathbf{x})+v^2=t^2u^2-(t^2+2t)uv+v^2,$$
$$\operatorname{Tr}(\mathbf{r})=u\operatorname{Tr}(\mathbf{x})+2v=-(t^2+2t)u+2v,$$

we get the following two points of order 4 on $\mathcal{E}_T^{(12)}$:

$$(t^2u^2-(t^2+2t)uv+v^2,(t^2u^2-(t^2+2t)uv+v^2)(-(t^2+2t)u+2v)),$$
$$(t^2u^2-(t^2+2t)uv+v^2,-(t^2u^2-(t^2+2t)uv+v^2)(-(t^2+2t)u+2v)),$$

where

$$t=-\frac{4T^2(T^2+1)}{(T^2-1)^2},\quad t^2+2t=\frac{8T^2(T^2+1)(T^4+4T^2-1)}{(T^2-1)^4},$$
$$v=\frac{1}{\mu}=\frac{1-T^2}{2T},\quad u=\frac{v}{\lambda}=-\frac{(T^2-1)^2}{8T^3}.$$

Substituting these expressions for t, u, and v in the above formulas, we obtain the order 4 points

$$\left(-\frac{3T^2+1}{T^2-1},-\frac{8T^3(3T^2+1)}{(T^2-1)^3}\right),\ \left(-\frac{3T^2+1}{T^2-1},\frac{8T^3(3T^2+1)}{(T^2-1)^3}\right).$$

REMARK 8.3. In Theorem 8.1 we do *not* assume that $\operatorname{char}(K)\neq 3$!

9. Elliptic curves with rational points of order 10

Let E be an elliptic curve over a field K with $\mathrm{char}(K) \neq 2$. Suppose that $E(K)$ contains exactly one point of order 2 and a point P of order 5. We may assume that the first coordinate of P is 0. Since P is divisible by 2, Theorem 4.6 and Remark 4.7 tell us that E is K-isomorphic to the elliptic curve

$$E_{t,y_0} : y^2 = (x+1)\left(x^2 + \left(t^2 + 2y_0\right)x + y_0^2\right), \tag{31}$$

where $t^2 + 4y_0 \neq 0$ and $1 - t^2 - 2y_0 + y_0^2 \neq 0$. We may also assume that $P = (0, y_0)$. Clearly P has order 5 if and only if there exists a $Q \in E(K)$ such that $2P = -Q$ and $2Q = P$. Using equation (20) with $p = t^2 + 2y_0$ and $q = y_0^2$ and the equality $x(Q) = t - y_0$ obtained in Remark 4.7, we can write the equalities $2P = -Q$ and $2Q = P$ in the following equivalent form:

$$\frac{(t^2 + 2y_0 - y_0^2)^2}{4y_0^2} - 1 = t - y_0.$$

Multiplying both sides by y_0^2 and removing parentheses, we get

$$t^4 + y_0^4 + 4t^2 y_0 - 4ty_0^2 - 2t^2 y_0^2 = 0,$$

$$(t^2 - y_0^2)^2 + 4ty_0(t - y_0) = 0.$$

If $t = y_0$, then by Remark 4.7, we obtain $Q_{1,t} = (0, -y_0)$, $P = -Q_{1,t}$, and so $3P = 0$, which is impossible. Dividing both sides of the above equation by $t - y_0$, we get

$$(t - y_0)(t + y_0)^2 + 4ty_0 = 0.$$

To obtain a rational parametrization of this equation, let us put $y_0 = ut$ with $u \in K \setminus \{0, 1\}$. Then

$$t^3(1-u)(1+u)^2 + 4t^2 u = 0,$$

hence

$$t = \frac{4u}{(u-1)(u+1)^2}, \quad y_0 = \frac{4u^2}{(u-1)(u+1)^2}.$$

Since

$$t^2 + 2y_0 = \frac{16u^2}{(u-1)^2(u+1)^4} + \frac{8u^2}{(u-1)(u+1)^2} = \frac{8u^2(u^3 + u^2 - u + 1)}{(u-1)^2(u+1)^4},$$

equation (31) transforms to

$$y^2 = (x+1)\left(x^2 + \frac{8u^2(u^3 + u^2 - u + 1)}{(u-1)^2(u+1)^4}x + \frac{16u^4}{(u-1)^2(u+1)^4}\right).$$

Since

$$t^2 + 4y_0 = \frac{16u^3(u^2 + u - 1)}{(u-1)^2(u+1)^4} \text{ and } 1 - t^2 - 2y_0 + y_0^2 = \frac{(u^2 - 4u - 1)(u-1)}{(u+1)^3},$$

the conditions $t^2 + 4y_0 \neq 0$ and $1 - t^2 - 2y_0 + y_0^2 \neq 0$ are equivalent to $u \neq 0, u \neq \pm 1$, $u^2 + u \neq 1$, and $u^2 - 4u - 1 \neq 0$. Moreover, $t^2 + 4y_0$ is not a square in K if and only if $u(u^2 + u - 1)$ is not a square in K.

We have proved the following statement.

THEOREM 9.1. *The following conditions on E are equivalent.*

(i) *$E(K)$ contains exactly one point of order* 2 *and a point of order* 10.

(ii) *There exists* $u \in K \setminus \{0, \pm 1, (-1 \pm \sqrt{5})/2, 2 \pm \sqrt{5}\}$ *such that* $u(u^2 + u - 1)$ *is not a square in* K *and* E *is isomorphic over* K *to the elliptic curve*

$$\mathcal{E}_u^{(10)} : y^2 = (x+1)\left(x^2 + \frac{8u^2(u^3+u^2-u+1)}{(u-1)^2(u+1)^4}x + \frac{16u^4}{(u-1)^2(u+1)^4}\right).$$

REMARK 9.2. $\mathcal{E}_u^{(10)}(K)$ contains a point $\left(0, 4u^2/(u-1)(u+1)^2\right)$ of order 5 and exactly one point $(-1, 0)$ of order 2.

REMARK 9.3. In Theorem 9.1 we do *not* assume that $\mathrm{char}(K) \neq 5$!

10. Elliptic curves in characteristic 2

In this section we assume that K is a field of characteristic 2 and $\bar{K}$ is its algebraic closure. It is known [**9**, Appendix A] that E is *ordinary* (i.e., $E(\bar{K})$ contains a point of order 2) if and only if $j(E) \neq 0$. Let E be an elliptic curve over K defined by the equation

$$y^2 + xy = x^3 + a_2x^2 + a_6,$$

where

$$a_2, a_6 \in K;\ a_6 \neq 0,\ j(E) = \frac{1}{a_6} \neq 0$$

As above, E has the only one infinite point $\infty = (0:1:0)$, which is taken as the zero of the group law on E. It is known [**9**, Appendix A] that E is ordinary. In addition, every ordinary elliptic curve $\tilde{E}$ over K is isomorphic to E for suitable $a_2 \in K$ and $a_6 = 1/j(\tilde{E})$ [**9**, Appendix A], [**11**, Sect. 2.8]. If $P = (x_0, y_0) \in E(\bar{K})$, then

$$-P = (x_0, y_0 + x_0) \in E(\bar{K}). \tag{32}$$

It follows that

$$W_3 = (0, \sqrt{a_6}) = \left(0, \frac{1}{\sqrt{j(E)}}\right) \in E(\bar{K})$$

is the only point of order 2 in $E(\bar{K})$.

This implies the following assertion (that may be also extracted from [**4**]).

PROPOSITION 10.1. *An elliptic curve* E *over a field* K *of characteristic* 2 *has a* K*-rational point of order* 2 *if and only if* $j(E)$ *is a nonzero square in* K.

Our first goal is to find explicitly both halves of P. So, let $Q = (x_1, y_1) \in E(\bar{K})$ with $2Q = P$. Clearly, $Q \neq W$, i.e., $x_1 \neq 0$, and therefore the tangent line $\mathcal{L}$ to E at Q is *not* vertical, i.e, may be written in the form

$$y = lx + m;\ l, m \in \bar{K}.$$

(If $Q \in E(K)$, then both l and m lie in K.) Since $2Q = P$, the line $\mathcal{L}$ contains $-P$. Restricting (as usual) the equation of E to $\mathcal{L}$, we get the equation

$$g(x) = x^3 + a_2x + a_6 + (lx+m)^2 + x(lx+m) = 0.$$

We know that $x = x_1$ is a multiple root of the monic cubic polynomial $g(x)$ and $x = x_0$ is a root of $g(x)$. This implies that

$$g(x) = (x + x_1)^2(x + x_0).$$

It follows that

$$x^3 + (a_2 + l^2 + l)x^2 + mx + (a_6 + m^2) = x^3 + x_0x^2 + x_1^2x + x_1^2x_0,$$

i.e,

$$(a_2 + l^2 + l) = x_0, \; m = x_1^2, \; (a_6 + m^2) = x_1^2 x_0.$$

This implies that

$$l^2 + l = x_0 + a_2.$$

Since $-P = (x_0, y_0 + x_0)$ lies on $L : y = lx + m$, we obtain

(33) $m = (y_0 + x_0) - lx_0 = y_0 + (l + 1)x_0, \; x_1 = \sqrt{m}, \; y_1 = lx_1 + m = l\sqrt{m} + m.$

If we replace l by $l + 1$, then we should replace m by $m + x_0$ and x_1 by $x_1 + \sqrt{x_0}$. We also obtain the following formulas:

$$\sqrt{a_6} + m = x_1\sqrt{x_0}, \; \sqrt{a_6} + y_0 + (l + 1)x_0 = x_1\sqrt{x_0}. \tag{34}$$

EXAMPLE 10.2. Suppose that $x_0 = 0$, i.e., P is a point of order 2. Then

$$m = \sqrt{a_6}, \; x_1 = \sqrt[4]{a_6}, \; y_1 = l\sqrt[4]{a_6} + \sqrt{a_6}.$$

If, in addition, $a_2 = 0$, then $l = 0$ or 1, and we get two halves $Q_1 = (\sqrt[4]{a_6}, \sqrt{a_6})$ and $Q_1 = (\sqrt[4]{a_6}, \sqrt[4]{a_6} + \sqrt{a_6})$ of P. This implies that Q_1 and Q_2 are (the only) points of order 4.

The following assertion may be extracted from [**4**, Prop. 1.1].

THEOREM 10.3. *Suppose that $E(K)$ contains a point of order 2, i.e., there exists $\beta \in K$ such that $a_6 = \beta^2$. Then a point $P = (x_0, y_0) \in E(K)$ is divisible by 2 in $E(K)$ if and only if the following conditions hold.*

(i) *x_0 is a square in K, i.e., there exists $r \in K$ such that $r^2 = x_0$.*
(ii) *There exists $l \in K$ such that*

$$l^2 + l = x_0 + a_2.$$

(iii) *If $x_0 = 0$, i.e., P is a point of order 2, then a_6 is a fourth power in K.*

PROOF. Let $P = (x_0, y_0)$ is a K-point on E.

Assume that $P \in 2E(K)$, i.e., there exists a point $Q = (x_1, y_1)$ on E such that $2Q = P$ and $x_1, y_1 \in K$. Then in the notation above $l, m \in K$, because the tangent line to E at K-point Q is defined over K. It follows that (ii) holds. If $x_0 = 0$, then it follows from Example 10.2 that conditions (i)-(iii) hold. So, we may assume that $x_0 \neq 0$. It follows from (34) that $\beta + m = x_1\sqrt{x_0}$. This implies that

$$\sqrt{x_0} = \frac{\beta + m}{x_1}$$

lies in K, i.e., (i) holds.

Now assume that conditions (i)-(iii) hold and $Q = (x_1, y_1) \in E(\bar{K})$ satisfies $2Q = P$. If $x_0 = 0$, then the explicit formulas of Example 10.2 tell us that $Q \in E(K)$. Suppose that $x_0 \neq 0$. Let $\mathcal{L} : y = lx + m$ be the equation of the tangent equation to E at Q. The condition (ii) implies that $l \in K$. Since $\mathcal{L}$ contains the K-point $-P$, m also lies in K. Now equation (34) tells us that $\beta + m = x_1 r$ with nonzero $r = \sqrt{x_0} \in K$, and therefore

$$x_1 = \frac{\beta + m}{r}$$

also lies in K. This implies that $y_1 = lx_1 + m$ also lies in K. □

COROLLARY 10.4. *An elliptic curve over a field K of characteristic 2 has a K-rational point of order 4 if and only if there exists a nonzero $\gamma \in K$ such that E is K-isomorphic to the elliptic curve*

$$\mathbf{E}_{4,\gamma} : y^2 + xy = x^3 + \gamma^4.$$

PROOF. The result follows almost immediately from Theorem 10.3 combined with Example 10.2. Indeed, if an elliptic curve E over K has a K-point of order 4, then it has a K-point of order 2 and therefore is ordinary, i.e., is K-isomorphic to

$$y^2 + xy = x^3 + a_2x^2 + a_6,$$

where

$$a_2, a_6 \in K;\ a_6 \neq 0,\ j(E) = \frac{1}{a_6} \neq 0$$

and a_6 is a square in K. In addition, $P = (0, \sqrt{a_6}) \in E(K)$ is a point of order 2 that actually lies in $2E(K)$. It follows from Theorem 10.3 that there exist $\beta, l \in K$ such that

$$\beta^4 = a_6,\ l^2 + l = a_2.$$

One should only notice that the change of variables $x \mapsto x, y \mapsto y + lx$ establishes a K-isomorphism between elliptic curves $y^2 + xy = x^3 + a_2x^2 a_6$ and

$$y^2 + xy = x^3 + (a_2 + l^2 + l)x^2 + a_6 = x^3 + \beta^4.$$

□

COROLLARY 10.5. *Let E be an elliptic curve over K such that $E(K)$ contains a point of order 4. Then $j(E)$ is a nonzero fourth power in K.*

Conversely, if c is a nonzero fourth power in F, then there exists an elliptic curve E over K such that $j(E) = c$ and $E(K)$ contains a point of order 4. Such an E is unique up to K-isomorphism.

PROOF. Suppose $E(K)$ contains a point of order 4. By Corollary 10.4, there exists a nonzero $\gamma \in K$ such that E is K-isomorphic to

$$\mathbf{E}_{4,\gamma} : y^2 + xy = x^3 + \gamma^4.$$

Clearly,

$$j(E) = j(\mathbf{E}_{4,\gamma}) = \frac{1}{\gamma^4} = \left(\frac{1}{\gamma}\right)^4.$$

This proves the first assertion.

Conversely, if $\delta \in K$ satisfies $\delta^4 = c$, then we put $\gamma = 1/\delta$ and consider the elliptic curve $\mathbf{E}_{4,\gamma}$ over K. Recall that $\mathbf{E}_{4,\gamma}(K)$ contains a point of order 4 and

$$j(\mathbf{E}_{4,\gamma}) = \frac{1}{\gamma^4} = \left(\frac{1}{\gamma}\right)^4 = \delta^4 = c.$$

This proves the second assertion. In order to prove the uniqueness, let us assume that E is an elliptic curve over K such that $j(E) = c$ and $E(K)$ contains a point of order 4. Then, thanks to Corollary 10.4, there exists nonzero $\gamma \in K$ such that E is K-isomorphic to $\mathcal{E}_{4,\gamma}$. This implies that

$$c = j(E) = j(\mathbf{E}_{4,\gamma}) = \frac{1}{\gamma^4}.$$

It follows that $\gamma = \sqrt[4]{1/c}$. □

THEOREM 10.6. *An elliptic curve over a field K of characteristic 2 has a K-rational point of order 8 if and only if there exists $t \in K \setminus \{0,1\}$ such that E is K-isomorphic to the elliptic curve*

$$\mathbf{E}_{8,t} : y^2 + xy = x^3 + \left(\frac{t}{t^2+1}\right)^8 .$$

PROOF. If an elliptic curve E over K has a K-point of order 8, then it has a K-point of order 4, and therefore by Corollary 10.4 is K-isomorphic to

$$\mathbf{E}_{4,\gamma} : y^2 + xy = x^3 + \gamma^4,$$

where γ is a nonzero element of K. A point Q of order 4 in $E(K)$ has x-coordinate $\gamma \neq 0$. Since Q is divisible by 2 in $E(K)$, Theorem 10.3 tells us that there exist $l, r \in K$ such that

$$l^2 + l = \gamma = r^2.$$

Now the parametrization

$$l = \frac{1}{t^2+1}, \quad r = \frac{t}{t^2+1}$$

of the conic

$$l^2 + l = r^2$$

gives us the formula

$$\gamma = \left(\frac{t}{t^2+1}\right)^2 .$$

□

REMARK 10.7. Let t, s be two nonzero elements of K. Then

$$t + \frac{1}{t} = s + \frac{1}{s}$$

if and only if $s = t$ or $s = t^{-1}$. This implies that $\mathbf{E}_{8,t} = \mathbf{E}_{8,s}$ if and only if $s = t^{\pm 1}$. (Hereafter both s and t are different from 1.) On the other hand,

$$j(\mathbf{E}_{8,t}) = \left(t + \frac{1}{t}\right)^{-8}, \quad j(\mathbf{E}_{8,s}) = \left(s + \frac{1}{s}\right)^{-8}.$$

This implies that $\mathbf{E}_{8,t}$ and $\mathbf{E}_{8,s}$ are isomorphic over $\bar{K}$ if and only if $s = t^{\pm 1}$, i.e., if and only if $\mathbf{E}_{8,t} = \mathbf{E}_{8,s}$.

REMARK 10.8. Let us find explicitly a point of order 8 in $\mathbf{E}_{8,t}(K)$. We have the order 4 point

$$P = (x_0, y_0) = (\gamma, \gamma^2) \in \mathbf{E}_{8,t}(K)$$

and the equalities

$$l = \frac{1}{t^2+1}, \quad r = \frac{t}{t^2+1}, \quad x_0 = \gamma = \left(\frac{t}{t^2+1}\right)^2 = l^2 + l = r^2, \quad y_0 = \gamma^2 = \left(\frac{t}{t^2+1}\right)^4 .$$

Let $Q = (x_1, y_1) \in \mathbf{E}_{8,t}(K)$ satisfy $2Q = P$ and l be the slope of the tangent line $y = lx + m$ to E at Q. Then

$$(y_0 + x_0) - lx_0 = m = x_1^2.$$

This implies that

$$(\gamma^2 + \gamma) - \frac{\gamma}{t^2+1} = m = x_1^2,$$

i.e.,

$$x_1 = (\gamma + \sqrt{\gamma}) + \frac{\sqrt{\gamma}}{t+1} = \left(\frac{t}{t^2+1}\right)^2 + \frac{t}{t^2+1} + \frac{t}{(t+1)^3} =$$

$$\frac{t^2 + t(t^2+1) + t(t+1)}{(t+1)^4} = \frac{t^3}{(t+1)^4}.$$

We have

$$m = x_1^2 = \frac{t^6}{(t+1)^8},\ y_1 = lx_1 + m = \frac{1}{t^2+1}\frac{t^3}{(t+1)^4} + \frac{t^6}{(t+1)^8} = \frac{t^3}{(t+1)^6} + \frac{t^6}{(t+1)^8} =$$

$$\frac{t^3(t^2+1)+t^6}{(t+1)^8}.$$

To summarize:

$$Q = \left(\frac{t^3}{(t+1)^4}, \frac{t^6+t^5+t^3}{(t+1)^8}\right) \in \mathcal{E}_{8,t}(K)$$

is a point of order 8.

COROLLARY 10.9. *Let q be a power of 2 and $\mathbb{F}_q$ a finite field that consists of q elements. Let N be a power of 2 and $\Sigma(q,N)$ the set of $\mathbb{F}_q$-isomorphism classes of elliptic curves E over $\mathbb{F}_q$ such that $E(\mathbb{F}_q)$ contains a point of order N. Then*

$$|\Sigma(q,4)| = q-1,\ |\Sigma(q,8)| = \frac{q}{2} - 1.$$

PROOF. We have $j(\mathbf{E}_{4,\gamma}) = 1/\gamma^4$. Since γ may take any nonzero value in $\mathbb{F}_q$, it follows from Corollary 10.4 that $|\Sigma(q,4)| = q-1$.

It follows from Theorem 10.6 combined with Remark 10.7 that there is a bijection between $\Sigma(q,8)$ and the set of all unordered pairs $\{(t,t^{-1}) \mid t \in \mathbb{F}_q \setminus \{0,1\}\}$. Since $t \neq t^{-1}$ for each $t \neq 0,1$, the set of such pairs consists of $(q-2)/2$ elements. □

EXAMPLE 10.10. Let us take $q = 4$ and $K = \mathbb{F}_4$. Then there is exactly one elliptic curve E over $\mathbb{F}_4$ (up to an $\mathbb{F}_4$-isomorphism) such that $E(\mathbb{F}_4)$ contains a point of order 8, namely,

$$\mathbf{E}_{4,\rho} = \mathbf{E}_{4,\rho+1} : y^2 + xy = x^3 + 1,$$

where $\rho \in \mathbb{F}_4 \setminus \mathbb{F}_2$ satisfies

$$\rho^2 + \rho + 1 = 0, \rho^3 = (\rho+1)^3 = 1, \rho^{-1} = \rho + 1,\ \rho + \rho^{-1} = 1.$$

The group $\mathbf{E}_{4,\rho}(\mathbb{F}_4)$ contains a point Q of order 8, namely (Remark 10.8),

$$Q = \left(\frac{\rho^3}{(\rho+1)^4}, \frac{\rho^6+\rho^5+\rho^3}{(\rho+1)^8}\right) = (\rho,\rho).$$

This implies that the order of the finite group $\mathbf{E}_{4,\rho}(\mathbb{F}_4)$ is divisible by 8. On the other hand, the Hasse bound tells us that the order of $\mathbf{E}_{4,\rho}(\mathbb{F}_4)$ does not exceed $4 + 2\sqrt{4} + 1 < 8 \cdot 2$. This implies that $\mathbf{E}_{4,\rho}(\mathbb{F}_4)$ has order 8 and therefore coincides with its cyclic subgroup of order 8 generated by Q.

References

[1] B. M. Bekker and Yu. G. Zarhin, *Division by 2 of rational points on elliptic curves* (Russian, with Russian summary), Algebra i Analiz **29** (2017), no. 4, 196–239, DOI 10.1090/spmj/1512; English transl., St. Petersburg Math. J. **29** (2018), no. 4, 683–713. MR3708868

[2] J. W. S. Cassels, *Diophantine equations with special reference to elliptic curves*, J. London Math. Soc. **41** (1966), 193–291, DOI 10.1112/jlms/s1-41.1.193. MR0199150

[3] A. W. Knapp, *Elliptic curves*, Mathematical Notes, vol. 40, Princeton University Press, Princeton, NJ, 1992. MR1193029

[4] K. Kramer, *Two-descent for elliptic curves in characteristic two*, Trans. Amer. Math. Soc. **232** (1977), 279–295, DOI 10.2307/1998941. MR0441977

[5] D. S. Kubert, *Universal bounds on the torsion of elliptic curves*, Proc. London Math. Soc. (3) **33** (1976), no. 2, 193–237, DOI 10.1112/plms/s3-33.2.193. MR0434947

[6] E. F. Schaefer, *2-descent on the Jacobians of hyperelliptic curves*, J. Number Theory **51** (1995), no. 2, 219–232, DOI 10.1006/jnth.1995.1044. MR1326746

[7] N. Schappacher and R. Schoof, *Beppo Levi and the arithmetic of elliptic curves*, Math. Intelligencer **18** (1996), no. 1, 57–69, DOI 10.1007/BF03024818. MR1381581

[8] A. Silverberg, *Explicit families of elliptic curves with prescribed mod N representations*, Modular forms and Fermat's last theorem (Boston, MA, 1995), Springer, New York, 1997, pp. 447–461. MR1638488

[9] J. H. Silverman, *The arithmetic of elliptic curves*, 2nd ed., Graduate Texts in Mathematics, vol. 106, Springer, Dordrecht, 2009. MR2514094

[10] S. Lang, *Elliptic functions*, 2nd ed., Graduate Texts in Mathematics, vol. 112, Springer-Verlag, New York, 1987. With an appendix by J. Tate. MR890960

[11] L. C. Washington, *Elliptic curves*, 2nd ed., Discrete Mathematics and its Applications (Boca Raton), Chapman & Hall/CRC, Boca Raton, FL, 2008. Number theory and cryptography. MR2404461

Department of Mathematics and Mechanics, St. Petersburg State University, Universitetsky prospekt, 28, Peterhof, St. Petersburg, 198504, Russia

Email address: bekker.boris@gmail.com

Department of Mathematics, Pennsylvania State University, University Park, Pennsylvania 16802

Email address: zarhin@math.psu.edu

Contemporary Mathematics
Volume **724**, 2019
https://doi.org/10.1090/conm/724/14583

The weighted moduli space of binary sextics

Lubjana Beshaj and Scott Guest

ABSTRACT. We use the weighted moduli height to investigate the distribution of fine moduli points in the moduli space of genus two curves. We show that for any genus two curve with equation $y^2 = f(x)$, its weighted moduli height $\mathfrak{h}(\mathfrak{p}) \leq 2^3\sqrt{3 \cdot 5 \cdot 7} \cdot H(f)$, where $H(f)$ is the minimal naive height of the curve. Based on the weighted moduli height $\mathfrak{h}$ we create a database of genus two curves defined over $\mathbb{Q}$ with small $\mathfrak{h}$ and show that for small such height ($\mathfrak{h} < 5$) about 30% of points are fine moduli points.

1. Introduction

In [**37**] authors studied heights of algebraic curves and compared the naive height with the moduli height. Their examples focused mostly on curves of genus two. In [**5**], the authors created a database of rational points in the moduli space of genus two curves $\mathcal{M}_2$ with moduli height $\mathfrak{h} \leq 20$ and considered the problem of what percentage of these points have corresponding genus 2 curves defined over $\mathbb{Q}$. It is widely believed that for almost all points $\mathfrak{p} \in \mathcal{M}_2(\mathbb{Q})$, there is no genus two representative defined over $\mathbb{Q}$. In the language of arithmetic geometry this means that for almost all genus two curves the field of moduli is not a minimal field of definition. However, results in [**5**] were somewhat surprising. For all $\mathfrak{h} \leq 20$ roughly 30% of the points the field of moduli was a field of definition. There were a couple interpretations to this.

Firstly, the points in $\mathcal{M}_2(\mathbb{Q})$ were represented using the absolute invariants i_1, i_2, i_3 of genus two curves. These are rational functions in terms of the Igusa invariants J_2, J_4, J_6, and J_{10}, see [**23**], [**5**], [**28**], but they are not defined when $J_2 = 0$. To handle the case $J_2 = 0$ another compactification of the moduli space was used. When using two different sets of invariants however, there is no natural means to make statements about interesting questions regarding the moduli space of genus two curves.

Secondly, it is possible that the "majority" of moduli points for which the field of moduli is a field of definition are concentrated around the "center" of the moduli space. In other words, as the moduli height $\mathfrak{h}$ goes to ∞, the percentage of curves for which the field of moduli is a field of definition tends to zero. However, computations in [**5**] didn't quite show this trend.

2010 *Mathematics Subject Classification.* Primary 14H10,14H45.

In [**29**] a "universal" equation of genus two curves was discovered. Here universal means that it works for any tuple of arithmetic invariants (J_2, J_4, J_6, J_{10}) including the case when $J_2 = 0$ or the case when the automorphism group has order > 2, which was treated separately in Mestre's original algorithm [**31**] and has not been implemented in most computational algebra packages. This raised the question on whether we could perform computations done in [**5**], but now using $[J_2, J_4, J_6, J_{10}]$ and the weighted projective space $\mathbb{WP}^3_{(2,4,6,10)}$ instead of the coordinate (i_1, i_2, i_3) used in [**5**]?

The main question became whether there was any measure of height for the weighted projective spaces which could be used for $\mathbb{WP}^3_{(2,4,6,10)}$ in order to perform a similar analysis as in [**5**]. Luckily, [**30**] came out which does precisely what we needed, defines a height on weighted projective spaces. Using this height for $\mathbb{WP}^3_{(2,4,6,10)}$ we were able to determine a unique tuple (J_2, J_4, J_6, J_{10}) for any point $\mathfrak{p} \in \mathbb{WP}^3_{(2,4,6,10)}$ and compile a database of genus two curves based on this weighted projective height.

In this database every point is represented by a unique minimal tuple of arithmetic invariants (J_2, J_4, J_6, J_{10}). Such database is valid for curves over any field of characteristic different from two. In the case of characteristic two we can do exactly the same thing by adding another invariant J_8 as explained in [**23**]. The disadvantage of using the weighted moduli height is that since powers of invariants are involved, one can not perform computations for very big heights. In this paper we only go for heights up to $\mathfrak{h} = 4$.

This paper is organized as follows. In section 2.1, we start with giving a brief definition of a weighted projective space and then define the height function on this space, following closely the approach in [**30**].

In section 3 we explain how we build the database. We start with creating a database of normalized tuples $\mathfrak{p} = [J_2 : J_4 : J_6 : J_{10}] \in \mathbb{WP}^3_{(2,4,6,10)}(\mathbb{Q})$ with weighted height $\mathfrak{h} \leq 4$ and $J_{10} \neq 0$. Then, in subsection 3.1 we explain how for each point $\mathfrak{p} \in \mathcal{M}_2(\mathbb{Q})$ we can compute the equation of the curve over the rationals, when such exists. In [**29**] the authors do exactly this and construct for every point $\mathfrak{p} \in \mathcal{M}_2$ the equation of the corresponding genus 2 curve. While we know that each $\mathfrak{p}$ is defined over $\mathbb{Q}$ if it corresponds to a curve defined over $\mathbb{Q}$, the converse in general is not true. It is then necessary to determine if a given moduli point actually corresponds to a curve over $\mathbb{Q}$. We say a moduli point $\mathfrak{p} \in \mathcal{M}_2(\mathbb{Q})$ is a fine point if and only we can find a representative curve $\mathcal{X}_\mathfrak{p}$ defined over $\mathbb{Q}$, otherwise we call it a coarse point or an obstruction point.

For each such tuple $\mathfrak{p} = [J_2 : J_4 : J_6 : J_{10}] \in \mathbb{WP}^3_{(2,4,6,10)}(\mathbb{Q}) \setminus \{J_{10} = 0\}$, we also compute the automorphism group of the corresponding curve, using algorithms presented in [**5**]. We then present the data on the distribution of fine points in the moduli space of genus two curves. Namely, we compare the number of moduli points below a given weighted moduli height to the number of fine points below that height for all weighted moduli heights $\mathfrak{h} < 5$. Finally, we give a table of specific results for all moduli points with $\mathfrak{h} < 1$.

Notation: Throughout this paper by a curve we mean a smooth, irreducible algebraic curve. Unless otherwise noted a curve $\mathcal{X}$ means the isomorphism class of $\mathcal{X}$ over some field k. With $\mathbb{WP}^3_{(2,4,6,10)}(\mathbb{Q})$ we denote the weighted projective space with weights $(2, 4, 6, 10)$ and $\mathfrak{h}(\mathfrak{p})$ we denote the weighted moduli height of a point $\mathfrak{p} \in \mathbb{WP}^3_{(2,4,6,10)}(\mathbb{Q})$ which we define in the preliminaries.

2. Preliminaries

Let $\mathcal{X}$ be a genus 2 curve defined over a field k with characteristic 0 or a prime $p \neq 2$. It is a well known fact that the affine equation of $\mathcal{X}$ can be given by

$$\mathcal{X} : y^2 = f(x) = a_6 \sum_{i=1}^{6} (x - \alpha_i) = a_6 x^6 + a_5 x^5 + \cdots + a_0 \tag{1}$$

where $\{\alpha_i\}_{i=1}^6$ are the ramification points of the map $\mathcal{X} \to \mathbb{P}^1$. The isomorphism classes of genus 2 curves are on one to one correspondence with the orbits of the $GL_2(k)$-action on the space of binary sextics. Igusa proved that the invariant ring $\mathbb{R}_6$ is generated by the Igusa arithmetic invariants J_2, J_4, J_6, J_{10}; see [**5, 23**] for details. Note that Igusa J-invariants $\{J_{2i}\}$, for $i = 1, 2, 3, 5$, are homogenous polynomials of degree $2i$ in $k[a_0, \dots, a_6]$ and J_{10} is the discriminant of the curve. Since in this paper we are working with smooth, irreducible curves than we take $J_{10} \neq 0$. The definitions of these invariants can be found in several papers such as [**5, 23, 39**].

2.1. The weighted projective space. In this subsection we will follow closely the notation in [**30**]. Let k be a field of characteristic zero and $(q_0, \dots, q_n)$ a fixed tuple of positive integers called **weights**. Consider the action of $k^\star$ on $\mathbb{A}^{n+1}$ as follows

$$\lambda * (x_0, \dots, x_n) = (\lambda^{q_0} x_0, \dots, \lambda^{q_n} x_n)$$

for $\lambda \in k^*$. The quotient of this action is called a **weighted projective space** and denoted by $\mathbb{WP}^n_{(q_0,\dots,q_n)}$. It is the projective variety $Proj\,(k[x_0, ..., x_n])$ associated to the graded ring $k[x_0, \dots, x_n]$ where the variable x_i has degree q_i for $i = 0, \dots, n$.

In [**30**] they define the concept of height on the weighted projective spaces. In this subsection we will mostly focus on the space $\mathbb{WP}^3_{(2,4,6,10)}(\mathcal{O}_K) \setminus \{J_{10} = 0\}$. Let $\mathfrak{p} = [J_2, J_4, J_6, J_{10}] \in \mathbb{WP}^3_{(2,4,6,10)}(\mathcal{O}_K) \setminus \{J_{10} = 0\}$ be a given weighted moduli point in the weighted projective space. We define the tuple $\mathfrak{p}$ to be a **normalized weighted moduli point** if the weighted greatest common divisor of its coordinates is one, where the weighted greatest common divisor is defined as follows

$$\operatorname{wgcd}(J_{2i}) = \prod_{v \in M_K} \prod_{\substack{d^{q_i} | J_i \\ d \in \mathcal{O}_K}} |d|_v$$

for all valuation v in the set of valuations M_K and divisors $d \in \mathcal{O}_K$ such that for all $i = 2, 4, 6, 10$ we have $d^i | J_i$. Note that if $K = \mathbb{Q}$ then,

$$\operatorname{wgcd}(J_{2i}) = \prod_{\substack{p^{q_i} | J_i \\ p \in \mathbb{Z}}} |p|$$

for all primes $p \in \mathbb{Z}$. The normalization of the point $\mathfrak{p} \in \mathbb{WP}^3_{\mathfrak{w}}(K)$ is

$$\mathfrak{q} = \frac{1}{\operatorname{wgcd}(\mathfrak{p})} \star \mathfrak{p}.$$

and from [**30**, Lemma. 1] we have that since $w = (2, 4, 6, 10)$ and $d = \gcd(2, 4, 6, 10) = 2$, then this normalization is unique up to a multiplication by a d-th root of unity. In other words, for every normalized point $\mathfrak{p} = [J_2, J_4, J_6, J_{10}] \in \mathbb{WP}^3_{\mathfrak{w}}(\mathbb{Q})$, there is another normalized point $\mathfrak{p}' = [-J_2, J_4, -J_6, -J_{10}]$ equivalent to $\mathfrak{p}$. Moreover, $\mathfrak{p}$ and $\mathfrak{p}'$ are isomorphic over the Gaussian integers, see [**30**, Cor. 2] for more details.

REMARK 1. *Notice that if the first coordinate $J_2 = 0$ then the normalization in the space $\mathbb{WP}^3_{(2,4,6,10)}(\mathbb{Q})$ is unique up to a multiplication by a $d = \gcd(4,6,10)$-th root of unity. For example, the pair of points $\mathfrak{p}$ and $\mathfrak{p}'$ in Table 1 are twists of each other, where $\mathfrak{p}' = \sqrt{-1}\cdot\mathfrak{p}$. In the coming section we will see why these are the only tuples of height $h = 1$ for which this is true.*

TABLE 1. Twists of height $\mathfrak{h} = 1$.

#	$\mathfrak{p}$	$\mathfrak{p}' = \sqrt{-1}\cdot\mathfrak{p}$
1	[0, -1, -1, -1]	[0, -1, 1, 1]
2	[0, -1, -1, 1]	[0, -1, 1, -1]
3	[0, -1, 0, -1]	[0, -1, 0, 1]
4	[0, 0, -1, -1]	[0, 0, 1, 1]
5	[0, 0, -1, 1]	[0, 0, 1, -1]
6	[0, 0, 0, -1]	[0, 0, 0, 1]
7	[0, 1, -1, -1]	[0, 1, 1, 1]
8	[0, 1, -1, 1]	[0, 1, 1, -1]
9	[0, 1, 0, -1]	[0, 1, 0, 1]

Next, we will see when a point $\mathfrak{p} \in \mathbb{WP}^3_w(\mathbb{Q})$ is normalized over $\overline{\mathbb{Q}}$. The point $\mathfrak{p}$ is called **absolutely normalized**, i.e. normalized over $\overline{\mathbb{Q}}$ if $\overline{\text{wgcd}}(\mathfrak{p}) = 1$ where

$$\overline{\text{wgcd}}(\mathfrak{p}) = \prod_{\substack{\lambda^{q_i} | J_i \\ \lambda \in \overline{\mathbb{Q}}}} |\lambda|$$

such that for all $i = 2,4,6,10$, $\lambda^i \in \mathbb{Z}$. We can compute the absolutely normalization of a point in the weighted projective space the same way as we computed the normalization of that point.

Two absolutely normalized tuples are twists of each other if and only if they are isomorphic over the Gaussian integers.

The **weighted multiplicative height** of $P = (x_0, \dots, x_n) \in \mathbb{P}^n_w(K)$ over K is defined as

$$\mathfrak{h}_K(P) = \frac{1}{\text{wgcd}(\mathfrak{p})} \prod_{v \in M_K} \max\left\{ |x_0|_v^{1/q_0}, \dots, |x_n|_v^{1/q_n} \right\} \tag{2}$$

for all valuations v is the set of valuations M_K and the **absolute weighted multiplicative height** over $\overline{\mathbb{Q}}$

$$\tilde{\mathfrak{h}}(P) = \frac{1}{\overline{\text{wgcd}}(\mathfrak{p})} \prod_{v \in M_K} \max\left\{ |x_0|_v^{1/q_0}, \dots, |x_n|_v^{1/q_n} \right\}. \tag{3}$$

Note that if we consider points $\mathfrak{p} \in \mathbb{WP}^3_w(\mathbb{Q})$ then it is clear that $\mathfrak{p}$ will have a representative over the integers. With such a representative for the coordinates of $\mathfrak{p}$, the non-Archimedean absolute values give no contribution to the height and the only one that matters is the Archimedean absolute value, i.e. just the maximum of the coordinates of the given point.

Next, we show how we can calculate the absolute minimal tuple and absolute minimum height of all zero-dimensional loci in $\mathcal{M}_2$. All such curves are defined over $\mathbb{Q}$, so we consider their respective weighted moduli points as existing in $\mathbb{WP}^3_{\mathfrak{w}}(\mathbb{Z})$.

EXAMPLE 1. *Let $\mathcal{X}$ be the genus two curve with equation $y^2 = x^6 - 1$. Its weighted moduli point is*

$$\mathfrak{p}_1 = [240, 1620, 119880, 46656] = \left[2^4 \cdot 3 \cdot 5, 2^2 \cdot 3^4 \cdot 5, 2^3 \cdot 3^4 \cdot 5 \cdot 37, 2^6 \cdot 3^6\right].$$

which is a minimal tuple since there is no $p \in \mathbb{Z}$ such that $p^{2i} | J_{2i}$, for $i = 1, 2, 3, 5$. This tuple has three twists for values of $\lambda = 2, 3, 6$. The corresponding twists are

$$\begin{aligned}
\mathfrak{p}_2 &= \frac{1}{2} \star \mathfrak{p} = [120, 405, 14985, 1458] = \left[2^3 \cdot 3 \cdot 5, 3^4 \cdot 5, 3^4 \cdot 5 \cdot 37, 2 \cdot 3^6\right] \\
\mathfrak{p}_3 &= \frac{1}{3} \star \mathfrak{p} = [80, 180, 4440, 192] \quad = \left[2^4 \cdot 5, 2^2 \cdot 3^2 \cdot 5, 2^3 \cdot 3 \cdot 5 \cdot 37, 2^6 \cdot 3\right] \\
\mathfrak{p}_4 &= \frac{1}{6} \star \mathfrak{p} = [40, 45, 555, 6] \quad = \left[2^3 \cdot 5, 3^2 \cdot 5, 3 \cdot 5 \cdot 37, 2 \cdot 3\right]
\end{aligned}$$

The heights are

$$\mathfrak{h}(\mathfrak{p}_1) = 4\sqrt{15},\ \mathfrak{h}(\mathfrak{p}_2) = 2\sqrt{30},\ \mathfrak{h}(\mathfrak{p}_3) = 4\sqrt{5},\ \mathfrak{h}(\mathfrak{p}_4) = 2\sqrt{10}.$$

The absolute minimal tuple is

$$\mathfrak{p} = [40, 45, 555, 6] = \left[2^3 \cdot 5, 3^2 \cdot 5, 3 \cdot 5 \cdot 37, 2 \cdot 3\right].$$

Hence, the absolute minimum height is

$$\bar{\mathfrak{h}}(\mathfrak{p}) = 2\sqrt{10}.$$

EXAMPLE 2. *Let $\mathcal{X}$ be the genus two curve with equation $y^2 = x^5 - x$. Its weighted moduli point is*

$$\mathfrak{p}_1 = [-40, -80, 320, -256] = \left[-1 \cdot 2^3 \cdot 5, -1 \cdot 2^4 \cdot 5, 2^6 \cdot 5, -1 \cdot 2^8\right],$$

which is a minimal tuple since there is no $p \in \mathbb{Z}$ such that $p^{2i} | J_{2i}$, for $i = 1, 2, 3, 5$. This tuple has a sole twist for the value $\lambda = 2$. This twist is

$$\mathfrak{p}_2 = [-20, -20, 40, -8] = \left[-1 \cdot 2^2 \cdot 5, -1 \cdot 2^2 \cdot 5, 2^3 \cdot 5, -1 \cdot 2^3\right].$$

The heights are

$$\mathfrak{h}(\mathfrak{p}_1) = 2\sqrt{10},\ \mathfrak{h}(\mathfrak{p}_2) = 2\sqrt{5}.$$

The absolute minimal tuple is

$$\mathfrak{p} = [-20, -20, 40, -8] = \left[-1 \cdot 2^2 \cdot 5, -1 \cdot 2^2 \cdot 5, 2^3 \cdot 5, -1 \cdot 2^3\right].$$

Hence, the absolute minimum height is

$$\bar{\mathfrak{h}}(\mathfrak{p}) = 2\sqrt{5}.$$

EXAMPLE 3. *Let $\mathcal{X}$ be the genus two curve with equation $y^2 = x^6 - x$. Its weighted moduli point is*

$$\mathfrak{p}_1 = [0, 0, 0, 3125] = \left[0, 0, 0, 5^5\right].$$

which is a minimal tuple since there is no $p \in \mathbb{Z}$ such that $p^{2i} | J_{2i}$, for $i = 1, 2, 3, 5$. This tuple has a sole twist for the value $\lambda = 5$. This twist is

$$\mathfrak{p}_2 = [0, 0, 0, 1]$$

The heights are

$$\mathfrak{h}(\mathfrak{p}_1) = \sqrt[10]{3125},\ \mathfrak{h}(\mathfrak{p}_2) = 1$$

The absolute minimal tuple is

$$\mathfrak{p} = [0, 0, 0, 1] .$$

Hence, the absolute minimum height is

$$\bar{\mathfrak{h}}(\mathfrak{p}) = 1.$$

In [**37**] was defined the moduli height of genus g algebraic curves. Consider $\mathfrak{p} \in \mathcal{M}_g$, where $\mathcal{M}_g$ is the moduli space of smooth irreducible algebraic curves of genus $g \geq 2$. It is known that $\mathcal{M}_g$ is a quasi projective variety of dimension $3g-3$ and is embedded in $\mathbb{P}^{3g-2}$. The moduli height is the usual height in the projective space $\mathbb{P}^{3g-2}$, i.e. if $P \in \mathbb{P}^{3g-2}(\mathbb{Q})$ then

$$H(P) = \max \left\{ |x_0|, \ldots, |x_n| \right\} .$$

Moreover, for $g = 2$ where a moduli point is given by $\mathfrak{p} = [J_2, \ldots, J_{10}] \in \mathcal{M}_2(\mathbb{Q})$ then

$$H(\mathfrak{p}) = \max \left\{ |J_2|, \ldots, |J_{10}| \right\} .$$

If we consider $\mathfrak{p} \in \mathbb{WP}^3_w(\mathbb{Q})$ then the weighted moduli height is

$$\mathfrak{h}(\mathfrak{p}) = \max \left\{ |J_2|^{1/2}, \ldots, |J_{10}|^{1/10} \right\} .$$

Hence, by definition for a given moduli point the weighted moduli height is always smaller than the moduli height. And this makes computing databases of genus two curves based on the weighted moduli height much more convenient.

2.2. Naive height versus weighted moduli height. Let $\mathcal{X}$ be a genus two curve with affine equation given as

$$y^2 = a_6 x^6 + a_5 x^5 + \cdots + a_0 .$$

The naive height $H(\mathcal{X})$ of a curve (resp. polynomial) is the maximum of the absolute value of the coefficients of the curve (resp. polynomial); see [**37**] for details. In [**37**] and [**5**] were used both the naive height and the moduli height to analyze points in $\mathcal{M}_2(\mathbb{Q})$. Even though using the weighted moduli height makes more sense in analyzing points in $\mathcal{M}_2(\mathbb{Q})$, since a weighted moduli point determines uniquely an isomorphism class of the curves, it would be interesting to see how the weighted moduli height relates to the naive height. In [**37**] it is given an inequality how the naive height is compared to the moduli height. In this subsection we capture previous results and also compare the naive height to the weighted moduli height.

In [**37**] the naive height of a curve is compared with the moduli height and showed that for a genus 2 curve with equation $y^2 = f(x)$ the moduli height is bounded as follows

$$\tilde{\mathfrak{h}}(f) \leq 2^{28} \cdot 3^9 \cdot 5^5 \cdot 7 \cdot 11 \cdot 13 \cdot 17 \cdot 43 \cdot H(f)^{10} .$$

Note that there is a typo in the computations in [**37**] and the above coefficient should be $2^{24} \cdot 3^7 \cdot 5^4 \cdot 7 \cdot 11 \cdot 13 \cdot 43$. Next, we consider how the weighted moduli height is related to the naive height of the curve.

LEMMA 1. *Let $\mathcal{X}$ be a genus 2 curve and $\mathfrak{p}$ its corresponding point in $\mathbb{WP}^3_w(\mathbb{Q})$. Then the weighted moduli height is bounded as follows*

$$\mathfrak{h}(\mathfrak{p}) \leq 2^3 \sqrt{3 \cdot 5 \cdot 7} \cdot H(\mathcal{X})$$

PROOF. We assume that the equation of the curve is given as $y^2 = f(x)$. Recall that each J_{2i} is a degree $2i$ polynomial evaluated at f, i.e degree $2i$ polynomial given in $k[a_0, \cdots, a_6]$. Then, from Lemma 15 in [**37**] we have

$$H(J_{2i}(f)) \leq c_0 \cdot H(J_{2i}) \cdot H(f)^{2i}$$

where $c_0 = \binom{d+n}{n}$ is the number of monomials of a degree d homogenous polynomial in $n+1$ variables. Computations of $H(J_{2i})$ is done in Maple and we get

$$\begin{aligned}
H(J_2) &\leq 2^6 \cdot 3 \cdot 5 \cdot 7 \cdot H(f)^2 \\
H(J_4) &\leq 2^3 \cdot 3^5 \cdot 5^2 \cdot 7 \cdot H(f)^4 \\
H(J_6) &\leq 2^5 \cdot 3^5 \cdot 5 \cdot 7 \cdot 11 \cdot 37 \cdot H(f)^6 \\
H(J_{10}) &\leq 2^9 \cdot 3^5 \cdot 5 \cdot 7 \cdot 11 \cdot 13 \cdot H(f)^{10}
\end{aligned}$$

The weighted moduli height of $\mathfrak{p}$ is computed as follows

$$\mathfrak{h}(\mathfrak{p}) = \max\{H(J_2(f))^{1/2}, \dots, H(J_{10}(f))^{1/10}\} \leq 2^3\sqrt{3 \cdot 5 \cdot 7} \cdot H(f).$$

This concludes the proof. $\square$

Hence, we get that the height of the given curve with corresponding moduli point $\mathfrak{p} = [J_2 : J_4 : J_6 : J_{10}]$ is as follows

$$H(\mathcal{X}) \geq \frac{1}{2^3\sqrt{3 \cdot 5 \cdot 7}} \cdot \mathfrak{h}(\mathfrak{p})$$

but we don't know if this lower bound is strict and if it can ever be achieved. In [**37**] are displayed all genus 2 curves (up to isomorphism) with naive height $H = 1$. In [**37**, Tables 1-4] there are displayed 230 isomorphism classes of genus two curves with naive height one. There are 184 genus two curves with naive height one and automorphism group $\mathbb{Z}_2$, 30 with automorphism group isomorphic to V_4, 11 with automorphism group D_4, two curves with group D_6, and three with automorphism group respectively 10, 24, and 48.

In an upcoming paper we will check how the weighted moduli height of those curves in [**37**, Tables 1-4] is compared to their naive height and see if this lower bound is sharp for any of those curves.

3. Rational points in the weighted moduli space

Next we will build a database of all isomorphism classes of genus 2 curves with bounded weighted moduli height. We start with creating a database of normalized tuples in $\mathbb{WP}^3_w(\mathbb{Q}) \setminus \{J_{10} = 0\}$ of height $\leq h$, for a given h. Then, we can compute the equation of the curve for a given $\mathfrak{p} \in \mathbb{WP}^3_w(\mathbb{Q}) \setminus \{J_{10} = 0\}$ using methods explained in the upcoming subsection 3.1.

Let h be a positive integer. The number of points $\mathfrak{p} \in \mathbb{W}P^3_{(2,4,6,10)}(\mathbb{Q}) \setminus \{J_{10} = 0\}$ of weighted moduli height $\leq h$ is given by the following lemma.

LEMMA 2. *For any given positive integer h there are at most*

$$2h^{10}(h^2+1)(2h^4+1)(2h^6+1)$$

normalized points in $\mathbb{W}P^3_{(2,4,6,10)}(\mathbb{Q}) \setminus \{J_{10} = 0\}$ of weighted moduli height $\leq h$.

PROOF. Let $\mathcal{X}$ be a genus two curve and $\mathfrak{p} = [J_2, J_4, J_6, J_{10}] \in \mathbb{W}P^3_{(2,4,6,10)}(\mathbb{Q}) \setminus \{J_{10} = 0\}$ the moduli point representing $\mathcal{X}$ of height $\mathfrak{h}(\mathfrak{p}) = h$. From the definition of weighted moduli height for every $i = 1, 2, 3, 5$ we have that

$$|J_{2i}|^{1/2i} \leq h$$

Hence, for a given height h, every element J_{2i} of $\mathfrak{p}$ satisfies

$$-h^{2i} \leq J_{2i} \leq h^{2i} \text{ for } i = 1, 2, 3, 5.$$

Since, $J_{10} \neq 0$ then there are $2h^{10}$ possible integer values for J_{10}. In [**30**, Corr.2] it is proved that for every normalized point $\mathfrak{p} = [J_2, J_4, J_6, J_{10}] \in \mathbb{WP}^3_{(2,4,6,10)}(\mathbb{Q})$, there is another normalized point $\mathfrak{p}' = [-J_2, J_4, -J_6, -J_{10}]$ equivalent to $\mathfrak{p}$. Since, we only want to count equivalence classes then we let $J_2 \geq 0$ and therefore there are h^2+1 possible integer values for J_2. Lastly, for $i = 2, 3$ there are $(2h^{2i}+1)$-possible integer values for J_{2i}.

But, up to this point we didn't count the points $\mathfrak{p} \in \mathbb{WP}^3_{(2,4,6,10)}(\mathbb{Q})$ that satisfy the property that there exists a $\lambda \in \mathbb{Q}^\star$ such that

$$\lambda \star (J_2, \ldots, J_{10}) = (\lambda^2 J_2, \ldots, \lambda^{10} J_{10}). \tag{4}$$

But there is not a rigorous mathematical way of counting this points. Hence, the number of normalized points $\mathfrak{p} \in \mathbb{WP}^3_{(2,4,6,10)}(\mathbb{Q})$ will be at most

$$2h^{10}(h^2+1)(2h^4+1)(2h^6+1).$$

This concludes the proof. □

However, for computational reasons it is more convenient to consider the space $\mathbb{WP}^3_{(1,2,3,5)}(\mathbb{Q})$ instead of $\mathbb{W}P^3_{(2,4,6,10)}(\mathbb{Q})$. In this case the above lemma becomes as follows.

LEMMA 3. *For any given positive integer h there are at most*

$$2h^5(h+1)(2h^2+1)(2h^3+1)$$

normalized points in $\mathbb{WP}^3_{(1,2,3,5)}(\mathbb{Q})$ of weighted moduli height $\leq h$.

TABLE 2. Points of bounded height.

height	# of tuples	# of points in $\mathbb{WP}^3_{(1,2,3,5)}(\mathbb{Q})$
1	36	27
2	29 376	24 423
3	2 031 480	1 776 549
4	43 591 680	39 206 865

The proof is similar to the previous lemma. In Table 2 for weighted moduli height $\mathfrak{h} \leq 4$ we display how many possible tuples are there in $\mathbb{WP}^3_{(1,2,3,5)}(\mathbb{Q})$ and then in the next column we display how many of this tuples give us isomorphism classes of genus two curves. And this results agree with Lemma 3.

Then, in the upcoming Table 3 we display all points $\mathfrak{p} \in \mathbb{WP}^3_{(1,2,3,5)}(\mathbb{Q}) \setminus \{J_{10} = 0\}$ with weighted moduli heights $\mathfrak{h} = 1$.

TABLE 3. Tuples with weighted moduli height $\mathfrak{h} = 1$.

#	$\mathfrak{p} = [J_2 : J_4 : J_6 : J_{10}]$	#	$\mathfrak{p} = [J_2 : J_4 : J_6 : J_{10}]$	#	$\mathfrak{p} = [J_2 : J_4 : J_6 : J_{10}]$
1	[0, -1, 0, 1]	10	[1, 0, 1, 1]	19	[0, 1, 1, 1]
2	[0, 1, 0, 1]	11	[1, -1, -1, 1]	20	[1, 0, 1, -1]
3	[0, -1, 1, 1]	12	[1, 1, -1, 1]	21	[1, -1, -1, -1]
4	[0, 0, 0, 1]	13	[1, 1, 1, -1]	22	[1, 1, -1, -1]
5	[0, 0, 1, -1]	14	[1, -1, 1, -1]	23	[1, -1, 0, -1]
6	[0, 0, 1, 1]	15	[1, 1, 1, 1]	24	[1, 1, 0, -1]
7	[1, 0, -1, 1]	16	[1, 0, -1, -1]	25	[1, 1, 0, 1]
8	[1, 0, 0, -1]	17	[0, -1, 1, -1]	26	[1, -1, 0, 1]
9	[1, 0, 0, 1]	18	[0, 1, 1, -1]	27	[1, -1, 1, 1]

Note that in Table 3 the points 1 to 15 are fine moduli points, i.e. with methods that we will explain in the upcoming subsection we can compute their equations defined over $\mathbb{Q}$. We have already seen points 1 to 6 and 17 to 19 and their twists in the previous section in Table 1.

3.1. Equations of curves from points in $\mathbb{W}P^3_{(2,4,6,10)}(\mathbb{Q})$. In [**29**] for every point $\mathfrak{p} = [J_2 : J_4 : J_6 : J_{10}] \in \mathbb{W}P^3_{(2,4,6,10)} \setminus \{J_{10} = 0\}$ was constructed a genus two curve defined over a field of definition. We recapture that result below:

Let $\mathfrak{p} = [J_2, J_4, J_6, J_{10}] \in \mathbb{WP}^3_{\mathfrak{w}}(\mathcal{O}_K)$ for some number field K and $\mathcal{O}_K$ its ring of integers. There is a genus-two curve corresponding to $\mathfrak{p}$ as follows:

i) If $J_2 \cdot J_{10} \neq 0$, then there is a genus 2 curve $C_{(\alpha,\beta)}$ given by

$$C_{(\alpha,\beta)} : y^2 = \sum_{i=0}^{6} a_i(\alpha, \beta) x^i$$

with coefficients given in [**29**, Eq. 3.9 and Eq. 3.10] and a pair (α, β) satisfying

$$\alpha^2 + \lambda_6 \beta^2 \sigma = \gamma,$$

where λ_6, σ, and γ are given in terms of some ρ and κ, which are themselves determined by $\mathfrak{p}$. Their equations are given in [**29**]. Moreover, $C_{(\alpha,\beta)}$ is defined over its field of moduli K, i.e. $a_i(\alpha, \beta) \in K$, $i = 0, \ldots, 6$, if and only if K-rational α and β exist. For the explicit equation of $C_{(\alpha,\beta)}$ we refer the reader directly to [**29**, Theorem 1].

ii) If $J_2 = 0$ and $J_4 \cdot J_6 \cdot J_{10} \neq 0$, then there is a genus 2 curve given by setting $\rho = \kappa \neq 0$ in the previous equation.

iii) If $J_2 = J_6 = 0$ and $J_4 \cdot J_{10} \neq 0$, there is only one genus-two curve given by

$$\begin{aligned} y^2 =& (4\nu + 1)(2\nu - 1)x^6 + 2(1 - \nu)(4\nu + 3)x^5 - 15(1 - \nu)x^4 \\ &+ 20(1 - \nu)^2 x^3 + 5(2\nu - 3)(1 - \nu)^2 x^2 + 6(1 - \nu)^3 x - (1 - \nu)^3, \end{aligned}$$

where

$$\nu = \frac{J_4^5}{2^2 3^5 5^5 J_{10}^2}.$$

iv) If $J_2 = J_4 = 0$ and $J_6 \cdot J_{10} \neq 0$, there is only one genus-two curve given by
$$\begin{aligned} y^2 =& 5x^6 + 12(1-\mu)x^5 - 15(1-\mu)x^4 - 80(1-\mu)^2x^3 \\ &+ 15(4\mu-7)(1-\mu)^2x^2 - 60(1-\mu)^3x + (4\mu-13)(1-\mu)^3, \end{aligned}$$
where
$$\mu = \frac{J_6^5}{2^4 3^4 5^5 J_{10}^3}.$$

v) If $J_2 = J_4 = J_6 = 0$ and $J_{10} \neq 0$, there is only one genus-two curve given by
$$y^2 = x^6 - x.$$

Now that we have seen how to compute equations of curves we turn back to one of our main goals for this paper: Investigate what proportion of points in $\mathbb{W}P^3_{(2,4,6,10)}(\mathbb{Q})$ are defined over $\mathbb{Q}$. The following remark is believed to be true.

REMARK 2. *For almost all points $\mathfrak{p} \in \mathcal{M}_2(\mathbb{Q})$, the field of moduli is not a field of definition. In other words, the probability that a point $\mathfrak{p} \in \mathcal{M}_2(\mathbb{Q})$ has a representative curve defined over $\mathbb{Q}$ is zero.*

However, this was contradicted by previous results in [**5**] where absolute invariants i_1, i_2, i_3 were used to identify the isomorphism classes of curves. At first we believed, this was because such invariants are not everywhere defined and to build the database two different set of invariants were used, see [**5**] for more details. But, as we will see below we get the same contradicting results using points in $\mathbb{WP}^3_{(1,2,3,5)}(\mathbb{Q})$ with minimal weighted moduli height. The following result is well known by work of Cardona/Quer [**12**] or Shaska; see [**40**].

LEMMA 4. *Every genus 2 curve with automorphism group of order > 2 is defined over its field of moduli. Moreover, its equation over its field of moduli it's given in* [**20**, Lemma 6].

TABLE 4. The number of points in $\mathbb{WP}^3_{(1,2,3,5)}(\mathbb{Q})$ for each automorphism group.

$\mathfrak{h}$	# pts	D_4	D_6	V_4	C_2	C_{10}	#fine pts.	$\frac{\text{\#fine pts.}}{\text{\#pts.}}$
$0 < \mathfrak{h} \leq 1$	27	0	0	0	26	1	15	0.555
$1 < \mathfrak{h} \leq 2$	24 396	0	0	0	24 396	0	9 423	0.386
$2 < \mathfrak{h} \leq 3$	1 752 126	0	0	2	1 752 124	0	596 818	0.340
$3 < \mathfrak{h} \leq 4$	37 430 316	1	0	14	37 430 301	0	11 365 256	0.303

Hence, we are left to check for curves with automorphism group of order two. In Table 4 given below we display the number of points with height $\mathfrak{h} \leq 5$ for each possible group of automorphisms and the number of points that are defined over the rationals. We stop at $\mathfrak{h} = 5$ since then the computations get to big. As we can see most of the curves are curves with automorphism group $\mathbb{Z}_2$, as expected. From the data we notice that the percentage of the curves defined over the rationals gets smaller as the weighted moduli height gets higher and as we can see approximately 30% of the points are defined over the rationals. This results agree with previous results from [**5**] and we expect the ratio to converges to zero as the height gets bigger.

Acknowledgment

We want to thank Prof. Tony Shaska for sharing his code for the universal equation of genus two curves from [**29**] and his helpful discussions and suggestions.

References

[1] C. Bagiński, J. J. Etayo, G. Gromadzki, and E. Martínez, *On real forms of a Belyi action of the alternating groups*, Albanian J. Math. **10** (2016), no. 1, 3–10. MR3490049

[2] Edoardo Ballico, *Symmetric tensor rank and the identification of a point using linear spans of an embedded variety*, Albanian J. Math. **8** (2014), no. 1, 9–21. MR3270073

[3] Marie José Bertin and Odile Lecacheux, *Automorphisms of certain Niemeier lattices and elliptic fibrations*, Albanian J. Math. **11** (2017), no. 1, 13–33. MR3724575

[4] L. Beshaj, *Minimal weierstrass equations for genus 2 curves* (201612), available at 1612.08318.

[5] L. Beshaj, R. Hidalgo, S. Kruk, A. Malmendier, S. Quispe, and T. Shaska, *Rational points in the moduli space of genus two*, Higher genus curves in mathematical physics and arithmetic geometry, Contemp. Math., vol. 703, Amer. Math. Soc., Providence, RI, 2018, pp. 83–115, DOI 10.1090/conm/703/14132. MR3782461

[6] L. Beshaj and M. Polak, *On the weighted projective space of binary octavics*, Algebraic curves and cryptography, 2018, pp. 164–179.

[7] Lubjana Beshaj, *Reduction theory of binary forms*, Advances on superelliptic curves and their applications, NATO Sci. Peace Secur. Ser. D Inf. Commun. Secur., vol. 41, IOS, Amsterdam, 2015, pp. 84–116. MR3525574

[8] T. Shaska and L. Beshaj, *The arithmetic of genus two curves*, Information security, coding theory and related combinatorics, NATO Sci. Peace Secur. Ser. D Inf. Commun. Secur., vol. 29, IOS, Amsterdam, 2011, pp. 59–98. MR2963126

[9] Lubjana Beshaj and Fred Thompson, *Equations for superelliptic curves over their minimal field of definition*, Albanian J. Math. **8** (2014), no. 1, 3–8. MR3253208

[10] Andrew R. Booker, Jeroen Sijsling, Andrew V. Sutherland, John Voight, and Dan Yasaki, *A database of genus-2 curves over the rational numbers*, LMS J. Comput. Math. **19** (2016), no. suppl. A, 235–254, DOI 10.1112/S146115701600019X. MR3540958

[11] S. Allen Broughton, *Quasi-platonic $PSL_2(q)$-actions on closed Riemann surfaces*, Albanian J. Math. **9** (2015), no. 1, 31–61. MR3438844

[12] Gabriel Cardona and Jordi Quer, *Field of moduli and field of definition for curves of genus 2*, Computational aspects of algebraic curves, Lecture Notes Ser. Comput., vol. 13, World Sci. Publ., Hackensack, NJ, 2005, pp. 71–83, DOI 10.1142/9789812701640_0006. MR2181874

[13] Harris B. Daniels, *Siegel functions, modular curves, and Serre's uniformity problem*, Albanian J. Math. **9** (2015), no. 1, 3–29. MR3338631

[14] Barry H. Dayton, *Algorithms for real numerical varieties with application to parameterizing quadratic surface intersection curves*, Albanian J. Math. **7** (2013), no. 2, 61–75. MR3109478

[15] Anand Deopurkar, Maksym Fedorchuk, and David Swinarski, *Gröbner techniques and ribbons*, Albanian J. Math. **8** (2014), no. 1, 55–70. MR3293901

[16] José A. Díaz-García and Ramón Gutiérrez-Sánchez, *Matrix multivariate Pearson II-Riesz distribution*, Albanian J. Math. **9** (2015), no. 1, 63–75. MR3438845

[17] Simon Foucart and Tatyana Sorokina, *Generating dimension formulas for multivariate splines*, Albanian J. Math. **7** (2013), no. 1, 25–35. MR3069881

[18] G. Frey and T. Shaska, *Curves, jacobians, and cryptography*, Algebraic curves and their applications, 2018, pp. 3–48.

[19] T. Alden Gassert, *A note on the monogeneity of power maps*, Albanian J. Math. **11** (2017), no. 1, 3–12. MR3659215

[20] J. Gutierrez and T. Shaska, *Hyperelliptic curves with extra involutions*, LMS J. Comput. Math. **8** (2005), 102–115, DOI 10.1112/S1461157000000917. MR2135032

[21] Jaime Gutierrez, D. Sevilla, and T. Shaska, *Hyperelliptic curves of genus 3 with prescribed automorphism group*, Computational aspects of algebraic curves, 2005, pp. 109–123. MR2182037

[22] Ruben Hidalgo and Tony Shaska, *On the field of moduli of superelliptic curves*, Higher genus curves in mathematical physics and arithmetic geometry, Contemp. Math., vol. 703, Amer. Math. Soc., Providence, RI, 2018, pp. 47–62, DOI 10.1090/conm/703/14130. MR3782459

[23] Jun-ichi Igusa, *Arithmetic variety of moduli for genus two*, Ann. of Math. (2) **72** (1960), 612–649, DOI 10.2307/1970233. MR0114819

[24] Jun-ichi Igusa, *On Siegel modular forms of genus two*, Amer. J. Math. **84** (1962), 175–200, DOI 10.2307/2372812. MR0141643

[25] Jun-ichi Igusa, *Modular forms and projective invariants*, Amer. J. Math. **89** (1967), 817–855, DOI 10.2307/2373243. MR0229643

[26] David Kalaj and Ken-ichi Sakan, *Quasiconformal harmonic mappings onto a convex domain revisited*, Albanian J. Math. **7** (2013), no. 2, 51–60. MR3109477

[27] Reynald Lercier, Christophe Ritzenthaler, and Jeroen Sijsling, *Explicit Galois obstruction and descent for hyperelliptic curves with tamely cyclic reduced automorphism group*, Math. Comp. **85** (2016), no. 300, 2011–2045, DOI 10.1090/mcom3032. MR3471117

[28] A. Malmendier and T. Shaska, *The Satake sextic in F-theory*, J. Geom. Phys. **120** (2017), 290–305, DOI 10.1016/j.geomphys.2017.06.010. MR3712162

[29] Andreas Malmendier and Tony Shaska, *A universal genus-two curve from Siegel modular forms*, SIGMA Symmetry Integrability Geom. Methods Appl. **13** (2017), Paper No. 089, 17, DOI 10.3842/SIGMA.2017.089. MR3731039

[30] J. Mandili and T. Shaska, *Computing heights on weighted projective spaces*, Algebraic curves and their applications, 2018, pp. 153–163.

[31] Jean-François Mestre, *Construction de courbes de genre* 2 *à partir de leurs modules* (French), Effective methods in algebraic geometry (Castiglioncello, 1990), Progr. Math., vol. 94, Birkhäuser Boston, Boston, MA, 1991, pp. 313–334. MR1106431

[32] Rezart Muço, Nejme Pjero, Ervin Ruci, and Eustrat Zhupa, *Classifying families of superelliptic curves*, Albanian J. Math. **8** (2014), no. 1, 23–35. MR3270074

[33] Monika Polak and Vasyl Ustimenko, *On stream cipher based on family of graphs* $\widetilde{D(n,q)}$ *of increasing girth*, Albanian J. Math. **8** (2014), no. 1, 37–45. MR3293899

[34] Monika Polak and Eustrat Zhupa, *Graph based linear error correcting codes*, Albanian J. Math. **10** (2016), no. 1, 37–45. MR3589800

[35] Nguyen Tien Quang, Che Thi Kim Phung, and Ngo Sy Tung, *Abelian crossed modules and strict Picard categories*, Albanian J. Math. **7** (2013), no. 1, 37–48. MR3069882

[36] T. Shaska, *Some remarks on the hyperelliptic moduli of genus 3*, Comm. Algebra **42** (2014), no. 9, 4110–4130, DOI 10.1080/00927872.2013.791305. MR3200084

[37] T. Shaska and L. Beshaj, *Heights on algebraic curves*, Advances on superelliptic curves and their applications, NATO Sci. Peace Secur. Ser. D Inf. Commun. Secur., vol. 41, IOS, Amsterdam, 2015, pp. 137–175. MR3525576

[38] T. Shaska and F. Thompson, *Bielliptic curves of genus 3 in the hyperelliptic moduli*, Appl. Algebra Engrg. Comm. Comput. **24** (2013), no. 5, 387–412, DOI 10.1007/s00200-013-0209-9. MR3118614

[39] T. Shaska and H. Völklein, *Elliptic subfields and automorphisms of genus 2 function fields*, Algebra, arithmetic and geometry with applications (2000), 703–723.

[40] Tony Shaska, *Genus 2 curves with* (3,3)*-split Jacobian and large automorphism group*, Algorithmic number theory (Sydney, 2002), Lecture Notes in Comput. Sci., vol. 2369, Springer, Berlin, 2002, pp. 205–218, DOI 10.1007/3-540-45455-1_17. MR2041085

[41] Elke Wolf, *Eigenvalues of composition combined with differentiation*, Albanian J. Math. **10** (2016), no. 1, 11–19. MR3589798

[42] Mohamed Zaatra, *A characterization of some Laguerre-Hahn orthogonal polynomials of class one*, Albanian J. Math. **10** (2016), no. 1, 21–36. MR3589799

[43] Albertas Zinevi˘ cius, *On the congruent number problem over integers of real number fields*, Albanian J. Math. **8** (2014), no. 1, 49–53. MR3293900

Department of Mathematical Sciences, United States Military Academy, West Point, New York 10996

Email address: lubjana.beshaj@westpoint.edu

Department of Mathematics and Statistics, Oakland University, Rochester, Michigan 48309

Email address: sguest@oakland.edu

Contemporary Mathematics
Volume **724**, 2019
https://doi.org/10.1090/conm/724/14584

A family of nonnormal double planes associated to hyperelliptic curves

Timothy J. Ford

ABSTRACT. Let C be the affine hyperelliptic curve defined by $y^2 = g(x)$, where $g(x)$ is a polynomial of degree at least three over a field k. Starting with an example that originally appeared in an article by the author and F. DeMeyer, a nonnormal rational affine double plane $X \to \mathbb{A}^2_k$ is constructed together with a one-to-one homomorphism from the subgroup of torsion elements in the Picard group of C to the Brauer group of X. This construction is generalized to the situation where C is an arbitrary affine variety.

1. Introduction

One theme of these proceedings is the application of divisor class groups of algebraic curves, with particular emphasis on hyperelliptic curves. The motivation for the main theorem of this article is a construction that permits us to compute the Brauer group invariant for a large class of algebraic surfaces in terms of the Picard group of certain hyperelliptic curves. This example, which originally appeared in [**5**], shows that given an affine hyperelliptic curve C over a field k defined by an equation of the form $y^2 = g(x)$, there is an integral nonnormal rational affine double plane $X \to \mathbb{A}^2_k$ together with an embedding $\operatorname{tors}(\operatorname{Pic}(C)) \to \mathrm{B}(X)$ from the subgroup of torsion elements in the Picard group of C to the Brauer group of X. See Example 2.3 below for the construction of C and X, where it is shown that the affine coordinate ring $\mathcal{O}(X)$ can be embedded as a subring of $k[x, y]$. The Azumaya $\mathcal{O}(X)$-algebras are described by pullbacks of matrix algebras. Hence, the existence of curves C such that $\operatorname{Pic}(C)$ contains nontrivial torsion provides examples of subrings of $k[x, y]$ which have nontrivial Azumaya algebras.

Before continuing we review for the benefit of the reader the definitions of the Brauer group and Picard group of a commutative ring R. For an introduction to these topics, the reader is referred to [**7**]. If A is an R-algebra and A^o denotes the opposite ring, then A can be made into a left $A \otimes_R A^o$-module by the action $a \otimes b \cdot x = axb$. By definition, the algebra A is R-separable if A is projective as a left $A \otimes_R A^o$-module. If A is both central and separable, then A is called an Azumaya R-algebra. The Brauer group of R, denoted $\mathrm{B}(R)$, parametrizes the Azumaya R-algebras up to Brauer equivalence. The importance of the Brauer group invariant lies in its subtle connections to the algebraic, arithmetic, and topological properties of R. The Picard group of R, written $\operatorname{Pic}(R)$, parametrizes the projective R-modules of

2010 *Mathematics Subject Classification.* Primary 14F22; Secondary 16K50, 14C20, 14J26.

constant rank one. When R is an integrally closed noetherian integral domain, the Picard group can be identified with a subgroup of the class group of Weil divisors, $\mathrm{Cl}(R)$. If R is a regular noetherian integral domain, then $\mathrm{Pic}(R) \to \mathrm{Cl}(C)$ is an isomorphism (see, for example, [**7**, Theorem 6.5.11]). If R is not regular, then from a computational point of view, the Picard group tends to be more elusive than the class group. It is fair to say that the family of rings for which the two invariants $\mathrm{B}(R)$, $\mathrm{Pic}(R)$ are either known or not difficult to compute is quite small. In the last fifty years much progress has been made on the problem of getting a better understanding of these two invariants, especially when R is the coordinate ring of an algebraic variety of low dimension. Curves and surfaces receive a lot of attention because in many cases the computations can be completely carried out.

As stated above, Example 2.3 below is an elaboration of an example from [**5**] which exhibits a class of algebraic surfaces for which the Brauer group invariant is computed in terms of the Picard group of certain hyperelliptic curves. To carry out this computation, the connecting homomorphism in a Mayer-Vietoris sequence is employed. This is but one example appearing in the literature that relates the group of divisors on a hyperelliptic curve to the Brauer group of an algebraic surface. We take the opportunity here to mention a few more. This is not intended to be an exhaustive list. In [**3**] an algebraic surface X is constructed such that X contains a nonrational singular point P corresponding to the blowing-down of a nonsingular hyperelliptic curve C on a rational surface Y. It is shown that the Brauer group of the local ring $\mathcal{O}_{X,P}$ contains a subgroup corresponding to the subgroup of torsion elements in the class group $\mathrm{Cl}(C)$. A similar but slightly more involved computation is the subject of [**8**]. In [**6**] the Brauer groups of two classes of double planes $\pi : X \to \mathbb{A}^2$ related to hyperelliptic curves C are studied. In the first case, the surface X is a ruled surface of the form $C \times \mathbb{A}^1$. In the second case, the ramification locus of π is a hyperelliptic curve C. In both sets of examples, the class groups of C and X are exploited to construct Azumaya algebras on X as well as on the open complements on X and $\mathbb{A}^2$ of the ramification locus of π. Finally, if $C \subseteq \mathbb{P}^2$ is a nonsingular plane hyperelliptic curve and $U = \mathbb{P}^2 - C$ is the open complement, then the Brauer group of U is the subject of the papers [**10**], [**11**], and [**16**]. In this example, torsion elements in $\mathrm{Cl}(C)$ give rise to nontrivial Azumaya $\mathcal{O}(U)$-algebras.

Of the examples mentioned in the previous paragraph, only the double plane X associated to a hyperelliptic curve C that arises in Example 2.3 is nonnormal. In all of the other instances, the surface is either nonsingular or normal. For a nonnormal algebraic variety X, relatively few examples exist for which $\mathrm{B}(X)$ is known. Comparatively little is known about the structure of an Azumaya $\mathcal{O}(X)$-algebra, if X is nonnormal. The main theorem of this article (Theorem 3.1) shows that for an arbitrary algebraic variety C there is associated a variety X such that $\mathrm{B}(X)$ can be computed in terms of the torsion subgroup of $\mathrm{Pic}(C)$. Here C is not assumed to be irreducible, and in general the variety X that we construct is integral but nonnormal. The Azumaya algebras on X are described in terms of pullback diagrams involving rings of matrices. Hence, Theorem 3.1 allows us to exhibit a large class of nonnormal algebraic varieties X for which the Brauer group is nontrivial and moreover Azumaya $\mathcal{O}(X)$-algebras are presented that generate $\mathrm{B}(X)$. The Picard group of X is described in Proposition 3.3.

2. Background material

As a general reference for all unexplained notation and terminology, [**7**] is recommended. Because all of the examples constructed below involve rings that are defined as pullbacks, we begin with a quick review of the necessary properties associated to cartesian squares of rings. Let

$$\begin{array}{ccc} A & \xrightarrow{h_2} & A_2 \\ {\scriptstyle h_1}\downarrow & & \downarrow{\scriptstyle f_2} \\ A_1 & \xrightarrow{f_1} & A_3 \end{array} \tag{2.1}$$

be a commutative diagram of homomorphisms of rings. Then (2.1) is called a *cartesian square* if $A \cong \{(x_1, x_2) \in A_1 \oplus A_2 \mid f_1(x_1) = f_2(x_2)\}$ and h_1, h_2 are induced by the coordinate projections. In this case we say A is the *pullback* of f_1 and f_2. For example, if B is a ring, I a two-sided ideal of B, and $A \subseteq B$ a subring such that $I \subseteq A$, then

$$\begin{array}{ccc} A & \xrightarrow{h_2} & A/I \\ {\scriptstyle h_1}\downarrow & & \downarrow{\scriptstyle f_2} \\ B & \xrightarrow{f_1} & B/I \end{array} \tag{2.2}$$

is a cartesian square where all of the maps are the natural maps (see [**7**, Exercise 1.4.20]). Propositions 2.1 and 2.2 contain some useful facts about finitely generated k-algebras and rings that are defined by pullback diagrams.

PROPOSITION 2.1. *Let*

$$\begin{array}{ccc} A & \xrightarrow{h_2} & A_2 \\ {\scriptstyle h_1}\downarrow & & \downarrow{\scriptstyle f_2} \\ A_1 & \xrightarrow{f_1} & A_3 \end{array}$$

be a cartesian square of rings.

(1) *The kernel of h_2 is isomorphic to the kernel of f_1.*
(2) *If f_1 is onto, then h_2 is onto.*
(3) *If f_1 is onto, and f_2 makes A_3 into a finitely generated left A_2-module, then h_1 makes A_1 into a finitely generated left A-module.*

PROOF. Parts (1) and (2) are left to the reader (see [**7**, Exercise 1.4.18]).

(3): Since f_1 is onto, it follows that h_2 is onto. Hence A_3 is finitely generated as an A-module. Let $\{z_1, \dots, z_n\}$ be a subset of A_1 such that $\{f_1(z_1), \dots, f_1(z_n)\}$ is a set of generators for A_3 as an A-module. Then $A_1 = Az_1 + \cdots + Az_n + I$, where I denotes the kernel of f_1. Identify A with $\{(a_1, a_2) \in A_1 \oplus A_2 \mid f_1(a_1) = f_2(a_2)\}$. Then $I \oplus (0)$ is a subset of A. Hence I is in the image of h_1. This proves A_1 is generated as an A-module by $z_1, \dots, z_n, 1$. □

A proof of Proposition 2.2 can be found in [**1**, Proposition 7.8].

PROPOSITION 2.2. *Let k be a commutative noetherian ring and B a finitely generated commutative k-algebra. Let A be a k-subalgebra of B with the property that B is a finitely generated A-module. Then A is a finitely generated k-algebra.*

EXAMPLE 2.3. As mentioned in Section 1, we present here an example that originally appeared in [**5**] which associates to an affine hyperelliptic curve C an affine double plane X. This example is also explored in [**7**, Section 14.3.3]. Both of these references have the added hypothesis that the ground field k is algebraically closed and has characteristic zero. In the following, these restrictions on k are relaxed. This example provides the motivation for Theorem 3.1 below.

Before continuing, we remark that the terminology "hyperelliptic curve" appearing in the title of this article is applied in a very loose sense. It will describe any affine plane curve defined by an equation of the form $y^2 = g(x)$. This includes the possibilities that C is reducible, or nonnormal. Therefore the genus of an irreducible component of C can be any nonnegative integer.

Let k be any field of characteristic p, where $p = 0$ is allowed. Let $g(x)$ be any polynomial in $k[x]$. The hyperelliptic curve mentioned in the title of this article is the affine plane curve C defined by $y^2 - g(x) = 0$. Let

$$S = k[x, y^2, y(y^2 - g(x))]$$

which is a subring of the polynomial ring $k[x, y]$. The double plane mentioned in the title that is associated to C is the surface X with affine coordinate ring S. To realize X as an affine double plane, define a k-algebra homomorphism

$$k[u, v, w] \xrightarrow{\theta} S = k[x, y^2, y(y^2 - g(x))]$$

by $u \mapsto x$, $v \mapsto y^2$, $w \mapsto y(y^2 - g(x))$. Then θ is onto, and one checks that the kernel of θ is the principal ideal generated by $w^2 - v(v - g(u))^2$. Thus, S is isomorphic to $k[u, v, w]/(w^2 - v(v - g(u))^2)$ which is a ramified quadratic extension of $k[u, v]$. Hence, S is a quadratic extension of $k[x, y^2]$.

If we set $\bar{S} = k[x, y]$, then $\bar{S}$ is integral over S and the quotient field of S is $k(x, y)$. It follows that $\bar{S}$ is the integral closure of S in $k(x, y)$. Let

$$c = S : \bar{S} = \{\alpha \in \bar{S} \mid \alpha\bar{S} \subseteq S\}$$

be the conductor ideal from $\bar{S}$ to S. One computes that $c = (y^2 - g(x), y(y^2 - g(x)))$. Notice that c is an ideal of $\bar{S}$ contained in S. As an ideal in $\bar{S}$ we have $c = (y^2 - g(x))$. Then $S/c \cong k[x]$ and $\bar{S}/c \cong k[x, y]/(y^2 - g(x))$. As in Eq. (2.2) there is a cartesian square

$$(2.3)\qquad \begin{array}{ccc} S & \xrightarrow{h_2} & k[x] \\ {\scriptstyle h_1}\downarrow & & \downarrow{\scriptstyle \phi_2} \\ k[x, y] & \xrightarrow{\phi_1} & k[x, y]/(y^2 - g(x)) = \mathcal{O}(C) \end{array}$$

of rings with ϕ_1 onto and ϕ_2 one-to-one. We are in the context of Theorem 3.1 below from which we conclude that there is an exact sequence of abelian groups

$$(2.4)\qquad 0 \to \operatorname{tors}(\operatorname{Pic}(C)) \xrightarrow{\partial_1} \mathrm{B}(S) \to \mathrm{B}(k[x, y]) \oplus \mathrm{B}(k[x])$$

and a sequence of abelian groups

$$(2.5)\qquad 0 \to \operatorname{tors}(\operatorname{Pic}(C)) \xrightarrow{\partial_1} \mathrm{B}(S) \to \mathrm{B}(k) \to 0$$

which is split exact if $p = 0$ and split exact modulo p-torsion subgroups if $p > 0$. For instance, if the degree of $g(x)$ is at least three and C is nonsingular, the group $\operatorname{tors}(\operatorname{Pic}(C))$ corresponds to a subgroup of the jacobian variety of C. Since the jacobian variety is an abelian variety (see [**14**]), for any $n \geq 2$ that is invertible in

k, there is a finite extension field F/k such that $\mathrm{Pic}(C \times F)$ contains an element of order n. Upon extension of the base field of X to F, it follows that the Brauer group $\mathrm{B}(X \times F)$ also contains an element of order n. Thus Eq. (2.4) shows that $\mathrm{B}(S)$ is nontrivial for a sufficiently general choice of $g(x)$. The reader is referred to [**2**], [**4**] and other articles in these proceedings for a more specific treatment of the subject of torsion in $\mathrm{Cl}(C)$. For an example when C is not irreducible, see Example 4.3.

3. The main theorem

The proof of Theorem 3.1 is based on the Mayer-Vietoris sequence for Brauer groups that was constructed by Knus and Ojanguren (see [**7**, Section 14.2]). The cartesian square of rings appearing in Theorem 3.1 is a generalization of Eq. (2.3). Thus, Theorem 3.1 is a generalization of Example 2.3 to the situation where the affine coordinate ring of C is an arbitrary finitely generated commutative k-algebra.

THEOREM 3.1. *Let k be a field of characteristic p, where $p = 0$ is allowed. Let R be a finitely generated commutative k-algebra of Krull dimension $m \geq 0$. Let $r > 0$ and*

$$0 \to I \to k[u_1, \dots, u_r] \xrightarrow{f_1} R \to 0$$

any finite presentation of R. Let $V = \{v_1, \dots, v_m\}$ be a set of indeterminates, X any subset of V, and $f_2 : k[X] \to R$ any one-to-one homomorphism of k-algebras. Consider the cartesian square of commutative k-algebras

$$\begin{array}{ccc} S & \xrightarrow{h_2} & k[X] \\ {\scriptstyle h_1}\downarrow & & \downarrow{\scriptstyle f_2} \\ k[u_1, \dots, u_r] & \xrightarrow{f_1} & R \end{array}$$

where S is the pullback of f_1 and f_2.

(1) *There is a homomorphism ∂_1 such that the sequence of abelian groups*

$$0 \to \mathrm{tors}(\mathrm{Pic}(R)) \xrightarrow{\partial_1} \mathrm{B}(S) \xrightarrow{(h_1,h_2)} \mathrm{B}(k[u_1, \dots, u_r]) \oplus \mathrm{B}(k[X])$$

is exact.

(2) *If k has characteristic $p = 0$, then there is a split exact sequence*

$$0 \to \mathrm{tors}(\mathrm{Pic}(R)) \xrightarrow{\partial_1} \mathrm{B}(S) \to \mathrm{B}(k) \to 0.$$

If k has characteristic $p > 0$, then the sequence is split exact modulo p-torsion subgroups.

PROOF. (1): It follows from Step 2 of the proof of Theorem 14.2.12 and Corollary 14.2.15 of [**7**] that there is a one-to-one homomorphism $\partial_1 : \mathrm{tors}(\mathrm{Pic}(R)) \to \mathrm{B}(S)$ of abelian groups. We sketch the key ingredients of the proof. To define ∂_1, let L be an R-progenerator module of constant rank one such that the class of L in the Picard group has finite order. An application of a theorem of Bass (see [**7**, Corollary 14.2.2]) implies there is an R-module isomorphism $\tau : L \otimes_R R^{(t)} \cong R^{(t)}$ for some $t > 0$. Then τ induces an R-algebra automorphism on the ring of matrices

$\bar{\tau} : M_t(R) \cong M_t(R)$. In fact, $\bar{\tau}$ is conjugation by τ. There is a cartesian square of rings

$$\begin{array}{ccccc} A & & \longrightarrow & & M_t(k[X]) \\ \downarrow & & & & \downarrow \\ M_t(k[u_1, \dots, u_r]) & \longrightarrow & M_t(R) & \xrightarrow{\bar{\tau}} & M_t(R) \end{array}$$

where A is the pullback. Then A is an Azumaya S-algebra, and $\partial_1(L)$ is the Brauer class containing A. The proof that ∂_1 is one-to-one follows from the identities $\operatorname{Pic}(k[u_1, \dots, u_r]) = \operatorname{Pic}(k[X]) = \operatorname{Pic}(k) = (0)$. Now we show exactness at $\mathrm{B}(S)$. For notational simplicity, we write $R_1 = k[u_1, \dots, u_r]$ and $R_2 = k[X]$. By [**7**, Theorem 14.2.11] every Azumaya S-algebra Λ is defined by a pullback diagram

$$\begin{array}{ccccc} \Lambda & & \longrightarrow & & \Lambda_2 \\ \downarrow & & & & \downarrow{\scriptstyle \phi_2} \\ \Lambda_1 & \xrightarrow{\phi_1} & R \otimes_{R_1} \Lambda_1 & \xrightarrow{\sigma} & R \otimes_{R_2} \Lambda_2 \end{array}$$

where each Λ_i is an Azumaya R_i-algebra and σ is an isomorphism of R-algebras. We have $R_i \otimes_S \Lambda \cong \Lambda_i$. Let Λ be an Azumaya S-algebra of rank t^2 and assume Λ is split by $R_1 \oplus R_2$. Then $\Lambda_i = R_i \otimes_S \Lambda$ is the endomorphism ring of an R_i-progenerator module. By Serre's Theorem (see [**15**, Theorem 4.62] or [**12**]), a finitely generated projective module over R_1 or R_2 is free. Therefore, Λ_i is isomorphic to the ring of matrices $M_t(R_i)$. Reversing the above steps shows that in this case, Λ is in the image of ∂_1.

(2): Since h_2 is onto, we know S has a maximal ideal $\mathfrak{m}$ such that $S/\mathfrak{m} \cong k$. Because S is a k-algebra, the natural homomorphism $\mathrm{B}(k) \to \mathrm{B}(S)$ splits. Consider the commutative diagram

$$\begin{array}{ccccccccc} 0 & \longrightarrow & \mathrm{B}(k) & \longrightarrow & \mathrm{B}(S) & \longrightarrow & \frac{\mathrm{B}(S)}{\mathrm{B}(k)} & \longrightarrow & 0 \\ & & \downarrow & & \downarrow{\scriptstyle (h_1,h_2)} & & \downarrow & & \\ 0 & \longrightarrow & \mathrm{B}(k) \oplus \mathrm{B}(k) & \longrightarrow & \mathrm{B}(R_1) \oplus \mathrm{B}(R_2) & \longrightarrow & \frac{\mathrm{B}(R_1) \oplus \mathrm{B}(R_2)}{\mathrm{B}(k) \oplus \mathrm{B}(k)} & \longrightarrow & 0 \end{array}$$

where the rows are split exact sequences. By a theorem of Auslander and Goldman (see [**7**, Theorem 13.6.3]), the natural maps $\mathrm{B}(k) \to \mathrm{B}(R_1)$ and $\mathrm{B}(k) \to \mathrm{B}(R_2)$ are isomorphisms if $p = 0$. In this case we see that the kernel of (h_1, h_2) is isomorphic to $\mathrm{B}(S)/\mathrm{B}(k)$. If $p > 0$, then by a theorem of R. Hoobler (see [**7**, Theorem 13.6.4]), the natural maps $\mathrm{B}(k) \to \mathrm{B}(R_1)$ and $\mathrm{B}(k) \to \mathrm{B}(R_2)$ are isomorphisms modulo p-torsion subgroups. □

Remark 3.2. Let R and S be as in Theorem 3.1. Since h_1 is an embedding of S into the polynomial ring $k[u_1, \dots, u_r]$, S is an integral domain. As we see in Example 4.4 below, in general S is not a finitely generated k-algebra. Assume R is of Krull dimension m. By E. Noether's Normalization Lemma, there is a one-to-one homomorphism $f_2 : k[v_1, \dots, v_m] \to R$ such that R is a finitely generated module over the ring $k[v_1, \dots, v_m]$. Propositions 2.1 and 2.2 show that in this case the homomorphism h_1 makes $k[u_1, \dots, u_r]$ into a finitely generated S-module and S is a finitely generated k-algebra.

Using information obtained in the proof of Theorem 3.1 together with the Mayer-Vietoris sequence of Milnor ([**7**, Theorem 14.2.10]), we compute the group of units of S and the Picard group of S.

PROPOSITION 3.3. *Let k, R and S be as in Theorem 3.1. Then*

(1) $S^* = k^*$, *and*

(2) *there is an exact sequence of abelian groups*

$$1 \to k^* \to R^* \xrightarrow{\partial_0} \operatorname{Pic}(S) \to 0.$$

PROOF. In our context, the Mayer-Vietoris sequence of Milnor is

$$\begin{aligned}(3.1)\quad 1 \to S^* \to (k[v_1,\dots,v_m])^* \times (k[X])^* \xrightarrow{(f_1,1/f_2)} R^* \xrightarrow{\partial_0} \\ \operatorname{Pic}(S) \to \operatorname{Pic}(k[v_1,\dots,v_m]) \oplus \operatorname{Pic}(k[X]) \to \operatorname{Pic}(R).\end{aligned}$$

We have $(k[v_1,\dots,v_m])^* = (k[X])^* = k^*$ and $\operatorname{Pic}(k[v_1,\dots,v_m]) = \operatorname{Pic}(k[X]) = \operatorname{Pic}(k) = (0)$. The map $(f_1, 1/f_2) : k^* \times k^* \to R^*$ is the difference map which is defined by $(a,b) \mapsto ab^{-1}$. The kernel and image of the difference map are both isomorphic to k^*. The sequence (3.1) breaks up into the isomorphism of Part (1) and the short exact sequence of Part (2). □

4. Examples and applications

We include in this section some examples and applications of Theorem 3.1. We begin by generalizing Example 2.3 to curves that are not necessarily hyperelliptic.

EXAMPLE 4.1. Let k be a field of characteristic p. Let f be a nonconstant polynomial in $k[x,y]$ defining the affine plane curve $C = Z(f)$. Let $\phi_1 : k[x,y] \to k[x,y]/(f)$ be the natural map. By Noether's Normalization Lemma, there is a one-to-one homomorphism of k-algebras $\phi_2 : k[z] \to k[x,y]/(f)$ such that $k[x,y]/(f)$ is finitely generated as a $k[z]$-module. Let

$$S = \{(u,v) \in k[x,y] \times k[z] \mid \phi_1(u) = \phi_2(v)\}$$

be the pullback. Then the cartesian square

$$\begin{array}{ccc} S & \xrightarrow{h_2} & k[z] \\ {\scriptstyle h_1}\downarrow & & \downarrow{\scriptstyle \phi_2} \\ k[x,y] & \xrightarrow{\phi_1} & \mathcal{O}(C) = k[x,y]/(f) \end{array}$$

commutes, where ϕ_1 and h_2 are onto, and ϕ_2 and h_1 are one-to-one. By Theorem 3.1, there is a sequence of abelian groups

$$0 \to \operatorname{tors}(\operatorname{Pic}(C)) \xrightarrow{\partial_1} \mathrm{B}(S) \to \mathrm{B}(k) \to 0$$

which is split exact modulo p-torsion subgroups. By Propositions 2.1 and 2.2, h_1 makes $k[x,y]$ into a finitely generated S-module and S is a finitely generated k-algebra.

REMARK 4.2. Suppose k has characteristic zero and S is a ring constructed using a plane curve $C \subseteq \mathbb{A}^2$ as in Example 4.1 such that S is a finitely generated k-subalgebra of $k[x,y]$ and $k[x,y]$ is a finitely generated module over S. Then the Brauer group functor can be used to show that the noetherian integral domain S is in general nonnormal. For instance, in [**13**], H. Li proves that if k is one of the fields $\mathbb{R}$ or $\mathbb{C}$, and $k \subseteq S \subseteq k[x,y]$ is a tower of subrings such that S is normal and the inclusion $S \subseteq k[x,y]$ is integral, then the natural maps $\mathrm{B}(k) \to \mathrm{B}(S) \to \mathrm{B}(k[x,y])$ are isomorphisms. By Example 4.1, we know that $\mathrm{B}(k) \to \mathrm{B}(S)$ is not an isomorphism when $\operatorname{tors}(\operatorname{Pic}(C)) \neq (0)$. If $k = \mathbb{C}$, then a general plane curve in $\mathbb{P}^2$ is nonsingular (see for example [**9**, Exercise I.5.15]) and the dimension of its jacobian is greater than or equal to one. Therefore, for a sufficiently general C, the Brauer group of S will be nontrivial.

EXAMPLE 4.3. In the context of Example 2.3, let k be a field in which 2 is invertible and $g(x) = (x^2-1)^2 \in k[x]$. The hyperelliptic curve C defined by the equation $y^2 = (x^2-1)^2$ is the union of two conics. Following [**7**, Exercise 14.2.20], one shows that $\operatorname{Pic}(C)$ is isomorphic to k^*, the group of invertible elements in k. If $S = k[x, y^2, y(y^2 - g(x))]$, then by Eq. (2.4), $\mathrm{B}(S)$ contains a subgroup isomorphic to $\mu(k)$, where $\mu(k)$ denotes the group of all roots of unity in k.

The next example shows that the ring S in Theorem 3.1 can be a k-algebra that is not finitely generated.

EXAMPLE 4.4. Let k be a field. Consider the cartesian square of commutative k-algebras

$$\begin{array}{ccc} S & \xrightarrow{h_2} & k \\ {\scriptstyle h_1}\downarrow & & \downarrow{\scriptstyle f_2} \\ k[x,y] & \xrightarrow{f_1} & k[x,y]/(y) \end{array}$$

where f_1 and f_2 are the natural maps and S is the pullback. Then S can be identified with the subring of $k[x,y]$ consisting of all polynomials $p(x,y)$ such that $p(x,0)$ is in k. Hence S can be identified with $k[y, xy, x^2y, x^3y, \dots]$. Notice that S is not a finitely generated k-algebra.

In Example 4.5 we show that any finite cyclic group is the Brauer group of an affine variety.

EXAMPLE 4.5. Let k be a field of characteristic p, where $p = 0$ is allowed. Let $m \geq 2$ and $S = k[x_0, \dots, x_m]$. Let $f \in S$ be a homogeneous irreducible polynomial of degree $n \geq 2$ and R the subring of homogeneous elements in $S[f^{-1}]$ of degree 0. Then R is a finitely generated k-algebra, R is a regular noetherian integral domain of Krull dimension m, R is birational to $k[x_1, \dots, x_m]$, $\operatorname{Pic}(R) = \operatorname{Cl}(R) \cong \mathbb{Z}/n$, and $R^* = k^*$. For proofs, see [**7**, Example 6.5.12].

Now assume n, the degree of f, is relatively prime to p. Since R is a finitely generated k-algebra, for some $r > 0$, there is a finite presentation of R

$$0 \to I \to k[u_1, \dots, u_r] \xrightarrow{f_1} R \to 0$$

where I is the kernel of f_1.

Let $V = \{v_1, \dots, v_m\}$ be a set of indeterminates and X any subset of V. Since $\dim(R) = m$, there is a one-to-one homomorphism $f_2 : k[X] \to R$ of k-algebras. Consider the cartesian square of commutative k-algebras

$$\begin{array}{ccc} S & \xrightarrow{h_2} & k[X] \\ {\scriptstyle h_1}\downarrow & & \downarrow{\scriptstyle f_2} \\ k[u_1, \dots, u_r] & \xrightarrow{f_1} & R \end{array}$$

where S is the pullback. By Theorem 3.1,

$$0 \to \operatorname{Pic}(R) \to \mathrm{B}(S) \to \mathrm{B}(k) \to 0$$

is split exact modulo subgroups of p-torsion. If we assume $p = 0$ and k is algebraically closed, then $\mathrm{B}(S) \cong \mathbb{Z}/n$. If we pick $f_2 : k[V] \to R$ such that R is a finitely generated module over $k[V]$, then S is a finitely generated k-algebra (Remark 3.2).

Our final example, Example 4.6, shows that any finite abelian group is the Brauer group of an affine variety.

EXAMPLE 4.6. Let k be an algebraically closed field of characteristic zero. Let $n_1, \dots, n_s$ be integers, each of which is greater than or equal to 2. Following the construction in Example 4.5, for each i let R_i be a finitely generated k-algebra of dimension $\dim(R_i) = m_i \geq 2$ such that $\operatorname{Pic}(R_i) \cong \mathbb{Z}/n_i$. For each i let X_i be a set of m_i indeterminates, and $f_{2,i} : k[X_i] \to R_i$ a one-to-one homomorphism of k-algebras. For some $r > 0$, there is a finite presentation

$$0 \to I \to k[u_1, \dots, u_r] \xrightarrow{f_1} R_1 \oplus \cdots \oplus R_s \to 0$$

of the k-algebra $R_1 \oplus \cdots \oplus R_s$, where I is the kernel of f_1. Consider the cartesian square of commutative k-algebras

$$\begin{array}{ccc} S & \xrightarrow{h_2} & k[X_1] \oplus \cdots \oplus k[X_s] \\ {\scriptstyle h_1}\downarrow & & \downarrow{\scriptstyle (f_{2,1},\dots,f_{2,s})} \\ k[u_1, \dots, u_r] & \xrightarrow{f_1} & R_1 \oplus \cdots \oplus R_s \end{array}$$

where S is the pullback. Then S is an integral domain with the property that $\mathrm{B}(S) \cong \operatorname{Pic}(R_1 \oplus \cdots \oplus R_s) \cong \mathbb{Z}/n_1 \oplus \cdots \oplus \mathbb{Z}/n_s$. By Noether's Normalization Lemma, for each i we can pick $f_{2,i}$ so that R_i is finitely generated as a module over $k[X_i]$. In this case S is a finitely generated k-algebra (Remark 3.2).

References

[1] M. F. Atiyah and I. G. Macdonald, *Introduction to commutative algebra*, Addison-Wesley Publishing Co., Reading, Mass.-London-Don Mills, Ont., 1969. MR0242802

[2] Abbey Bourdon and Pete L. Clark, *Torsion points and galois representations on CM elliptic curves*, preprint. arXiv:1612.03229 [math.NT].

[3] Djordje N. Bulj, Timothy J. Ford, and Drake M. Harmon, *Generically trivial Azumaya algebras on a rational surface with a non rational singularity*, Comm. Algebra **41** (2013), no. 11, 4333–4338, DOI 10.1080/00927872.2012.695837. MR3169521

[4] Harris B. Daniels, *Torsion subgroups of rational elliptic curves over the compositum of all D_4 extensions of the rational numbers*, J. Algebra **509** (2018), 535–565, DOI 10.1016/j.jalgebra.2018.02.038. MR3812213

[5] F. R. DeMeyer and T. J. Ford, *On the Brauer group of superspaces and subrings of $k[x, y]$*, Brauer groups in ring theory and algebraic geometry (Wilrijk, 1981), Lecture Notes in Math., vol. 917, Springer, Berlin-New York, 1982, pp. 211–221. MR657431

[6] Timothy J. Ford, *The Brauer group of an affine double plane associated to a hyperelliptic curve*, Comm. Algebra **45** (2017), no. 4, 1416–1442, DOI 10.1080/00927872.2016.1175608. MR3576667

[7] Timothy J. Ford, *Separable algebras*, Graduate Studies in Mathematics, vol. 183, American Mathematical Society, Providence, RI, 2017. MR3618889

[8] Timothy J. Ford and Drake M. Harmon, *The Brauer group of an affine rational surface with a non-rational singularity*, J. Algebra **388** (2013), 107–140, DOI 10.1016/j.jalgebra.2013.04.022. MR3061681

[9] Robin Hartshorne, *Algebraic geometry*, Springer-Verlag, New York-Heidelberg, 1977. Graduate Texts in Mathematics, No. 52. MR0463157

[10] Bill Jacob, *Division algebras which only ramify along a hyperelliptic curve in* $\mathbf{P}^2$, J. Algebra **242** (2001), no. 2, 720–728, DOI 10.1006/jabr.2001.8843. MR1848967

[11] Boris È. Kunyavskiĭ, Louis H. Rowen, Sergey V. Tikhonov, and Vyacheslav I. Yanchevskiĭ, *Division algebras that ramify only on a plane quartic curve*, Proc. Amer. Math. Soc. **134** (2006), no. 4, 921–929, DOI 10.1090/S0002-9939-05-08106-2. MR2196022

[12] T. Y. Lam, *Serre's problem on projective modules*, Springer Monographs in Mathematics, Springer-Verlag, Berlin, 2006. MR2235330

[13] Hongnian Li, *Brauer groups over affine normal surfaces*, K-theory and algebraic geometry: connections with quadratic forms and division algebras (Santa Barbara, CA, 1992), Proc. Sympos. Pure Math., vol. 58, Amer. Math. Soc., Providence, RI, 1995, pp. 309–317. MR1327306

[14] David Mumford, *Abelian varieties*, Tata Institute of Fundamental Research Studies in Mathematics, No. 5, Published for the Tata Institute of Fundamental Research, Bombay; Oxford University Press, London, 1970. MR0282985

[15] Joseph J. Rotman, *An introduction to homological algebra*, Pure and Applied Mathematics, vol. 85, Academic Press, Inc. [Harcourt Brace Jovanovich, Publishers], New York-London, 1979. MR538169

[16] Michel Van den Bergh, *Division algebras on* $\mathbf{P}^2$ *of odd index, ramified along a smooth elliptic curve are cyclic* (English, with English and French summaries), Algèbre non commutative, groupes quantiques et invariants (Reims, 1995), Sémin. Congr., vol. 2, Soc. Math. France, Paris, 1997, pp. 43–53. MR1601190

Department of Mathematics, Florida Atlantic University, Boca Raton, Florida 33431

Email address: ford@fau.edu

Contemporary Mathematics
Volume **724**, 2019
https://doi.org/10.1090/conm/724/14585

On the discriminant of certain quadrinomials

Shuichi Otake and Tony Shaska

ABSTRACT. We give an explicit formula for the discriminant $\Delta_f(x)$ of the quadrinomials of the form $f(x) = x^n + t(x^2 + ax + b)$. The proof uses Bezoutians of polynomials.

1. Introduction

In [**43**] Selmer studied polynomials $f(x) = x^n \pm x \pm 1$ and proved that $x^n - x - 1$ is irreducible for all ≥ 2, while $x^n + x + 1$ is irreducible for $n \not\equiv 2 \bmod 3$. In the process he proved that the discriminant of $f(x)$ is (up to a multiplication by -1) $\Delta = \pm\left(n^n \pm (n-1)^{n-1}\right)$, see [**43**] for precise formulas. He also noticed that polynomials $f(x)$ have very small discriminants. He checked his results for $n \leq 20$. Other mathematicians have considered trinomials $f(x) = x^n + ax + b$ for various reasons; see [**34**], [**48**]. In general, finding conditions on the coefficients of a polynomial to have minimal discriminant is a difficult problem. It is related to reduction of binary forms discussed in [**7**] and their heights [**44**]. More generally it is a special case of finding conditions on the coefficients of a binary form such that the corresponding point in the weighted moduli space of invariants is normalized; see [**32**].

This paper is the first of hopefully others to come to determine for what a, b, t the quadrinomial

$$f(x) = x^n + t(x^2 + ax + b) \tag{1}$$

has minimal discriminant, is reduced in the sense of [**7**], or has minimal naive height. There have been plenty of efforts to determine such formulas for certain classes of polynomials. In [**24**], [**45**] the authors focus on the computation of the discriminant of a trinomial which has been carried out in different ways. In this paper, we determine explicitly a formula for the discriminant of $f(x)$ by using the approach of Bezoutians.

We prove (see Thm. 1) that the discriminant $\Delta_f(x)$ of the polynomials in Eq. (1) is given by the formula

$$\Delta = (-1)^{m_1} t^{n-1}\left((n-2)^{n-2}(a^2 - 4b)t^2 + \gamma_{\mathbf{c}}\, t - n^n b^{n-1}\right), \tag{2}$$

2010 *Mathematics Subject Classification.* Primary 12Y05, 12E10.

such that

$$\gamma_{\mathbf{c}} = \sum_{k=0}^{m_0} (-1)^{n+k} n^k (n-1)^{n-2k-4} (n-2)^k a^{n-2k-4} b^k \cdot S_k,$$

and $m_0 = \lfloor (n-3)/2 \rfloor$, $m_1 = \lceil (n-3)/2 \rceil$ for $a \neq 0$ or $a = 0$ and n even, and $\gamma_{\mathbf{c}} = 0$ for $a = 0$ and n odd, and S_k given as in (4). While the method and the proofs are tedious and technical we avoid some of the details of the proof of Thm. 1 and direct the reader to the extended version of this paper on [**38**].

Notation: Throughout this paper n is a positive integer, Δ denotes the discriminant of a polynomial $f(x)$ with respect to the variable x. The symbol $\lfloor \cdot \rfloor$ is the floor function and $\lceil \cdot \rceil$ is the ceiling function. $B_n(f,g)$ denotes the level n Bezoutian of two polynomials f and g. For a polynomial $f \in \mathbb{R}[x]$ the symbol N_f denotes the number of real roots of $f(x)$. We denote by $\binom{n}{m}$ $(n, m \in \mathbb{Z}_{\geq 0})$ the binomial coefficient.

2. Preliminaries

Let F be a field of characteristic zero and $f_1(x)$, $f_2(x)$ be polynomials over F. Then, for any integer n such that $n \geq \max\{\deg f_1, \deg f_2\}$, we put

$$B_n(f_1, f_2) := \frac{f_1(x)f_2(y) - f_1(y)f_2(x)}{x-y} = \sum_{i,j=1}^{n} \alpha_{ij} x^{n-i} y^{n-j} \in F[x,y],$$

$$M_n(f_1, f_2) := (\alpha_{ij})_{1 \leq i,j \leq n}.$$

The $n \times n$ matrix $M_n(f_1, f_2)$ is called the Bezoutian of f_1 and f_2. When $f_2 = f_1'$, the formal derivative of f_1 with respect to the indeterminate x, we often write $B_n(f_1) := B_n(f_1, f_1')$, $M_n(f_1) := M_n(f_1, f_1')$ and the matrix $M_n(f_1)$ is called the Bezoutian of f_1.

We denote by N_{f_1} the number of distinct real roots of $f_1(x)$ and $\Delta(f_1)$ the discriminant of $f_1(x)$. Moreover, for any real symmetric matrix M, we denote by $\sigma(M)$ the index of inertia of M. The following are the list of important properties of Bezoutians.

PROPOSITION 1. *Notations as above, we have*

(i) $M_n(f_1, f_2)$ *is an* $n \times n$ *symmetric matrix over* F. $(M_n(f_1, f_2) \in Sym_n(F))$.
(ii) $B_n(f_1, f_2)$ $(M_n(f_1, f_2))$ *is linear in* f_1 *and* f_2*, separately.*
(iii) $B_n(f_1, f_2) = -B_n(f_2, f_1)$ $(M_n(f_1, f_2) = -M_n(f_2, f_1))$.
(iv) $N_{f_1} = \sigma(M_n(f_1))$.
(v) $\Delta(f_1) = \dfrac{1}{\mathrm{led}(f_1)^2} \det M_n(f_1)$, *where* $\mathrm{led}(f_1)$ *is the leading coefficient of* f_1. *In particular, if* f_1 *is monic, we have* $\Delta(f_1) = \det M_n(f_1)$.
(vi) Let λ, μ, ν *be integers such that* $\lambda \geq \mu > \nu \geq 0$. *Then* $M_\lambda(x^\mu, x^\nu) = (m_{ij})_{1 \leq i,j \leq \lambda}$, *where*

$$m_{ij} = \begin{cases} 1 & i+j = 2\lambda - (\mu+\nu) + 1 \quad (\lambda - \mu + 1 \leq i,j \leq \lambda - \nu), \\ 0 & \textit{otherwise.} \end{cases}$$

PROOF. For (i)–(iv), see [**22**, Theorem 8.25, 9.2]. For (v), see equation (4) of [**21**, p.217]. For (vi), see [**39**, Lemma 2] □

Let $n \geq 3$ be an integer, $a, b \in \mathbb{R}$, $\boldsymbol{c} = (b, a, 1) \in \mathbb{R}^3$, and

$$g_{\boldsymbol{c}}(x) = x^2 + ax + b \in \mathbb{R}[x], \quad f_{\boldsymbol{c}}(t; x) = x^n + t g_{\boldsymbol{c}}(x) \in \mathbb{R}(t)[x]. \tag{3}$$

We will denote by

$$S_k = \binom{n-k-3}{k}\left((n-1)^3 a^4 - \frac{n(n-1)\left\{5n^2 - (6k+23)n + 10k + 24\right\}}{n-k-3} a^2 b\right) + 4n^2(n-2)\binom{n-k-4}{k} b^2, \tag{4}$$

Our main result is the following:

THEOREM 1. *For any $n \geq 5$ and k $(0 \leq k \leq m_0)$, the discriminant of*

$$f(x) = x^n + t\left(x^2 + ax + b\right)$$

is given by the formula

$$\Delta = (-1)^{m_1} t^{n-1}\left((n-2)^{n-2}(a^2 - 4b)t^2 + \gamma_{\boldsymbol{c}}\, t - n^n b^{n-1}\right), \tag{5}$$

such that $m_0 = \lfloor (n-3)/2 \rfloor$, $m_1 = \lceil (n-3)/2 \rceil$ and

$$\gamma_{\boldsymbol{c}} = \sum_{k=0}^{m_0} (-1)^{n+k} n^k (n-1)^{n-2k-4} (n-2)^k a^{n-2k-4} b^k S_k,$$

for $a \neq 0$ or $a = 0$ and n even, and $\gamma_{\boldsymbol{c}} = 0$ for $a = 0$ and n odd.

The rest of the paper will be focused on proving the theorem.

REMARK 1. *Notice that if $n = 3$, then*

$$\Delta\left(f_{\boldsymbol{c}}(t; x)\right) = t^2\left((a^2 - 4b)t^2 - (4a^3 - 18ab)t - 27b^2\right).$$

By a direct computation, we have that the discriminant when $n = 4$ is given by

$$\Delta\left(f_{\boldsymbol{c}}(t; x)\right) = -t^3\left\{(4a^2 - 16b)t^2 + (27a^4 - 144a^2 b + 128b^2)t - 256b^3\right\}$$

and we get Thm. 1 for $n = 4$.

In the following, we put

$$\begin{aligned}
A_{\boldsymbol{c}}(t) &= (a_{ij}^{(\boldsymbol{c})}(t))_{1 \leq i,j \leq n} = M_n(f_{\boldsymbol{c}}(t; x)) \in \mathrm{Sym}_n(\mathbb{R}(t)), \\
B_{\boldsymbol{c}} &= (b_{ij}^{(\boldsymbol{c})})_{1 \leq i,j \leq 2} = M_2(g_{\boldsymbol{c}}(x)) \in \mathrm{Sym}_2(\mathbb{R}), \\
P_{\boldsymbol{c}}(t) &= \det A_{\boldsymbol{c}}(t) = \Delta(f_{\boldsymbol{c}}(t; x)).
\end{aligned}$$

To prove Thm. 1, we will use the equality $\Delta(f_{\boldsymbol{c}}(t; x)) = \det A_{\boldsymbol{c}}(t)$. Here, let $Q_m(k; c) = (q_{ij})_{1 \leq i,j \leq m}$, $R_m(k, l; c) = (r_{ij})_{1 \leq i,j \leq m}$ be $m \times m$ elementary matrices such that

$$Q_m(k; c) = \begin{bmatrix} 1 & & & & & \\ & \ddots & & & & \\ & & 1 & & & \\ & & & c & & \\ & & & & 1 & \\ & & & & & \ddots & \\ & & & & & & 1 \end{bmatrix}, \quad R_m(k, l; c) = \begin{bmatrix} 1 & & & & & \\ & \ddots & & & & \\ & & 1 & & c & \\ & & & \ddots & & \\ & & & & 1 & \\ & & & & & \ddots & \\ & & & & & & 1 \end{bmatrix},$$

where $q_{kk} = c$, $r_{kl} = c$, respectively and, for any $m \times m$ matrices $M_1, M_2, \cdots, M_l$, put $\prod_{k=1}^{l} M_k = M_1 M_2 \cdots M_l$. Moreover, put $l_k = n + k$. Then, as is the case in [**39**], we inductively define the matrix $A_{\mathbf{c}}(t)_k$ $(1 \le k \le n-2)$ as follows;

(1) $A_{\mathbf{c}}(t)_1 = (a_{ij}^{(\mathbf{c})}(t)_1)_{1 \le i,j \le n} = {}^t S_{\mathbf{c}}(t)_1 A_{\mathbf{c}}(t) S_{\mathbf{c}}(t)_1$, where

$$S_{\mathbf{c}}(t)_1 = Q_n(1; 1/\sqrt{n}) \prod_{k=0}^{1} R_n(1, l_k - 1; -a_{1,l_k-1}^{(\mathbf{c})}(t)/\sqrt{n}).$$

(2) Let

$$n_0 = \begin{cases} (n-1)/2, & n : \text{odd}, \\ n/2, & n : \text{even}. \end{cases}$$

Then, $A_{\mathbf{c}}(t)_k = (a_{ij}^{(\mathbf{c})}(t)_k)_{1 \le i,j \le n} = {}^t S_{\mathbf{c}}(t)_k A_{\mathbf{c}}(t)_{k-1} S_{\mathbf{c}}(t)_k$, where

$$S_{\mathbf{c}}(t)_k = \begin{cases} \displaystyle\prod_{m=l_0-k+1}^{n} R_n\left(l_0 - k, m; -\frac{a_{km}^{(\mathbf{c})}(t)_{k-1}}{-(n-2)t}\right) & (2 \le k \le n_0) \\ \displaystyle R_n\left(l_0 - k, k; -\frac{a_{kk}^{(\mathbf{c})}(t)_{k-1}}{-2(n-2)t}\right) \prod_{m=k+1}^{n} R_n\left(l_0 - k, m; -\frac{a_{km}^{(\mathbf{c})}(t)_{k-1}}{-(n-2)t}\right) & \\ & (n_0 < k \le n-2). \end{cases}$$

By the definition of $A_{\mathbf{c}}(t)_k$ $(1 \le k \le n-2)$, we have $\det A_{\mathbf{c}}(t) = n \det A_{\mathbf{c}}(t)_{n-2}$. Therefore, to compute the polynomial $P_{\mathbf{c}}(t) = \det A_{\mathbf{c}}(t)$, let us first compute the matrix $A_{\mathbf{c}}(t)_{n-2}$ concretely.

3. Computation of the matrix $A_{\mathbf{c}}(t)_{n-2}$

Suppose $n \ge 5$. Then, by Proposition 4 and equation (5) in [**39**], we have

(6)

$$A_{\mathbf{c}}(t) = \begin{bmatrix} n & 0 & \cdots & \cdots & \cdots & 0 & 2t & at \\ 0 & & & & & -(n-2)t & -(n-1)at & -nbt \\ \vdots & & & & \iddots & -(n-1)at & -nbt & 0 \\ \vdots & & & \iddots & \iddots & -nbt & 0 & \vdots \\ \vdots & & \iddots & \iddots & & \iddots & \vdots & \vdots \\ 0 & -(n-2)t & -(n-1)at & -nbt & \iddots & & 0 & 0 \\ 2t & -(n-1)at & -nbt & 0 & \cdots & 0 & 2t^2 & at^2 \\ at & -nbt & 0 & \cdots & \cdots & 0 & at^2 & (a^2-2b)t^2 \end{bmatrix}$$

and hence

(7)

$$A_{\mathbf{c}}(t)_1 = \begin{bmatrix} 1 & 0 & \cdots & \cdots & \cdots & 0 & 0 & 0 \\ 0 & & & & & -(n-2)t & -(n-1)at & -nbt \\ \vdots & & & & \iddots & -(n-1)at & -nbt & 0 \\ \vdots & & & \iddots & \iddots & -nbt & 0 & \vdots \\ \vdots & & \iddots & \iddots & \iddots & \iddots & \vdots & \vdots \\ 0 & -(n-2)t & -(n-1)at & -nbt & \iddots & & 0 & 0 \\ 0 & -(n-1)at & -nbt & 0 & \cdots & 0 & 2\left(1-\frac{2}{n}\right)t^2 & \left(1-\frac{2}{n}\right)at^2 \\ 0 & -nbt & 0 & \cdots & \cdots & 0 & \left(1-\frac{2}{n}\right)at^2 & \left\{\left(1-\frac{1}{n}\right)a^2 - 2b\right\}t^2 \end{bmatrix}.$$

Here, similar to $A_{\mathbf{c}}(t)_k$ $(2 \le k \le n-2)$, we inductively define the matrix $W(t)_k = (w_{ij}(t)_k)_{1\le i,j\le n}$ $(2 \le k \le n-2)$ as follows;

(1) $W(t)_1 = (w_{ij}(t)_1)_{1\le i,j\le n}$, where

$$w_{ij}(t)_1 = \begin{cases} qt & i+j=n,\ 2\le i,j\le n-2, \\ rt & i+j=n+1,\ 2\le i,j\le n-1, \\ st & i+j=n+2,\ 2\le i,j\le n, \\ 0 & \text{otherwise.} \end{cases}$$

(2) $W(t)_k = (w_{ij}(t)_k)_{1\le i,j\le n}$ $(2 \le k \le n-2) = {}^tS(t)_k W(t)_{k-1} S(t)_k$, where

$$S(t)_k = \begin{cases} \displaystyle\prod_{m=n-k+1}^{n} R_n\left(n-k, m; -\frac{w_{km}(t)_{k-1}}{qt}\right) & (2 \le k \le n_0), \\ \displaystyle R_n\left(n-k, k; -\frac{w_{kk}(t)_{k-1}}{-2qt}\right) \prod_{m=k+1}^{n} R_n\left(n-k, m; -\frac{w_{km}(t)_{k-1}}{qt}\right) & \\ & (n_0 < k \le n-2). \end{cases}$$

Let us now compute the matrix $W(t)_{n-2}$ for any integer $n \ge 7$.

LEMMA 1. *Let $n \ge 7$ and $\{x_m\}$, $\{y_m\}$ be sequences defined by*

$$x_0 = 0,\ x_1 = -qt,\ x_2 = rt,\ x_{m+2} = -\frac{r}{q}x_{m+1} - \frac{s}{q}x_m\ (m \ge 1),$$
$$y_0 = y_1 = 0,\ y_{m+1} = -\frac{s}{q}x_m\ (m \ge 1)$$

and put

$$n_1 = \begin{cases} n_0 - 1 & n : \textit{odd}, \\ n_0 - 2 & n : \textit{even}. \end{cases}$$

Then, for any integer k such that $2 \le k \le n_1$, we have

$$w_{ij}(t)_k = \begin{cases} 0 & (i,j) = (k,\ell) \textit{ or } (\ell,k)\ (n-k+1 \le \ell \le n), \\ x_{\ell-n+k+2} & (i,j) = (k+1,\ell) \textit{ or } (\ell,k+1)\ (n-k \le \ell \le n-1), \\ y_{\ell-n+k+2} & (i,j) = (k+2,\ell) \textit{ or } (\ell,k+2)\ (n-k \le \ell \le n-1), \\ -sx_k/q & (i,j) = (k+1,n) \textit{ or } (n,k+1), \\ -sy_k/q & (i,j) = (k+2,n) \textit{ or } (n,k+2), \\ w_{ij}(t)_{k-1} & \textit{otherwise.} \end{cases}$$

PROOF. Let us prove this lemma by induction on k. First, suppose $k = 2$. Then, by the definition of $W(t)_2$, we have

$$W(t)_2 = \begin{bmatrix}
0 & \cdots & \cdots & \cdots & \cdots & \cdots & 0 & 0 & 0 & 0 \\
\vdots & & & & & & & qt & 0 & 0 \\
\vdots & & & & & & qt & rt & (qs-r^2)t/q & -rst/q \\
\vdots & & & & & \iddots & rt & st & -rst/q & -s^2t/q \\
\vdots & & & & \iddots & \iddots & st & 0 & 0 & 0 \\
\vdots & & & \iddots & \iddots & \iddots & \iddots & \iddots & \iddots & \vdots \\
0 & 0 & qt & rt & st & \iddots & \iddots & \iddots & & \vdots \\
0 & qt & rt & st & 0 & \iddots & \iddots & & & \vdots \\
0 & 0 & (qs-r^2)t/q & -rst/q & 0 & \iddots & & & & \vdots \\
0 & 0 & -rst/q & -s^2t/q & 0 & \cdots & \cdots & \cdots & \cdots & 0
\end{bmatrix},$$

which implies the claim of Lem. 1 for $k = 2$ since

$$x_2 = rt,\ x_3 = \frac{(qs - r^2)t}{q},\ y_2 = st,\ y_3 = -\frac{rst}{q}.$$

Next, suppose Lem. 1 is true for $k = 2, \cdots, m-1$ $(m - 1 < n_1)$. Then, since

$$W(t)_{m-1} = \begin{bmatrix}
0 & \cdots & \cdots & \cdots & \cdots & \cdots & \cdots & \cdots & \cdots & \cdots & \cdots & 0 & 0 \\
\vdots & & & & & & & & & \iddots & \iddots & \vdots & \vdots \\
\vdots & & & & & & & & qt & 0 & \cdots & 0 & 0 \\
\vdots & & & & & & & qt & x_2 & x_3 & \cdots & x_m & -sx_{m-1}/q \\
\vdots & & & & & & \iddots & rt & y_2 & y_3 & \cdots & y_m & -sy_{m-1}/q \\
\vdots & & & & & \iddots & \iddots & st & 0 & 0 & \cdots & 0 & 0 \\
\vdots & & & & \iddots & \iddots & \iddots & \iddots & \iddots & & & \iddots & \vdots \\
\vdots & & & qt & rt & st & \iddots & \iddots & & & \iddots & & \vdots \\
\vdots & & qt & x_2 & y_2 & 0 & \iddots & & & \iddots & & & \vdots \\
\vdots & \iddots & 0 & x_3 & y_3 & 0 & & & \iddots & & & & \vdots \\
\vdots & \iddots & \vdots & \vdots & \vdots & \vdots & & \iddots & & & & & \vdots \\
0 & \cdots & 0 & x_m & y_m & 0 & \iddots & & & & & & \vdots \\
0 & \cdots & 0 & -sx_{m-1}/q & -sy_{m-1}/q & 0 & \cdots & \cdots & \cdots & \cdots & \cdots & \cdots & 0
\end{bmatrix},$$

we have

$$W(t)_m = \left[\begin{array}{ccccccccccccc}
0 & \cdots & \cdots & \cdots & \cdots & \cdots & \cdots & \cdots & \cdots & \cdots & \cdots & 0 & 0 \\
\vdots & & & & & & & & & \iddots & \iddots & \vdots & \vdots \\
\vdots & & & & & & & & qt & 0 & \cdots & 0 & 0 \\
\vdots & & & & & & & qt & 0 & 0 & \cdots & 0 & 0 \\
\vdots & & & & & & \iddots & rt & x'_2 & x'_3 & \cdots & x'_m & x'_{m+1} \\
\vdots & & & & & \iddots & \iddots & st & y'_2 & y'_3 & \cdots & y'_m & y'_{m+1} \\
\vdots & & & & \iddots & \iddots & \iddots & \iddots & \iddots & & & \iddots & \vdots \\
\vdots & & & qt & rt & st & \iddots & \iddots & & & \iddots & & \vdots \\
\vdots & & qt & 0 & x'_2 & y'_2 & \iddots & & & \iddots & & & \vdots \\
\vdots & \iddots & 0 & 0 & x'_3 & y'_3 & & & \iddots & & & & \vdots \\
\vdots & \iddots & \vdots & \vdots & \vdots & \vdots & & \iddots & & & & & \vdots \\
0 & \cdots & 0 & 0 & x'_m & y'_m & \iddots & & & & & & \vdots \\
0 & \cdots & 0 & 0 & x'_{m+1} & y'_{m+1} & \cdots & \cdots & \cdots & \cdots & \cdots & \cdots & 0
\end{array}\right],$$

where

$$x'_\ell = y_\ell - \frac{x_\ell}{qt} \cdot rt = -\frac{r}{q} x_\ell - \frac{s}{q} x_{\ell-1} = x_{\ell+1} \ (2 \leq \ell \leq m),$$

$$y'_\ell = -\frac{x_\ell}{qt} \cdot st = -\frac{s}{q} x_\ell = y_{\ell+1} \ (2 \leq \ell \leq m),$$

$$x'_{m+1} = -\frac{s}{q} y_{m-1} + \left(-\frac{r}{q}\right) \cdot \left(-\frac{s}{q} x_{m-1}\right) = -\frac{s}{q}\left(-\frac{r}{q} x_{m-1} - \frac{s}{q} x_{m-2}\right) = -\frac{s}{q} x_m$$

$$y'_{m+1} = \left(-\frac{s}{q}\right) \cdot \left(-\frac{s}{q} x_{m-1}\right) = -\frac{s}{q} y_m.$$

This completes the proof of Lem. 1. □

PROPOSITION 2. *For any integer $n \geq 5$, we have*

$$w_{ij}(t)_{n-2} = \begin{cases} qt & i + j = n \ (2 \leq i, j \leq n-2), \\ x_{n-1} & (i,j) = (n-1, n-1), \\ y_{n-1} & (i,j) = (n-1, n) \text{ or } (n, n-1), \\ (-s/q) y_{n-2} & (i,j) = (n,n), \\ 0 & \text{otherwise.} \end{cases}$$

PROOF. Since Prop. 2 has been proved for $n = 5, 6$ in [**38**, p. 6, Example 2], we suppose $n \geq 7$. First, by solving the recurrence relation given in Lem. 1, we have

$$x_m = \begin{cases} \frac{(-1)^m m r^{m-1}}{2^{m-1} q^{m-2}} t & (r^2 - 4qs = 0), \\ \frac{(-1)^m \left\{ \left(r + \sqrt{r^2 - 4qs}\right)^m - \left(r - \sqrt{r^2 - 4qs}\right)^m \right\}}{2^m q^{m-2} \sqrt{r^2 - 4qs}} t & (r^2 - 4qs \neq 0). \end{cases} \tag{8}$$

Note that

$$\frac{\left(r + \sqrt{r^2 - 4qs}\right)^m - \left(r - \sqrt{r^2 - 4qs}\right)^m}{2^m q^{m-2} \sqrt{r^2 - 4qs}} = \frac{2}{2^m q^{m-2}} \sum_{k=0}^{\lfloor \frac{m-1}{2} \rfloor} \binom{m}{2k+1} r^{m-2k-1} (r^2 - 4qs)^k$$

and hence, by putting $r^2 - 4qs = 0$, we have

$$\frac{(-1)^m \left\{ \left(r + \sqrt{r^2 - 4qs}\right)^m - \left(r - \sqrt{r^2 - 4qs}\right)^m \right\}}{2^m q^{m-2} \sqrt{r^2 - 4qs}} t = \frac{(-1)^m m r^{m-1}}{2^{m-1} q^{m-2}} t,$$

the solution for the case of $r^2 - 4qs = 0$. Here, suppose n is odd. Then, we have $n_1 = (n-3)/2$ and hence by Lem. 1,

$$w_{ij}(t)_{n_1} = \begin{cases} 0 & (i,j) = ((n-3)/2, \ell) \text{ or } (\ell, (n-3)/2) \; ((n+5)/2 \leq \ell \leq n), \\ x_{\ell-(n-1)/2} & (i,j) = ((n-1)/2, \ell) \text{ or } (\ell, (n-1)/2) \; ((n+3)/2 \leq \ell \leq n-1), \\ y_{\ell-(n-1)/2} & (i,j) = ((n+1)/2, \ell) \text{ or } (\ell, (n+1)/2) \; ((n+3)/2 \leq \ell \leq n-1), \\ -sx_{(n-3)/2}/q & (i,j) = ((n-1)/2, n) \text{ or } (n, (n-1)/2), \\ -sy_{(n-3)/2}/q & (i,j) = ((n+1)/2, n) \text{ or } (n, (n+1)/2), \\ w_{ij}(t)_{(n-5)/2} & \text{otherwise.} \end{cases}$$

Therefore, we have the next expression of the matrix $W(t)_{n_1}$;

$$W(t)_{n_1} = \begin{bmatrix} 0 & \cdots & \cdots & \cdots & \cdots & \cdots & \cdots & \cdots & 0 & 0 \\ \vdots & & & & & & \iddots & & \vdots & \vdots \\ \vdots & & & & & qt & 0 & \cdots & 0 & 0 \\ \vdots & & & & qt & x_2 & x_3 & \cdots & x_{(n-1)/2} & -sx_{(n-3)/2}/q \\ \vdots & & & qt & rt & y_2 & y_3 & \cdots & y_{(n-1)/2} & -sy_{(n-3)/2}/q \\ \vdots & & qt & x_2 & y_2 & 0 & & & & \vdots \\ \vdots & \iddots & 0 & x_3 & y_3 & & & & & \vdots \\ \vdots & & \vdots & \vdots & \vdots & & & & & \vdots \\ 0 & \cdots & 0 & x_{(n-1)/2} & y_{(n-1)/2} & & & & & \vdots \\ 0 & \cdots & 0 & -sx_{(n-3)/2}/q & -sy_{(n-3)/2}/q & \cdots & \cdots & \cdots & \cdots & 0 \end{bmatrix}.$$

Then, since $W(t)_{n_1+1} = {}^tS(t)_{n_1+1} W(t)_{n_1} S(t)_{n_1+1}$, where

$$S(t)_{n_1+1} = \prod_{m=(n+3)/2}^{n} R_n\left(\frac{n+1}{2}, m; -\frac{w_{(n-1)/2,m}(t)_{n_1}}{qt}\right)$$

and $w_{n-1,(n+1)/2}(t)_{n_1+1} = w_{(n+1)/2,n-1}(t)_{n_1+1} = x_{(n+1)/2}$, we have

$$w_{n-1,n-1}(t)_{n_1+1} = \frac{\left(r + \sqrt{r^2 - 4qs}\right)^{n-1} - \left(r - \sqrt{r^2 - 4qs}\right)^{n-1}}{2^{n-1} q^{n-3} \sqrt{r^2 - 4qs}} t = x_{n-1}.$$

Similarly, we have

$$w_{n-1,n}(t)_{n_1+1} = -\frac{s}{q} x_{n-2} = y_{n-1},$$

see [**38**] for details. Moreover, since $w_{n,(n+1)/2}(t)_{n_1+1} = w_{(n+1)/2,n}(t)_{n_1+1} = -sx_{(n-1)/2}/q$, we have

$$w_{n,n}(t)_{n_1+1} = -\frac{s}{q}\left(-\frac{s}{q} x_{n-3}\right) = -\frac{s}{q} y_{n-2}.$$

Therefore, we can express the matrix $W(t)_{n_0}$ $(= W(t)_{n_1+1})$ as follows;

$$
W(t)_{n_0} = \begin{bmatrix}
0 & \cdots & \cdots & \cdots & \cdots & \cdots & \cdots & \cdots & 0 & 0 \\
\vdots & & & & & & \iddots & & \vdots & \vdots \\
\vdots & & & & & qt & 0 & \cdots & 0 & 0 \\
\vdots & & & & qt & 0 & 0 & \cdots & 0 & 0 \\
\vdots & & & qt & * & * & * & \cdots & * & * \\
\vdots & & qt & 0 & * & * & * & \cdots & * & * \\
\vdots & \iddots & 0 & 0 & * & * & * & \cdots & * & * \\
\vdots & & \vdots & \vdots & \vdots & \vdots & \vdots & \ddots & \vdots & \vdots \\
0 & \cdots & 0 & 0 & * & * & * & \cdots & x_{n-1} & y_{n-1} \\
0 & \cdots & 0 & 0 & * & * & * & \cdots & y_{n-1} & (-s/q)y_{n-2}
\end{bmatrix}.
$$

Then, by the definition of the matrix W_k $(n_0 < k \leq n-2)$, we have

$$
W(t)_{n-2} = \begin{bmatrix}
0 & \cdots & \cdots & \cdots & \cdots & \cdots & \cdots & 0 & 0 \\
\vdots & & & & & \iddots & & \vdots & \vdots \\
\vdots & & & & qt & 0 & \cdots & 0 & 0 \\
\vdots & & & qt & 0 & 0 & \cdots & 0 & 0 \\
\vdots & & qt & 0 & 0 & 0 & \cdots & 0 & 0 \\
\vdots & \iddots & 0 & 0 & 0 & 0 & \cdots & 0 & 0 \\
\vdots & & \vdots & \vdots & \vdots & \vdots & \ddots & \vdots & \vdots \\
0 & & 0 & 0 & 0 & 0 & \cdots & x_{n-1} & y_{n-1} \\
0 & \cdots & 0 & 0 & 0 & 0 & \cdots & y_{n-1} & (-s/q)y_{n-2}
\end{bmatrix},
$$

which completes the proof of Prop. 2 for odd n.

Next, suppose n is even. Then, we have $n_1 = (n-4)/2$ and hence by Lem. 1, we have

$$
w_{ij}(t)_{n_1} = \begin{cases}
0 & (i,j) = ((n-4)/2, \ell) \text{ or } (\ell, (n-4)/2) \ ((n+6)/2 \leq \ell \leq n), \\
x_{\ell-n/2} & (i,j) = ((n-2)/2, \ell) \text{ or } (\ell, (n-2)/2) \ ((n+4)/2 \leq \ell \leq n-1), \\
y_{\ell-n/2} & (i,j) = (n/2, \ell) \text{ or } (\ell, n/2) \ ((n+4)/2 \leq \ell \leq n-1), \\
-sx_{(n-4)/2}/q & (i,j) = ((n-2)/2, n) \text{ or } (n, (n-2)/2), \\
-sy_{(n-4)/2}/q & (i,j) = (n/2, n) \text{ or } (n, n/2), \\
w_{ij}(t)_{(n-6)/2} & \text{otherwise.}
\end{cases}
$$

Thus, we have the next expression of the matrix $W(t)_{n_1}$;

$$W(t)_{n_1}=\left[\begin{array}{ccccccccccc}
0 & \cdots & \cdots & \cdots & \cdots & \cdots\ \cdots & \cdots & \cdots & \cdots & 0 & 0\\
\vdots & & & & & & & \iddots & & \vdots & \vdots\\
\vdots & & & & & & qt & 0 & \cdots & 0 & 0\\
\vdots & & & & & qt & x_2 & x_3 & \cdots & x_{(n-2)/2} & -sx_{(n-4)/2}/q\\
\vdots & & & & qt & rt & y_2 & y_3 & \cdots & y_{(n-2)/2} & -sy_{(n-4)/2}/q\\
\vdots & & & qt & rt & st & 0 & 0 & \cdots & 0 & 0\\
\vdots & & qt & x_2 & y_2 & 0 & 0 & & & & \vdots\\
\vdots & \iddots & 0 & x_3 & y_3 & 0 & & & & & \vdots\\
\vdots & & \vdots & \vdots & \vdots & \vdots & & & & & \vdots\\
0 & \cdots & 0 & x_{(n-2)/2} & y_{(n-2)/2} & 0 & & & & & \vdots\\
0 & \cdots & 0 & -sx_{(n-4)/2}/q & -sy_{(n-4)/2}/q & 0 & \cdots & \cdots & \cdots & \cdots & 0
\end{array}\right].$$

Then, for any integer ℓ $((n+4)/2 \leq \ell \leq n)$, let us put

$$V(t)_\ell = (v_{ij}(t)_\ell)_{1\leq i,j\leq n} = {}^tR(t)_\ell W(t)_{n_1} R(t)_\ell,$$

where

$$R(t)_\ell = \prod_{m=(n+4)/2}^{\ell} R_n\left((n+2)/2, m; -\frac{w_{(n-2)/2,m}(t)_{n_1}}{qt}\right).$$

Note that we have $V(t)_n = W(t)_{n_1+1}$.

CLAIM . *For any integer ℓ' $((n+4)/2 \leq \ell' \leq n-1)$, we have*

$$v_{ij}(t)_{\ell'} = \begin{cases} 0 & (i,j) = ((n-2)/2, \ell) \text{ or } (\ell, (n-2)/2)\ ((n+4)/2 \leq \ell \leq \ell'),\\ x_{\ell-(n-2)/2} & (i,j) = (n/2, \ell) \text{ or } (\ell, n/2)\ ((n+2)/2 \leq \ell \leq \ell'),\\ y_{\ell-(n-2)/2} & (i,j) = ((n+2)/2, \ell) \text{ or } (\ell, (n+2)/2)\ ((n+2)/2 \leq \ell \leq \ell'),\\ \left(-x_{\ell-n/2}/qt\right)^2 y_2 & (i,j) = (\ell,\ell)\ ((n+4)/2 \leq \ell \leq \ell'),\\ (-x_{m-n/2}/qt) & \\ \qquad \times y_{\ell-(n-2)/2} & (i,j) = (\ell, m) \text{ or } (m,\ell)\ ((n+4)/2 \leq \ell < m \leq \ell'),\\ w_{ij}(t)_{(n-4)/2} & \text{otherwise.} \end{cases}$$

PROOF OF CLAIM. To ease notation, let us put $k = (n-2)/2$. Then, by definition,

$$V(t)_{(n+4)/2}=\left[\begin{array}{ccccccccccc}
0 & \cdots & \cdots & \cdots & \cdots & \cdots & \cdots & \cdots & \cdots & 0 & 0\\
\vdots & & & & & & & \iddots & & \vdots & \vdots\\
\vdots & & & & & & qt & 0 & \cdots & 0 & 0\\
\vdots & & & & & qt & 0 & x_3 & \cdots & x_k & -sx_{k-1}/q\\
\vdots & & & & qt & x_2 & x_3 & y_3 & \cdots & y_k & -sy_{k-1}/q\\
\vdots & & & qt & x_2 & y_2 & y_3 & 0 & \cdots & 0 & 0\\
\vdots & & qt & 0 & x_3 & y_3 & \left(-x_2/qt\right)^2 y_2 & & & & \vdots\\
\vdots & \iddots & 0 & x_3 & y_3 & 0 & & & & & \vdots\\
\vdots & & \vdots & \vdots & \vdots & \vdots & & & & & \vdots\\
0 & \cdots & 0 & x_k & y_k & 0 & & & & & \vdots\\
0 & \cdots & 0 & -sx_{k-1}/q & -sy_{k-1}/q & 0 & \cdots & \cdots & \cdots & \cdots & 0
\end{array}\right]$$

and we get Claim for $V(t)_{(n+4)/2}$. Here, suppose Claim is true for $V(t)_{\ell'-1}$ and hence

$$V(t)_{\ell'-1} = \left[\begin{array}{ccccccccccc}
0 & \cdots & \cdots & \cdots & \cdots & \cdots & \cdots & \cdots & 0 & 0 & \cdots \\
\vdots & & & & & \iddots & & & \vdots & \vdots & \\
\vdots & & & & qt & 0 & 0 & \cdots & 0 & x_{\ell'-k-1} & \cdots \\
\vdots & & & qt & x_2 & x_3 & x_4 & \cdots & x_{\ell'-k-1} & y_{\ell'-k-1} & \cdots \\
\vdots & & qt & x_2 & y_2 & y_3 & y_4 & \cdots & y_{\ell'-k-1} & 0 & \cdots \\
\vdots & \iddots & 0 & x_3 & y_3 & (-x_2/qt)^2 y_2 & (-x_3/qt)\, y_3 & \cdots & (-x_{\ell'-k-2}/qt) y_3 & 0 & \cdots \\
\vdots & & 0 & x_4 & y_4 & (-x_3/qt)\, y_3 & (-x_3/qt)^2 y_2 & \cdots & (-x_{\ell'-k-2}/qt) y_4 & 0 & \cdots \\
\vdots & & \vdots & \vdots & \vdots & \vdots & \vdots & \ddots & \vdots & \vdots & \\
0 & & 0 & x_{\ell'-k-1} & y_{\ell'-k-1} & (-x_{\ell'-k-2}/qt) y_3 & (-x_{\ell'-k-2}/qt) y_4 & \cdots & (-x_{\ell'-k-2}/qt)^2 y_2 & 0 & \cdots \\
0 & \cdots & x_{\ell'-k-1} & y_{\ell'-k-1} & 0 & 0 & 0 & \cdots & 0 & 0 & \cdots \\
\vdots & & \vdots & \vdots & \vdots & \vdots & \vdots & & \vdots & \vdots &
\end{array}\right].$$

Thus, by a direct computation, we have

$$V(t)_{\ell'} = \left[\begin{array}{ccccccccccc}
0 & \cdots & \cdots & \cdots & \cdots & \cdots & \cdots & \cdots & 0 & 0 & \cdots \\
\vdots & & & & & \iddots & & & \vdots & \vdots & \\
\vdots & & & & qt & 0 & 0 & \cdots & 0 & x_{\ell'-k} & \cdots \\
\vdots & & & qt & x_2 & x_3 & x_4 & \cdots & x_{\ell'-k} & y_{\ell'-k} & \cdots \\
\vdots & & qt & x_2 & y_2 & y_3 & y_4 & \cdots & y_{\ell'-k} & 0 & \cdots \\
\vdots & \iddots & 0 & x_3 & y_3 & (-x_2/qt)^2 y_2 & (-x_3/qt)\, y_3 & \cdots & (-x_{\ell'-k-1}/qt) y_3 & 0 & \cdots \\
\vdots & & 0 & x_4 & y_4 & (-x_3/qt)\, y_3 & (-x_3/qt)^2 y_2 & \cdots & (-x_{\ell'-k-1}/qt) y_4 & 0 & \cdots \\
\vdots & & \vdots & \vdots & \vdots & \vdots & \vdots & \ddots & \vdots & \vdots & \\
0 & & 0 & x_{\ell'-k} & y_{\ell'-k} & (-x_{\ell'-k-1}/qt) y_3 & (-x_{\ell'-k-1}/qt) y_4 & \cdots & (-x_{\ell'-k-1}/qt)^2 y_2 & 0 & \cdots \\
0 & \cdots & x_{\ell'-k} & y_{\ell'-k} & 0 & 0 & 0 & \cdots & 0 & 0 & \cdots \\
\vdots & & \vdots & \vdots & \vdots & \vdots & \vdots & & \vdots & \vdots &
\end{array}\right]$$

and we get Claim by induction on ℓ'. □

By above Claim,

$$V(t)_{n-1} = \left[\begin{array}{cccccccccc}
0 & \cdots & \cdots & \cdots & \cdots & \cdots & \cdots & \cdots & 0 & 0 \\
\vdots & & & & & \iddots & & & \vdots & \vdots \\
\vdots & & & & qt & 0 & 0 & \cdots & 0 & -sx_{k-1}/q \\
\vdots & & & qt & x_2 & x_3 & x_4 & \cdots & x_{k+1} & -sy_{k-1}/q \\
\vdots & & qt & x_2 & y_2 & y_3 & y_4 & \cdots & y_{k+1} & 0 \\
\vdots & \iddots & 0 & x_3 & y_3 & (-x_2/qt)^2 y_2 & (-x_3/qt)\, y_3 & \cdots & (-x_k/qt) y_3 & 0 \\
\vdots & & 0 & x_4 & y_4 & (-x_3/qt)\, y_3 & (-x_3/qt)^2 y_2 & \cdots & (-x_k/qt) y_4 & 0 \\
\vdots & & \vdots & \vdots & \vdots & \vdots & \vdots & \ddots & \vdots & \vdots \\
0 & & 0 & x_{k+1} & y_{k+1} & (-x_k/qt) y_3 & (-x_k/qt) y_4 & \cdots & (-x_k/qt)^2 y_2 & 0 \\
0 & \cdots & -sx_{k-1}/q & -sy_{k-1}/q & 0 & 0 & 0 & \cdots & 0 & 0
\end{array}\right]$$

and hence, by the definition of $V(t)_n$, we have

$$W(t)_{n_1+1} = V(t)_n =$$

$$\left[\begin{array}{cccccccccc}
0 & \cdots & \cdots & \cdots & \cdots & \cdots & \cdots & \cdots & 0 & 0 \\
\vdots & & & & & \iddots & & & \vdots & \vdots \\
\vdots & & & & qt & 0 & 0 & \cdots & 0 & 0 \\
\vdots & & & qt & x_2 & x_3 & x_4 & \cdots & x_{k+1} & -sx_k/q \\
\vdots & & qt & x_2 & y_2 & y_3 & y_4 & \cdots & y_{k+1} & -sy_k/q \\
\vdots & \iddots & 0 & x_3 & y_3 & (-x_2/qt)^2 y_2 & (-x_3/qt)\, y_3 & \cdots & (-x_k/qt) y_3 & (x_{k-1}/q^2t^2) y_2 y_3 \\
\vdots & & 0 & x_4 & y_4 & (-x_3/qt)\, y_3 & (-x_3/qt)^2 y_2 & \cdots & (-x_k/qt) y_4 & (x_{k-1}/q^2t^2) y_2 y_4 \\
\vdots & & \vdots & \vdots & \vdots & \vdots & \vdots & \ddots & \vdots & \vdots \\
0 & & 0 & x_{k+1} & y_{k+1} & (-x_k/qt) y_3 & (-x_k/qt) y_4 & \cdots & (-x_k/qt)^2 y_2 & (x_{k-1}/q^2t^2) y_2 y_{k+1} \\
0 & \cdots & 0 & -sx_k/q & -sy_k/q & (x_{k-1}/q^2t^2) y_2 y_3 & (x_{k-1}/q^2t^2) y_2 y_4 & \cdots & (x_{k-1}/q^2t^2) y_2 y_{k+1} & (x_{k-1}/q^2t^2)^2 y_2^3
\end{array}\right].$$

Therefore, by the definition of the matrix $W(t)_{n_1+2} = W(t)_{n_0}$, we have

$$w_{n-1,n-1}(t)_{n_0} = x_{n-1}.$$

Similarly, we have

$$w_{n-1,n}(t)_{n_0} = -\frac{s}{q}\, y_{n-2}.$$

Therefore, we can express the matrix $W(t)_{n_0}$ as follows;

$$W(t)_{n_0} = \begin{bmatrix} 0 & \cdots & \cdots & \cdots & \cdots & \cdots & \cdots & 0 & 0 \\ \vdots & & & & & \iddots & & \vdots & \vdots \\ \vdots & & & & qt & 0 & \cdots & 0 & 0 \\ \vdots & & & qt & 0 & 0 & \cdots & 0 & 0 \\ \vdots & & qt & 0 & * & * & \cdots & * & * \\ \vdots & \iddots & 0 & 0 & * & * & \cdots & * & * \\ \vdots & & \vdots & \vdots & \vdots & \vdots & \ddots & \vdots & \vdots \\ 0 & & 0 & 0 & * & * & \cdots & x_{n-1} & y_{n-1} \\ 0 & \cdots & 0 & 0 & * & * & \cdots & y_{n-1} & (-s/q)y_{n-2} \end{bmatrix}.$$

Then, by the definition of the matrix W_k $(n_0 < k \leq n-2)$, we have

$$W(t)_{n-2} = \begin{bmatrix} 0 & \cdots & \cdots & \cdots & \cdots & \cdots & \cdots & 0 & 0 \\ \vdots & & & & & \iddots & & \vdots & \vdots \\ \vdots & & & & qt & 0 & \cdots & 0 & 0 \\ \vdots & & & qt & 0 & 0 & \cdots & 0 & 0 \\ \vdots & & qt & 0 & 0 & 0 & \cdots & 0 & 0 \\ \vdots & \iddots & 0 & 0 & 0 & 0 & \cdots & 0 & 0 \\ \vdots & & \vdots & \vdots & \vdots & \vdots & \ddots & \vdots & \vdots \\ 0 & & 0 & 0 & 0 & 0 & \cdots & x_{n-1} & y_{n-1} \\ 0 & \cdots & 0 & 0 & 0 & 0 & \cdots & y_{n-1} & (-s/q)y_{n-2} \end{bmatrix},$$

which completes the proof of Prop. 2 for even n. □

In the end, let us apply Prop. 2 to the matrix $A_{\mathbf{c}}(t)_1$ $(n \geq 5)$. Then, by the definition of the matrix $A_{\mathbf{c}}(t)_k$ $(2 \leq k \leq n-2)$ and Prop. 2, we have

$$(9)\quad A_{\mathbf{c}}(t)_{n-2} = \begin{bmatrix} 1 & 0 & \cdots & \cdots & \cdots & 0 & 0 & 0 \\ 0 & & & & & -(n-2)t & 0 & 0 \\ \vdots & & & & \iddots & \iddots & 0 & 0 \\ \vdots & & & \iddots & \iddots & \iddots & \vdots & \vdots \\ \vdots & & \iddots & \iddots & \iddots & \iddots & \vdots & \vdots \\ 0 & -(n-2)t & \iddots & \iddots & \iddots & & 0 & 0 \\ 0 & 0 & 0 & \cdots & \cdots & 0 & a^{(\mathbf{c})}_{n-1,n-1}(t)_{n-2} & a^{(\mathbf{c})}_{n-1,n}(t)_{n-2} \\ 0 & 0 & 0 & \cdots & \cdots & 0 & a^{(\mathbf{c})}_{n,n-1}(t)_{n-2} & a^{(\mathbf{c})}_{n,n}(t)_{n-2} \end{bmatrix},$$

where

$$\begin{cases} a^{(\mathbf{c})}_{n-1,n-1}(t)_{n-2} = 2\,(1-2/n)\,t^2 + \bar{x}_{n-1} \\ a^{(\mathbf{c})}_{n-1,n}(t)_{n-2} = a^{(\mathbf{c})}_{n,n-1}(t)_{n-2} = (1-2/n)\,at^2 - (nb\bar{x}_{n-2})/(n-2) \\ a^{(\mathbf{c})}_{n,n}(t)_{n-2} = \left\{(1-1/n)\,a^2 - 2b\right\} t^2 + (n^2b^2\bar{x}_{n-3})/(n-2)^2. \end{cases}$$

Here, for any integer $m \geq 0$, we denote

$$\bar{x}_m = \frac{(-1)^m (A^m - B^m)}{2^m\{-(n-2)\}^{m-2}\sqrt{(n-1)^2a^2 - 4n(n-2)b}}t,$$

$$A = -(n-1)a + \sqrt{(n-1)^2a^2 - 4n(n-2)b},$$

$$B = -(n-1)a - \sqrt{(n-1)^2a^2 - 4n(n-2)b}.$$

4. Proof of theorems

By a direct computation, we have

(10)

$$\det \begin{bmatrix} a^{(\mathbf{c})}_{n-1,n-1}(t)_{n-2} & a^{(\mathbf{c})}_{n-1,n}(t)_{n-2} \\ a^{(\mathbf{c})}_{n,n-1}(t)_{n-2} & a^{(\mathbf{c})}_{n,n}(t)_{n-2} \end{bmatrix} = \frac{(n-2)(a^2-4b)}{n}t^4 + \left(\frac{2nb^2\bar{x}_{n-3}}{n-2}\right.$$
$$\left. + \frac{\{(n-1)a^2 - 2nb\}\,\bar{x}_{n-1}}{n} + 2ab\bar{x}_{n-2}\right)t^2 + \frac{n^2b^2}{(n-2)^2}\left(\bar{x}_{n-1}\bar{x}_{n-3} - \bar{x}^2_{n-2}\right).$$

Here, let us put

$$\alpha(t) = \left\{\frac{2nb^2\bar{x}_{n-3}}{n-2} + \frac{\{(n-1)a^2 - 2nb\}\,\bar{x}_{n-1}}{n} + 2ab\bar{x}_{n-2}\right\}t^2,$$

$$\beta(t) = \frac{n^2b^2}{(n-2)^2}\left(\bar{x}_{n-1}\bar{x}_{n-3} - \bar{x}^2_{n-2}\right).$$

LEMMA 2. *Suppose $n \geq 5$ and S_k as in Eq. (4). Then, we have $\beta(t) = -\frac{n^{n-1}b^{n-1}}{(n-2)^{n-3}}t^2$ and*

$$\alpha(t) = \sum_{k=0}^{m_0} \frac{(-1)^{n+k}n^k(n-1)^{n-2k-4}(n-2)^k a^{n-2k-4}b^k S_k}{n(n-2)^{n-3}}t^3$$

for $a \neq 0$, or $a = 0$ and n even, otherwise $\alpha(t) = 0$.

PROOF. We first prove the second equality. Then, by the definition of $\bar{x}_m$, and straight computation we have

$$\beta(t) = -\frac{n^{n-1}b^{n-1}}{(n-2)^{n-3}}t^2,$$

which is the second equality. For the first equality, let us first suppose $a = 0$. Then, we have

$$\begin{aligned}\bar{x}_m &= \frac{(-1)^m\left\{\{-4n(n-2)b\}^{m/2} - (-1)^m\{-4n(n-2)b\}^{m/2}\right\}}{2^m\{-(n-2)\}^{m-2}\{-4n(n-2)b\}^{1/2}}t \\ &= \begin{cases} \dfrac{2\{-4n(n-2)b\}^{(m-1)/2}}{2^m(n-2)^{m-2}}t & (m : \text{odd}), \\ 0 & (m : \text{even}) \end{cases} \\ &= \begin{cases} \dfrac{(-nb)^{(m-1)/2}}{(n-2)^{(m-3)/2}}t & (m : \text{odd}), \\ 0 & (m : \text{even}) \end{cases}\end{aligned}$$

and hence we have

$$\alpha(t)=\left\{\frac{2nb^2\bar{x}_{n-3}}{n-2}-2b\bar{x}_{n-1}\right\}t^2=\begin{cases}\dfrac{(-1)^{n/2}4n^{(n-2)/2}b^{n/2}}{(n-2)^{(n-4)/2}}t^3 & (n:\text{even}),\\ 0 & (n:\text{odd}),\end{cases}$$

which is the claim of Lem. 2 for the case $a=0$ since for even n $(n\geq 6)$,

$$\begin{aligned}&\frac{\sum_{k=0}^{m_0}(-1)^{n+k}n^k(n-1)^{n-2k-4}(n-2)^ka^{n-2k-4}b^kS_k}{n(n-2)^{n-3}}t^3\\ &=\frac{(-1)^{(3n-4)/2}n^{(n-4)/2}(n-2)^{(n-4)/2}b^{(n-4)/2}\cdot 4n^2(n-2)b^2}{n(n-2)^{n-3}}t^3\\ &=\frac{(-1)^{n/2}4n^{(n-2)/2}b^{n/2}}{(n-2)^{(n-4)/2}}t^3.\end{aligned}$$

Next, we suppose $a\neq 0$ and let us put

$$\begin{aligned}&F=(n-1)^2a^3-2n(2n-3)ab,\ G=(n-1)a^2-2nb,\\ &H=(n-1)^2a^2-4n(n-2)b.\end{aligned}$$

Then, we have

$$\begin{aligned}&\frac{2nb^2\bar{x}_{n-3}}{n-2}+\frac{\{(n-1)a^2-2nb\}\,\bar{x}_{n-1}}{n}+2ab\bar{x}_{n-2}\\ &=\frac{8n^2(n-2)b^2+\{(n-1)a^2-2nb\}\,A^2+4n(n-2)abA}{2^{n-1}n(n-2)^{n-3}\sqrt{H}}A^{n-3}t\\ &\qquad-\frac{8n^2(n-2)b^2+\{(n-1)a^2-2nb\}\,B^2+4n(n-2)abB}{2^{n-1}n(n-2)^{n-3}\sqrt{H}}B^{n-3}t\end{aligned}$$

and

$$8n^2(n-2)b^2+\{(n-1)a^2-2nb\}\,A^2+4n(n-2)abA=2F\sqrt{H}+2GH.$$

Hence, by putting

$$A^{n-3}=\left\{-(n-1)a+\sqrt{(n-1)^2a^2-4n(n-2)b}\right\}^{n-3}=I+J\sqrt{H},$$

we have

$$\frac{2nb^2\bar{x}_{n-3}}{n-2}+\frac{\{(n-1)a^2-2nb\}\,\bar{x}_{n-1}}{n}+2ab\bar{x}_{n-2}=\frac{-FI+GHJ}{2^{n-3}n(n-2)^{n-3}}t.$$

Then, by using Lemma 14 in [**38**, p. 18, Lemma 4], we have

$$I=\sum_{k=0}^{m_0}(-1)^{n+k+1}2^{n-4}n^k(n-1)^{n-2k-3}(n-2)^k\frac{n-3}{n-k-3}\binom{n-k-3}{k}a^{n-2k-3}b^k$$

and

$$J=\sum_{k=0}^{m_0}(-1)^{n+k}2^{n-4}n^k(n-1)^{n-2k-4}(n-2)^k\binom{n-k-4}{k}a^{n-2k-4}b^k,$$

which implies,

$$\begin{aligned}-FI+GHJ &= -\{(n-1)^2a^3-2n(2n-3)ab\}I\\ &\quad +\{(n-1)^3a^4-2n(n-1)(3n-5)a^2b+8n^2(n-2)b^2\}J\\ &= \sum_{k=0}^{m_0}(-1)^{n+k}2^{n-3}n^k(n-1)^{n-2k-4}(n-2)^k a^{n-2k-4}b^k\\ &\quad\times\Bigg\{(n-1)^3\binom{n-k-3}{k}a^4\\ &\qquad -\frac{n(n-1)\left\{5n^2-(6k+23)n+10k+24\right\}}{n-k-3}\binom{n-k-3}{k}a^2b\\ &\qquad +4n^2(n-2)\binom{n-k-4}{k}b^2\Bigg\}\end{aligned}$$

Therefore, we finally obtain

$$\begin{aligned}\alpha(t) &= \left\{\frac{2nb^2\bar{x}_{n-3}}{n-2}+\frac{\left\{(n-1)a^2-2nb\right\}\bar{x}_{n-1}}{n}+2ab\bar{x}_{n-2}\right\}t^2=\frac{-FI+GHJ}{2^{n-3}n(n-2)^{n-3}}t^3\\ &= \sum_{k=0}^{m_0}\frac{(-1)^{n+k}n^k(n-1)^{n-2k-4}(n-2)^k a^{n-2k-4}b^k S_k}{n(n-2)^{n-3}}t^3,\end{aligned}$$

where S_k as in Eq. (4). This completes the proof. □

Next we are ready to prove Thm. 1.

Proof of Thm. 1. By the definition of $A_{\mathbf{c}}(t)_{n-2}$ and equations (9), (10), we have

$$\begin{aligned}\Delta\left(f_{\mathbf{c}}(t;x)\right) &= \det A_{\mathbf{c}}(t)=n\cdot\det A_{\mathbf{c}}(t)_{n-2}\\ &= n\cdot(-1)^{\lceil (n-3)/2\rceil}(n-2)^{n-3}t^{n-3}\cdot\det\begin{bmatrix} a^{(\mathbf{c})}_{n-1,n-1}(t)_{n-2} & a^{(\mathbf{c})}_{n-1,n}(t)_{n-2}\\ a^{(\mathbf{c})}_{n,n-1}(t)_{n-2} & a^{(\mathbf{c})}_{n,n}(t)_{n-2}\end{bmatrix}.\end{aligned}$$

Therefore, by Lem. 2, we have

$$\Delta\left(f_{\mathbf{c}}(t;x)\right)=(-1)^{m_1}t^{n-1}\Big\{(n-2)^{n-2}(a^2-4b)t^2+\gamma_{\mathbf{c}}t-n^nb^{n-1}\Big\},$$

where

$$\gamma_{\mathbf{c}}=\begin{cases}\displaystyle\sum_{k=0}^{m_0}(-1)^{n+k}n^k(n-1)^{n-2k-4}(n-2)^k a^{n-2k-4}b^k S_k & (a\neq 0,\text{ or } a=0,\ n:\text{even}),\\ 0 & (a=0,\ n:\text{odd}),\end{cases}$$

which completes the proof. □

5. Concluding remarks

It is object of further investigation if such result could be generalized to polynomials $f(x)=x^n+t\cdot g(x)$, where $g(x)$ is a general cubic. As a quick application of Thm. 1 we get that for $b\neq 0$ and $\frac{(n-1)^2a^2}{4n(n-2)}\leq b$ the polynomial $f(t,x)$ has no real roots, for any real number $t>0$; see [**38**, Cor. 1]. It would be interesting to check whether out approach would work to generalize for any polynomial of type $f(t,x) = x^n + t \cdot g(x)$, for $\deg g \geq 3$.

It is a quite open problem to determine for what integer values of a, b, and t the quadrinomial $f(x)$ is irreducible or the discriminant Δ_f has a minimal value. Moreover, it is our intention to study in the future for what conditions on a, b, t the quadrinomial is reduced in the sense of [**7**].

Another motivation of looking at the discriminant of such family of polynomials comes from our efforts to construct superelliptic Jacobians with large endomorphism rings. We have checked computationally that for small n and $f(t,x) \in \mathbb{Q}[t,x]$, for almost all t the Galois group $\mathrm{Gal}_{\mathbb{Q}}(f,x)$ is isomorphic to S_n. Due to results of Zarhin this implies that the superelliptic curve $y^m = f(t,x)$ has large endomorphism ring; see [**20**] for related matters. Hence, curves $y^n = f(x)$, where $f(x)$ is as above, are smooth curves whose Jacobians are expected to have large endomorphism rings. This remains the focus of further investigation.

References

[1] C. Bagiński, J. J. Etayo, G. Gromadzki, and E. Martínez, *On real forms of a Belyi action of the alternating groups*, Albanian J. Math. **10** (2016), no. 1, 3–10. MR3490049

[2] Edoardo Ballico, *Symmetric tensor rank and the identification of a point using linear spans of an embedded variety*, Albanian J. Math. **8** (2014), no. 1, 9–21. MR3270073

[3] Oz Ben-Shimol, *On Galois groups of prime degree polynomials with complex roots*, Algebra Discrete Math. **2** (2009), 99–107. MR2589076

[4] Victor Beresnevich, Vasili Bernik, and Friedrich Götze, *Integral polynomials with small discriminants and resultants*, Adv. Math. **298** (2016), 393–412, DOI 10.1016/j.aim.2016.04.022. MR3505745

[5] Marie José Bertin and Odile Lecacheux, *Automorphisms of certain Niemeier lattices and elliptic fibrations*, Albanian J. Math. **11** (2017), no. 1, 13–33. MR3724575

[6] L. Beshaj, R. Hidalgo, S. Kruk, A. Malmendier, S. Quispe, and T. Shaska, *Rational points in the moduli space of genus two*, Higher genus curves in mathematical physics and arithmetic geometry, Contemp. Math., vol. 703, Amer. Math. Soc., Providence, RI, 2018, pp. 83–115, DOI 10.1090/conm/703/14132. MR3782461

[7] Lubjana Beshaj, *Reduction theory of binary forms*, Advances on superelliptic curves and their applications, NATO Sci. Peace Secur. Ser. D Inf. Commun. Secur., vol. 41, IOS, Amsterdam, 2015, pp. 84–116. MR3525574

[8] Lubjana Beshaj, *Integral binary forms with minimal height*, ProQuest LLC, Ann Arbor, MI, 2016. Thesis (Ph.D.)–Oakland University. MR3579531

[9] Lubjana Beshaj, *Minimal integral Weierstrass equations for genus 2 curves*, Higher genus curves in mathematical physics and arithmetic geometry, Contemp. Math., vol. 703, Amer. Math. Soc., Providence, RI, 2018, pp. 63–82, DOI 10.1090/conm/703/14131. MR3782460

[10] Lubjana Beshaj and Fred Thompson, *Equations for superelliptic curves over their minimal field of definition*, Albanian J. Math. **8** (2014), no. 1, 3–8. MR3253208

[11] A. Bialostocki and T. Shaska, *Galois groups of prime degree polynomials with nonreal roots*, Computational aspects of algebraic curves, 2005, pp. 243–255. MR2182043

[12] David W. Boyd, Greg Martin, and Mark Thom, *Squarefree values of trinomial discriminants*, LMS J. Comput. Math. **18** (2015), no. 1, 148–169, DOI 10.1112/S1461157014000436. MR3303602

[13] S. Allen Broughton, *Quasi-platonic $PSL_2(q)$-actions on closed Riemann surfaces*, Albanian J. Math. **9** (2015), no. 1, 31–61. MR3438844

[14] Harris B. Daniels, *Siegel functions, modular curves, and Serre's uniformity problem*, Albanian J. Math. **9** (2015), no. 1, 3–29. MR3338631

[15] Barry H. Dayton, *Algorithms for real numerical varieties with application to parameterizing quadratic surface intersection curves*, Albanian J. Math. **7** (2013), no. 2, 61–75. MR3109478

[16] Anand Deopurkar, Maksym Fedorchuk, and David Swinarski, *Gröbner techniques and ribbons*, Albanian J. Math. **8** (2014), no. 1, 55–70. MR3293901

[17] Karl Dilcher and Kenneth B. Stolarsky, *Resultants and discriminants of Chebyshev and related polynomials*, Trans. Amer. Math. Soc. **357** (2005), no. 3, 965–981, DOI 10.1090/S0002-9947-04-03687-6. MR2110427

[18] Carrie Finch and Lenny Jones, *On the irreducibility of* $\{-1, 0, 1\}$*-quadrinomials*, Integers **6** (2006), A16, 4. MR2247810

[19] V. Flammang, *The Mahler measure of trinomials of height 1*, J. Aust. Math. Soc. **96** (2014), no. 2, 231–243, DOI 10.1017/S1446788713000633. MR3194146

[20] G. Frey and T. Shaska, *Curves, Jacobians, and cryptography*, Algebraic curves and their applications, 2018.

[21] G. Frobenius, *Ueber das Trägheitsgesetz der quadratischen Formen* (German), J. Reine Angew. Math. **114** (1895), 187–230, DOI 10.1515/crll.1895.114.187. MR1580376

[22] Paul A. Fuhrmann, *A polynomial approach to linear algebra*, 2nd ed., Universitext, Springer, New York, 2012. MR2894784

[23] T. Alden Gassert, *A note on the monogeneity of power maps*, Albanian J. Math. **11** (2017), no. 1, 3–12. MR3659215

[24] Gary R. Greenfield and Daniel Drucker, *On the discriminant of a trinomial*, Linear Algebra Appl. **62** (1984), 105–112, DOI 10.1016/0024-3795(84)90089-2. MR761061

[25] Ruben Hidalgo and Tony Shaska, *On the field of moduli of superelliptic curves*, Higher genus curves in mathematical physics and arithmetic geometry, Contemp. Math., vol. 703, Amer. Math. Soc., Providence, RI, 2018, pp. 47–62, DOI 10.1090/conm/703/14130. MR3782459

[26] Jonas Jankauskas, *On the reducibility of certain quadrinomials*, Glas. Mat. Ser. III **45(65)** (2010), no. 1, 31–41, DOI 10.3336/gm.45.1.03. MR2646435

[27] David Joyner and Tony Shaska, *Self-inversive polynomials, curves, and codes*, Higher genus curves in mathematical physics and arithmetic geometry, Contemp. Math., vol. 703, Amer. Math. Soc., Providence, RI, 2018, pp. 189–208, DOI 10.1090/conm/703/14138. MR3782467

[28] David Kalaj and Ken-ichi Sakan, *Quasiconformal harmonic mappings onto a convex domain revisited*, Albanian J. Math. **7** (2013), no. 2, 51–60. MR3109477

[29] Kiran S. Kedlaya, *A construction of polynomials with squarefree discriminants*, Proc. Amer. Math. Soc. **140** (2012), no. 9, 3025–3033, DOI 10.1090/S0002-9939-2012-11231-6. MR2917075

[30] Wilhelm Ljunggren, *On the irreducibility of certain trinomials and quadrinomials*, Math. Scand. **8** (1960), 65–70, DOI 10.7146/math.scand.a-10593. MR0124313

[31] Andreas Malmendier and Tony Shaska, *A universal genus-two curve from Siegel modular forms*, SIGMA Symmetry Integrability Geom. Methods Appl. **13** (2017), Paper No. 089, 17, DOI 10.3842/SIGMA.2017.089. MR3731039

[32] J. Mandili and T. Shaska, *Computation of heights on weighted projective spaces*, Algebraic curves and their applications, 2018.

[33] W. H. Mills, *The factorization of certain quadrinomials*, Math. Scand. **57** (1985), no. 1, 44–50, DOI 10.7146/math.scand.a-12105. MR815429

[34] Shigefumi Mori, *The endomorphism rings of some abelian varieties. II*, Japan. J. Math. (N.S.) **3** (1977), no. 1, 105–109. MR529440

[35] Rezart Muço, Nejme Pjero, Ervin Ruci, and Eustrat Zhupa, *Classifying families of superelliptic curves*, Albanian J. Math. **8** (2014), no. 1, 23–35. MR3270074

[36] Shuichi Otake, *Counting the number of distinct real roots of certain polynomials by Bezoutian and the Galois groups over the rational number field*, Linear Multilinear Algebra **61** (2013), no. 4, 429–441, DOI 10.1080/03081087.2012.689983. MR3005628

[37] Shuichi Otake, *A Bezoutian approach to orthogonal decompositions of trace forms or integral trace forms of some classical polynomials*, Linear Algebra Appl. **471** (2015), 291–319, DOI 10.1016/j.laa.2015.01.005. MR3314338

[38] Shuichi Otake and Tony Shaska, *Bezoutians and the discriminant of a certain quadrinomials* (2018), available at `1806.09008`.

[39] Shuichi Otake and Tony Shaska, *Some remarks on the non-real roots of polynomials* (2018), available at `1802.02708`.

[40] Monika Polak and Eustrat Zhupa, *Graph based linear error correcting codes*, Albanian J. Math. **10** (2016), no. 1, 37–45. MR3589800

[41] Nguyen Tien Quang, Che Thi Kim Phung, and Ngo Sy Tung, *Abelian crossed modules and strict Picard categories*, Albanian J. Math. **7** (2013), no. 1, 37–48. MR3069882

[42] John Riordan, *Combinatorial identities*, John Wiley & Sons, Inc., New York-London-Sydney, 1968. MR0231725

[43] Ernst S. Selmer, *On the irreducibility of certain trinomials*, Math. Scand. **4** (1956), 287–302, DOI 10.7146/math.scand.a-10478. MR0085223

[44] T. Shaska and L. Beshaj, *Heights on algebraic curves*, Advances on superelliptic curves and their applications, NATO Sci. Peace Secur. Ser. D Inf. Commun. Secur., vol. 41, IOS, Amsterdam, 2015, pp. 137–175. MR3525576

[45] Richard G. Swan, *Factorization of polynomials over finite fields*, Pacific J. Math. **12** (1962), 1099–1106. MR0144891

[46] Elke Wolf, *Eigenvalues of composition combined with differentiation*, Albanian J. Math. **10** (2016), no. 1, 11–19. MR3589798

[47] Mohamed Zaatra, *A characterization of some Laguerre-Hahn orthogonal polynomials of class one*, Albanian J. Math. **10** (2016), no. 1, 21–36. MR3589799

[48] Yuri G. Zarhin, *Galois groups of Mori trinomials and hyperelliptic curves with big monodromy*, Eur. J. Math. **2** (2016), no. 1, 360–381, DOI 10.1007/s40879-015-0048-2. MR3454107

Department of Applied Mathematics, Waseda University, Japan
Email address: shuichi.otake.8655@gmail.com

Department of Mathematics and Statistics, Oakland University, Rochester, Michigan 48309
Email address: shaska@oakland.edu

Contemporary Mathematics
Volume **724**, 2019
https://doi.org/10.1090/conm/724/14586

Semistable types of hyperelliptic curves

Tim Dokchitser, Vladimir Dokchitser, Céline Maistret, and Adam Morgan

ABSTRACT. In this paper, we explore three combinatorial descriptions of semistable types of hyperelliptic curves over local fields: dual graphs, their quotient trees by the hyperelliptic involution, and configurations of the roots of the defining equation ('cluster pictures'). We construct explicit combinatorial one-to-one correspondences between the three, which furthermore respect automorphisms and allow to keep track of the monodromy pairing and the Tamagawa group of the Jacobian. We introduce a classification scheme and a naming convention for semistable types of hyperelliptic curves and types with a Frobenius action. This is the higher genus analogue of the distinction between good, split and non-split multiplicative reduction for elliptic curves. Our motivation is to understand L-factors, Galois representations, conductors, Tamagawa numbers and other local invariants of hyperelliptic curves and their Jacobians.

1. Introduction

Suppose K is a field with a discrete valuation, say of odd residue characteristic, and C/K is a hyperelliptic curve of genus g,

$$C : y^2 = f(x).$$

Our motivation is the study of the arithmetic of C and its Jacobian, including its minimal model, Tamagawa number, L-factor, conductor and other invariants related to the Birch–Swinnerton-Dyer conjecture. It would be desirable to have a classification of reduction types in the fashion of Kodaira types for elliptic curves, which moreover would take into consideration non-algebraically closed residue fields. In order to do so, for semistable curves, this paper develops a correspondence between three natural combinatorial objects attached to C that control its arithmetic. The correspondence is explicit and gives a simple way to pass between these objects in practice.

1.1. Correspondence. First, C has semistable reduction over some finite extension F/K: it has a model over the ring of integers of F with stable special fibre $\bar{C}$. Thus, $\bar{C}$ has only ordinary double points as singularities, and $\mathrm{Aut}(\bar{C})$ is

2010 *Mathematics Subject Classification.* Primary 11G20; Secondary 14H45, 05C22.
Key words and phrases. Hyperelliptic curves, hyperelliptic graphs, BY trees, cluster pictures, Tamagawa group, semistable reduction.
This research was supported by EPSRC grants EP/M016838/1 and EP/M016846/1. All relevant data is contained within the article; no supporting data is held elsewhere.
The second author was supported by a Royal Society University Research Fellowship.

finite (assuming $g \geq 2$ for the moment). Associated to $\bar{C}$ is its dual graph G, with a vertex for each geometric irreducible component, decorated with its genus and an edge for each intersection. It is often referred to as a 'semistable type' of C; e.g., in genus 2 there are seven types (omitting genus 0 markings):

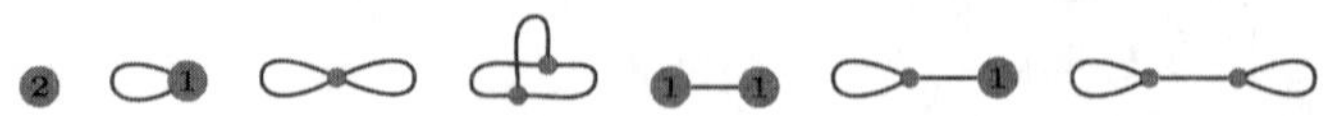

Second, G has an involution ι that comes from the hyperelliptic involution $y \mapsto -y$ on C, and the topological quotient $G/\langle\iota\rangle$ is a tree, say T. It has genus markings on the vertices as well, and a natural 2-colouring: colour points over which $G \to T$ is 2:1 yellow, and the branch locus blue. In genus 2, the corresponding trees are (yellow edges drawn squiggly and yellow vertices hollow)

Third, the set $X \subset \bar{K}$ of the $2g+1$ or $2g+2$ roots of the defining polynomial $f(x)$ gives another natural combinatorial invariant — how the roots 'cluster' together. Call a non-empty subset $\mathfrak{s} \subset X$ a *cluster* if it is of the form $X \cap$ (some disc in $\bar{K}$), and view X abstractly as a finite set with a collection Σ of clusters $\mathfrak{s} \subset X$, a *cluster picture*. Different presentations $y^2 = f(x)$ of the same curve may give different cluster pictures; however there is an equivalence relation induced by Möbius transformations of the roots. When $|X| = 6$ ($g = 2$), there are seven equivalence classes, represented by

The leftmost one illustrates the 6 roots being all equidistant, in the next one the last two roots are closer to each other than to the other four, and so on.

The three sets of 7 pictures raise the obvious question, to which the answer turns out to be 'Yes'. There is an established combinatorial notion of a 'hyperelliptic graph'. In this paper, we introduce 'BY (blue/yellow) trees' and 'cluster pictures', again in a combinatorial fashion, and formally define genus and equivalence. We then prove

THEOREM 1.1 (Main correspondence). *There is an explicit genus-preserving one-to-one correspondence between*

- *Hyperelliptic graphs up to isomorphism,*
- *BY trees up to isomorphism,*
- *Cluster pictures up to equivalence.*

In order to work with a fixed model of a hyperelliptic curve, it is also natural to ask for a graph-theoretic counterpart of a cluster picture that determines it up to isomorphism, rather than up to equivalence. The right notions turn out to be *open BY tree* / *open hyperelliptic graph*, with one extra edge or ι-orbit of edges with no endpoint ('going off to ∞'). An open hyperelliptic graph G has a unique largest hyperelliptic subgraph, its *core* $\tilde{G}$, and G, G' are called *equivalent* if $\tilde{G} \cong \tilde{G}'$; similarly for BY trees $T, \tilde{T}$.

The refined version of Theorem 1.1 is

THEOREM 1.2 (Open correspondence). *There is an explicit genus-preserving and equivalence-preserving one-to-one correspondence between*

- *Open hyperelliptic graphs up to isomorphism,*
- *Open BY trees up to isomorphism,*
- *Cluster pictures up to isomorphism.*

The construction is summarized in Table 4.20.

For example, here are open hyperelliptic graphs with core , open BY trees with core , and the corresponding cluster pictures that form one full equivalence class:

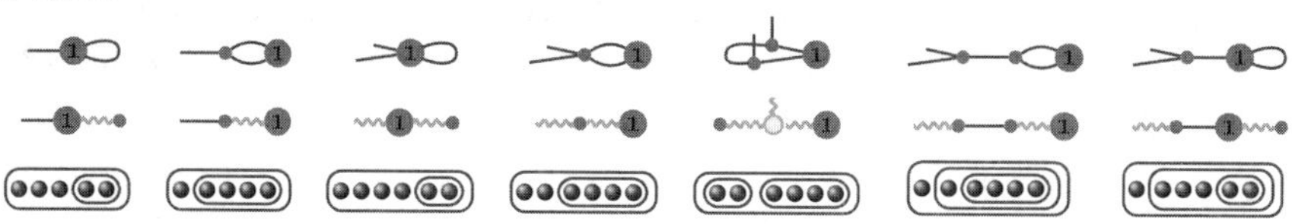

1.2. Invariants. The dual graph G of a semistable curve carries several important invariants, notably

- A *metric* (edges have length),
- *Automorphisms* (e.g. coming from the action of Galois),
- Homology *lattice* $\Lambda = H_1(G,\mathbb{Z})$,
- Symmetric positive-definite *pairing*[1] on Λ induced by the metric,
- *Tamagawa group*[2] $\Phi(G) = \Lambda^\vee/\Lambda$.

All of these have counterparts for BY trees and cluster pictures:

- The analogue of an (open or not) *metric* hyperelliptic graph G / BY tree T (edges have length) is a *metric cluster* picture Σ, with distances between clusters.
- For a BY tree T, the counterpart of $\Lambda_G = H_1(G,\mathbb{Z})$ is the relative homology group $\Lambda_T = H_1(T,T_b,\mathbb{Z})$ with respect to the blue part $T_b \subset T$. For a cluster picture Σ, Λ_Σ is, essentially, a permutation module on certain clusters ('even but not übereven').
- In all three settings, the metric determines a pairing on Λ. When all edge lengths of G are integers, we call G *integral* and this notion also transports to T and Σ as well. In this case, the pairing is $\mathbb{Z}$-valued, and we can define the *Tamagawa group* $\Phi = \Lambda^\vee/\Lambda$.
- The group $\operatorname{Aut} G$ corresponds to the group $\operatorname{Aut} T$ of automorphisms of T together with a choice of sign for each yellow component, and to the group $\operatorname{Aut}\Sigma$ of permutations of clusters in Σ with a choice of (compatible) signs for clusters of even size.

THEOREM 1.3. *The correspondence in Theorem* 1.2 *extends to the metric case. Suppose G, T and Σ correspond to one another. Choose a section $s: G/\langle\iota\rangle \to G$. Then there are canonical isomorphisms*

$$\operatorname{Aut}(\tilde{G}) \cong \operatorname{Aut}(\tilde{T}) \cong \operatorname{Aut}(\Sigma)$$

[1] the monodromy pairing of Grothendieck [**10**]

[2] also known in graph theory as the Jacobian (see §7.2), Picard group or the sandpile group of G; for curves over local fields this is the group of connected components of the Néron model of the Jacobian, and the size of its Frobenius invariants is called the Tamagawa number

and canonical Aut$(\cdot)$*-equivariant isomorphisms*

$$\Lambda_{\tilde{G}} \cong \Lambda_{\tilde{T}} \cong \Lambda_{\Sigma}$$

as lattices with a pairing. If G is integral, they induce isomorphisms

$$\Phi(\tilde{G}) \cong \Phi(\tilde{T}) \cong \Phi(\Sigma).$$

The Tamagawa group $\Phi(\tilde{G})$ is also equivariantly isomorphic to the graph-theoretic Jacobian of G, see Proposition 7.10. As one application, we get a simple description of the 2-torsion in $\Phi(\tilde{G})$.

COROLLARY 1.4. *Let G be an integral hyperelliptic graph of genus ≥ 2, $G^{\langle\iota\rangle}$ the set of fixed points of the involution ι, and $\mathcal{W}$ the set of those connected components of $G^{\langle\iota\rangle}$ that contain a point of integer distance from a vertex. Then we have isomorphisms of* Aut G*-modules*

$$\Phi(G)[2] \cong \begin{cases} 0 & \text{if } \mathcal{W} = \emptyset \text{ and } \operatorname{rk} H_1(G,\mathbb{Z}) \text{ even,} \\ \mathbb{F}_2 & \text{if } \mathcal{W} = \emptyset \text{ and } \operatorname{rk} H_1(G,\mathbb{Z}) \text{ odd,} \\ \ker(\mathbb{F}_2[\mathcal{W}] \xrightarrow{\text{sum}} \mathbb{F}_2) & \text{otherwise.} \end{cases}$$

(Here 'sum' *denotes the sum of the coefficients map.)*

There is an algorithm due to Betts [**1**] that computes, for a BY tree T, the Tamagawa group $\Phi(T)$ and the group of invariants $\Phi(T)^{F=1}$ for $F \in \operatorname{Aut} T$. By the theorem above, it gives a way to compute $\Phi(G)$ and $\Phi(\Sigma)$ for hyperelliptic graphs and cluster pictures as well.

1.3. Hyperelliptic curves in odd residue characteristic. The present paper is purely combinatorial, and was motivated by its geometric counterpart [**9**] that studies the arithmetic and geometry of hyperelliptic curves over local fields. We briefly sketch the results for the interested reader.

Let K be a local field of odd residue characteristic p, with valuation val_K on the separable closure $\bar{K}$. As before, let C/K be a hyperelliptic curve

$$C : y^2 = f(x)$$

of genus ≥ 2. Suppose it is semistable over some finite Galois extension F/K. The dual graph G_C of the special fibre of the minimal regular model of C/F is naturally a metric hyperelliptic graph, and it has a Galois action[3]

$$\mathrm{Gal}(\bar{K}/K) \to \operatorname{Aut} G_C.$$

The cluster picture $\Sigma_C = (X, \Sigma)$ is given by the collections of roots that are cut out by p-adic discs,

$$X = \{\text{roots of } f \text{ in } \bar{K}\}, \qquad \Sigma = \{\mathfrak{s} = X \cap D \mid D \subset \bar{K} \text{ disc}, \mathfrak{s} \neq \emptyset\}.$$

It carries a metric, with the distance between clusters determined by the p-adic distances between the roots:

$$\delta(\mathfrak{s}, \mathfrak{s}') = \text{diameter}(\mathfrak{s}) + \text{diameter}(\mathfrak{s}') - 2 \cdot \text{diameter}(\mathfrak{s} \cup \mathfrak{s}'),$$

[3]This is slightly non-trivial, and relies on uniqueness of the minimal regular model of C over F

where for $X' \subseteq X$, $\text{diameter}(X') = \min_{r,r' \in X'}(\text{val}_K(r - r'))$. An automorphism $\sigma \in \text{Gal}(\bar{K}/K)$ acts on X and permutes the clusters. Moreover, one can assign a natural sign $\epsilon_\sigma(\mathfrak{s}) = \pm 1$ for every even cluster $\mathfrak{s}$. This gives a homomorphism

$$\text{Gal}(\bar{K}/K) \to \text{Aut}\,\Sigma_C.$$

Having produced both the hyperelliptic graph G_C and the cluster picture Σ_C, we can now state

THEOREM 1.5 ([**9**]). *The core of the hyperelliptic graph associated to the cluster picture Σ_C by Theorem 1.2 is isomorphic to the dual graph G_C. The isomorphism preserves the metric and the* $\text{Gal}(\bar{K}/K)$*-action.*

Thus, from a cluster picture of C/K, which is an elementary invariant constructed from the roots of $f(x)$, we recover the semistable model of C/F together with the Galois action. This allows us to determine some of the main arithmetic invariants of C, such as

- Necessary and sufficient conditions for C/K to be semistable;
- The Galois representation $H^1_{\text{ét}}(C/\bar{K}, \mathbb{Q}_l)$;
- The conductor of C;
- Equations for the minimal regular model of C/F;
- The Tamagawa number of the Jacobian of C over F.[4]

1.4. Classification of semistable types and naming convention. Recall that one of the main motivations for studying hyperelliptic graphs, BY trees and cluster pictures was to produce a classification of semistable types of hyperelliptic curves. The results in this paper allow one to produce such a classification in any genus; the tables in §9 classify objects up to genus 3 and objects together with an automorphism in genus 1 and 2. Keeping track of an automorphism is important, as the action of Frobenius on the dual graph of the special fiber of the minimal regular model of hyperellptic curves is key to the study of their arithmetic. One standard application of such classifications is that they enable one to use a systematic case by case analysis. For example, [**15**] employs the one in the present paper to prove a general result on the parity of ranks of Jacobians of genus 2 curves.

We end by proposing a naming convention for BY trees or, equivalently, hyperelliptic graphs and cluster pictures. In the context of semistable curves of genus 1 or 2, our notation is compatible with that of Kodaira [**14**] and Namikawa–Ueno [**16**]. As an illustration, the seven types in genus 2 shown previously get the names

$$2 \qquad 1_n \qquad \text{I}_{n,m} \qquad \text{U}_{n,m,k} \qquad 1{\times}1 \qquad 1{\times}\text{I}_n \qquad \text{I}_n{\times}\text{I}_m$$

(with the same ordering). Moreover, we also allow an arbitrary automorphism to be encoded in the type. For example, for type 1_n above, there are two choices of automorphisms giving

$$1_n^+ \qquad 1_n^-.$$

These correspond to a genus 2 curve whose reduction has one split or non-split node, respectively.

Related work. Hyperelliptic graphs and their link to hyperelliptic curves are well-known [**3, 7, 8, 13**]. Our definition (3.2) is as in [**7, 13**]; it is stronger than that

[4] Note also that Corollary 1.4 fully describes the 2-torsion of the Tamagawa group.

of [**8**] which does not require condition (3).[5] Cluster pictures (and, to some extent, BY trees) appear implicitly in [**6**] in setting of rigid geometry. As far as we are aware, neither the correspondence nor an analysis of automorphism groups (which is important when the residue field is not algebraically closed) have been studied.

1.5. Layout. §2 recalls terminology about graphs and properties of their homology, relative homology and the natural pairing coming from edge lengths. §3 introduces hyperelliptic graphs, BY trees and cluster pictures. §4 and §5 state and prove the one-to-one correspondence between hyperelliptic graphs, BY trees and cluster pictures. §6 shows that it preserves the lattice Λ as a module under automorphism groups. §7 discusses Tamagawa groups of hyperelliptic graphs. The naming convention is addressed in §8. The classification of semistable types is discussed in §9.

1.6. Notation. Throughout the paper we use the following notation:

G	hyperelliptic graph / open hyperelliptic graph (§3.1)
T	BY tree / open BY tree (§3.2)
Σ	cluster picture (§3.3)
$\mathfrak{s}$	cluster (element of Σ)
g	genus function on vertices of G or T / clusters in Σ, and the total genus of G, T or Σ (Definitions 3.8, 3.23, 3.38)
Λ	homology lattice of G, T or Σ (Definitions 3.16, 3.31, 3.48)
$\Lambda^\vee$	$= \mathrm{Hom}(\Lambda, \mathbb{Z})$, the dual lattice
δ	distance function on G, T or Σ (Definitions 3.15, 3.30, 3.45)
$\tilde{G}, \tilde{T}$	core of an open hyperelliptic graph/BY tree (Definitions 3.12, 3.25)
T_b, T_y	blue/yellow part of T
ι	hyperelliptic involution on a hyperelliptic graph G
s	continuous section $G/\langle\iota\rangle \to G$ to the quotient map (§4.3)
$\mathbb{Z}[X]$	free abelian group on X

2. Background on metric graphs

2.1. Graphs. In what follows, the word *graph* refers to a topological space G homeomorphic to a finite (combinatorial) graph. It comes with a set $V(G)$ of *vertices* (containing all points $x \in G$ of degree $\neq 2$) and *edges* $E(G)$. *Graph isomorphisms* are homotopy classes of homeomorphisms that preserve vertices and edges. Loops and multiple edges are allowed, though note that in the topological setting the action of $\mathrm{Aut}\, G$ might permute multiple edges and reverse direction of loops.

By a *metric graph* we mean a topological graph G along with a function $l : E(G) \to \mathbb{R}_{>0}$ which assigns a length to each edge. This may be extended into a metric on G. We write $\delta(v, v')$ for the shortest distance between $v, v' \in V(G)$. We require isomorphisms and automorphisms of metric graphs to preserve lengths.

---

[5]Condition (3) of 3.2 is needed for geometric reasons, for otherwise there are no hyperelliptic curves with such special fibres; see [**13**] Thm 1.2 and compare 36 types in genus 3 of [**8**] with 32 in Table 9.4. From the point of view of our 1-1 correspondence, it is forced automatically by cluster pictures.

2.2. Homology of graphs. For a topological space X, $H_i(X)$ denotes the i-th singular homology group of the space X with coefficients in $\mathbb{Z}$. For a subspace $A \subseteq X$, $H_i(X, A)$ denotes the i-th relative (singular) homology group, again with coefficients in $\mathbb{Z}$. Most calculations will be carried out via simplicial homology. This will give the same answer where used: see [**11**, Theorem 2.27]. We refer to Section 2 of op. cit. for more details and proofs of everything outlined below.

Let G be a graph. We make G into a Δ-complex by taking the 0-simplices (resp. 1-simplices) to be the set of vertices (resp. edges, along with their endpoint(s)) of G and write $C_0(G)$ (resp. $C_1(G)$) for the free $\mathbb{Z}$-module on the 0-simplices (resp. 1-simplices) of G. For each choice of orientation on the edges of G, which for a non-loop edge e amounts to a choice of 'nose' e_+ and 'tail' e_-, we have an associated boundary map $d : C_1(G) \to C_0(G)$ sending a non-loop edge e to $e_+ - e_-$, and sending loops to 0. We then have

$$H_1(G) = \operatorname{Ker}(d) \subseteq C_1(G).$$

Given two choices of orientation, there is a canonical isomorphism between the associated homology groups and so $H_1(G)$ is independent of the choice of orientation. For the rest of this section, fix an orientation on G.

2.2.1. *Action of automorphisms.* Let $\operatorname{Aut} G$ be the group of automorphisms of G. Then $\operatorname{Aut} G$ acts on $C_0(G)$ via its action on the set of vertices, and on $C_1(G)$ via its action on the set of edges save that now we add signs to this action to take account of the orientation. Explicitly, if $\sigma \in \operatorname{Aut} G$ maps the (unsigned) non-loop edge e to e' then we define the action on $e \in C_1(G)$ as

$$\sigma(e) = \begin{cases} e' & \sigma(e_+) = e'_+ \\ -e' & \sigma(e_+) = e'_-. \end{cases}$$

The action on loops is defined similarly, introducing a minus sign if σ maps a loop with its positive orientation to a loop with its negative orientation. This action commutes with the boundary map and induces an action of $\operatorname{Aut} G$ on $H_1(G)$. Whilst the action on $C_1(G)$ depends on the choice of orientation, the induced action on $H_1(G)$ does not (upon canonically identifying the homology groups arising from different choices of orientation).

2.2.2. *Length pairing on homology of metric graphs.* Suppose now that G is a metric graph with associated length function l. Define the *length pairing* on $C_1(G)$ by setting

$$\langle e, e' \rangle = \begin{cases} l(e) & e = e,' \\ 0 & e \neq e', \end{cases}$$

and extending bilinearly. This is independent of the orientation on the edges of G. The induced pairing on $H_1(G)$ is positive definite (since the pairing on $C_1(G)$ is) and invariant under the action of $\operatorname{Aut} G$ (since the same is true of the pairing on $C_1(G)$, for any choice of orientation).

In particular, $H_1(G)$ is a finitely generated, torsion free $\mathbb{Z}$-module which carries a canonical action of $\operatorname{Aut} G$ and, in the metric case, a positive definite real valued, invariant pairing.

2.2.3. *Relative homology.* Let H be a subgraph of G (that is, a closed subspace of G which is a union of edges and vertices of G). In the metric case, we put the induced metric on H so that it is also a metric graph. Then H has a natural structure of a Δ-complex inherited from that on G. Having fixed an orientation on the edges of G, we have an induced orientation on the edges of H. The boundary operator $d : C_1(G) \to C_0(G)$ then maps $C_1(H)$ into $C_0(H)$ and we have

$$H_1(G, H) = \ker \left(\frac{C_1(G)}{C_1(H)} \longrightarrow \frac{C_0(G)}{C_0(H)} \right).$$

Writing $\mathrm{Aut}_H\, G$ for the subgroup of automorphisms of G preserving H, the action of $\mathrm{Aut}\, G$ on $C_1(G)$ defined above induces an action of $\mathrm{Aut}_H\, G$ on $H_1(G, H)$. Again, this is independent of the choice of orientation on G.

In the metric case, define the *relative length pairing* on $C_1(G)$ by setting

$$\langle e, e' \rangle = \begin{cases} l(e) & e = e', e \notin H, \\ 0 & \text{otherwise}, \end{cases}$$

and extending bilinearly. This induces a positive definite pairing on $H_1(G, H)$ which is invariant under the action of $\mathrm{Aut}_H\, G$. In particular, as with $H_1(G)$, $H_1(G, H)$ is a finitely generated, torsion free (as $C_1(H)$ is a direct summand of $C_1(G)$) $\mathbb{Z}$-module equipped with an action of $\mathrm{Aut}_H\, G$ and, in the metric case, a positive definite real valued, invariant pairing.

For an example illustrating all the definitions above, see Example 3.32.

2.2.4. *Subdivision of edges.* It will often be convenient to define a new Δ-complex structure on G by 'subdividing' certain edges. That is, we take points $x_1, ..., x_m$ lying on edges of G and define a Δ-complex structure on G by taking the set of 0-simplices to be the set $V(G) \cup \{x_1, ..., x_m\}$, and redefining the set of 1-simplices accordingly. Let us temporarily denote G, along with the new Δ-complex structure as G^*, and fix an orientation on the 1-simplices of G^* (not necessarily induced from the orientation on the 1-simplices of G). Then there is a natural map $C_1(G) \to C_1(G^*)$ sending an edge $e \in C_1(G)$ to the (signed) sum over its subdivisions. This map induces an isomorphism $H_1(G) \xrightarrow{\sim} H_1(G^*)$. If the set $\{x_1, ..., x_m\}$ is preserved by an automorphism σ of G, then σ induces an action on $H_1(G^*)$ by the same formula as in the case of G and the isomorphism $H_1(G) \cong H_1(G^*)$ defined above preserves this action. Moreover, in the metric case, each 1-simplex of G^* has an associated length and we thus obtain a length pairing on $H_1(G^*)$ by the same formula as for G. Since the length of an edge of G is the sum of the lengths of its subdivisions, the isomorphism above identifies the pairings also. Consequently, we will frequently subdivide edges without further comment, with the caveat that when computing actions of automorphisms, we use subdivisions which are preserved by the automorphisms of interest.

The above discussion applies equally well to the relative homology group of G with respect to a subgraph H (along with its automorphism action and, in the metric case, pairing), provided that when defining the simplicial complex structure on G by subdividing edges, we give H the simplicial complex structure inherited from the new one on G as before.

3. Hyperelliptic graphs, BY trees and cluster pictures

3.1. Hyperelliptic graphs.

DEFINITION 3.2 (Hyperelliptic graph). Let G be a connected graph equipped with

- $g : V(G) \to \mathbb{Z}_{\geq 0}$, a function that assigns a *genus* to every vertex,
- $\iota : G \to G$ an involution (graph isomorphism of order ≤ 2).

TABLE 3.1. All type names, hyperelliptic graphs, BY trees and balanced cluster pictures in genus 0, 1 and 2.

Type name	G	T	Σ
		Genus 0	
0			
		Genus 1	
1			
$\mathrm{U}_{n,m}$			
		Genus 2	
2			
1_n			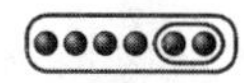
$\mathrm{I}_{n,m}$			
$\mathrm{U}_{n,m,k}$		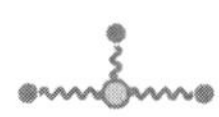	
$\mathrm{I}_n \times \mathrm{I}_m$			
$1 \times \mathrm{I}_n$			
1×1	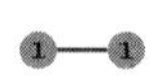		

We say G (or, more precisely, (G, g, ι)) is a *hyperelliptic graph* if

(1) vertices of non-zero genus are ι-invariant;
(2) genus 0 vertices have degree ≥ 3;
(3) $2g(v) + 2 \geq \#$ ι-invariant edges at v, for every vertex v;
(4) the topological quotient $G/\langle\iota\rangle$ is contractible (that is, a tree).

In addition, it is convenient to declare two exceptions

(3.3) 

(with ι=reflection in the x-axis and vertices of genus 0) to be hyperelliptic graphs as well, although they violate (2).

DEFINITION 3.4 (Open version). An *open* hyperelliptic graph G is a connected graph G with at least one open[6] edge, a genus function $g : V(G) \to \mathbb{Z}_{\geq 0}$ and an involution ι, satisfying conditions (1)-(3) of Definition 3.2 for all vertices, and such that the quotient $G/\langle\iota\rangle$ is a tree with a unique open edge (in particular, G has either a unique open edge, or a pair of open edges swapped by ι). In addition, we declare the graph

(3.5) 

(with vertex of genus 0 and ι=reflection in the x-axis) to be an open hyperelliptic graph, though it violates (2).

REMARK 3.6. To stress the distinction from the open case, we will sometimes also refer to hyperelliptic graphs (as in Definition 3.2) as *closed hyperelliptic graphs*.

[6]that is, open on one end; such graphs are called 'graphs with legs' in [7]

REMARK 3.7. Any open hyperelliptic graph is homeomorphic to a connected component of $\bar{G} \setminus I$ for some hyperelliptic graph $\bar{G} = (\bar{G}, g, \iota)$ for $I \subset V(\bar{G})$ an ι-orbit of vertices, together with the induced genus function and action of ι. In other words, G is a hyperelliptic graph with an extra 'open edge/pair of open edges'. The missing vertex/pair of vertices are referred to as ∞ or ∞^+, ∞^-.

We draw hyperelliptic graphs as follows, with numbers indicating the genus $g(v)$ when it is positive (and omitted when $g(v) = 0$).

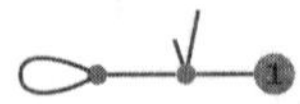

(closed) hyperelliptic graph open hyperelliptic graph

DEFINITION 3.8 (Genus). For both closed and open hyperelliptic graphs, the *genus of G* is given by

$$g(G) = \operatorname{rk} H_1(G) + \sum_{v \in V(G)} g(v).$$

DEFINITION 3.9 (Isomorphism). Two hyperelliptic graphs (closed or open) are *isomorphic* if there is a homeomorphism between them that preserves the defining data (vertices, edges, genus markings, commutes with ι). We write $\operatorname{Aut} G$ for the group of *automorphisms* of G. (Recall that they are considered up to homotopy; see §2.1.)

REMARK 3.10. In every hyperelliptic graph G that is not the exceptional circle graph from (3.3), ι is the unique involution in $\operatorname{Aut} G$ that fixes all the vertices of positive genus, and such that the quotient $G/\langle\iota\rangle$ is a tree (see e.g. [**12**] Prop 1.4). In particular, it is central, and graph isomorphisms commute with ι automatically.

EXAMPLE 3.11. Table 3.1 (2nd column) lists all hyperelliptic graphs of genus 0,1 and 2, and Table 4.1 open ones of genus 0 and 1, up to isomorphism.

DEFINITION 3.12 (Core). A *(closed) hyperelliptic subgraph* H of a (open or closed) hyperelliptic graph G is a (closed) hyperelliptic graph H such that

- As a topological space, H is a union of vertices and edges of G, closed in G, and closed under ι.
- The vertices of H are exactly those vertices of G that are in H, except for those that have genus 0 and degree 2 in H. The latter become points on the edges of H. (When H is the second exceptional graph (3.3), we declare its two ι-invariant points to be genus 0 vertices.)
- The genus of a vertex of H is the same as its genus in G.

The *core* $\tilde{G}$ of an open hyperelliptic graph G is its maximal closed hyperelliptic subgraph. By Proposition 5.7 below, it is unique, has the same genus as G, and can be easily obtained from G by removing a few vertices and edges near ∞.

DEFINITION 3.13 (Equivalence). We say that two open hyperelliptic graphs are *equivalent* if they have isomorphic cores.

EXAMPLE 3.14. Take a single vertex of genus 1 with a loop, and let ι reverse the direction of the loop:

$\tilde{G}$ = 

It is a (closed) hyperelliptic graph $\tilde{G}$ of genus 1+1=2. There are, up to isomorphism, 7 open hyperelliptic graphs G with $\tilde{G}$ as their core:

Graphs #1,#2,#5 have $\operatorname{Aut} G = \langle 1, \iota \rangle \cong C_2$, and the other four $\operatorname{Aut} G \cong C_2^2$.

DEFINITION 3.15 (Metric version). A *metric hyperelliptic graph* G (closed or open) is a hyperelliptic graph equipped with an ι-invariant length function $\delta : E(G) \to \mathbb{R}_{>0}$ on the edges (excluding the open edge). In this case, we write $\delta(v, v')$ for the shortest distance between $v, v' \in V(G)$. We require isomorphisms and automorphisms of metric graphs to preserve δ, and for a hyperelliptic subgraph $H \subset G$ we require[7] $\delta_H(v, v') = \delta_G(v, v')$ for the vertices of H. Similarly, we say that two open metric hyperelliptic graphs are equivalent if there is an isomorphism between their cores which preserves distance.

DEFINITION 3.16 (The homology lattice Λ). Let G be a closed or open hyperelliptic graph. We set $\Lambda_G = H_1(G)$. Recall from §2.2 that a (closed) hyperelliptic graph comes with a natural action of $\operatorname{Aut} G$ and, if G is a metric hyperelliptic graph, a non-degenerate, real valued, $\operatorname{Aut} G$-invariant pairing. The same is also true of the open case. Indeed, in Lemma 6.4 we will show that if G is an open hyperelliptic graph with core $\tilde{G}$, then $H_1(G)$ is canonically isomorphic to $H_1(\tilde{G})$. Since automorphisms of G induce automorphisms of the core via restriction, we may use the closed case to equip Λ_G with an action of $\operatorname{Aut} G$ and, in the metric case, with a non-degenerate, real valued, $\operatorname{Aut} G$-invariant pairing.

EXAMPLE 3.17. Consider the following open hyperelliptic graph G:

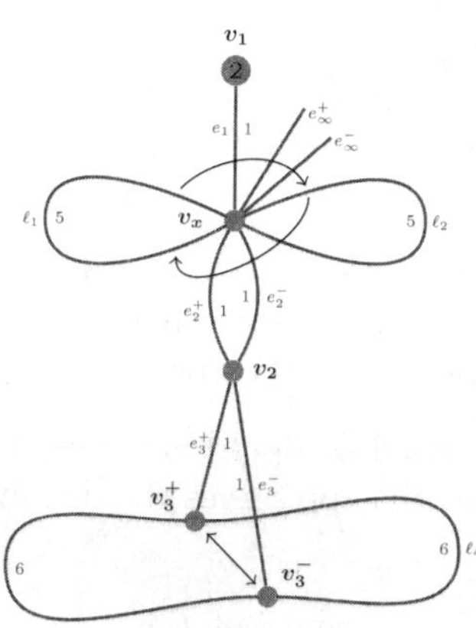

The graph has vertices $v_1, v_x, v_2, v_3^+, v_3^-$ of genera $2, 0, 0, 0, 0$; loops ℓ_1, ℓ_2 of lengths $5, 5$; edges $\ell_3, \ell_4, e_1, e_2^+, e_2^-, e_3^+, e_3^-$ of lengths $6, 6, 1, 1, 1, 1, 1$ and two open edges e_∞^+, e_∞^- going off to infinity from v_x. The involution ι fixes v_1, v_2, v_x, e_1, swaps v_3^+ with v_3^-, e_2^+ with e_2^-, e_3^+ with e_3^- and e_∞^+ with e_∞^-, and reverses the directions of $\ell_1, \ell_2, \ell_3, \ell_4$. G admits an automorphism $\sigma \in \operatorname{Aut}_G$ of order 4, that fixes all vertices except v_3^+ and v_3^-, fixes the edges e_1, e_2^+, e_2^- pointwise, swaps e_3^+ and e_3^-, reverses the directions of ℓ_3 and ℓ_4, and has an order four action on the loops, sending $\ell_1 \to -\ell_2 \to -\ell_1 \to \ell_2$ (where $-\ell$ denotes ℓ with the opposite orientation).

The lattice Λ_G is

$$\Lambda_G = H_1(G) = \langle \ell_1, \ell_2, \ell_3 + e_3^+ - e_3^-, \ell_4 + e_3^+ - e_3^-, e_2^+ - e_2^- \rangle \simeq \mathbb{Z}^5,$$

where we have picked an orientation for each edge and loop (edges $e_2^\pm, e_3^\pm$ going bottom-to-top, ℓ_3, ℓ_4 right-to-left, and the loops ℓ_1, ℓ_2 oriented clockwise) and where $+$ means concatenation and $-x$ is x with the opposite orientation. The loops in the above basis have lengths $5, 5, 8, 8, 2$, and have trivial intersections, except for the third and fourth basis elements whose intersection has length 2. Thus the

[7] When H is the exceptional circle graph from (3.3), the vertices might change, and so we require instead that the total length of the circle is the same in H and in G

length-pairing (see Section 2.2.2) and the action of σ on Λ_G are

$$\langle\cdot,\cdot\rangle = \begin{pmatrix} 5 & 0 & 0 & 0 & 0 \\ 0 & 5 & 0 & 0 & 0 \\ 0 & 0 & 8 & 2 & 0 \\ 0 & 0 & 2 & 8 & 0 \\ 0 & 0 & 0 & 0 & 2 \end{pmatrix}, \qquad \sigma = \begin{pmatrix} 0 & -1 & 0 & 0 & 0 \\ 1 & 0 & 0 & 0 & 0 \\ 0 & 0 & -1 & 0 & 0 \\ 0 & 0 & 0 & -1 & 0 \\ 0 & 0 & 0 & 0 & 1 \end{pmatrix}.$$

3.2. BY trees.

DEFINITION 3.18 (BY tree). A BY tree is a finite tree T with a genus function $g : V(T) \to \mathbb{Z}_{\geq 0}$ on vertices and a 2-colouring blue/yellow on vertices and edges such that

(1) yellow vertices have genus 0, degree ≥ 3, and only yellow edges;
(2) blue vertices of genus 0 have at least one yellow edge;
(3) at every vertex, $2g(v) + 2 \geq \#$ blue edges at v.

Note that all leaves are blue.

In diagrams, yellow edges are drawn squiggly and yellow vertices hollow for the benefit of viewing them in black and white.

EXAMPLE 3.19. Table 3.1 (third column) lists all BY trees of genus 0,1 and 2 up to isomorphism.

NOTATION 3.20. As a topological space, $T = T_b \coprod T_y$ with T_b the *blue part*, and T_y the *yellow part.* (Thus $T_b \subset T$ is a closed subset.)

DEFINITION 3.21 (Open version). An *open BY tree* T is a finite tree T with a unique open edge, a genus function $g : V(T) \to \mathbb{Z}_{\geq 0}$ on vertices and a 2-colouring blue/yellow on vertices and edges, satisfying conditions (1), (2) and (3) of Definition 3.18.

In other words, as for hyperelliptic graphs, an open BY tree T is a BY tree with one 'missing' vertex, that we will refer to as ∞. Again, we sometimes refer to BY trees of Definition 3.18 as *closed,* to distinguish them from the open ones.

DEFINITION 3.22 (Isomorphic BY trees). Two (closed or open) BY trees are *isomorphic* if there is a homeomorphism between them that preserves the defining data (vertices, edges, genus markings, colouring).

DEFINITION 3.23. For a closed or open BY tree T, the *genus* of T is

$$g(T) = \operatorname{rk} H_1(T, T_b) + \sum_{v \in V(T)} g(v),$$

where the first term is the (singular) relative homology group (see §2.2.3).

REMARK 3.24. The relative homology sequence

$$0 = H_1(T) \longrightarrow H_1(T, T_b) \longrightarrow H_0(T_b) \longrightarrow H_0(T) = \mathbb{Z}$$

gives an isomorphism $H_1(T, T_b) \cong \tilde{H}_0(T_b)$ where the latter group is the reduced homology of T_b in degree zero. In particular, $\operatorname{rk} H_1(T, T_b)$ is equal to one less than the number of connected components of T_b.

Definition 3.25 (Core). A *(closed) BY subtree* T' of a (closed or open) BY tree T is a (closed) BY tree T' such that

- As a topological space, T' is a union of vertices and edges of T, and is closed in T.
- The vertices of T' are exactly those vertices of T that are in T', except for those of genus 0 that in T' have degree 2 and incident edges of the same colour as the vertex. These exceptional vertices become points on the edges of T'.
- The genus of a vertex of T' is the same as its genus on T.

The *core* $\tilde{T}$ of an open BY tree T is its maximal closed BY subtree. Again, we will see later (Proposition 5.7) that it is unique, has the same genus, and is obtained from T by removing a few vertices and edges near ∞.

Definition 3.26 (Equivalence). We say that two open BY trees are *equivalent* if they have isomorphic cores.

Definition 3.27. An *isomorphism of (closed or open) BY trees* $T \to T'$ is a pair (α, ϵ) where

- α is a graph isomorphism $T \to T'$ (in the open case, $T \cup \{\infty\} \to T' \cup \{\infty\}$ with $\alpha(\infty) = \infty$) that preserves the genera of the vertices and the colours, and
- $\epsilon(Z) = \pm 1$ is a collection of signs for every connected component Z of the yellow part $T_y \subset T$.

Equivalently, ϵ is a collection of signs $\epsilon(v) \in \{\pm 1\}$ and $\epsilon(e) \in \{\pm 1\}$ for every yellow vertex and yellow edge, such that $\epsilon(v) = \epsilon(e)$ whenever e ends at v. The isomorphisms are composed by a cocycle rule

$$(\alpha, \epsilon_\alpha) \circ (\beta, \epsilon_\beta) = \big(\alpha \circ \beta, \bullet \mapsto \epsilon_\beta(\bullet)\epsilon_\alpha(\beta(\bullet))\big).$$

An *automorphism* of T is an isomorphism from T to itself. We write $\operatorname{Aut} T$ for the group of automorphisms. (As all ϵ may be chosen to be $+1$, this extended notion of an isomorphism does not affect the earlier definition of BY trees being isomorphic.)

Example 3.28. Take a tree $\tilde{T}$ on 2 blue vertices, one of genus 0, and one of genus 1, with one yellow edge between them:

$\tilde{T}$ = 

It is a (closed) BY tree of genus 1+1=2. There are, up to isomorphism, 7 open BY trees T with $\tilde{T}$ as their core:

Trees #1,#2,#5 have $\operatorname{Aut} T = \operatorname{Aut} \tilde{T} \cong C_2$, and the other four $\operatorname{Aut} T \cong C_2^2$.

Remark 3.29. If $T_y^{(1)}, \ldots, T_y^{(r)}$ are the connected components of $T_y \subset T$, then, by definition,

$$\operatorname{Aut} T = \operatorname{Aut}_0(T) \ltimes (\mathbb{Z}/2\mathbb{Z})^r,$$

where $\operatorname{Aut}_0(T)$ consists of those elements for which $\epsilon(T_y^{(j)}) = +1$ for all j. Equivalently, $\operatorname{Aut}_0(T)$ is the group of (homotopy classes of) homeomorphisms $T \to T$ that preserve $V(T), E(T), T_b, T_y$ and g.

DEFINITION 3.30 (Metric version). A *metric (open or not) BY tree* is one with a length function $\delta : E(T) \to \mathbb{R}_{>0}$ on the edges (excluding the open one). We denote by $\delta(v, v')$ the distance between $v, v' \in V(T)$, and we require isomorphisms/automorphisms of metric trees to preserve δ. Similarly, we say that two open metric BY trees are equivalent if there is an isomorphism between their cores which preserves distance.

DEFINITION 3.31 (The lattice Λ). Let T be a (open or not) BY tree. We set $\Lambda_T = H_1(T, T_b)$. In the closed case, as detailed in Section 2.2, it comes with a natural action of $\mathrm{Aut}_0\, T$ and, if T is a metric BY tree, a non-degenerate, real valued, $\mathrm{Aut}_0 T$-invariant pairing. We extend the action of $\mathrm{Aut}_0\, T$ to an action of the full automorphism group $\mathrm{Aut}\, T$ as follows. Let $\sigma = (\sigma_0, \epsilon_\sigma)$ be an element of $\mathrm{Aut}\, T$ and let e be a yellow edge, viewed as an element of $C_1(T)$. Then we set $\sigma(e) = \epsilon(Z)\sigma_0(e)$ where Z is the connected component of T_y containing e and the action of σ_0 on $C_1(T)$ is as in Section 2.2. This induces the sought action of $\mathrm{Aut}\, T$ on $H_1(T, T_b)$.

We will show in Lemma 6.4 that if T is an open BY tree with core $\tilde{T}$, then $H_1(T, T_b)$ is canonically isomorphic to $H_1(\tilde{T}, \tilde{T}_b)$. Since automorphisms of T induce automorphisms of the core via restriction, we may use the closed case to equip Λ_T with an action of $\mathrm{Aut}\, T$ and, in the metric version, a non-degenerate, real valued, $\mathrm{Aut}\, T$-invariant pairing also.

EXAMPLE 3.32. Consider the following open BY tree T:

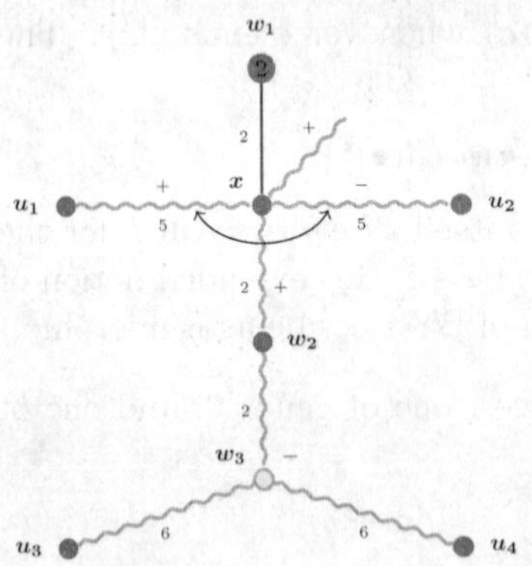

The graph T has vertices $u_1, \ldots u_4, w_1, w_2, w_3, x$ of genera $0,0,0,0,2,0,0,0$. There is an edge from x going off to infinity. T admits an automorphism $\sigma = (\alpha, \epsilon) \in \mathrm{Aut}_T$ of order 4, where α swaps u_1 and u_2, and fixes all the other vertices and the sign function ϵ is given by

$$\epsilon(u_1x) = 1, \quad \epsilon(u_2x) = -1, \quad \epsilon(w_2x) = 1,$$

$$\epsilon(w_3) = \epsilon(w_3w_2) = \epsilon(u_3w_3) = \epsilon(u_4w_3) = -1,$$

where yz denotes the edge between the vertices y and z.

The lattice $\Lambda_T = H_1(T, T_b)$ is the relative homology of T with respect to its blue part. In other words it consists of 1-chains in T whose boundary is blue (see Section 2.2.3). Writing $[a, b]$ for the shortest path from a to b, and

$$c_1 = [u_1, x], \quad c_2 = [u_2, x], \quad c_3 = [u_3, w_2], \quad c_4 = [u_4, w_2], \quad c_5 = [w_2, x],$$

the lattice is given by

$$\Lambda_T = H_1(T, T_b) = \langle c_1, c_2, c_3, c_4, c_5 \rangle \simeq \mathbb{Z}^5.$$

The cycles have lengths $5, 5, 8, 8, 2$, and have trivial intersections, except for the third and fourth basis elements whose intersection has length 2. Clearly α swaps c_1 and c_2 and fixes the other basis vectors, while the sign ϵ in 1 on c_1 and c_5, and

-1 on c_2, c_3, c_4. Thus the length-pairing (see Section 2.2.2) and the action of σ on Λ_T are

$$\langle \cdot,\cdot\rangle = \begin{pmatrix} 5 & 0 & 0 & 0 & 0\\ 0 & 5 & 0 & 0 & 0\\ 0 & 0 & 8 & 2 & 0\\ 0 & 0 & 2 & 8 & 0\\ 0 & 0 & 0 & 0 & 2\end{pmatrix}, \qquad \sigma = \begin{pmatrix} 0 & -1 & 0 & 0 & 0\\ 1 & 0 & 0 & 0 & 0\\ 0 & 0 & -1 & 0 & 0\\ 0 & 0 & 0 & -1 & 0\\ 0 & 0 & 0 & 0 & 1\end{pmatrix}.$$

3.3. Cluster pictures.

DEFINITION 3.33 (Cluster picture). Let X be a finite set and $\Sigma \subset \mathcal{P}(X)$ a collection of non-empty subsets of X; elements of Σ are called *clusters*. Then Σ (or (X,Σ)) is a *cluster picture* if

(1) Every singleton ('root') is a cluster, and X is a cluster.
(2) Two clusters are either disjoint or contained in one another.

We say that Σ has *genus* g if $|X| \in \{2g+1, 2g+2\}$.

Two cluster pictures (X,Σ_X) and (Y,Σ_Y) are *isomorphic* if there is a bijection $X \to Y$ that takes Σ_X to Σ_Y. (So we may take $X = \{1,...,n\}$, and just consider its cluster pictures, up to S_n-permutations.)

REMARK 3.34. Let $X = \{r_1,...,r_n\} \subset K$ be a finite subset of a field with a (non-trivial) valuation. Then the non-empty subsets of X that are cut out by discs in $\bar{K}$ form a cluster picture. Conversely, every cluster picture arises in this way: for any K that has at least n elements in its residue field one can find $X \subset K$ that realises it.

EXAMPLE 3.35. Let $X = \{1,2,3,4,5,6\}$ and let

$$\Sigma = \Big\{\{1\},...,\{6\},\{5,6\},X\Big\}.$$

Thus, apart from the required singletons and X, there is one extra cluster $\mathfrak{s} = \{5,6\}$. We draw Σ with ovals around every $\mathfrak{s} \in \Sigma$ with $|\mathfrak{s}| > 1$:

$\Sigma \;=$ 

It is a cluster picture of genus 2. In the language of the last remark, it is realised, for example, by $\{1,2,3,4,-p,p\} \subset \mathbb{Q}_p$ $(p \geq 5)$.

DEFINITION 3.36 (Children). If $\mathfrak{s}' \subsetneq \mathfrak{s}$ is a maximal subcluster, we write $\mathfrak{s}' < \mathfrak{s}$ and refer to $\mathfrak{s}'$ as a *child* of $\mathfrak{s}$, and $\mathfrak{s}$ as the *parent* of $\mathfrak{s}'$.

DEFINITION 3.37 (Types of clusters). A cluster $\mathfrak{s}$ is *proper* if $|\mathfrak{s}| > 1$ or $|\mathfrak{s}| = |X| = 1$, a *twin* if $|\mathfrak{s}| = 2$, and it is *odd/even* if its size is odd/even. A proper cluster is *übereven* if it has no odd children.

DEFINITION 3.38 (Genus). A non-übereven cluster has *genus* $g = g(\mathfrak{s})$ if it has $2g+1$ or $2g+2$ odd children; übereven clusters are declared to have genus 0.

DEFINITION 3.39 (Balanced). A cluster picture Σ is *balanced* if $|X|$ is even, X is the only cluster of size $> \frac{|X|}{2}$, and there are either 0 or 2 clusters of size $\frac{|X|}{2}$.

EXAMPLE 3.40. Table 3.1 (last column) lists all balanced cluster pictures of genus 0,1 and 2 up to isomorphism.

DEFINITION 3.41. An *isomorphism* $\Sigma \to \Sigma'$ is an equivalence class of pairs (α, ϵ) with $\alpha : \Sigma \to \Sigma'$ a bijection that preserves cluster sizes and inclusions, and $\epsilon(\mathfrak{s}) = \pm 1$ a collection of signs for even clusters $\mathfrak{s} \in \Sigma$, such that

$$\epsilon(\mathfrak{s}') = \epsilon(\mathfrak{s}) \qquad \text{for } \mathfrak{s} \text{ übereven}, \mathfrak{s}' < \mathfrak{s}.$$

Here we say pairs (α, ϵ) and (α', ϵ') are equivalent if α and α' induce the same map between the sets of proper clusters of Σ and Σ', and $\epsilon = \epsilon'$. We compose isomorphisms by a cocycle rule

$$(\alpha, \epsilon_\alpha) \circ (\beta, \epsilon_\beta) = \big(\alpha \circ \beta, \mathfrak{s} \mapsto \epsilon_\beta(\mathfrak{s})\epsilon_\alpha(\beta(\mathfrak{s}))\big).$$

An *automorphism* of Σ is an isomorphism from Σ to itself, and we write $\operatorname{Aut}\Sigma$ for the group of automorphisms. (As all ϵ may be chosen to be $+1$, this extended notion of an isomorphism does not affect the definition of being isomorphic.) Equivalently, one may think of an element of $\operatorname{Aut}\Sigma$ as a pair $(\sigma, \epsilon_\sigma)$ where σ is a permutation of the proper clusters of Σ and ϵ_σ is a collection of signs as above, composition of two such being given by the same formula. We will frequently take this viewpoint without further comment in what follows.

REMARK 3.42. As for BY trees (Remark 3.29), we have

$$\operatorname{Aut}\Sigma = \operatorname{Aut}_0(\Sigma) \ltimes (\mathbb{Z}/2\mathbb{Z})^r,$$

where $\operatorname{Aut}_0(\Sigma)$ consists of elements for which $\epsilon(\mathfrak{s}) = +1$ for all even clusters $\mathfrak{s} \in \Sigma$. The number r is the number of equivalence classes of even clusters for the equivalence relation generated by '$\mathfrak{s}' \sim \mathfrak{s}$ if $\mathfrak{s}$ is übereven and $\mathfrak{s}' < \mathfrak{s}$'.

DEFINITION 3.43 (Equivalence). We say that cluster pictures (X_1, Σ_1) and (X_2, Σ_2) are *equivalent* if (X_2, Σ_2) is isomorphic to a cluster picture obtained from (X_1, Σ_1) in a finite number of the following steps $(X, \Sigma) \to (X', \Sigma')$.

(i) ('add cocluster') $X' = X$, $\Sigma' = \Sigma \cup \{X \setminus \mathfrak{s}\}$ for some $\mathfrak{s} < X$, $|X| = 2g+2$.
(ii) ('remove cocluster') $X' = X$, $\Sigma' = \Sigma \setminus \{\mathfrak{s}\}$ for some $\mathfrak{s} < X$ with $|\mathfrak{s}| \geq 2$, when X has exactly two children, $|X| = 2g+2$.
(iii) ('$2g+1 \to 2g+2$') $X' = X \coprod \{r\}$, $\Sigma' = \Sigma \cup \{X'\} \cup \{\{r\}\}$, when $|X| = 2g+1$.
(iv) ('$2g+2 \to 2g+1$') $X = X' \coprod \{r\}$, $\Sigma = \Sigma' \cup \{X\} \cup \{\{r\}\}$, when $|X| = 2g+2$.

In general, note that

- the genus of a cluster picture is preserved under equivalence;
- (ii)=(i)$^{-1}$ and (iv)=(iii)$^{-1}$, hence this is an equivalence relation;
- (i) does nothing when X has only 2 children;

EXAMPLE 3.44. There are, up to isomorphism, 7 cluster pictures $\tilde{\Sigma}$ in the equivalence class of Σ from Example 3.35:

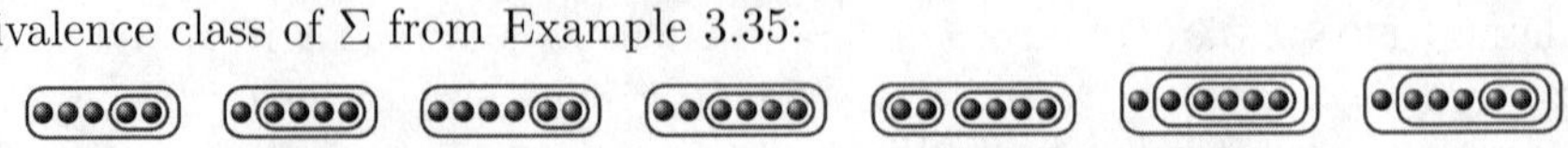

Cluster pictures #1,#2,#5 have $\operatorname{Aut}\Sigma \cong C_2$, and the other four $\operatorname{Aut}\Sigma \cong C_2^2$. All 7 have trivial $\operatorname{Aut}_0(\Sigma)$, so the automorphisms come from choices of signs on even clusters.

As for hyperelliptic graphs and BY trees, we have a metric version:

DEFINITION 3.45 (Metric version). A cluster picture (X, Σ) is *metric* when every pair of proper clusters $\mathfrak{s} < \mathfrak{t}$ is assigned a *distance* $\delta(\mathfrak{s}, \mathfrak{t}) = \delta(\mathfrak{t}, \mathfrak{s}) \in \mathbb{R}_{>0}$. The distance function clearly extends to every pair of proper clusters: if $\mathfrak{s}$ and $\mathfrak{t}$ are distinct proper clusters with least common ancestor $\mathfrak{u}$ (no child of $\mathfrak{u}$ contains both $\mathfrak{s}$ and $\mathfrak{t}$, but $\mathfrak{u}$ does), so that

$$\mathfrak{s} < \mathfrak{s}_1 < ... < \mathfrak{s}_{k-1} < \mathfrak{u} > \mathfrak{t}_{m-1} > ... > \mathfrak{t}_1 > \mathfrak{t},$$

we let $\delta(\mathfrak{s}, \mathfrak{t})$ be the sum of the $k+m$ distances between adjacent clusters in the chain; we let $\delta(\mathfrak{s}, \mathfrak{s})$=0.

An *isomorphism as metric cluster pictures* is one that preserves δ.

We say that two metric cluster pictures are *equivalent* if one is (up to isomorphism) obtained from the other by a finite number of 'metric versions' of the moves (i)-(iv) of Definition 3.43, in which we allow any metric δ' on (X', Σ') such that

- $\delta'(\mathfrak{t}, \mathfrak{t}') = \delta(\mathfrak{t}, \mathfrak{t}')$ for every $\mathfrak{t}, \mathfrak{t}' \in \Sigma \cap \Sigma'$ with (for moves (i)-(ii)) $\mathfrak{t}, \mathfrak{t}' \neq X$,
- $\delta'(\mathfrak{t}, X \setminus \mathfrak{s}) = \delta(\mathfrak{t}, X)$ for all $\mathfrak{t} \in \Sigma' \setminus \{X, X \setminus \mathfrak{s}\}$ in move (i),
- $\delta'(\mathfrak{t}, X) = \delta(\mathfrak{t}, \mathfrak{s})$ for all $\mathfrak{t} \in \Sigma' \setminus \{X\}$ in move (ii).

In diagrams, distances are shown using subscripts on clusters: a cluster gets a subscript indicating the distance to its parent (cf. Example 3.49).

REMARK 3.46. The clusters arising in Remark 3.34 are naturally metric cluster pictures. In this setting, for a cluster $\mathfrak{s}$ we define its 'depth' as $\delta(\mathfrak{s}) = \min_{r,r' \in \mathfrak{s}}(\mathrm{val}_K(r - r'))$, and for a pair $\mathfrak{s} < \mathfrak{s}'$ set the distance to be given by the 'relative depth', $\delta(\mathfrak{s}, \mathfrak{s}') = \delta(\mathfrak{s}) - \delta(\mathfrak{s}')$. This is the same as the formula given in §1.3 in terms of the diameter.

We now define the lattice Λ_Σ for a cluster picture (X, Σ). We need one preliminary piece of notation.

NOTATION 3.47. Let (X, Σ) be a cluster picture. Let

$$\mathcal{E}_\Sigma = \{\text{even, non-übereven clusters } \mathfrak{s} \neq X\}.$$

Further, for $\mathfrak{s} \in \mathcal{E}_\Sigma$, write $\hat{\mathfrak{s}}$ for the smallest non-übereven cluster strictly containing $\mathfrak{s}$. If no such cluster exists, we set $\hat{\mathfrak{s}} = X$.

DEFINITION 3.48 (The lattice Λ). Let (X, Σ) be a cluster picture. If X is not an übereven cluster, let

$$\Lambda_\Sigma = \mathbb{Z}[\mathcal{E}_\Sigma] = \Big\{ \sum_{\mathfrak{s} \in \mathcal{E}_\Sigma} \lambda_\mathfrak{s} \mathfrak{s} \;\Big|\; \lambda_\mathfrak{s} \in \mathbb{Z} \Big\}.$$

If X is übereven, let

$$\Lambda_\Sigma = \Big\{ \sum_{\mathfrak{s} \in \mathcal{E}_\Sigma} \lambda_\mathfrak{s} \mathfrak{s} \in \mathbb{Z}[\mathcal{E}_\Sigma] \;\Big|\; \sum_{\hat{\mathfrak{s}} = X} \lambda_\mathfrak{s} = 0 \Big\}.$$

Further, define an action of $\sigma = (\sigma_0, \epsilon_\sigma) \in \mathrm{Aut}\,\Sigma$ on Λ_Σ by

$$\sigma \cdot \mathfrak{s} = \epsilon_\sigma(\mathfrak{s}) \sigma_0(\mathfrak{s}).$$

In the metric case, define a pairing on Λ_Σ by setting, for $\mathfrak{s}_1, \mathfrak{s}_2 \in \mathcal{E}_\Sigma$,

$$\langle \mathfrak{s}_1, \mathfrak{s}_2 \rangle = \begin{cases} 2\delta(\mathfrak{s}_1 \wedge \mathfrak{s}_2, \hat{\mathfrak{s}}_1) & \hat{\mathfrak{s}}_1 = \hat{\mathfrak{s}}_2, \\ 0 & \hat{\mathfrak{s}}_1 \neq \hat{\mathfrak{s}}_2, \end{cases}$$

where $\mathfrak{s}_1 \wedge \mathfrak{s}_2$ denotes the least common ancestor of $\mathfrak{s}_1$ and $\mathfrak{s}_2$.

EXAMPLE 3.49. Consider the following cluster picture Σ:

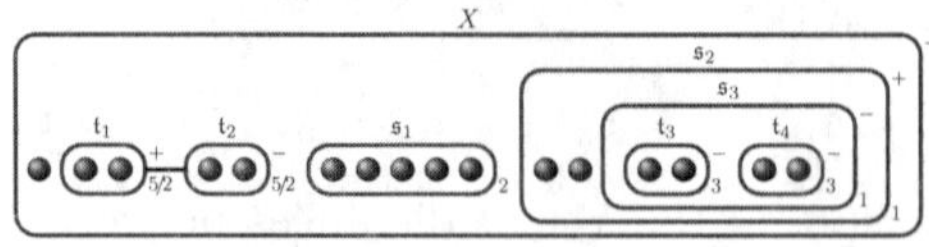

The cluster picture Σ has clusters $\mathfrak{t}_1, \ldots, \mathfrak{t}_4$ (twins), $\mathfrak{s}_1, \mathfrak{s}_2, \mathfrak{s}_3$, X of genera $0,0,0,0,2,0,0,0$ with distances between children and parents $\delta(\mathfrak{t}_1, X) = \delta(\mathfrak{t}_2, X) = 5/2$, $\delta(\mathfrak{t}_3, \mathfrak{s}_3) = \delta(\mathfrak{t}_4, \mathfrak{s}_3) = 3$, $\delta(\mathfrak{s}_3, \mathfrak{s}_2) = 1, \delta(\mathfrak{s}_2, X) = 1$, $\delta(\mathfrak{s}_1, X) = 2$.

This cluster picture admits an automorphism $\sigma = (\alpha, \epsilon) \in \mathrm{Aut}_\Sigma$ of order 4, where σ swaps $\mathfrak{t}_1$ and $\mathfrak{t}_2$ (which is indicated by the black line between the two leftmost twins) and fixes all other clusters. The sign function ϵ is

$$\epsilon(\mathfrak{t}_1) = 1, \quad \epsilon(\mathfrak{t}_2) = -1, \quad \epsilon(\mathfrak{s}_2) = 1, \quad \epsilon(X) = 1,$$
$$\epsilon(\mathfrak{t}_3) = \epsilon(\mathfrak{t}_4) = \epsilon(\mathfrak{s}_3) = -1,$$

indicated by the $+$ and $-$ on top of the respective even clusters.

Here the set of even non-übereven clusters (that are not X) is $\mathcal{E}_\Sigma = \{\mathfrak{t}_1, \mathfrak{t}_2, \mathfrak{t}_3, \mathfrak{t}_4, \mathfrak{s}_2\}$, and since X is not übereven,

$$\Lambda_\Sigma = \langle \mathfrak{t}_1, \mathfrak{t}_2, \mathfrak{t}_3, \mathfrak{t}_4, \mathfrak{s}_2 \rangle \simeq \mathbb{Z}^5.$$

By definition

$$\hat{\mathfrak{t}}_1 = \hat{\mathfrak{t}}_2 = \hat{\mathfrak{s}}_2 = X \qquad \text{and} \qquad \hat{\mathfrak{t}}_3 = \hat{\mathfrak{t}}_4 = \mathfrak{s}_2,$$

and

$$\mathfrak{t}_1 \wedge \mathfrak{t}_2 = \mathfrak{t}_1 \wedge \mathfrak{s}_2 = \mathfrak{t}_2 \wedge \mathfrak{s}_2 = X \qquad \text{and} \qquad \mathfrak{t}_3 \wedge \mathfrak{t}_4 = \mathfrak{s}_3.$$

Thus by definition of the pairing on Λ_Σ and since the action of σ on Λ_Σ is a signed permutation given by α and ϵ,

$$\langle \cdot, \cdot \rangle = \begin{pmatrix} 5 & 0 & 0 & 0 & 0 \\ 0 & 5 & 0 & 0 & 0 \\ 0 & 0 & 8 & 2 & 0 \\ 0 & 0 & 2 & 8 & 0 \\ 0 & 0 & 0 & 0 & 2 \end{pmatrix}, \qquad \sigma = \begin{pmatrix} 0 & -1 & 0 & 0 & 0 \\ 1 & 0 & 0 & 0 & 0 \\ 0 & 0 & -1 & 0 & 0 \\ 0 & 0 & 0 & -1 & 0 \\ 0 & 0 & 0 & 0 & 1 \end{pmatrix}.$$

4. One-to-one correspondence (open case)

In this section we explain how to pass between cluster pictures, hyperelliptic graphs and BY trees. We construct maps $\underline{G}$, $\underline{T}$ and $\underline{\Sigma}$ (which become inverse to each other upon passing to isomorphism classes) between objects according to the following diagram:

$$\{\text{open hyperelliptic graphs}\} \underset{\underline{G}}{\overset{\underline{T}}{\rightleftarrows}} \{\text{open BY trees}\} \underset{\underline{T}}{\overset{\underline{\Sigma}}{\rightleftarrows}} \{\text{cluster pictures}\}.$$

In addition, for $\underline{F}$ one of $\underline{G}$, $\underline{T}$ or $\underline{\Sigma}$, and X an object on which it is defined, we construct a map $\underline{F} : \mathrm{Aut}\, X \to \mathrm{Aut}\, \underline{F}(X)$, which turns out to be an isomorphism in each case. Here there is a subtlety: for an open hyperelliptic graph G, the map $\mathrm{Aut}\, G \to \mathrm{Aut}\, \underline{T}(G)$ depends on a choice of section $G/\langle\iota\rangle \to G$. The exact dependence of the map on the choice is examined in Proposition 4.7. By contrast, all other maps on automorphism groups are canonical. The constructions are summarised in Table 4.20 and illustrated in Examples 4.21 and 4.22.

TABLE 4.1. Open hyperelliptic graphs, open BY trees and cluster pictures in genus 0 and 1 up to isomorphism.

G	T	Σ	Aut	Λ
	Core 0 (genus 0)			
			C_1	0
			C_2	0
	Core 1 (genus 1)			
			C_1	0
			C_2	0
			C_2	0
	Core $\mathrm{U}_{n,m}$ (genus 1)			
			C_2	$\mathbb{Z}$
			$C_2 \times C_2$	$\mathbb{Z}$
		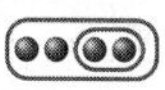	$C_2 \times C_2$	$\mathbb{Z}$
			$C_2 \times C_2$	$\mathbb{Z}$

The main result is the following.

THEOREM 4.2. *The maps $\underline{G}$, $\underline{T}$ and $\underline{\Sigma}$ defined in Constructions* 4.4, 4.8, 4.13 *and* 4.15 *give a genus preserving one-to-one correspondence between isomorphism classes of (either metric or not)*

(i) *Cluster pictures,*
(ii) *Open hyperelliptic graphs,*
(iii) *Open BY trees.*

Moreover, the associated maps on automorphism groups (see Constructions 4.4, 4.8, 4.13 *and* 4.15 *and the preceding paragraph) are isomorphisms.*

PROOF. Combine Proposition 4.11 ('hyperelliptic graphs $\leftrightarrow$ BY trees') with Proposition 4.19 ('BY trees $\leftrightarrow$ cluster pictures'). □

4.1. Hyperelliptic graphs $\leftrightarrow$ BY trees. We begin with the maps between open hyperelliptic graphs and open BY trees, as well as the associated maps on automorphism groups. In fact, the constructions apply equally well in the closed version. Since both versions will be relevant later, we cover both here. Before detailing the constructions, we briefly discuss the notion of a 'section to the quotient map' for a hyperelliptic graph.

REMARK 4.3 (Sections to the quotient map). Let G be a (open or not) hyperelliptic graph and $\pi : G \to G/\langle\iota\rangle$ be the quotient map. In order to construct the map between automorphisms of G and automorphisms of the associated BY tree $T = G/\langle\iota\rangle$, it will be necessary to choose a continuous map $s : T \to G$ such that $\pi \circ s = \mathrm{id}$. That is, a continuous section to π (we henceforth refer to s as a *section*, the continuity being understood). More concretely, such a choice amounts to the following. Write $G_y = \pi^{-1}(T_y)$ and choose a decomposition $G_y = G_y^+ \coprod G_y^-$, such that $\pi : G_y^{\pm} \to T_y$ is a homeomorphism (that is, a choice of 'top' and 'bottom' above every connected component of T_y). Such a choice determines a section s by sending $x \in T_b$ to its unique preimage in G, and $x \in T_y$ to its unique preimage in G_y^+. Conversely, a section s determines such a decomposition by taking G_y^+ to be $s(T_y)$.

CONSTRUCTION 4.4 ($\underline{T}(G)$).
Objects: Let (G, g, ι) be a (closed) hyperelliptic graph, without loss of generality considered to be metric. Define

$$T = \underline{T}(G) = G/\langle\iota\rangle,$$

the topological quotient. It is a tree by Definition 3.2 (4), and we colour the branch locus T_b of the quotient map $\pi : G \to T$ blue and $T_y = T \backslash T_b$ yellow. In other words, under π, blue points have one preimage and yellow points have two preimages. We make T into a graph as follows:

Write $V(G) = \{v_1, ..., v_k, v_{k+1}, \iota(v_{k+1}), \dots, v_n, \iota(v_n)\}$, with the first k vertices ι-invariant, followed by the pairs swapped by ι. Then $v_1, ..., v_k$ give vertices $\bar{v}_1, ..., \bar{v}_k \in V(T)$, each $\bar{v}_j$ declared blue of genus $g(v_j)$, and each pair $\{v_j, \iota(v_j)\}$ for $j > k$ gives one vertex $\bar{v}_j \in V(T)$, declared yellow of genus 0. Next, take an edge of G, say of length d. It is either mapped to another edge by ι, is ι-invariant, or is ι-anti-invariant. Then

- each ι-invariant edge vw gives a blue edge $\bar{v}\bar{w}$ of length $2d$;
- each ι-anti-invariant edge $v\iota(v)$ (allowing for $\iota(v) = v$ in the case of loops) gives a yellow edge of length d from $\bar{v}$ to an extra blue genus 0 leaf of T;
- each swapped pair of edges vw and $\iota(v)\iota(w)$ gives a yellow edge $\bar{v}\bar{w}$ of length $2d$.

Finally, if G is an open (possibly metric) hyperelliptic graph, we define $\underline{T}(G)$ in the same way (treating w as ∞ for the open edge(s)).
Automorphisms: Let G, T and $\pi : G \to T$ be as above. To define the map $\operatorname{Aut} G \to \operatorname{Aut} T$ we begin by choosing a section $s : T \to G$ to π. Now given $\sigma \in \operatorname{Aut} G$ we construct an automorphism $\underline{T}(\sigma) = (\underline{T}(\sigma)_0, \epsilon_{\underline{T}(\sigma)})$ of T as follows (if we wish to record the choice of section, we write $\underline{T}(\sigma; s)$). Since σ necessarily commutes with the hyperelliptic involution ι (see Remark 3.10), it induces a graph-theoretic automorphism $\bar{\sigma}$ of T which preserves genus and colour by construction. We set $\underline{T}(\sigma)_0 = \bar{\sigma}$. Next, let Z be a connected component of T_y. Then we define

$$\epsilon_{T(\sigma;s)}(Z) = \begin{cases} 1 & s(\bar{\sigma}(Z)) = \sigma(s(Z)), \\ -1 & \text{else.} \end{cases}$$

PROPOSITION 4.5. *If G is a hyperelliptic graph, then $\underline{T}(G)$ is a BY tree. The same holds in the open case.*

PROOF. It is easy to check the claim in genus 0 and 1 (see Tables 3.1, 4.1), so assume $g > 1$ from now on. We check the three conditions of Definition 3.18:

(1) *Yellow vertices have genus 0, degree ≥ 3, and only yellow edges*: all yellow vertices come from pairs $v_j, \iota(v_j)$ consisting of two distinct vertices of G swapped by ι, and are declared to have genus 0 in the construction. Moreover, yellow vertices have only yellow outgoing edges, because blue edges only come from edges between ι-invariant vertices. Finally, $G \to T$ is a two-to-one topological cover at $\bar{v}_j$, and so $\deg \bar{v}_j = \deg v_j \geq 3$ by Definition 3.2 (2).

(2) *Blue vertices of genus 0 have at least one yellow edge*: blue vertices either come from loops, in which case they have a yellow edge by construction, or from ι-invariant vertices v_j. Suppose $g(v_j) = 0$ but $\bar{v}_j$ has only blue outgoing edges. Then $\deg v_j \geq 3$ by 3.2 (2), so v_j has at least three ι-invariant outgoing edges, contradicting Definition 3.2 (3).

(3) *At every vertex, $2g(v)+2 \geq$ # blue edges at v*: if v is yellow, or a blue leaf coming from a loop, then it has only yellow edges and there is nothing to prove. If v is blue, the blue edges from $\bar{v}$ are in one-to-one correspondence with the ι-invariant edges from v, and the claim follows from Definition 3.2 (3). □

We now examine the dependence of the map $T(-,s) : \operatorname{Aut} G \to \operatorname{Aut} T$ constructed above on the choice of section s. For this it will be useful to have the following definition.

DEFINITION 4.6. Let G be a hyperelliptic graph (open or not), let $T = \underline{T}(G)$ be the associated BY tree, and let $\pi : G \to T$ be the quotient map. Further, let s and s' be two sections to π. We define the automorphism $\psi_{s,s'} = (\mathrm{id}, \epsilon_{s,s'})$ of T by setting, for a component Z of T_y,

$$\epsilon_{s,s'}(Z) = \begin{cases} 1 & s(Z) = s'(Z) \\ -1 & s(Z) \neq s'(Z). \end{cases}$$

PROPOSITION 4.7. *Let G be a hyperelliptic graph (possibly open), let $T = \underline{T}(G)$ be the associated BY tree, and let $\pi : G \to T$ be the quotient map. Then for each section $s : T \to G$ to π, the map $\underline{T}(-;s)$ defines a homomorphism* $\operatorname{Aut} G \to \operatorname{Aut} T$. *Moreover, if s and s' are two choices of section then for all $\sigma \in \operatorname{Aut} G$ we have*

$$\underline{T}(\sigma; s') = \psi_{s,s'} \circ \underline{T}(\sigma; s) \circ \psi_{s,s'}^{-1}.$$

In particular, the map $\underline{T}(-;s)$ depends on s only up to conjugation.

PROOF. It is immediate from the construction that for any automorphism σ of G, $\underline{T}(\sigma; s)$ is indeed an automorphism of T and it is clear that $\underline{T}(-;s)$ is a homomorphism. A straightforward calculation verifies the statement concerning the dependence on the choice of section. □

CONSTRUCTION 4.8 ($\underline{G}(T)$).
Objects: Let T be a BY tree, viewed as a topological space. Let G be the topological space given by glueing two disjoint copies T^+ and T^- of T along their common blue parts. Then G comes with a natural map $\pi : G \to T$ making it into a two-to-one cover of T ramified along T_b, as well as an involution ι (swapping the elements of the fibres of π over points in T_y) such that $G/\langle\iota\rangle = T$. When T has genus ≥ 2, we make G into a graph as follows:

- A blue vertex $\bar{v} \in V(T)$ which is not a genus 0 leaf gives an ι-invariant vertex $v \in V(G)$ of genus $g(\bar{v})$.
- a yellow vertex $\bar{v} \in V(T)$ gives vertices v^+ and v^- in $V(G)$ (where $v^+ \in T^+$ and $v^- \in T^-$) swapped by ι, of genus 0,
- a (necessarily blue) genus 0 leaf $\bar{v}$ of T with an edge from $\bar{v}$ to a blue vertex $\bar{w}$, of length d, gives an ι-anti-invariant loop on w of length d,
- a (blue) genus 0 leaf $\bar{v}$ of T joined by an edge of length d to a yellow vertex $\bar{w}$ gives an ι-anti-invariant edge between w^+ and w^- of length d,

- a blue edge between (necessarily blue vertices) $\bar{v}$ and $\bar{w}$ of length d gives an ι-invariant edge between v and w of length $d/2$,
- a yellow edge between $\bar{v}$ and $\bar{w}$ of length d gives a pair of edges, one between v^+ and w^+ and one between v^- and w^-, swapped by ι, each of length $d/2$ (here, if $\bar{v}$ (resp. $\bar{w}$) is blue we set $v^+ = v^- = v$ (resp. $w^+ = w^- = w$)).

When T is an open BY tree, we use exactly the same construction (say, by adding a vertex at ∞, following the steps above, and removing the vertices above ∞).

Finally, when T has genus 0 or 1, we have to declare vertices of G slightly differently, and we refer to Tables 3.1, 4.1 for the correspondence.

Automorphisms: Write G_b for those points in G fixed by ι and G_y for $G \setminus G_b$. Further, write G_y^+ for the points in G_y which come from T^+ and G_y^- for the points coming from T^-. To ease notation, for $x \in G$ we also set $\bar{x} := \pi(x) \in T$.

Now let $\sigma = (\sigma_0, \epsilon_\sigma)$ be an automorphism of T. We define the automorphism $\underline{G}(\sigma)$ of G as follows:

- for $x \in G_b$ we set $\underline{G}(\sigma)(x)$ to be the unique point of G lying over $\sigma(\bar{x})$
- for $x \in G_y^+$ we set

$$\underline{G}(\sigma)(x) = \begin{cases} \pi^{-1}(\{\bar{x}\}) \cap G_y^+ & \epsilon(\bar{x}) = 1 \\ \pi^{-1}(\{\bar{x}\}) \cap G_y^- & \epsilon(\bar{x}) = -1. \end{cases}$$

- similarly, for $x \in G_y^-$ we set

$$\underline{G}(\sigma)(x) = \begin{cases} \pi^{-1}(\{\bar{x}\}) \cap G_y^- & \epsilon(\bar{x}) = 1 \\ \pi^{-1}(\{\bar{x}\}) \cap G_y^+ & \epsilon(\bar{x}) = -1. \end{cases}$$

(In the above, for $\bar{x} \in T$ yellow, we write $\epsilon_\sigma(\bar{x})$ for the value of ϵ_σ on the connected component of T_y containing $\bar{x}$.)

Note that σ commutes with the hyperelliptic involution ι.

PROPOSITION 4.9. *If T is a BY tree, then $\underline{G}(T)$ is a hyperelliptic graph. The same holds in the open case. Further, the map $\underline{T}(-)$ gives a homomorphism* $\operatorname{Aut} T \to \operatorname{Aut} G$.

PROOF. That $\underline{G}(T)$ is a hyperelliptic graph follows by reversing the argument of Proposition 4.5. The claim about automorphisms is clear. □

REMARK 4.10. As constructed, for a BY tree T, $G = \underline{G}(T)$ comes equipped with a canonical section $s : T \to G$ sending $\bar{x} \in T$ to x^+ (in the notation of Construction 4.8). In general, suppose that $s' : T \to G$ is any section. Then given $\sigma = (\sigma_0, \epsilon_\sigma) \in \operatorname{Aut} T$ we may define an automorphism $\underline{G}(\sigma; s') \in \operatorname{Aut} G$ by first defining it on the image of s' as

$$\underline{G}(\sigma; s')(s'(\bar{x})) = \begin{cases} s'(\sigma_0(\bar{x})) & \bar{x} \text{ blue or } \bar{x} \text{ yellow and } \epsilon_\sigma(\bar{x}) = 1, \\ \iota(s'(\sigma_0(\bar{x}))) & \text{else,} \end{cases}$$

and then extending to the whole of G by insisting that $\underline{G}(\sigma; s')$ commutes with the hyperelliptic involution. Then $\underline{G}(\sigma; s)$ agrees with $\underline{G}(\sigma)$ as defined in Construction 4.8. Moreover, writing $\phi_{s,s'}$ for the automorphism of G such that $s' = \phi_{s,s'} \circ s$, we have

$$\underline{G}(\sigma; s') = \phi_{s,s'} \circ \underline{G}(\sigma; s) \circ \phi_{s,s'}^{-1}$$

for all $\sigma \in \operatorname{Aut} T$.

The following proposition establishes the correspondence between hyperelliptic graphs and BY trees.

PROPOSITION 4.11. *Let G be a hyperelliptic graph, possibly open and/or metric and let $T = \underline{T}(G)$ be the corresponding BY tree. Then*

(1) *We have an equality of genera $g(G) = g(T)$,*
(2) *For any choice of section s, the map $\underline{T}(-;s)$ gives an isomorphism $\operatorname{Aut} G \xrightarrow{\sim} \operatorname{Aut} T$.*
(3) *We have $\underline{G}(T) \cong G$ (as metric graphs if G is metric).*

Conversely, let T be a BY tree, possibly open and/or metric, and let $G = \underline{G}(T)$ be the corresponding hyperelliptic graph. Then

(1)′ *We have an equality of genera $g(T) = g(G)$,*
(2)′ *The map $\underline{G}(-)$ gives an isomorphism $\operatorname{Aut} T \xrightarrow{\sim} \operatorname{Aut} G$,*
(3)′ *We have $\underline{T}(G) \cong T$ (as metric BY trees if T is metric).*

PROOF. First let G be a hyperelliptic graph and T the associated BY tree. For concreteness we consider the closed non-metric case, the argument being identical in the other cases. It is clear from the constructions that we have $\underline{G}(T) \cong G$ (non-canonically). We will show later in Proposition 6.6 that there is an isomorphism $H_1(G) \cong H_1(T, T_b)$. Since $\pi : G \to T$ is also a bijection on the vertices of positive genus, we have $g(T) = g(G)$ (cf. Definitions 3.8, 3.23). We have now established (1) and (3).

To show (2), since changing the section s serves to conjugate $\underline{T}(-;s)$, it suffices to prove the result for a single section s, which we now fix. If $\sigma \in \operatorname{Aut} G$ is such that $\underline{T}(\sigma;s)$ is the trivial automorphism of T, then σ acts trivally on the quotient $G/\langle\iota\rangle$ and also preserves the section s. Such an automorphism is easily seen to be the identity so $\underline{T}(-;s)$ is injective. To show surjectivity it now suffices to show that $|\operatorname{Aut} T| \leq |\operatorname{Aut} G|$. By (3), it suffices to show that $\underline{G}(-) : \operatorname{Aut} T \to \operatorname{Aut} \underline{G}(T)$ is injective which we do independently below. Modulo this remaining claim, this completes the proof of (1), (2) and (3).

Now let T be a BY tree (say closed and non-metric) and $G = \underline{G}(T)$. Again, it is clear from the construction that we have $\underline{T}(G) \cong T$ (and now the isomorphism is canonical) so (3)′ is proven. Part (1)′ now follows from (3)′ and (1).

We now show (2)′. Let $\sigma = (\sigma_0, \epsilon_\sigma)$ be an automorphism of T. If $\underline{G}(\sigma)$ is trivial then it is immediate from the definition that $\underline{G}(\sigma)$ is an automorphism of G that commutes with the hyperelliptic involution and acts trivially on $s(T)$. Since such automorphisms are necessarily trivial, $\underline{G}(\sigma)$ is injective (which also completes the proof of (2)). As above, the isomorphism of (3)′ along with the injectivity of $\underline{T}(-) : G \to \underline{T}(G)$ shown previously forces $\underline{G}(\sigma)$ to be an isomorphism. □

REMARK 4.12. It is tempting to define 'signed hyperelliptic graphs' as pairs (G, s) where G is a hyperelliptic graph (possibly open) and $s : G/\langle\iota\rangle \to G$ is a section to the quotient map, with isomorphisms between two such pairs (G, s) and (G', s') required to take s onto s'. It is easy to see that two such pairs (G, s) and (G', s') are isomorphic if and only if G and G' are, and that $\underline{G}$ and $\underline{T}$ give a one-to-one correspondence between signed hyperelliptic graphs and BY trees in which there are no choices involved in identifying automorphism groups (automorphisms of signed hyperelliptic graphs which forget the section must be allowed though). Additionally, the isomorphism $G \cong \underline{G}(\underline{T}(G))$ becomes canonical. We have decided against setting up the correspondence this way, however, as for our application to hyperelliptic curves, the hyperelliptic graphs we obtain do not come with a natural section s and so it is convenient to allow an arbitrary choice.

4.2. Cluster pictures $\leftrightarrow$ BY trees. We now construct the maps between cluster pictures and open BY trees.

CONSTRUCTION 4.13 ($\underline{T}(\Sigma)$).
Objects: Let (X,Σ) be a cluster picture. Define $T=\underline{T}(\Sigma)$ to be the open BY tree whose vertices are

- one vertex $v_{\mathfrak{s}}$ for every proper cluster $\mathfrak{s}$ which is not a twin, coloured yellow if $\mathfrak{s}$ is übereven and blue otherwise.
- one blue vertex (a leaf) $v_{\mathfrak{t}}$ for every twin $\mathfrak{t}$.

For the edges

- for every pair $\mathfrak{s}'<\mathfrak{s}$ (see Definition 3.36) with $\mathfrak{s}'$ proper, link $v'_{\mathfrak{s}}$ and $v_{\mathfrak{s}}$ by an edge, yellow if $\mathfrak{s}'$ is even and blue otherwise.
- add one open edge $v_X\infty$, yellow if X is even and blue otherwise.

In the metric version, set the length to be $\delta(\mathfrak{s},\mathfrak{s}')$ for blue edges and $2\delta(\mathfrak{s},\mathfrak{s}')$ for yellow edges.

Finally, define the genus of a vertex $v_{\mathfrak{s}}$ to be the genus of the cluster $\mathfrak{s}$ as in Definition 3.38.

Automorphisms: Let $\sigma=(\sigma_0,\epsilon_\sigma)$ be an element of $\operatorname{Aut}\Sigma$, where σ_0 is viewed as a permutation of the proper clusters. Then we define an automorphism $\underline{T}(\sigma)=(\underline{T}(\sigma)_0,\epsilon_{\underline{T}(\sigma)})$ of $T=\underline{T}(\Sigma)$ as follows. For a vertex $v_{\mathfrak{s}}$ of T, set $\underline{T}(\sigma)_0(v_{\mathfrak{s}})=v_{\sigma(\mathfrak{s})}$. To define $\epsilon_{\underline{T}(\sigma)}$, for a yellow component T_y of T, pick a yellow edge from $v_{\mathfrak{s}}$ to $v_{\mathfrak{s}'}$ in T_y where we take $v_{\mathfrak{s}'}$ to be nearer to ∞ (since yellow vertices have only yellow edges, each component of T_y has at least one yellow edge). Then $\mathfrak{s}$ is an even cluster and we set $\epsilon_{\underline{T}(\sigma)}(T_y)=\epsilon_\sigma(\mathfrak{s})$. The compatibility of signs on even clusters as in Definition 3.33 ensures that this is well defined.

PROPOSITION 4.14. *Let (X,Σ) be a cluster picture. Then*

(1) *$\underline{T}(\Sigma)$ is an open BY tree,*
(2) *The map $\underline{T}(-)$ defines a homomorphism $\operatorname{Aut}\Sigma\to\operatorname{Aut}\underline{T}(\Sigma)$.*

PROOF. (1) Let us check the conditions of a BY tree (Definition 3.21). When $|X|\le 2$, this is easy to check by hand. Otherwise:

Yellow vertices have genus 0, degree ≥ 3, and only yellow edges: übereven clusters have genus 0 and at least 2 children, giving at least 2 edges and an edge to the parent or to ∞. So there are at least 3 outgoing edges, all of which are yellow since all children of an übereven cluster are even.

Blue vertices of genus 0 have at least one yellow edge: every even cluster has a yellow edge to its parent or ∞, and every odd cluster of size >2 has either positive genus or at least one even child that gives a yellow edge.

At every vertex, $2g(v)+2\ge\#$ blue edges at v: every vertex v comes either from a twin or from a cluster $\mathfrak{s}$ of size >2. In the former case, v has no blue edges. In the latter,

- if $\mathfrak{s}$ is odd, it has an odd number ($=2g(\mathfrak{s})+1$) of odd children, and

$$\begin{aligned}2g(\mathfrak{s})+2 &= 1+\#\{\text{odd children}\}\ge 1+\#\{\text{odd proper children}\}\\ &= \#\text{ blue edges at }v.\end{aligned}$$

• if $\mathfrak{s}$ is even, it has an even number ($=2g(\mathfrak{s})+2$) of odd children, and

$$\begin{aligned} 2g(\mathfrak{s})+2 &= \#\{\text{odd children}\} \geq \#\{\text{odd proper children}\} \\ &= \#\text{ blue edges at } v. \end{aligned}$$

(2) If $\sigma = (\sigma_0, \epsilon_\sigma)$ is an automorphism of Σ then σ_0 preserves inclusion and the size of proper clusters by definition. Thus $\underline{T}(\sigma)$ preserves adjacency, colour and genus and it is now clear that $\underline{T}(\sigma)$ is an automorphism of $\underline{T}(\Sigma)$. That the map $\underline{T}(-)$ is a homomorphism follows from the way we have defined composition of automorphisms for cluster pictures and BY trees (cf. Definitions 3.27, 3.41). □

The construction in the opposite direction is as follows.

CONSTRUCTION 4.15 ($\underline{\Sigma}(T)$).
Objects: Let T be an open (possibly metric) BY tree. Define a partial order on the vertices of T by setting $v \preceq v'$ if v' lies on the unique shortest path from v to ∞.

For each blue vertex $v \in T_b$, let $\deg_{T_b}(v)$ be the number of blue edges at v (i.e. the degree of v in T_b) and set $m_v = 2g(v)+2-\deg_{T_b}(v)$, which is non-negative by Definition 3.21 (3). Take m_v singletons $x_{v,1},\dots,x_{v,m_v}$ and define $X_v := \{x_{v,1},\dots,x_{v,m_v}\}$. Now take $X = \bigcup_{v\in T_b} X_v$.

Further, for every vertex v of T (of any colour), set

$$\mathfrak{s}_v = \bigcup_{v'\preceq v,\ v' \text{ blue}} X_{v'}$$

and define the subset $\Sigma \subseteq \mathcal{P}(X)$ as

$$\Sigma = \bigcup_{v\in T}\{\mathfrak{s}_v\} \cup \bigcup_{x\in X}\{x\}.$$

Set $\underline{\Sigma}(T) = (X,\Sigma)$, the cluster picture associated to T. In the metric case, for $e \in E(T)$, write

$$l(e) = \begin{cases} \delta(e) & e \text{ blue}, \\ \frac{1}{2}\delta(e) & e \text{ yellow}, \end{cases}$$

and extend to a distance function of T in the obvious way. Now for vertices $v, w \in T$, define $\delta(\mathfrak{s}_v,\mathfrak{s}_w) = l(v,w)$.

Automorphisms: Let $\sigma = (\sigma_0,\epsilon_\sigma) \in \operatorname{Aut}(T)$. Then we define an element $\underline{\Sigma}(\sigma) = (\underline{\Sigma}(\sigma)_0, \epsilon_{\underline{\Sigma}(\sigma)})$ of $\operatorname{Aut}\Sigma$ as follows. Noting that the map $v \mapsto \mathfrak{s}_v$ is a bijection between the vertices of T and the proper clusters of Σ, define a permutation $\underline{\Sigma}(\sigma)_0$ of the proper clusters of Σ by setting $\underline{\Sigma}(\sigma)_0(\mathfrak{s}_v) = \mathfrak{s}_{\sigma_0(v)}$. Since $\mathfrak{s}_v \subseteq \mathfrak{s}_{v'}$ if and only if $v \preceq v'$, this preserves inclusion. To define $\epsilon_{\underline{\Sigma}(\sigma)}$, let $\mathfrak{s} \in \Sigma$ be an even cluster. Then $\mathfrak{s} = \mathfrak{s}_v$ for a vertex v of T and we'll see in Corollary 4.18 (2) below that the edge from v towards ∞ is yellow. Writing Z for the connected component of T_y containing this edge, we define $\epsilon_{\underline{\Sigma}(\sigma)}(\mathfrak{s}) = \epsilon_\sigma(Z)$.

PROPOSITION 4.16. *Let T be an open BY tree. Then $\underline{\Sigma}(T)$ is a cluster picture. Moreover, $\underline{\Sigma}(-)$ defines a homomorphism* $\operatorname{Aut} T \to \operatorname{Aut}\underline{\Sigma}(T)$.

PROOF. In both cases this is clear from Construction 4.15. □

Given a vertex v of an open BY tree T, it is not obvious from Construction 4.15 how the size of the associated cluster $\mathfrak{s}_v$ relates to invariants of T and v. The following two results explain this, as well as showing that Construction 4.15 preserves genus.

PROPOSITION 4.17. *Let T be an open BY tree, e its unique open edge, and $\underline{\Sigma}(T) = (X, \Sigma)$ the associated cluster picture. Then*

$$|X| = \begin{cases} 2g(T)+2 & \text{if } e \text{ is yellow,} \\ 2g(T)+1 & \text{if } e \text{ is blue.} \end{cases}$$

In particular, $g(T) = g(\Sigma)$, and $|X|$ is even if and only if e is yellow.

PROOF. By Construction 4.15 we see that

$$|X| = \sum_{v \in T_b} \left(2g(v) + 2 - \deg_{T_b}(v)\right).$$

Now since yellow vertices have genus 0, we may split this sum as

$$|X| = \sum_{v \in T} 2g(v) + \sum_{v \in T_b} \left(2 - \deg_{T_b}(v)\right).$$

If e is yellow then T_b is a disjoint union of closed connected trees, hence

$$\sum_{v \in T_b} \left(2 - \deg_{T_b}(v)\right) = 2|V(T_b)| - 2|E(T_b)| = 2\#\{\text{connected comps. of } T_b\}.$$

On the other hand, if e is blue, it is counted one fewer times in the sum. We thus obtain

$$|X| = 2\left(\sum_{v \in T} g(v) + \#\{\text{connected comps. of } T_b\}\right) - \begin{cases} 0 & \text{if } e \text{ is yellow,} \\ 1 & \text{if } e \text{ is blue.} \end{cases}$$

Since $\mathrm{rk} H_1(T, T_b)$ is equal to one less than the number of connected components of T_b (see Remark 3.24) the result follows. $\square$

COROLLARY 4.18. *Let T be an open BY tree and $\Sigma = \underline{\Sigma}(T)$ the associated cluster picture. Fix a vertex $v \in T$ and (in the notation Construction 4.15) let $\mathfrak{s}_v \in \Sigma$ be the associated cluster. Further, denote by e_v the edge from v towards ∞. Then:*

(1) *We have*

$$|\mathfrak{s}_v| = \begin{cases} 2g(T_v)+2 & \text{if } e_v \text{ is yellow,} \\ 2g(T_v)+1 & \text{if } e_v \text{ is blue,} \end{cases}$$

where here T_v denotes the open BY tree generated by the vertices $v' \preceq v$ of T along with the open edge e_v,

(2) *The cluster $\mathfrak{s}_v$ is even if and only if e_v is yellow, and übereven if and only if v itself is yellow,*

(3) *We have an equality of genera $g(\mathfrak{s}_v) = g(v)$.*

PROOF. The claims are easy to check when T has genus 0 or 1, so assume the genus is at least 2.

(1). Apply Proposition 4.17 to the open BY tree T_v.

(2). That $\mathfrak{s}_v$ is even if and only if e_v is yellow is clear from (1). Next, since yellow vertices have only yellow edges and no associated singletons, it is clear that if v is yellow then $\mathfrak{s}_v$ is übereven. For the converse it is convenient to first observe that if v is blue then the number of odd children of $\mathfrak{s}_v$ is either $2g(v)+2$ or $2g(v)+1$, the former case occuring if and only if e_v is yellow. Indeed, by Construction 4.15 the number of children of $\mathfrak{s}_v$ of size 1 is given by $m_v = 2g(v) + 2 - \deg_{T_b}(v)$, whilst part (1) applied to the vertices adjacent to v shows that the number of odd proper children of $\mathfrak{s}_v$ is given by the number of blue edges at v, excluding the edge e_v towards infinity (should this be blue).

Since $g(v)$ is non negative, it now follows that if v is blue then $\mathfrak{s}_v$ cannot be übereven.

(3). If $v \in T$ is yellow then $\mathfrak{s}_v$ is übereven by (2) and both $g(v)$ and $g(\mathfrak{s}_v)$ are 0. Suppose now that $v \in T$ is blue. As above, the number of odd children of $\mathfrak{s}_v$ is either $2g(v)+1$ or $2g(v)+2$. It is now immediate from the definition of the genus of a cluster that $g(\mathfrak{s}_v) = g(v)$ as claimed. $\square$

The following Proposition completes the proof of Theorem 4.2.

PROPOSITION 4.19. *Let T be an open (possibly metric) BY tree and $(X,\Sigma) = \underline{\Sigma}(T)$ be the associated cluster picture. Then*

(1) *we have an equality of genera $g(T) = g(\Sigma)$.*
(2) *The map $\sigma \mapsto \underline{\Sigma}(\sigma)$ gives an isomorphism $\operatorname{Aut} T \xrightarrow{\sim} \operatorname{Aut} \Sigma$.*
(3) *We have $\underline{T}(\Sigma) \cong T$ (as metric BY trees if T is metric).*

Conversely, let (X,Σ) be a (possibly metric) cluster picture and let $T = \underline{T}(\Sigma)$ be the associated open BY tree. Then

(1)′ *we have an equality of genera $g(\Sigma) = g(T)$.*
(2)′ *The map $\sigma \mapsto \underline{T}(\sigma)$ gives an isomorphism $\operatorname{Aut} \Sigma \xrightarrow{\sim} \operatorname{Aut} T$.*
(3)′ *We have $\underline{\Sigma}(T) \cong \Sigma$ (as metric cluster pictures if Σ is metric).*

PROOF. We consider the non-metric case throughout, the metric case being an easy extension. First let T be an open BY tree and $(X,\Sigma) = \underline{\Sigma}(T)$ the associated cluster picture. Part (1) was shown previously in Proposition 4.17.

Next we show part (3). In the notation of Constructions 4.13 and 4.15, consider the map $f : T \to \underline{T}(\Sigma)$ sending $v \in T$ to $v_{\mathfrak{s}_v}$. It is clear from the constructions that this is a graph theoretic isomorphism. Moreover, by Construction 4.13 and Corollary 4.18 (2) we see that f preserves colour (of both edges and vertices). Finally, to see that f preserves genus, fix $v \in T$. Since for $\mathfrak{s} \in \Sigma$ we defined the genus of $v_{\mathfrak{s}}$ to be the genus of the cluster $\mathfrak{s}$, it suffices to show that $g(v) = g(\mathfrak{s}_v)$ for each $v \in T$, which is Corollary 4.18 (3).

To show (2), let $\sigma = (\sigma_0, \epsilon_\sigma) \in \operatorname{Aut} T$ and suppose that $\underline{\Sigma}(\sigma)$ is trivial in $\operatorname{Aut} \Sigma$. Then as $v \mapsto \mathfrak{s}_v$ is a bijection between the vertices of T and the proper clusters of Σ, σ_0 fixes every vertex of T. Moreover, if $\epsilon_{\underline{\Sigma}(\sigma)}$ is trivial on each even cluster then ϵ_σ must be trivial on each yellow edge and is then itself trivial. This shows that $\underline{\Sigma}(-)$ is injective and, in particular, that we have $|\operatorname{Aut} T| \leq |\operatorname{Aut} \Sigma|$. To show that $\underline{\Sigma}(-)$ is an isomorphism, it suffices to show that we also have the reverse inequality. In light of (3), it suffices to show that the map $\underline{T}(-) : \operatorname{Aut} \Sigma \to \operatorname{Aut} \underline{T}(\Sigma)$ is injective, which we do below.

We now turn to (1)′, (2)′ and (3)′, for which we fix a cluster picture (X,Σ) and let $T = \underline{T}(\Sigma)$ be the associated BY tree. We first show (3)′. In the notation of Constructions 4.13 and 4.15, we'll show that the map $h : \mathfrak{s} \mapsto \mathfrak{s}_{v_{\mathfrak{s}}}$ is an isomorphism of cluster pictures. It is clear that it gives a bijection on proper clusters which preserves inclusion. To complete the argument, we prove by induction that $|\mathfrak{s}| = |\mathfrak{s}_{v_{\mathfrak{s}}}|$ for all proper clusters $\mathfrak{s}$.

First suppose that $\mathfrak{s}$ is a minimal proper cluster, i.e. all its children are singletons. Then by the definition of the genus of $\mathfrak{s}$ we have

$$|\mathfrak{s}| = \begin{cases} 2g(\mathfrak{s})+1 & \mathfrak{s} \text{ odd,} \\ 2g(\mathfrak{s})+2 & \mathfrak{s} \text{ even.} \end{cases}$$

Now $v_\mathfrak{s}$ is blue and has no children in T by minimality of $\mathfrak{s}$. Moreover, its parent edge is yellow if $\mathfrak{s}$ is even and blue if $\mathfrak{s}$ is odd. In particular we have

$$\deg_{T_b}(v_\mathfrak{s}) = \begin{cases} 1 & \mathfrak{s} \text{ odd,} \\ 0 & \mathfrak{s} \text{ even,} \end{cases}$$

whence $m_{v_\mathfrak{s}} = |\mathfrak{s}_{v_\mathfrak{s}}| = |\mathfrak{s}|$ as desired (here, as in Construction 4.15, for any vertex v of T we set $m_v = 2g(v) + 2 - \deg_{T_b}(v)$).

Next, take a proper cluster $\mathfrak{s} \in \Sigma$ and suppose that we have $|\mathfrak{s}'| = |\mathfrak{s}_{v_{\mathfrak{s}'}}|$ for all proper clusters $\mathfrak{s}' \subsetneq \mathfrak{s}$. In particular, to show that $|\mathfrak{s}| = |\mathfrak{s}_{v_\mathfrak{s}}|$, it suffices to show that $\mathfrak{s}$ and $\mathfrak{s}_{v_\mathfrak{s}}$ both have the same number of children of size 1. Now by Corollary 4.18 and the discussion on genera in the proof of (3), it follows that $\mathfrak{s}$ and $\mathfrak{s}_{v_\mathfrak{s}}$ have the same genus and parity. Combining the inductive hypothesis with the observation that the genus and parity of a cluster together determine how many odd children it has (cf. Definition 3.38) completes the proof.

Part (1)′ now follows upon combining (1) and (3)′.

Finally, we show (2)′. In light of (3)′ and the injectivity of the map in (2) shown above, it suffices to show that $\sigma \mapsto \underline{T}(\sigma)$ is injective (and this also completes the proof that the map in (2) is an isomorphism). This follows by noting that $\mathfrak{s} \mapsto v_\mathfrak{s}$ is a bijection between proper clusters of Σ and vertices of T, and that even sized clusters give vertices whose parent edge is yellow, so triviality of $\epsilon_{\underline{T}(\sigma)}$ forces triviality of ϵ_σ. □

4.3. Summary of constructions and examples. The one-to-one correspondence given by Theorem 4.2 is easy to use in practice. Table 4.20 summarizes the constructions (it follows from Constructions 4.4, 4.8, 4.13 and Proposition 4.19). We illustrate how to use it in Examples 4.21 and 4.22.

As in Remark 4.3, the hyperelliptic graph described in the third column of Table 4.2 comes with the decomposition $G_y = G_y^+ \coprod G_y^-$, where G_y^+ consists of all edges and vertices denoted with a $+$. (Thus in order to construct automorphisms of a cluster picture or a BY tree from that of a hyperelliptic graph, it is first necessary to pick such a decomposition.)

Recall that in a BY tree, from every vertex v there is a shortest path towards ∞. The *parent edge* of v is the edge a incident to v on this path. Similarly, for a vertex v of a hyperelliptic graph, a *parent edge* is an edge incident to v on one of the shortest paths towards ∞.

TABLE 4.20. Dictionary for the correspondence (open case, genus ≥ 1).

Cluster picture Σ	Open BY tree T	Open hyperelliptic graph G
cluster $\mathfrak{s}$ with $\lvert\mathfrak{s}\rvert > 2$ of genus g	vertex $w_\mathfrak{s}$ not a genus 0 leaf of genus g with parent edge $a_\mathfrak{s}$	ι-orbit of vertices $\{v_\mathfrak{s}\}$ or $\{v_\mathfrak{s}^+, v_\mathfrak{s}^-\}$ of genus g with ι-orbit of parent edge(s) $p_\mathfrak{s}$ or $\{p_\mathfrak{s}^+, p_\mathfrak{s}^-\}$
• $\mathfrak{s}$ odd	• $w_\mathfrak{s}$ blue $a_\mathfrak{s}$ blue	• $v_\mathfrak{s}$ with parent edge $p_\mathfrak{s}$
• $\mathfrak{s}$ even non-übereven	• $w_\mathfrak{s}$ blue $a_\mathfrak{s}$ yellow	• $v_\mathfrak{s}$ with parent edges $p_\mathfrak{s}^+, p_\mathfrak{s}^-$
• $\mathfrak{s}$ übereven	• $w_\mathfrak{s}$ yellow $a_\mathfrak{s}$ yellow	• $v_\mathfrak{s}^+, v_\mathfrak{s}^-$ with parent edges $p_\mathfrak{s}^+, p_\mathfrak{s}^-$
$\mathfrak{s} < \mathfrak{s}'$	$a_\mathfrak{s} = w_\mathfrak{s} w_{\mathfrak{s}'}$	$p_\mathfrak{s} = v_\mathfrak{s} v_{\mathfrak{s}'}$ or $p_\mathfrak{s}^\pm = v_\mathfrak{s} v_{\mathfrak{s}'}$ or $v_\mathfrak{s}^\pm v_{\mathfrak{s}'}$ or $v_\mathfrak{s} v_{\mathfrak{s}'}^\pm$ or $v_\mathfrak{s}^\pm v_{\mathfrak{s}'}^\pm$
$\delta(\mathfrak{s}, \mathfrak{s}') = \begin{cases} 2d & \mathfrak{s} \text{ odd} \\ d & \mathfrak{s} \text{ even} \end{cases}$	length $2d$	length d
twin $\mathfrak{t}$ $\mathfrak{t} < \mathfrak{s}$ $\delta(\mathfrak{t}, \mathfrak{s}) = d/2$	genus 0 leaf $u_\mathfrak{t}$ with yellow edge $a_\mathfrak{t}$ to $w_\mathfrak{s}$ of length d	an edge $\ell_\mathfrak{t}$ $\ell_\mathfrak{t} = v_\mathfrak{s} v_\mathfrak{s}$ or $\ell_\mathfrak{t} = v_\mathfrak{s}^+ v_\mathfrak{s}^-$ of length d
$(\sigma, \epsilon) \in \operatorname{Aut} \Sigma$	$(\sigma, \epsilon) \in \operatorname{Aut} T$	$\sigma \in \operatorname{Aut} G$
$\sigma : \mathfrak{s} \mapsto \mathfrak{s}_2$	$\sigma : w_\mathfrak{s} \mapsto w_{\mathfrak{s}_2}$	$\sigma : v_\mathfrak{s} \mapsto v_{\mathfrak{s}_2}$ or $\sigma : v_\mathfrak{s}^\pm \mapsto v_{\mathfrak{s}_2}^{\pm\epsilon(v_\mathfrak{s}^*)}$ $\sigma : p_\mathfrak{s} \mapsto p_{\mathfrak{s}_2}$ or $\sigma : p_\mathfrak{s}^\pm \mapsto p_{\mathfrak{s}_2}^{\pm\epsilon(v_\mathfrak{s}^*)}$
$\epsilon(\mathfrak{s})$, for $\mathfrak{s}$ even	$\epsilon(a_\mathfrak{s})$, for $a_\mathfrak{s}$ yellow	$\epsilon(v_\mathfrak{s}^*) \in \{\pm 1\}$
$\sigma : \mathfrak{t} \mapsto \mathfrak{t}_2$, $\epsilon(\mathfrak{t})$	$\sigma : u_\mathfrak{t} \mapsto u_{\mathfrak{t}_2}$, $\epsilon(a_\mathfrak{t})$	$\sigma : \ell_\mathfrak{t} \mapsto \epsilon(\ell_\mathfrak{t})\ell_{\mathfrak{t}_2}$, $\epsilon(\ell_\mathfrak{t}) \in \{\pm 1\}$ (where $-\ell_\mathfrak{t}$ is $\ell_\mathfrak{t}$ with reversed orientation)

EXAMPLE 4.21. Consider the open hyperelliptic graph G and the open BY tree T together with the automorphisms σ_G and (α_T, ϵ_T) of Examples 3.17 and 3.32:

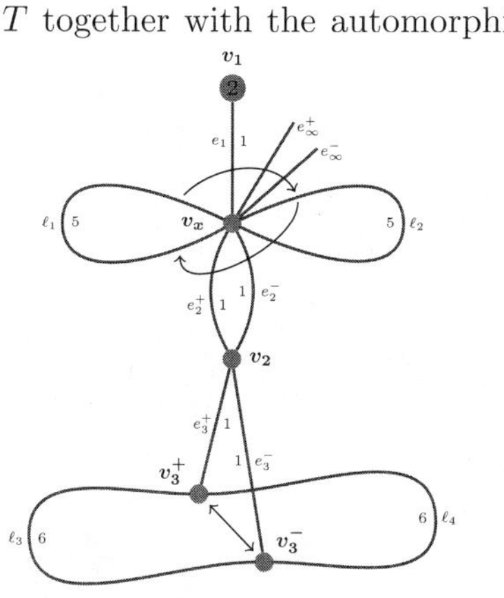

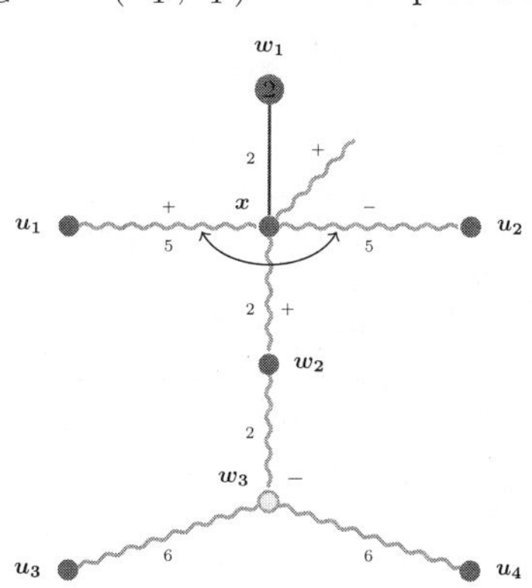

Following Table 4.20, from G we form the associated BY tree by creating:

- blue vertices w_1, x, w_2 of genera 2,0,0 corresponding to the ι-invariant vertices v_1, v_x, v_2 of G,
- a yellow vertex w_3 corresponding to $\{v_3^+, v_3^-\}$,
- blue genus 0 leaves u_1, u_2, u_3, u_4 corresponding to the loops ℓ_1, ℓ_2 and the ι-anti-invariant edges ℓ_3, ℓ_4,
- a blue edge from w_1 to x of length 2 corresponding to ι-invariant edge e_1,
- yellow edges from w_2 to x and w_3 to w_2 of lengths 2,2 corresponding to the ι-orbits $\{e_2^+, e_2^-\}$ and $\{e_3^+, e_3^-\}$,
- yellow edges from u_1 to x, u_2 to x, u_3 to w_3, u_4 to w_3 of lengths 5,5,6,6 corresponding to $\ell_1, \ell_2, \ell_3, \ell_4$,
- a yellow open edge from x to ∞ corresponding to the two open edges from v_x to ∞.

This construction precisely yields T.

To compute the automorphism (α, ϵ) corresponding to σ_G, consider the hyperelliptic graph G with the decomposition of its ι-permuted part $G_y = G_y^+ \coprod G_y^-$, where G_y^+ consists of $e_2^+, e_3^+, v_3^+, e_\infty^+$ and the top halves of ℓ_1, ℓ_2, ℓ_3 and ℓ_4 (call these $\ell_1^+, \ell_2^+, \ell_3^+, \ell_4^+$).

Then (α, ϵ) is given as follows: α acts on T by swapping u_1 and u_2, since σ_G swaps ℓ_1 and ℓ_2 and fixes all other ι-orbits, and

$$\begin{aligned}
\sigma_G(\ell_1^+) = \ell_2^+ &\implies \epsilon(u_1x) = 1,\\
\sigma_G(\ell_2^+) = \ell_1^- &\implies \epsilon(u_2x) = -1,\\
\sigma_G(e_\infty^+) = e_\infty^+ &\implies \epsilon(x\infty) = 1,\\
\sigma_G(e_2^+) = e_2^+ &\implies \epsilon(xw_2) = 1,\\
\sigma_G(v_3^+) = v_3^- &\implies \epsilon(u_3w_3) = \epsilon(u_4w_3) = \epsilon(w_2w_3) = \epsilon(w_3) = -1.
\end{aligned}$$

In other words, we obtain $(\alpha, \epsilon) = (\alpha_T, \epsilon_T)$.

From T, to form the associated hyperelliptic graph, we create:

- vertices v_1, v_x, v_2 fixed by ι of genera 2,0,0, corresponding to the blue vertices w_1, x, w_2,
- vertices v_3^+, v_3^- swapped by ι, of genera 0,0 corresponding to the yellow vertex w_3,
- two edges e_∞^+, e_∞^- swapped by ι, corresponding the open yellow edge from x to ∞,
- an edge from v_1 to v_x of length 1, corresponding to the blue edge from w_1 to x,
- two edges e_2^+, e_2^- swapped by ι, of lengths 1, corresponding to the yellow edge from the blue vertex w_2 to the blue vertex x,
- two edges e_3^+, e_3^- swapped by ι, of lengths 1, corresponding to the yellow edge from the yellow vertex w_3 to the blue vertex w_2,
- two ι-anti-invariant edges ℓ_1, ℓ_2 from v_x to itself of lengths 5, corresponding to the yellow edges from the blue vertex x to the genus 0 leaves u_1 and u_2,
- two ι-anti-invariant edges ℓ_3, ℓ_4 from v_3^+ to v_3^- of lengths 6, corresponding to the yellow edges from the yellow vertex w_3 to the genus 0 leaves u_3 and u_4.

We obtain precisely G.

The automorphism σ corresponding to $(\alpha_T, \epsilon_T) \in \mathrm{Aut}(T)$ is given by

- σ fixes v_1, v_x, v_2 and preserves $\{v_3^+, v_3^-\}$ since α_T fixes w_1, x, w_2 and w_3,
- σ fixes $e_\infty^+, e_\infty^-, e_2^+, e_2^-$ since α_T fixes w_2x, $x\infty$ and ϵ_T is 1 on both edges,
- σ swaps e_3^+ and e_3^-, v_3^+ and v_3^- and changes the orientation of ℓ_3 and ℓ_4 since α_T fixes w_3w_2, w_3, u_3w_3, u_4w_3 and ϵ_T is -1 on all of them,
- σ maps ℓ_1 to ℓ_2 since α_T maps u_1 to u_2 and $\epsilon_T(u_1x) = 1$,
- σ maps ℓ_2 to $-\ell_1$ since α_T maps u_2 to u_1 and $\epsilon_T(u_2x) = -1$ (where $-\ell$ denotes a loop ℓ with opposite orientation).

We then obtain $\sigma = \sigma_G$.

EXAMPLE 4.22. Consider the cluster picture Σ and the open BY tree T together with the automorphisms $(\alpha_\Sigma, \epsilon_\Sigma)$ and (α_T, ϵ_T) from Examples 3.49 and 3.32:

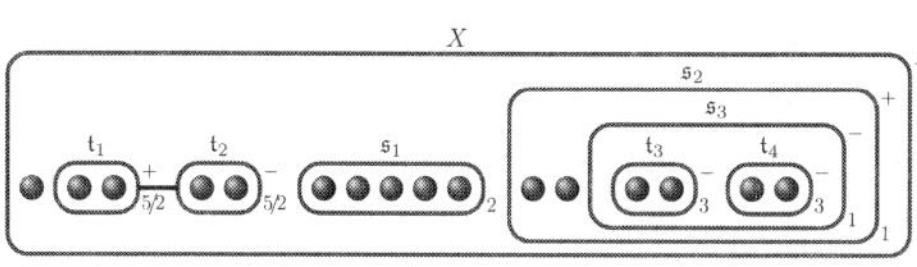

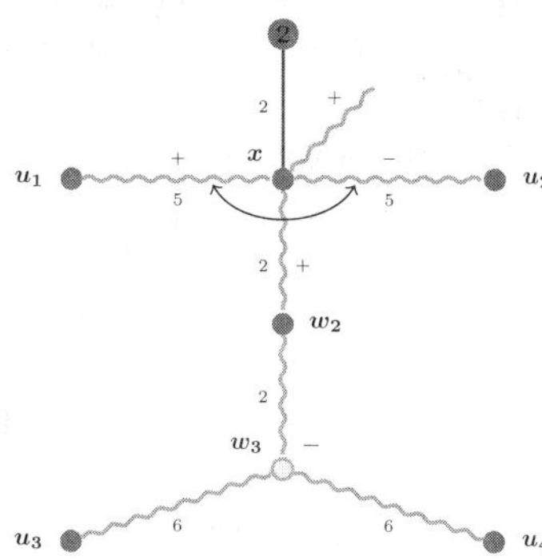

Following Table 4.20, from Σ we construct the associated BY tree by creating:

- blue vertices w_1, x, w_2 of genera 2, 0, 0, corresponding to the non-übereven clusters $\mathfrak{s}_1, X, \mathfrak{s}_2$ of size > 2,
- a yellow vertex w_3 of genus 0 corresponding to the übereven cluster $\mathfrak{s}_3$,
- blue genus 0 leaves u_1, u_2, u_3, u_4 corresponding to the twins $\mathfrak{t}_1, \mathfrak{t}_2, \mathfrak{t}_3, \mathfrak{t}_4$,
- a yellow open edge from x to ∞, since the top cluster X is even,
- a blue edge from w_1 to x of length 2, since $\mathfrak{s}_1$ is odd, $\mathfrak{s}_1 < X$ and $\delta(\mathfrak{s}_1, X) = 2$,
- yellow edges from u_1 to x, u_2 to x, w_2 to x, w_3 to w_2, u_3 to w_3 and u_4 to w_3 of lengths 5,5,2,2,6,6, since $\mathfrak{t}_1, \mathfrak{t}_2, \mathfrak{s}_2, \mathfrak{s}_3, \mathfrak{t}_3, \mathfrak{t}_4$ are even, $\mathfrak{t}_1 < X$, $\mathfrak{t}_2 < X$, $\mathfrak{s}_2 < X$, $\mathfrak{s}_3 < \mathfrak{s}_2$, $\mathfrak{t}_3 < \mathfrak{s}_3$, $\mathfrak{t}_4 < \mathfrak{s}_3$, and $\delta(\mathfrak{t}_1, X) = \frac{5}{2}$, $\delta(\mathfrak{t}_2, X) = \frac{5}{2}$, $\delta(\mathfrak{s}_3, \mathfrak{s}_2) = 1$, $\delta(\mathfrak{s}_2, X) = 1$, $\delta(\mathfrak{t}_3, \mathfrak{s}_3) = 3$ and $\delta(\mathfrak{t}_4, \mathfrak{s}_3) = 3$.

This yields T. The automorphism (α, ϵ) corresponding to $(\alpha_\Sigma, \epsilon_\Sigma)$ is given by

- α swaps u_1 and u_2 and fixes all other vertices, since α_Σ swaps $\mathfrak{t}_1$ and $\mathfrak{t}_2$ and fixes all other clusters,
- ϵ is 1 on the yellow edges $x\infty$, u_1x, w_2x, since $\epsilon_\Sigma(X) = \epsilon_\Sigma(\mathfrak{t}_1) = \epsilon_\Sigma(\mathfrak{s}_2) = 1$,
- ϵ is -1 on the yellow edges u_2x, w_3w_2, u_3w_3, u_4w_3, since $\epsilon_\Sigma(\mathfrak{t}_2) = \epsilon_\Sigma(\mathfrak{s}_3) = \epsilon_\Sigma(\mathfrak{t}_3) = \epsilon_\Sigma(\mathfrak{t}_4) = -1$,
- $\epsilon(w_3) = -1$ since it inherits the sign from its parent edge.

This shows precisely that $(\alpha, \epsilon) = (\alpha_T, \epsilon_T)$.

From T we construct the associated cluster picture by creating:

- clusters $\mathfrak{s}_1, \mathfrak{s}_2, \mathfrak{s}_3, \mathfrak{s}_x$ of size > 2 and genera 2,0,0,0 corresponding to vertices that are not genus 0 leaves w_1, w_2, w_3, w_x,
- twins $\mathfrak{t}_1, \mathfrak{t}_2, \mathfrak{t}_3, \mathfrak{t}_4$ corresponding to the genus 0 leaves u_1, u_2, u_3, u_4,
- $\mathfrak{s}_x$ is the top cluster X since $a_x = x\infty$,

- $\mathfrak{s}_1$ is an odd child of s_x of relative depth 2 since $a_1 = w_1w_x$ is blue of length 2,
- $\mathfrak{s}_2$ is an even non übereven child of s_x of relative depth 1 since w_2 is blue and $a_2 = w_2w_x$ is yellow of length 2,
- $\mathfrak{s}_3$ is an übereven child of $\mathfrak{s}_2$ of relative depth 1 since w_3 is yellow and $a_3 = w_3w_2$ is yellow of length 2,
- $\mathfrak{t}_1$ and $\mathfrak{t}_2$ are children of $\mathfrak{s}_x$ of relative depth $\frac{5}{2}$ since $a_{\mathfrak{t}_1} = u_1x$ and $a_{\mathfrak{t}_2} = u_2x$, both of length 5,
- $\mathfrak{t}_3$ and $\mathfrak{t}_4$ are children of $\mathfrak{s}_3$ of relative depth 3 since $a_{\mathfrak{t}_3} = u_3w_3$ and $a_{\mathfrak{t}_4} = u_4w_3$, both of length 6,
- $\mathfrak{s}_1$ has 5 roots since it is odd and of genus 2,
- $\mathfrak{s}_3$ has no roots outside $\mathfrak{t}_3, \mathfrak{t}_4$ since it is übereven,
- $\mathfrak{s}_2$ has 2 roots in addition to $\mathfrak{s}_3$ since it is even non-übereven of genus 0,
- $\mathfrak{s}_x$ has 1 root in addition to $\mathfrak{s}_1, \mathfrak{s}_2, \mathfrak{t}_1, \mathfrak{t}_2$ since it is even of genus 0.

This is precisely Σ.

The automorphism (α, ϵ) corresponding to (α_T, ϵ_T) is given by

- α swaps $\mathfrak{t}_1$ and $\mathfrak{t}_2$ and fixes all other clusters since α_T swaps u_1 and u_2 and fixes all other vertices,
- $\epsilon(X) = \epsilon(\mathfrak{t}_1) = \epsilon(\mathfrak{s}_2) = 1$, since ϵ_T is 1 on the parent edges of x, u_1 and w_2,
- $\epsilon(\mathfrak{t}_2) = \epsilon(\mathfrak{t}_3) = \epsilon(\mathfrak{t}_4) = \epsilon(\mathfrak{s}_3) = -1$, since ϵ_T is -1 on the parent edges of u_2, u_3, u_4, w_3.

This yields $(\alpha, \epsilon) = (\alpha_\Sigma, \epsilon_\Sigma)$.

5. One-to-one correspondence (closed case)

In this section we study the notion of equivalence for open hyperelliptic graphs, open BY trees and cluster pictures (see Definitions 3.13, 3.26 and 3.43). This enables us to prove a 'closed version' of the correspondences of Section 4. We also

- Explain how to explicitly obtain the core of an open hyperelliptic graph or open BY tree, and address the converse, namely which open BY trees have a specified core (Proposition 5.7, Corollary 5.10, Table 5.6);
- Identify a canonical representative in an equivalence class of BY trees, that corresponds to a balanced cluster picture (Remark 5.15);
- Describe 'principal clusters' in a cluster picture, that correspond to vertices in the core of the associated hyperelliptic graph;
- Interpret the moves for equivalence of cluster pictures (Definition 3.43) in terms of the associated BY tree (proof of Lemma 5.20).

The precise statement of the closed correspondence is as follows:

THEOREM 5.1. *There is a genus-preserving one-to-one correspondence, both in the metric and non-metric case, between*

(i) *Balanced cluster pictures up to isomorphism,*
(i′) *Cluster pictures up to equivalence,*
(ii) *Hyperelliptic graphs up to isomorphism,*
(ii′) *Open hyperelliptic graphs up to equivalence,*
(iii) *BY trees up to isomorphism,*
(iii′) *Open BY trees up to equivalence.*

Explicitly, the correspondence between hyperelliptic graphs and BY trees is given by the maps $\underline{G}$ and $\underline{T}$ of Section 4 and similarly, the correspondence between open BY *trees, open hyperelliptic graphs and cluster pictures is given by the maps* $\underline{G}, \underline{T}$ *and* $\underline{\Sigma}$. *Maps* $(ii') \to (ii)$ *and* $(iii') \to (iii)$ *are given by taking the core, whilst* $(i) \to (i)'$ *takes a balanced cluster picture to its equivalence class.*

In genus ≥ 2, *the correspondences (i')↔(ii)↔(iii) set up bijections between various invariants*[8] *as shown in Table* 5.3.

PROOF. The correspondence $(ii) \leftrightarrow (iii)$ was shown previously in Proposition 4.11. Lemma 5.5 combined with Proposition 5.7 (both shown below) and the open correspondence of Theorem 4.2 gives $(ii') \leftrightarrow (iii')$. The correspondences $(ii') \leftrightarrow (ii)$ and $(iii') \leftrightarrow (iii)$ follow from the definition of equivalence: Proposition 5.7 shows that the core exists and is unique, and Corollary 5.10 and Remark 5.11 show that each closed hyperelliptic graph (resp. closed BY tree) arises as the core of some open hyperelliptic graph (resp. open BY tree). Proposition 5.21 below shows that two cluster pictures are equivalent if and only if the associated BY trees are, which combined with the open correspondence of Theorem 4.2 gives $(i') \leftrightarrow (ii')$. Finally, we show in Lemma 5.25 that each equivalence class of cluster pictures contains a unique balanced one, giving $(i) \leftrightarrow (i')$.

The bijections between the invariants shown in Table 5.3 follow from the explicit description of the correspondences. □

The above result makes no mention of automorphism groups. Since equivalent objects need not have the same automorphism groups, the situation is more delicate (see e.g. Examples 3.14, 3.28 and 3.44). However, the correspondence for balanced cluster pictures is fairly clean:

THEOREM 5.2. *Let* (X, Σ) *be a balanced (possibly metric) cluster picture and* $\tilde{G}$ *be the core of the associated hyperelliptic graph* $\underline{G}(\Sigma)$. *Then the natural map* $\mathrm{Aut}(\Sigma) \to \mathrm{Aut}(\tilde{G})$, *given by applying* $\underline{G} \circ \underline{T}$ *and restricting to the core, is surjective. The kernel is trivial if* X *is übereven, and* C_2 *if* X *is non-übereven (generated by the trivial permutation with* $\epsilon(X) = -1$ *and all other signs* $+1$).

PROOF. This follows from Corollary 5.26 along with the comparison of equivalence between hyperelliptic graphs and BY trees (Lemma 5.5 and Proposition 4.11). □

5.1. Equivalence: BY trees, hyperelliptic graphs. Recall that the core of an open hyperelliptic graph (resp. open BY tree) is its maximal closed hyperelliptic subgraph (resp. closed BY subtree). Proposition 5.7 below shows that the core exists and is unique. Granted this, we single out vertices that come from the core:

DEFINITION 5.4 (Principal vertex).

(1) Let G be an open hyperelliptic graph of genus ≥ 2, with core $\tilde{G}$. A vertex v of G is *principal* if it corresponds to a vertex in the core (i.e. lies in the core and does not become a point on an edge upon removing $G \setminus \tilde{G}$).

(2) Let T be an open BY tree of genus ≥ 2, with core $\tilde{T}$. A vertex v of T is *principal* if it corresponds to a vertex in the core which is not a genus 0 leaf of $\tilde{T}$.

Still assuming Proposition 5.7, we now show that cores and principal vertices are preserved under the correspondence:

[8]The definition of cotwins and principal clusters is given in Definition 5.17.

TABLE 5.3. Dictionary for the one-to-one correspondence (closed case, genus ≥ 2).

Hyperelliptic graph	BY tree	Cluster picture
ι-invariant vertices of genus g	blue vertices of genus g that are not genus 0 leaves	principal non-übereven clusters of genus g
pairs of vertices $\{v, \iota(v)\}$	yellow vertices of genus 0	principal übereven clusters
loops of length 2δ	genus 0 leaves with the edge to a blue vertex of length 2δ	twins of distance δ to a principal non-übereven parent, and cotwins of distance δ to a non-übereven child, and X if $X = \mathfrak{s} \coprod \mathfrak{t}$ with $\mathfrak{s}$ even non-übereven, $\mathfrak{t}$ twin, of distance δ to each other
ι-anti-invariant edges of length 2δ, that are not loops	genus 0 leaves with the edge to a yellow vertex of length 2δ	twins of distance δ to a principal übereven parent, and cotwins of distance δ to an übereven child, and X if $X = \mathfrak{s} \coprod \mathfrak{t}$ with $\mathfrak{s}$ übereven, $\mathfrak{t}$ twin, of distance δ to each other
ι-invariant edges of length $\delta/2$	blue edges of length δ	odd principal clusters of distance δ to a principal parent, and X if $X = \mathfrak{s}_1 \coprod \mathfrak{s}_2$ with $\mathfrak{s}_1, \mathfrak{s}_2$ odd principal of distance δ to each other
pairs of edges $\{e, \iota(e)\}$ of length δ	yellow edges of length 2δ not incident to a genus 0 leaf	even principal clusters of distance δ to a principal parent, and X if $X = \mathfrak{s}_1 \coprod \mathfrak{s}_2$ with $\mathfrak{s}_1, \mathfrak{s}_2$ even principal of distance δ to each other

LEMMA 5.5. *The correspondences $\underline{G}$ and $\underline{T}$ between open hyperelliptic graphs and open BY trees preserve equivalence. Explicitly, if T is an open BY tree with core $\tilde{T}$, then the core of $\underline{G}(T)$ is isomorphic to $\underline{G}(\tilde{T})$. Similarly, for an open hyperelliptic graph G with core $\tilde{G}$, the core of $\underline{T}(G)$ is isomorphic to $\underline{T}(\tilde{G})$.*

Moreover, let G be an open hyperelliptic graph of genus ≥ 2, $T = \underline{T}(G)$ the associated BY tree and $\pi : G \to T$ the quotient map. Then π induces a bijection *between ι-orbits of principal vertices of G and principal vertices of T (here $\iota \in \mathrm{Aut}\, G$ is the hyperelliptic involution).*

PROOF. Recall that the maps $\underline{G}$ and $\underline{T}$ of Constructions 4.4 and 4.8 were defined for both open and closed objects. It follows from the explicit constructions that if G is an open hyperelliptic graph and G' a (closed) hyperelliptic subgraph, then $\underline{T}(G')$ is canonically a closed BY subtree of $\underline{T}(G)$, and that inclusion of subtrees is preserved by $\underline{T}$. Since the same is also true if we start with an open BY tree and consider the map $\underline{G}$ applied to subtrees of T, we obtain the result.

For the second part, it only remains to recall from Construction 4.4 that vertices of T either arise from ι-orbits of vertices of G, or from ι-anti-invariant edges, and that the latter are precisely the genus zero leaves of T. □

TABLE 5.6. Neighbourhoods of ∞ (genus ≥ 2).

Case	Connected component of ∞ of $G \setminus$ {principal vertices}	Connected component of ∞ of $T \setminus$ {principal vertices}	Configuration of maximal principal clusters of Σ
O	v	v	(0)
E	v	v	(E)
U	$v^{\pm}$	v	(U)
OE	v	v	● (E)
OU	$v^{\pm}$	v	● (U)
EE	v	v	●● (E)
EU	$v^{\pm}$	v	●● (U)
EO	v	v	● (0)
EOE	v	v	● ● (E)
EOU	$v^{\pm}$	v	● ● (U)
OxO	v w	v w	(0)(0)
ExE	v w	v w	(E)(E)
UxU	$v^{\pm}$ $w^{\pm}$	v w	(U)(U)
UxE	$v^{\pm}$ w	v w	(U)(E)
TxE	v	v	(●●)(E)
TxU	$v^{\pm}$	v	(●●)(U)

In every row the three entries correspond to one another under $\underline{G}, \underline{T}, \underline{\Sigma}$.
Column 2/3: labelled vertices are principal, others are not.
Column 2/3: dashed edges, and the leftmost vertex in EO/EOE/EOU, are not in the core.
Column 4: O/E/U denotes an odd/even/übereven maximal principal cluster.

PROPOSITION 5.7. *Let T be an open BY tree and G an open hyperelliptic graph.*
(1) The core $\tilde{T}$ of T (resp. $\tilde{G}$ of G) exists and is unique.
(2) $g(\tilde{T}) = g(T)$ and $g(G) = g(\tilde{G})$.

Now assume $g(T) \geq 2$ and $g(G) \geq 2$.

(3) There are 16 possibilities for the connected component of ∞ of $T \setminus \{$principal vertices$\}$, and 16 corresponding ones for hyperelliptic graphs. They are given in Table 5.6.
(4) In the notation of Table 5.6, *$\tilde{T}$ is obtained from T by starting at ∞ and removing*

- *(Cases O–EU) edge*
- *(Cases EO–EOU) edge, vertex, edge*
- *(Cases OxO–TxU) edge, vertex*

and $\tilde{G}$ from G by removing

- *(Cases O–U) edge(s)*
- *(Case EO) edges, vertex, edge*
- *(Case EOE, EOU) edges, vertex, edge, vertex*
- *(Cases OE–EU, OxO–TxU) edge(s), vertex/vertices.*

Here 'edge/edges' means removing the unique edge / ι-orbit of two edges from ∞ or the latest removed vertex; and 'vertex/vertices' means taking the vertex / ι-orbit of two vertices incident to the latest removed edge(s), and (a) when they have degree 1, removing them or (b) when they have degree 2, declaring them to not be vertices anymore (but interior points on the resulting merged edges instead).

PROOF. That the core of T (resp. G) exists and is unique follows by inspection in genus 0 and 1 (see Tables 3.1, 4.1) and, otherwise, from the explicit construction of the core detailed below. For that, we just do the $T \to \tilde{T}$ case, the $G \to \tilde{G}$ case being its translation via the correspondence between open hyperelliptic graphs and open BY trees.

Clearly, to get from T to $\tilde{T}$, the unique open edge (say, from v_0) needs to be removed. If v_0 becomes a valid BY tree vertex (see Definition 3.18), we are done. There are 7 such configurations depending on whether v_0 is a genus 0 leaf or not, and on the edge colours (Cases O–EU).

Otherwise, after the open edge is removed, v_0 must violate 3.18 (1) or (2). If it violates (1), v_0 is yellow (of genus 0) with exactly two other edges. There are 5 such configurations depending on the two adjacent vertices, and on the edge colours (Cases ExE–TxU). Declaring v_0 to not be a vertex gives a BY tree.

If v_0 violates (2), it must have become a blue genus 0 vertex with no yellow edges. Then it has one or two blue edges (by 3.18 (3)), and the removed edge was yellow (by (2)). When there are two blue edges, this is case OxO. Declaring v_0 to not be a vertex gives a BY tree. If there is one blue edge, we remove v_0 and this edge — these are Cases EO, EOE, EOU depending on the vertex adjacent to v_0.

The statement about the genus is clear, as no positive genus vertices are removed and the (relative) homology is unchanged. (See also Proposition 6.4.) □

An immediate corollary is the following.

COROLLARY 5.8. *Let T be an open BY tree with core $\tilde{T}$. Then the map* $\operatorname{Aut} T \to \operatorname{Aut} \tilde{T}$ *given by restriction of automorphisms has kernel C_2 in Cases*

EE-OxO and is injective in all other cases. Its image consists of all those automorphisms of $\tilde{T}$ fixing the point closest to ∞.

REMARK 5.9. The proof of Proposition 5.7 gives a straightforward way to compute the core in practice.

We now consider the opposite direction. That is, given a closed BY tree $\tilde{T}$, which open BY trees have $\tilde{T}$ as their core:

COROLLARY 5.10. *Let $\tilde{T}$ be a closed BY tree. Then an open BY tree T has core $\tilde{T}$ if and only if it is obtained from $\tilde{T}$ in one of the following ways:*

- *declaring a point on an edge of $\tilde{T}$ to be a vertex of genus 0 (and the same colour as the edge) and adding a yellow open edge at this vertex,*
- *adding a yellow open edge to a vertex of $\tilde{T}$,*
- *adding a blue open edge to a blue vertex v of $\tilde{T}$ which has $2g(v) + 2 >$ #blue edges at v,*
- *adding 'closed blue edge $\to$ genus 0 blue vertex $\to$ open yellow edge' to a blue vertex v of $\tilde{T}$ which has $2g(v) + 2 >$ #blue edges at v.*

PROOF. This is true by inspection in genus 0 and 1. Otherwise, it follows from Proposition 5.7 that the core of any open BY tree is obtained by removing either a single open edge, blue or yellow, or removing the configuration consisting of a closed blue edge followed by an open yellow edge. To prove the result, one just checks which conditions need to be satisfied at a point $x \in \tilde{T}$ in order for the graph given by glueing on one of these three configurations to be a valid open BY tree with core $\tilde{T}$. □

REMARK 5.11. Corollaries 5.8 and 5.10 have obvious analogues for hyperelliptic graphs via the correspondence. Since the statements are neater for BY trees and these are the ones we will use when comparing the notion of equivalence for BY trees/hyperelliptic graphs to that for cluster pictures, we have omitted them.

5.1.1. *Centres of BY trees.* Given a closed BY tree $\tilde{T}$, Corollary 5.10 can be viewed as describing the equivalence class of open BY trees with core (isomorphic to) $\tilde{T}$. In this subsection we single out a canonical representative in each equivalence class of open BY trees. To do this, we first single out a canonical 'centre' (either a vertex or edge) on a given closed BY tree. Glueing on an open yellow edge there gives the sought representative of the associated equivalence class of open BY trees.

The following purely graph theoretic lemma shows the existence of a 'centre' with respect to a weighting on the vertices of a tree. We omit the proof.

LEMMA 5.12. *Let T be a finite connected tree and $w : V(T) \to \mathbb{R}_{\geq 0}$ be a 'weight' function on the vertices of T such that each vertex of degree one or two has positive weight. For a subtree $T' \leq T$, set $w(T') = \sum_{v \in T'} w(v)$ and for each $v \in T$, define*

$$\phi(v) = \max\left\{w(T') \mid T' \textit{ is a connected component of } T \setminus \{v\}\right\}.$$

Then either

(1) $\min_{v \in T} \phi(v) < \frac{1}{2} w(T)$, *in which case the minimum is attained at a unique vertex of* T, *and all other vertices have* $\phi(v) > \frac{1}{2}\phi(T)$,

or

(2) $\min_{v \in T} \phi(v) = \frac{1}{2} w(T)$, *in which case the minimum is attained at precisely two vertices of* T, *and these vertices are adjacent.*

In case (1) *we call the minimising vertex the centre of* T *with respect to the weighting* ϕ. *In case* (2), *we define the centre to be the edge joining the two minimising vertices.*

DEFINITION 5.13. Let T be a closed BY tree. We define its *centre* to be the vertex or edge afforded by Lemma 5.12 applied to the weight function $w : V(T) \to \mathbb{Z}_{\geq 0}$ given by

$$w(v) = \begin{cases} 0 & v \text{ yellow}, \\ 2g(v) + 2 - \deg_{T_b}(v) & v \text{ blue}, \end{cases}$$

where here for a blue vertex v, $\deg_{T_b}(v)$ denotes the number of blue edges at v. Note that as w is invariant under all automorphisms of T, the centre of T is also.

REMARK 5.14. Let T be a closed BY tree. With w as in Definition 5.13 above, the same argument as in the proof of Proposition 4.17 gives

$$w(T) = 2g(T) + 2.$$

Similarly, for each $v \in T$ and each connected component T' of $T \setminus \{v\}$ we have

$$w(T') = \begin{cases} 2g(T') + 2 & \text{if the open edge of } T' \text{ is yellow}, \\ 2g(T') + 1 & \text{if the open edge of } T' \text{ is blue}. \end{cases}$$

REMARK 5.15. Glueing an open yellow edge to the centre of a closed BY tree $\tilde{T}$ gives (up to isomorphism) a canonical representative in the equivalence class of open BY trees having $\tilde{T}$ as their core. Letting T denote this representative, the natural map $\operatorname{Aut} T \to \operatorname{Aut} \tilde{T}$ given by restriction of automorphisms is surjective (this follows from Corollary 5.8 since all automorphisms of $\tilde{T}$ fix the centre). The kernel of the restriction homomorphism is trivial if the centre of $\tilde{T}$ is yellow, whilst if the centre of $\tilde{T}$ is blue then the kernel is isomorphic to C_2, generated by the automorphism of T which fixes all vertices, has sign -1 on (the component of T_y containing) the yellow open edge, and trivial sign on all other components of T_y (again see Corollary 5.8).

REMARK 5.16. Since equivalence is preserved by the correspondence between open BY trees and open hyperelliptic graphs, the construction above gives a canonical representative in each equivalence class of open hyperelliptic graphs.

5.2. Equivalence: Cluster pictures. We now turn to cluster pictures. We begin by describing the clusters that will correspond to principal vertices and to genus 0 leaves on the associated open BY tree,

5.2.1. *Principal clusters, twins and cotwins.*

DEFINITION 5.17. Let (X, Σ) be a cluster picture of genus $g \geq 2$. Recall that a cluster of size 2 is a *twin.*

A cluster $\mathfrak{s}$ is a *cotwin* if it has a child of size $2g$ whose complement is not a twin.

A proper cluster $\mathfrak{s}$ is *principal* if it is neither a twin nor a cotwin, and if $|\mathfrak{s}| = 2g + 2$ then $\mathfrak{s}$ has at least 3 children.

EXAMPLE 5.18. Each of the 7 pictures from Example 3.44 has exactly one principal cluster (smallest one of size ≥ 4), and either one twin or one cotwin.

REMARK 5.19. When $g(\Sigma) \geq 2$, either Σ has a unique maximal principal cluster $\mathfrak{s}$, of size $\geq 2g$, or $X = \mathfrak{s} \coprod \mathfrak{s}'$ is a union of two principal clusters. Marking a maximal principal cluster by O/E/U according to whether it is odd/even non-übereven/übereven gives 16 possible configurations, as listed in Table 5.6 (right column). Note that Σ has a cotwin if and only if it is in cases OE-EU or EOE-EOU.

LEMMA 5.20. *Let Σ be a cluster picture of genus ≥ 2, and $T = \underline{T}(\Sigma)$ the associated open BY tree, with core $\tilde{T}$. Then a cluster $\mathfrak{s} \in \Sigma$ is principal if and only if the associated vertex $v_{\mathfrak{s}}$ of T (see Construction 4.13) is a principal vertex. Moreover, $\mathfrak{s}$ corresponds to a genus 0 leaf in $\tilde{T}$ if and only if it is either a twin or a cotwin.*

PROOF. This follows from Construction 4.13 (describing the association $\mathfrak{s} \mapsto v_{\mathfrak{s}}$) and Table 5.6, which shows what is removed to obtain the core. Note that twins correspond to genus 0 leaves of T, each of which remains a genus 0 leaf in the core, whilst cotwins correspond to vertices of T which are not genus 0 leaves but become so when passing to the core. □

5.2.2. *Comparison with equivalence for open BY trees.* We now show that the maps $\underline{T}$ and $\underline{\Sigma}$ between cluster pictures and open BY trees preserve equivalence. Since (up to isomorphism) these maps are inverse to each other, it suffices to show the result for $\underline{T}$.

PROPOSITION 5.21. *Two cluster pictures (X, Σ), (X', Σ') are equivalent if and only if the corresponding BY trees $\underline{T}(\Sigma)$, $\underline{T}(\Sigma')$ are. The same holds in the metric case.*

PROOF. Again, this is true by inspection in genus 0 and 1 (see Tables 3.1, 4.1). Now suppose (X, Σ) is a cluster picture of genus ≥ 2, and $T = \underline{T}(\Sigma)$.

In the notation of Table 5.6, cluster pictures without clusters of size $2g$ or $2g+1$ fall into cases E, U, OxO, ExE, UxU and UxE. Possible moves (see Definitions 3.43, 3.45) between such cluster pictures are

$$\text{E/U} \underset{\text{(ii)}}{\overset{\text{(i)}}{\rightleftarrows}} \text{OxO/ExE/UxE/UxU.}$$

The corresponding BY tree in cases E and U has an open yellow edge attached to a principal vertex v. A principal child $\mathfrak{s}$ of X corresponds to an adjacent principal vertex $v_{\mathfrak{s}}$ and the moves (i) above are obtained by adding a 'cocluster' to $\mathfrak{s}$. The effect on T is to move the yellow edge to a point on the edge between v and $v_{\mathfrak{s}}$

(without changing the metric on the core in the metric case). The moves (ii) above are the inverse of this.

The only moves to and from cluster pictures that have a cluster of size $2g$ are

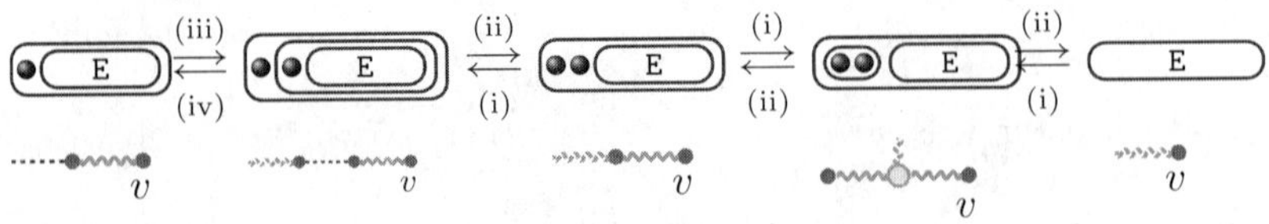

and

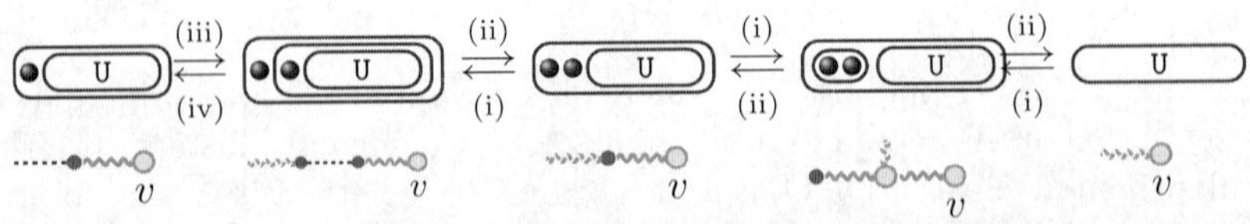

depending on whether the cluster of size $2g$ is non-übereven or übereven. By construction of T, moving along the chains transforms the 'tail at ∞' without altering the core, as shown above.

Finally, cluster pictures that have a cluster of size $2g+1$ but not of size $2g$ are cases O and EO. The only moves between them are

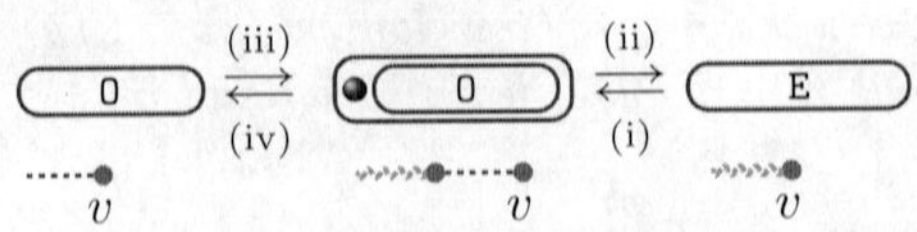

As before, these transform the tail at ∞ without altering the core, as shown.

This covers all possible moves between cluster pictures. Thus, equivalent cluster pictures yield BY trees with isomorphic cores (in other words, equivalent).

Conversely, if T are T' are open BY trees with the same core, the moves described above, the fact that BY trees are connected and Corollary 5.10 show that the associated cluster pictures are equivalent. □

REMARK 5.22. Incidentally, the proof of the proposition shows that every equivalence of cluster pictures can be broken up into moves (i)-(iv) *uniquely* (without going back). More precisely, fix an equivalence class of cluster pictures. Consider the graph whose vertices are cluster pictures in this class (up to isomorphism) and edges are given by moves (i)-(iv). Then this graph is a tree. See Table 5.23 for an example; here directions of arrows indicate moves (ii) and (iv).

TABLE 5.23. Example: An equivalence class of hyperelliptic graphs, open BY trees, and cluster pictures (Type $1\times \mathrm{I}_n$).

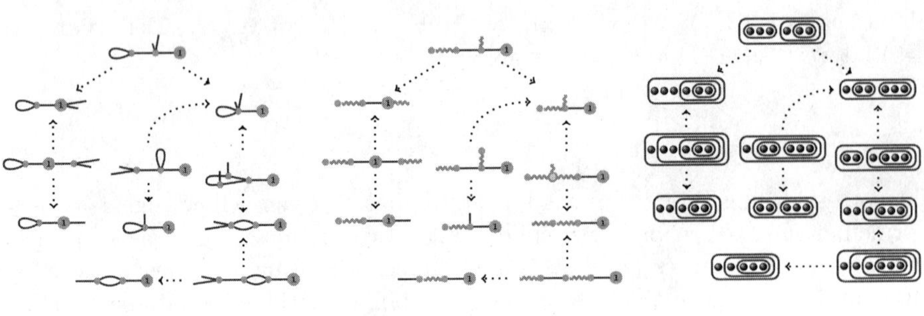

PROPOSITION 5.24. *Say that two proper clusters $\mathfrak{s}_1, \mathfrak{s}_2$ in a cluster picture (X, Σ) are adjacent if $\mathfrak{s}_1 < \mathfrak{s}_2$, $\mathfrak{s}_2 < \mathfrak{s}_1$, or $X = \mathfrak{s}_1 \coprod \mathfrak{s}_2$ with $\mathfrak{s}_1, \mathfrak{s}_2 < X$ and X is even. If (X, Σ) and (X', Σ') are equivalent cluster pictures of genus ≥ 2, then there is an adjacency preserving bijection*

$$\{principal\ \mathfrak{s} \in \Sigma\} \leftrightarrow \{principal\ \mathfrak{s}' \in \Sigma'\},$$
$$\{twins\ and\ cotwins\ \mathfrak{s} \in \Sigma\} \leftrightarrow \{twins\ and\ cotwins\ \mathfrak{s}' \in \Sigma'\}.$$

In the metric case, we can also insist that the bijection preserves distances between clusters.

PROOF. Let $T = \underline{T}(\Sigma)$ and let $\tilde{T}$ be its core. By Lemma 5.20, under the map $\mathfrak{s} \mapsto v_{\mathfrak{s}}$, principal clusters of Σ correspond to vertices of $\tilde{T}$ which are not genus 0 leaves, whilst twins and cotwins of Σ correspond to genus 0 leaves of $\tilde{T}$. First note that vertices $v_{\mathfrak{s}}$ and $v_{\mathfrak{s}'}$ are adjacent in the open BY tree T if and only if $\mathfrak{s} < \mathfrak{s}'$ or $\mathfrak{s}' < \mathfrak{s}$. That two vertices in the core $\tilde{T}$ are adjacent if and only if the corresponding clusters are now follows by consulting Table 5.6. In particular, any isomorphism from $\tilde{T}$ to the core of $\underline{T}(\Sigma')$ (one such necessarily exists by Proposition 5.21) induces a bijection as in the statement. □

5.2.3. *Centres and balanced cluster pictures.* Recall that a cluster picture (X, Σ) is balanced if $|X| = 2g + 2$ is even, there are either 0 or 2 clusters of size $g + 1$, and X is the only cluster of size $> g + 1$. (For instance, in Table 4.1, the second row in each of the three groups is balanced.)

LEMMA 5.25. *Every equivalence class of cluster pictures has (up to isomorphism) a unique balanced one. Under $\underline{T}$, it corresponds to the canonical representative of the associated equivalence class of open BY trees as defined in Remark 5.15.*

PROOF. Let Σ be a cluster picture of genus g, $T = \underline{T}(\Sigma)$ the associated BY tree, and $\tilde{T}$ its core. Then by Proposition 5.21, the cluster pictures equivalent to Σ are precisely those associated to the open BY trees obtained from $\tilde{T}$ by one of the operations of Corollary 5.10. Note that glueing an open blue edge to a vertex of $\tilde{T}$ results in a cluster picture of odd size (see Table 5.6) and such cluster pictures are not balanced. Similarly, glueing a closed blue edge (whose endpoint is blue of genus 0) followed by an open yellow edge onto a vertex of $\tilde{T}$ results in a cluster picture having a cotwin which again is not balanced.

Next, fix a vertex v of $\tilde{T}$ and consider the cluster picture (X', Σ') associated to the open BY tree obtained by glueing a yellow open edge to v. We have $|X'| = 2g+2$. Moreover, the children of X' are all of the form $\mathfrak{s}_{v'}$ for v' adjacent to v. For each such vertex, let $T_{v'}$ denote the connected component of $\tilde{T} \setminus v$ containing v'. By Remark 4.18, it follows that the size of $\mathfrak{s}_{v'}$ is equal to $w(T_{v'})$ where w is the weight function of Definition 5.13. It now follows from Lemma 5.12 that (X', Σ') is balanced if and only if v is the centre of $\tilde{T}$ (see also Remark 5.14).

Similarly, one sees that the cluster picture associated to the open BY tree obtained by glueing an open yellow edge to an existing edge of $\tilde{T}$ is balanced if and only if this edge is the centre of $\tilde{T}$. □

COROLLARY 5.26. *Let (X, Σ) be a balanced cluster picture, let $T = \underline{T}(\Sigma)$ be the associated open BY tree and let $\tilde{T}$ denote the core of T. Then the natural map $\operatorname{Aut}\Sigma \to \operatorname{Aut}\tilde{T}$, sending $\sigma \in \operatorname{Aut}\Sigma$ to the restriction of $\underline{T}(\sigma)$ to $\tilde{T}$, is surjective.*

Its kernel is trivial if X is übereven, and C_2 if X is non-übereven (generated by the trivial permutation with $\epsilon(X) = -1$ and all other signs $+1$).

PROOF. By Theorem 4.2, the map $\mathrm{Aut}\,\Sigma \to \mathrm{Aut}\,T$ sending σ to $\underline{T}(\sigma)$ is an isomorphism. The result now follows from Lemma 5.25 and the corresponding statement for BY trees discussed in Remark 5.15. □

6. The homology lattice Λ

In this section we study the lattices Λ attached to hyperelliptic graphs, BY trees and cluster pictures, along with their natural automorphism actions, and show that the correspondences identify them.

Action of automorphisms. Recall how to identify automorphism groups across the correspondence. Suppose G is a hyperelliptic graph, $T = \underline{T}(G)$ the associated BY tree, and $\Sigma = \underline{\Sigma}(T)$ the associated cluster picture. A choice of a section $s : G/\langle\iota\rangle \to G$ as in Construction 4.4 gives an isomorphism $\mathrm{Aut}\,G \to \mathrm{Aut}\,T$, which makes Λ_T an $\mathrm{Aut}\,G$-module. There is also a canonical isomorphism $\mathrm{Aut}\,T \to \mathrm{Aut}\,\Sigma$, independent of any choices, which makes Λ_Σ into an $\mathrm{Aut}\,G$-module as well. Finally, automorphisms of an open BY tree T act on the core $\tilde{T}$, and similarly for hyperelliptic graphs.

THEOREM 6.1 (Lattice correspondence). *If Σ is a cluster picture, then there are canonical* $\mathrm{Aut}\,\Sigma$*-equivariant isomorphisms*

$$\Lambda_{\widetilde{\underline{G}(\underline{T}(\Sigma))}} \cong \Lambda_{\underline{G}(\underline{T}(\Sigma))} \cong \Lambda_\Sigma \cong \Lambda_{\underline{T}(\Sigma)} \cong \Lambda_{\widetilde{\underline{T}(\Sigma)}}.$$

If T is a BY tree, then there are canonical $\mathrm{Aut}\,T$*-equivariant isomorphisms*

$$\Lambda_{\underline{G}(T)} \cong \Lambda_T \cong \Lambda_{\underline{\Sigma}(T)}.$$

If G is an open hyperelliptic graph, choose a section $s : G/\langle\iota\rangle \to G$. Then there are canonical $\mathrm{Aut}\,G$*-equivariant isomorphisms*

$$\Lambda_G \cong \Lambda_{\underline{T}(G)} \cong \Lambda_{\underline{\Sigma}(G)}.$$

For another section s', the two isomorphisms $\Lambda_G \cong \Lambda_{\underline{T}(G)}$ differ by $\psi_{s,s'}$ of Proposition 4.7. The isomorphism $\Lambda_{\underline{T}(G)} \cong \Lambda_{\underline{\Sigma}(G)}$ does not depend on the choice of s.

In the metric case, all isomorphisms preserve the pairings.

PROOF. This follows upon combining Lemmas 6.4 with Propositions 6.6 and 6.18 and Remark 6.7. □

COROLLARY 6.2. *Let (X, Σ) and (X', Σ') be equivalent cluster pictures. Then there is an isomorphism $\Lambda_\Sigma \cong \Lambda_{\Sigma'}$ which, in the metric case, preserves the respective pairings.*

PROOF. In both the metric and non-metric cases, the core $\tilde{T}$ of $\underline{T}(\Sigma)$ is an invariant of its equivalence class. Hence so is the associated lattice $\Lambda_{\tilde{T}}$. The result now follows from Theorem 6.1. □

REMARK 6.3. A proof of Corollary 6.2 without passing through the correspondence can be given by using Proposition 5.24.

6.1. Reduction to the closed case. We begin by showing that the homology groups of open hyperelliptic graphs (resp. open BY trees) are isomorphic to those of their core. As a consequence we will only consider the closed case after this subsection.

PROPOSITION 6.4. *Let G be an open hyperelliptic graph with core $\tilde{G}$. Then there is a canonical isomorphism*

$$H_1(G) \cong H_1(\tilde{G}),$$

equivariant for the action of $\operatorname{Aut} G$. *Similarly, if T is an open BY tree with core $\tilde{T}$ then there is a canonical isomorphism*

$$H_1(T, T_b) \cong H_1(\tilde{T}, \tilde{T}_b),$$

equivariant for the action of $\operatorname{Aut} T$. *(In the above, automorphisms of G (resp. T) act on $H_1(\tilde{G})$ (resp. $H_1(\tilde{T}, \tilde{T}_b)$) via their restriction to the core.)*

PROOF. It is easy to check the claim in genus 0 and 1 (see Tables 3.1, 4.1), so assume $g \geq 2$. From Table 5.6 we see that any open hyperelliptic graph admits a deformation retract onto its core. This induces the sought isomorphism on homology groups.

In the case of BY trees, Table 5.6 shows that T admits a deformation retract onto its core $\tilde{T}$ which induces a deformation retract from T_b to $\tilde{T}_b$. This induces maps $H_i(T, T_b) \to H_i(\tilde{T}, \tilde{T}_b)$ for each i. It also induces maps $H_i(T) \to H_i(\tilde{T})$ and $H_i(T_b) \to H_i(\tilde{T}_b)$ for each i which, being induced by deformation retracts, are isomorphsims. That the maps on relative homology groups are also isomorphisms now follows from the relative homology exact sequence and the 5-lemma.

The claim about the action of automorphisms is immediate since the deformation retracts act as identity on the core by definition. □

REMARK 6.5. It follows from Lemma 6.4 that the action of an automorphism of an open hyperelliptic graph G (resp. open BY tree T) on $H_1(G)$ (resp. $H_1(T, T_b)$) depends only on its restriction to the core.

6.2. Hyperelliptic graphs ↔ BY trees.

PROPOSITION 6.6. *Let G be a hyperelliptic graph, $T = \underline{T}(G)$ the associated BY tree and $s : T \to G$ a section to the quotient map $\pi : G \to T$. Then there is a canonical isomorphism*

$$H_1(G) \cong H_1(T, T_b),$$

equivariant for the action of $\operatorname{Aut} G$ *and, in the metric case, preserving the respective pairings. (In the above,* $\operatorname{Aut} G$ *acts on $H_1(T, T_b)$ via the isomorphism* $\operatorname{Aut} G \to \operatorname{Aut} T$ *determined by s (see Construction 4.4)).*

PROOF. We take the usual Δ-complex structure on T, so that the 0-simplices are the vertices and the 1-simplices are the edges. For the Δ-complex structure on G, we take the usual one, and then subdivide each ι-anti-invariant edge at the preimage of the associated vertex of T (which in each case is a genus 0 leaf). Define a map of complexes $C_\bullet(T) \to C_\bullet(G)$ given by $x \mapsto s(x) - \iota(s(x))$. Since the section s is continuous, this map is compatible with the boundary operators on each side (strictly speaking, we need to choose an orientation on the edges of T and G respectively to define the boundary operators; we do this in such a way that both ι and π are orientation-preserving). The kernel of this map of complexes is $C_\bullet(T_b)$ and, along with the quotient map $\pi : C_\bullet(G) \to C_\bullet(T)$ we obtain a short exact sequence of complexes

$$0 \to C_\bullet(T)/C_\bullet(T_b) \longrightarrow C_\bullet(G) \longrightarrow C_\bullet(T) \to 0.$$

Since $H_2(T) = 0 = H_1(T)$ (T is contractible) and $H_\bullet(T, T_b)$ is the homology of the leftmost complex, this sequence gives an isomorphism

$$H_1(T, T_b) \xrightarrow{\sim} H_1(G).$$

In the metric case, the pairings on $H_1(G)$ and $H_1(T, T_b)$ are induced by ones on $C_1(G)$ and $C_1(T)$ respectively. The scaling factors in Construction 4.4 are defined in such a way that the map in degree 1 in the short exact sequence above preserves these. Similarly, the compatibility with automorphism actions can be checked on the level of the map $C_1(T) \to C_1(G)$ (again, see Construction 4.4). □

REMARK 6.7. For each section $s : T \to G$, write $f_s : H_1(G) \to H_1(T, T_b)$ for the (inverse of the) isomorphism constructed in Proposition 6.6. Then given two sections s and s' one has

$$f_{s'} = \psi_{s,s'} \circ f_s$$

where $\psi_{s,s'} \in \operatorname{Aut} T$ is as in Definition 4.6.

REMARK 6.8. Another approach to proving the existence of the isomorphism of Proposition 6.6 is as follows. Writing G_b for the subgraph of G fixed by the hyperelliptic involution, one has $H_1(G) \cong \tilde{H}_0(G_b)$. This follows by enlarging the closed sets $T_b \subset T, G_b \subset G$ to their small open neighbourhoods $\tilde{T}_b, \tilde{G}_b$ and applying the Mayer–Vietoris sequence to the open sets $U = \tilde{G}_b \cup s(T_y)$ and $V = \tilde{G}_b \cup \iota(s(T_y))$ which cover G (note that $U \cap V$ is homotopic to G_b, whilst U and V individually are homotopic to the tree T). Since G_b and T_b are homeomorphic, we have $\tilde{H}_0(G_b) \cong \tilde{H}_0(T_b)$. The latter group is isomorphic to $H_1(T, T_b)$ via the relative homology sequence.

6.3. BY trees. In this subsection we give an explicit description of the first relative homology group of a (closed, possibly metric) BY tree with respect to its blue part. This will be necessary for establising the second isomorphism of Theorem 6.1 but may also be of independent interest. It will be convenient to work with rooted BY trees, i.e. BY trees with a distinguished point (which may be a vertex but could also be a point on an edge). Our description of the relative homology group will naturally be compatible with automorphisms of the BY tree which fix the root (but not general automorphisms).

6.3.1. *Rooted BY trees.*

DEFINITION 6.9. By a *rooted BY tree* we mean a pair (T, R) where T is a (closed, possibly metric) BY tree and R, the 'root', is a point on T (i.e. a vertex or a point on an edge). By an automorphism of a rooted BY tree we mean an automorphism of the underlying BY tree (complete with signs on yellow components) that preserves R. We write $\operatorname{Aut}_R T$ for the group of automorphisms of a rooted BY tree (T, R). Given a vertex $v \neq R$ of T, we refer to the unique edge of v in the direction of R as the *parent edge* of v.

REMARK 6.10. Every BY tree has a centre (in the sense of Definition 5.13), which is fixed by all automorphisms. Thus any BY tree can be made into a rooted BY tree in such a way that there is no difference between 'rooted' and 'non-rooted' automorphisms.

DEFINITION 6.11. Let T be an open BY tree and $\tilde{T}$ its core. Then we give $\tilde{T}$ the structure of a rooted BY tree by defining the root R to be the point on $\tilde{T}$ which is closest to ∞ in T.

REMARK 6.12. Let $\tilde{T}$ be the core of an open BY tree T of genus ≥ 2. Then whether or not the root R is a vertex depends on the type of the neighbourhood of infinity in T (cf. Table 5.6). Specifically, it is a vertex of $\tilde{T}$ in cases O–EOU, and lies on an edge otherwise. In cases O–E and OE–EOU, R is a blue vertex and in case U, R is a yellow vertex. In case OxO, R is a point on a blue edge and in cases ExE–TxU, R is a point on a yellow edge. Note that the automorphism group of an open BY tree $\tilde{T}$ maps surjectively to the automorphism group of its core, viewed as a rooted BY tree (the map being restriction of automorphisms). This is an isomorphism apart from cases EE-OxO, where the kernel is isomorphic to C_2 and acts trivially on $H_1(T, T_b)$ (see Corollary 5.8).

6.3.2. *Homology of a rooted BY tree.* Let (T, R) be a rooted (closed) BY tree with blue part T_b and yellow part T_y. For the purposes of computation, we make T into a Δ-complex in the usual way, save that, in the case that the root R lies on an edge, we subdivide this edge at the root R so that R becomes a 0-simplex. Moreover, we orient all edges so that they point towards R. Note that every automorphism of T is orientation preserving since it fixes the root.

For the rest of the section we adopt the following convention.

CONVENTION 6.13. If T' is a closed subtree of a rooted BY tree (T, R) then we take as the root of T' the point on T' closest to R in T (which is either R itself or a vertex of both T and T' (or both)).

DEFINITION 6.14. Let (T, R) be a rooted BY tree. For a connected component Y of T_y, let $\overline{Y}$ denote its closure in T, viewed as a rooted tree with root R_Y as in Convention 6.13. Write $\mathcal{L}_Y$ for the set of (non-root) leaves of $\overline{Y}$ and define $\mathcal{L}_T = \cup_Y \mathcal{L}_Y$ (note that this union is disjoint); equivalently,

$$\mathcal{L}_T = \{\text{blue vertices } v \neq R \text{ whose parent edge is yellow}\}.$$

For $v \in \mathcal{L}_T$, take Y for which $v \in \mathcal{L}_Y$ and define $\hat{v} = R_Y$. We then define the free $\mathbb{Z}$-module Π_T by

$$\Pi_T = \begin{cases} \mathbb{Z}[\mathcal{L}_T] & \text{if } R \text{ is blue,} \\ \left\{\sum_v \lambda_v v \in \mathbb{Z}[\mathcal{L}_T] \mid \sum_{\hat{v}=R} \lambda_v = 0\right\} & \text{if } R \text{ is yellow.} \end{cases}$$

In the metric case, we define a pairing on $\mathbb{Z}[\mathcal{L}_T]$ and Π_T by setting

$$\langle v_1, v_2 \rangle = \begin{cases} \delta(v_1 \wedge v_2, \hat{v}_1) & \text{if } \hat{v}_1 = \hat{v}_2, \\ 0 & \text{otherwise,} \end{cases}$$

where $v_1 \wedge v_2$ denotes the point (vertex or R (or both)) at which the unique paths in T from v_1 to R and v_2 to R meet. Automorphisms $(\sigma, \epsilon_\sigma) \in \mathrm{Aut}_R T$ act on $\mathbb{Z}[\mathcal{L}_T]$ and Π_T by setting, for a leaf $v_1 \in \overline{Y}$,

$$(\sigma, \epsilon_\sigma) \cdot v_1 = \epsilon_\sigma(Y)\sigma(v_1),$$

and extending linearly.

REMARK 6.15. For $v \in \mathcal{L}_T$, $\hat{v}$ is blue unless R is yellow and $\hat{v} = R$ (as all yellow vertices of T have only yellow edges).

REMARK 6.16. Note that the action of $\mathrm{Aut}_R T$ on $\mathbb{Z}[\mathcal{L}_T]$ is particularly simple, being given by signed permutations.

PROPOSITION 6.17. *Let (T, R) be a rooted BY tree. Then there is a canonical isomorphism*

$$\Pi_T \cong H_1(T, T_b),$$

equivariant for the action of $\mathrm{Aut}_R T$ *and, in the metric case, respecting the pairings.*

PROOF. First note that the map sending a connected component of T_b to its root gives a bijection between the set of connected components of T_b not containing R and the set $\mathcal{L}_T$. It now follows that

$$\mathrm{rk}\ \Pi_T = \mathrm{rk}\ H_1(T, T_b),$$

as Remark 3.24 shows that the rank of $H_1(T, T_b)$ is one less that the number of connected components of T_b.

Now consider the homomorphism $p : \mathbb{Z}[\mathcal{L}_T] \to C_1(T)/C_1(T_b)$ where for $v \in \mathcal{L}$, we define $p(v)$ as the shortest path in T from v to $\hat{v}$. Since this path is yellow by construction, p is injective. Note also that $C_1(T)/C_1(T_b)$ is a free $\mathbb{Z}$-module since $C_1(T)$ is the direct sum of $C_1(T_b)$ and $C_1(T_y)$. We claim that the image of p is a direct summand of $C_1(T)/C_1(T_b) = C_1(T_y)$. Indeed, for each $v \in \mathcal{L}_T$, its parent edge appears in $p(v)$ with multiplicity one, and does not appear in $p(v')$ for any $v \neq v' \in \mathcal{L}_T$. Thus the set $\{p(v) \mid v \in \mathcal{L}_T\}$ may be completed to a basis for $C_1(T_y)$ by adding in all yellow edges of T except the parent edges of vertices in $\mathcal{L}_T$.

Now denote by $\tilde{p}$ the restriction of p to Π_T, which is injective since p is. We claim that its image is contained in $H_1(T, T_b)$. Indeed, writing $d : C_1(T) \to C_0(T)$ for the boundary map as usual, for $v \in \mathcal{L}_T$ we have $d(p(v)) = \hat{v} - v$. Provided that $\hat{v}$ is blue (i.e. unless R is yellow and $\hat{v} = R$), we see that $d(p(v))$ lies in $C_0(T_b)$ in which case $p(v)$ is an element of $H_1(T, T_b)$. In particular, the claim holds for R blue. On the other hand, if R is yellow and $v, v' \in \mathcal{L}_T$ with $\hat{v} = \hat{v}' = R$, then $d(p(v - v')) = v' - v \in C_0(T_b)$. By the way we have defined Π_T when R is yellow, this proves the claim in this instance also.

Since Π_T is a direct summand of $\mathbb{Z}[\mathcal{L}_T]$ (it is the kernel of the homomorphism into $\mathbb{Z}$ sending $v \in \mathcal{L}_T$ with $\hat{v}$ yellow to 1, and all other elements of $\mathcal{L}_T$ to 0, which shows that $\mathbb{Z}[\mathcal{L}_T]/\Pi_T$ is torsion free), p is injective, and $p(\mathbb{Z}[\mathcal{L}_T])$ is a direct summand on $C_1(T_y)$, it follows that $\tilde{p}(\Pi_T)$ is a direct summand of $C_1(T_y)$ also. We are now in the following situation: we have inclusions of free, finite rank $\mathbb{Z}$-modules $\tilde{p}(\Pi_T) \subseteq H_1(T, T_b) \subseteq C_1(T_y)$, with $\tilde{p}(\Pi_T)$ a direct summand of $C_1(T_y)$ and $\mathrm{rk}\ \tilde{p}(\Pi_T) = \mathrm{rk}\ H_1(T, T_b)$. It now follows formally that $\tilde{p}(\Pi_T) = H_1(T, T_b)$ whence $\tilde{p}$ is an isomorphism. □

6.4. Cluster pictures ↔ BY trees.

PROPOSITION 6.18. *Let (X, Σ) be a cluster picture, $T = \underline{T}(\Sigma)$ the associated open BY tree and $\tilde{T}$ its core. Then there is a canonical isomorphism*

$$\Lambda_\Sigma \cong \Pi_{\tilde{T}},$$

equivariant for the action of $\mathrm{Aut}\,\Sigma = \mathrm{Aut}\,T$ *and, in the metric case, preserving the respective pairings. (Here, Π_T is as defined in Definition* 6.14.*)*

PROOF. We first claim that, under the correspondence between cluster pictures and open BY trees, the set $\mathcal{E}_\Sigma$ of Definition 3.47 (consisting of even, non- übereven clusters $\mathfrak{s} \neq X$) is identified with the set $\mathcal{L}_{\tilde{T}}$ (see Definition 6.14). Indeed, under the correspondence, even non-übereven clusters correspond to blue vertices of T whose parent edge is yellow. Now if $\mathfrak{s}$ is an even non-übereven cluster with associated vertex $v_\mathfrak{s}$, one easily checks from Table 5.6 that the parent edge is in the core if and only if $\mathfrak{s} \neq X$. Thus, as desired, $\mathcal{E}_\Sigma$ corresponds to the set of blue vertices in the core which are not the root, and whose parent edge is yellow.

It now follows that the map $\mathfrak{s} \leftrightarrow v_\mathfrak{s}$ induces an isomorphism between Λ_Σ and $\Pi_{\tilde{T}}$. Moreover, given $\mathfrak{s} \in \mathcal{E}_\Sigma$, corresponding to a leaf $v_\mathfrak{s}$ of a yellow component Y of $\tilde{T}$, one sees that $\hat{\mathfrak{s}}$ (cf. Definition 3.47) corresponds to the root R_Y of $\bar{Y}$. Indeed, this is immediate if $\mathfrak{s}$ is contained in some non-übereven cluster, and the case where no such cluster exists follows upon consulting Table 5.6. The claimed result now follows immediately from the definitions of Λ_Σ and $\Pi_{\tilde{T}}$, complete with pairing and action of automorphisms (using the identification of automorphism groups as given in Proposition 4.19). □

6.5. An example. Consider Examples 3.49, 3.32, 3.17, 4.21 and 4.22 with T and G open. We construct bases for $\Lambda_\Sigma, \Lambda_T, \Lambda_G$ that illustrate why the lattices are isomorphic.

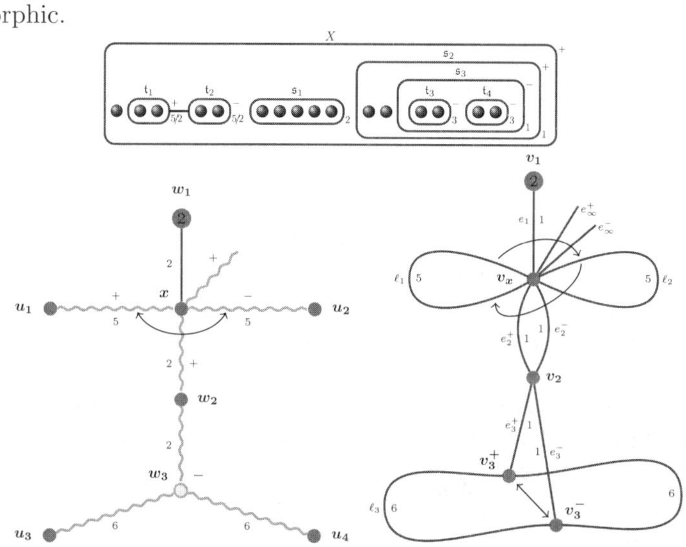

Recall that by definition

$$\Lambda_\Sigma = \mathbb{Z}\mathfrak{t}_1 + \mathbb{Z}\mathfrak{t}_2 + \mathbb{Z}\mathfrak{t}_3 + \mathbb{Z}\mathfrak{t}_4 + \mathbb{Z}\mathfrak{s}_2,$$

the basis vectors corresponding to the even, non-übereven clusters in Σ.

We produce a basis for $\Lambda_T = H_1(T, T_b)$ by looking at one yellow component at a time. For a given yellow component Y (with closure $\bar{Y}$), let z be the leaf of $\bar{Y}$ closest to ∞ and as basis vectors take the shortest paths from the other leaves to z. In our example, we have 4 yellow components and the basis vectors are $[u_1, x], [u_2, x], [u_3, w_2], [u_4, w_2], [w_2, x]$, which is precisely the basis in Example 3.32. In other words, the basis we've chosen is indexed by blue vertices with a yellow edge towards ∞, which exactly correspond to even, non übereven clusters of Σ (see Example 4.22).

To see why the two pairings coincide, consider for example the two paths $[u_3, w_2], [u_4, w_2]$. Their intersection is of length 2 which agrees with the distance $\delta(w_3, w_2)$. Since w_2 corresponds to $\mathfrak{s}_2 = \hat{\mathfrak{t}}_3 = \hat{\mathfrak{t}}_4$ and w_3 corresponds to $\mathfrak{s}_3 = \mathfrak{t}_3 \wedge \mathfrak{t}_4$,

$$2 = \langle [u_3, w_2], [u_4, w_2] \rangle = \delta(w_3, w_2) = \delta(\mathfrak{t}_3 \wedge \mathfrak{t}_4, \hat{\mathfrak{t}}_3) = \langle \mathfrak{t}_3, \mathfrak{t}_4 \rangle = 2.$$

Now consider the action of the automorphisms $\sigma_\Sigma = (\alpha_\Sigma, \epsilon_\Sigma)$ and $\sigma_T = (\alpha_T, \epsilon_T)$ on the lattices. Recall that $\sigma_\Sigma(\mathfrak{s}) = \epsilon_\Sigma(\mathfrak{s})\alpha_\Sigma(\mathfrak{s})$ by definition, e.g. $\sigma_\Sigma(\mathfrak{t}_2) = -\mathfrak{t}_1$ and that $\sigma_T([u, v]) = \epsilon_T(z)[\alpha_T(u), \alpha_T(v)]$, where $[u, v]$ is a path within one yellow component and z any point on that component, e.g. $\sigma_T([u_2, x]) = -[u_1, x]$.

A path in our basis of the form $[u, *]$ is sent to $[\alpha_T(u), *]$ which is another basis vector, e.g. $\alpha_T([u_2, x]) = [u_1, x]$. By construction, if u corresponds to the cluster $\mathfrak{s}$ then $\alpha_T(u)$ corresponds to $\alpha_\Sigma(\mathfrak{s})$ and $\epsilon_T(u) = \epsilon_\Sigma(\mathfrak{s})$, e.g. $\epsilon_T([u_2, x]) = -1 = \epsilon_\Sigma(\mathfrak{t}_2)$. It follows that the action of σ_T on Λ_T is the same as the action of σ_Σ on Λ_Σ.

Now consider the hyperelliptic graph G with the decomposition of its ι-permuted part $G_y = G_y^+ \coprod G_y^-$, where G_y^+ consists of $e_2^+, e_3^+, v_3^+, e_\infty^+$ and the top halves of ℓ_1, ℓ_2, ℓ_3 and ℓ_4 (call these $\ell_1^+, \ell_2^+, \ell_3^+, \ell_4^+$). To construct a basis for Λ_G in a systematic way

- construct G' from G_y by removing its edges towards ∞ and taking the closure. In our example $G' = G \setminus \{e_1, v_1, e_\infty^+, e_\infty^-\}$,
- the ι-invariant points remaining are v_x, v_2 and the mid-points of $\ell_1, \ell_2, \ell_3, \ell_4$,
- for ι-invariant points with an edge in G' towards ∞, create a loop by following G_y^+ towards ∞ to the next ι-invariant point and back via G_y^-; for our example we obtain the loops ℓ_1 (oriented clockwise) and ℓ_2 (anti-clockwise), and the loops $e_2^+ - e_2^-$, $\ell_3^+ + e_3^+ - e_3^- - \ell_3^-$, $\ell_4^+ + e_3^+ - e_3^- - \ell_4^-$, where we have oriented each edge and half-edge towards ∞. It is exactly the basis given in Example 3.17.

Under the $2:1$ map $G \to T$, these loops correspond to yellow paths from one blue vertex to another blue vertex which is closer to ∞. By construction, this gives the basis of Λ_T.

Since both pairings measure the length of the intersection of loops/paths, we get the same pairing on both spaces, e.g.

$$2 = \langle [u_3, w_2], [u_4, w_2] \rangle = \delta(w_3, w_2) = 2\delta(e_3^+) = \delta(e_3^+) + \delta(e_3^-) =$$

$$= \langle \ell_3^+ + e_3^+ - e_3^- - \ell_3^-, \ell_4^+ + e_3^+ - e_3^- - \ell_4^- \rangle = 2.$$

As in Example 4.21, the action of σ_G on $G/\langle \iota \rangle = T$ is that of α_T, in particular $\alpha_T([u_2, x]) = [u_1, x]$ corresponds to $\sigma_G(\ell_2) = \pm \ell_1$. Moreover $\epsilon_T([u_2, x]) = -1$ corresponds to $\sigma_G(\ell_2^+) = \ell_1^-$ so that $\sigma_T([u_2, x]) = -[u_1, x]$ corresponds precisely to $\sigma_G(\ell_2) = -\ell_1$.

7. Tamagawa groups of hyperelliptic graphs

In this section we study the Tamagawa group $\Phi(G)$ (see Definition 7.4) of a hyperelliptic graph G whose edge lengths are integers, and the corresponding group for BY trees and cluster pictures. In Proposition 7.10 we identify it with the graph-theoretic Jacobian of G along with automorphism action. We then give an explicit description of the 2-torsion in this group (Corollary 7.13).

7.1. Integral hyperelliptic graphs and Tamagawa groups.

DEFINITION 7.1. A (closed) metric hyperelliptic graph G is *integral* if all edge lengths are integers, unless G is the genus 1 circle graph from (3.3). In that exceptional case, we say that G is integral if the sum of lengths of its two edges is an integer.

A closed metric BY tree T is *integral* if $\underline{G}(T)$ is.

A metric cluster picture (X, Σ) is *integral* if the core of $\underline{G}(\underline{T}(\Sigma))$ is.

LEMMA 7.2. *A (closed) metric BY tree is integral if and only if all edges have integral length and all edges not incident to a genus 0 leaf have even length.*

A metric cluster picture (X, Σ) of genus ≥ 2 is integral if and only if

- $\delta(\mathfrak{s}, \mathfrak{s}') \in 2\mathbb{Z}$ *for* $\mathfrak{s}' < \mathfrak{s}$ *with* $\mathfrak{s}', \mathfrak{s}$ *principal,* $\mathfrak{s}'$ *odd,*
- $\delta(\mathfrak{s}, \mathfrak{s}') \in 2\mathbb{Z}$ *for* $\mathfrak{s}, \mathfrak{s}'$ *odd principal,* $X = \mathfrak{s} \bigsqcup \mathfrak{s}'$,
- $\delta(\mathfrak{s}, \mathfrak{s}') \in \mathbb{Z}$ *for* $\mathfrak{s}' < \mathfrak{s}$ *with* $\mathfrak{s}', \mathfrak{s}$ *principal,* $\mathfrak{s}'$ *even,*
- $\delta(\mathfrak{s}, \mathfrak{s}') \in \mathbb{Z}$ *for* $\mathfrak{s}, \mathfrak{s}'$ *even principal,* $X = \mathfrak{s} \bigsqcup \mathfrak{s}'$,
- $\delta(\mathfrak{s}, \mathfrak{t}) \in \frac{1}{2}\mathbb{Z}$ *for a twin* $\mathfrak{t} < \mathfrak{s}$,
- $\delta(\mathfrak{s}, \mathfrak{c}) \in \frac{1}{2}\mathbb{Z}$ *for* $\mathfrak{c}$ *a cotwin,* $\mathfrak{s} < \mathfrak{c}$ *of size* $2g$.

PROOF. This follows immediately from the (closed) case of the correspondence between cluster pictures, BY trees and hyperelliptic graphs as detailed in Section 5 (see in particular Table 5.3). □

REMARK 7.3. Given an integral hyperelliptic graph G, the paring on $H_1(G, \mathbb{Z})$ takes integer values. In particular, $H_1(G, \mathbb{Z})$ embeds in its abstract dual $H_1(G, \mathbb{Z})^\vee$ via $x \mapsto \langle x, - \rangle$. By using the correspondences of previous sections, it follows that if X is either an integral BY tree or an integral cluster picture, then Λ_X embeds into its abstract dual $\Lambda_X^\vee$ similarly.

DEFINITION 7.4. Let X be a hyperelliptic graph/BY tree/cluster picture and suppose that X is integral. Then we define the *Tamagawa group* $\Phi(X)$ as

$$\Phi(X) = \Lambda_X^\vee / \Lambda_X.$$

In each case, the action of $\operatorname{Aut} X$ on Λ_X induces an action on $\Phi(X)$.

THEOREM 7.5 (Tamagawa group correspondence). *Let (X, Σ) be an integral cluster picture, T (resp. G) the associated open BY tree (resp. open hyperelliptic graph) and $\tilde{T}$ (resp. $\tilde{G}$) its core. Then we have isomorphisms*

$$\Phi(\Sigma) \cong \Phi(\tilde{T}) \cong \Phi(\tilde{G}),$$

equivariant for the action of $\operatorname{Aut} \Sigma$.

PROOF. This follows immediately from Theorem 6.1. □

7.2. Jacobians of graphs. In this section, we show that the Tamagawa group of an integral hyperelliptic graph G of coincides with the Jacobian of a (combinatorial) graph $G_{\mathbb{Z}}$ canonically associated to G. Throughout this subsection, G has genus ≥ 2.

NOTATION 7.6. For an integral hyperelliptic graph G, we denote by $G_{\mathbb{Z}}$ the graph having the same underlying topological space as G, but whose set of vertices consists of those points on G which are an integer distance from the vertices of G. Equivalently, $G_{\mathbb{Z}}$ is the graph obtained by subdividing each edge e of G, say of length l, by adding $l-1$ vertices at intervals of unit distance along the edge, so as to obtain a new graph all of whose edge lengths are 1.

REMARK 7.7. We have $\operatorname{Aut} G_{\mathbb{Z}} = \operatorname{Aut} G$ and the discussion in Section 2.2.4 shows that $H_1(G_{\mathbb{Z}})$ is canonically isomorphic to $H_1(G)$, with the isomorphism preserving the respective pairings and automorphism actions.

In what follows we shall think of $G_{\mathbb{Z}}$ as being a finite combinatorial graph with unweighted edges (though possibly with loops and multiple edges) and disregard the genus marking. We now recall the definition of the Jacobian of such a graph.

DEFINITION 7.8. Let $\mathcal{G}$ be a finite combinatorial graph, possibly with loops and multiple edges (we reserve the letter 'G' for hyperelliptic graphs). Write $\operatorname{Div}(\mathcal{G})$ for the free $\mathbb{Z}$-module on the vertices $V(\mathcal{G})$ of $\mathcal{G}$ and $\operatorname{Div}^0(\mathcal{G})$ for the subgroup of $\operatorname{Div}(\mathcal{G})$ consisting of elements whose coefficients sum to zero. Contained in $\operatorname{Div}^0(\mathcal{G})$ is a certain full rank submodule $\operatorname{Prin}(\mathcal{G})$ consisting of 'principal divisors' which may be defined as follows:

For $v, v' \in V(\mathcal{G})$, set

$$v \cdot v' = \begin{cases} \deg(v) - 2 \,\#\text{ loops at } v & v = v', \\ -\#\text{ edges between } v \text{ and } v' & v \neq v', \end{cases}$$

and define a map $\alpha : \operatorname{Div}(\mathcal{G}) \to \operatorname{Div}(\mathcal{G})$ by, for $v \in V(\mathcal{G})$, setting

$$\alpha(v) = \sum_{v' \in V(\mathcal{G})} (v \cdot v')v',$$

and extending linearly. We then have $\operatorname{Prin}(\mathcal{G}) = \operatorname{im}(\alpha)$.

The *Jacobian* of $\mathcal{G}$ is then defined as

$$\operatorname{Jac}(\mathcal{G}) = \operatorname{Div}^0(\mathcal{G}) / \operatorname{Prin}(\mathcal{G}).$$

It is a finite abelian group and the action of $\operatorname{Aut} \mathcal{G}$ on $\operatorname{Div}(\mathcal{G})$ induces an action on $\operatorname{Jac}(\mathcal{G})$.

REMARK 7.9. The notion of the Jacobian of a graph appears in multiple places in the literature and is referred to by several different names, the most notable other ones being the sandpile group and the Picard group (see [**17**, Section 1.1] and the references therein for an overview of its occurence). Various equivalent definitions of the Jacobian also appear in the literature. The definition above is a slight variant of the one given in [**4**]. There the Jacobian is only defined for graphs without loops (but possibly with multiple edges). Our definition of $v \cdot v'$ above ensures that our definition of $\operatorname{Jac}(\mathcal{G})$ (along with automorphism action) agrees with that of the Jacobian of the graph obtained by removing all loop-edges from $\mathcal{G}$.

We also remark that in [**5**, Section 3.1] a generalisation of the Jacobian is defined for arbitrary metric graphs. In the case that G is an integral hyperelliptic graph the group $\operatorname{Jac}_{\mathbb{Z}}(G)$ in the notation of loc. cit. agrees with $\operatorname{Jac}(G_{\mathbb{Z}})$ as defined above. However, since the definition of the Jacobian of a metric graph is less elementary than that of a finite combinatorial graph, we have elected to work with $G_{\mathbb{Z}}$ rather than introduce $\operatorname{Jac}_{\mathbb{Z}}(G)$.

PROPOSITION 7.10. *Let G be an integral hyperelliptic graph of genus ≥ 2. Then there is a canonical isomorphism*

$$\mathrm{Jac}(G_{\mathbb{Z}}) \cong \Phi(G),$$

equivariant for the action of $\mathrm{Aut}\, G_{\mathbb{Z}} = \mathrm{Aut}\, G$.

REMARK 7.11. Several versions of this proposition, in various levels of generality (in particular, it is not specific to hyperelliptic graphs), appear in the literature though to the best of our knowledge the action of automorphism groups is not considered. We begin by reducing to the situation covered by [**4**, Theorem B.4] and deduce the compatibility of automorphisms from the explicit map defined there.

PROOF OF PROPOSITION 7.10. When defining $\Phi(G)$ as $H_1(G)^\vee/H_1(G)$ we are at liberty to choose the Δ-complex structure on G and we do so by taking the 0-simplices to consist of the vertices of G and the 1-simplices as the edges of $G_{\mathbb{Z}}$ (along with their endpoints). Note that if e is a loop-edge of $G_{\mathbb{Z}}$ then it generates an orthogonal direct summand of $H_1(G)$ and, having length 1, we have $\langle e, e\rangle = 1$. In particular we see that e does not contribute to the quotient $H_1(G)^\vee/H_1(G)$. Combining this observation with Remark 7.9, it suffices to prove the result under the assumption that $G_{\mathbb{Z}}$ contains no loops.

We are now in the situation covered by [**4**, Theorem B.4] and our choice of Δ-complex structure on G ensures that $H_1(G)$, as computed with this choice, coincides with their $\Lambda^1(G_{\mathbb{Z}})$. Following loc. cit., we now fix a base vertex $v \in V(G_{\mathbb{Z}})$ and define a map

$$f_v : \mathrm{Div}(G_{\mathbb{Z}}) \to H_1(G)^\vee/H_1(G) = \Phi(G)$$

as follows. Given $v' \in V(G_{\mathbb{Z}})$, pick a path $p_{v,v'}$ in $G_{\mathbb{Z}}$ from v to v' and set $f_v(v') = \langle p_{v,v'}, -\rangle$. Since $p_{v,v'}$ has integral length, its pairing with all elements of $H_1(G)$ is integral and so it defines a valid element of $H_1(G)^\vee$. Moreover, given two different choices of path from v to v', their difference is an element of $H_1(G)$ so f_v is independent of the choice of path $p_{v,v'}$. Restricting f_v to $\mathrm{Div}^0(G_{\mathbb{Z}})$ we obtain a map $f : \mathrm{Div}^0(G_{\mathbb{Z}}) \to \Phi(G)$ which does not depend on the choice of base vertex v. Then as asserted in loc. cit. (see [**2**, Proposition 7.2] for the proof), the map f induces the sought isomorphism $\mathrm{Jac}(G_{\mathbb{Z}}) \cong \Phi(G)$.

With the explicit map in hand, it is easy to check compatibility with automorphisms. Let $\sigma \in \mathrm{Aut}\, G$ and view it as an automorphism of $G_{\mathbb{Z}}$. Let $v, v' \in V(G_{\mathbb{Z}})$. Then f sends $v - v' \in \mathrm{Div}^0(G_{\mathbb{Z}})$ to $\langle p_{v,v'}, -\rangle$ where $p_{v,v'}$ is any path from v to v'. Now $\sigma(p_{v,v'})$ is a path from $\sigma(v)$ to $\sigma(v')$ and $\sigma(v) - \sigma(v') \in \mathrm{Div}^0(G_{\mathbb{Z}})$ is mapped by f to $\langle \sigma(p_{v,v'}), -\rangle$, which is the same as we obtain by acting by σ on $f(v - v')$. The result now follows since $\mathrm{Div}^0(G_{\mathbb{Z}})$ is generated by the elements $v - v'$ as v and v' range over the vertices of $G_{\mathbb{Z}}$. □

7.3. 2-torsion in the Tamagawa group. As an application of the correspondence between hyperelliptic graphs and BY trees, and the description of the group $H_1(T, T_b)$ for a BY tree T afforded by Proposition 6.17, we end this section by computing the 2-torsion in the Tamagawa group of a hyperelliptic graph.

The result for BY trees is the following.

THEOREM 7.12. *Let T be an integral BY tree of genus ≥ 2. Write S for the set of connected components of T_b, excluding the genus 0 leaves of T whose unique* *(necessarily yellow) edge has odd length;* $\operatorname{Aut} T$ *acts naturally on* S*. Then, as an* $\operatorname{Aut} T$*-module,*

$$\Phi(T)[2] \cong \begin{cases} 0 & S=\emptyset \text{ and } \operatorname{rk} H_1(T,T_b) \text{ even}, \\ \mathbb{Z}/2\mathbb{Z} & S=\emptyset \text{ and } \operatorname{rk} H_1(T,T_b) \text{ odd}, \\ \ker\left((\mathbb{Z}/2\mathbb{Z})[S] \xrightarrow{\text{sum}} \mathbb{Z}/2\mathbb{Z}\right) & \text{else}, \end{cases}$$

where 'sum' *denotes the sum of the coefficients map.*

PROOF. For the time being we will ignore the action of $\operatorname{Aut} T$, adding it back in at the end. Let $R \in T$ be a vertex and make T into a rooted BY tree by taking R to be the root. Now let Π_T be as in Definition 6.14, so that by Proposition 6.17 we have an isomorphism of $\mathbb{Z}$-lattices $\Pi_T \cong H_1(T,T_b)$ and, in particular, we have

$$\Phi(T) \cong \Pi_T^\vee/\Pi_T.$$

Noting that Π_T is torsion free as a $\mathbb{Z}$-module and applying the snake lemma to the commutative diagram with exact rows

$$\begin{array}{ccccccccc} 0 & \longrightarrow & \Pi_T & \longrightarrow & \Pi_T^\vee & \longrightarrow & \Phi(T) & \longrightarrow & 0 \\ & & \downarrow{\scriptstyle 2} & & \downarrow{\scriptstyle 2} & & \downarrow{\scriptstyle 2} & & \\ 0 & \longrightarrow & \Pi_T & \longrightarrow & \Pi_T^\vee & \longrightarrow & \Phi(T) & \longrightarrow & 0, \end{array}$$

it follows that we have

$$\Phi(T)[2] \cong \ker\left(\Pi_T/2\Pi_T \longrightarrow \Pi_T^\vee/2\Pi_T^\vee\right).$$

Suppose first that $S \neq \emptyset$, so that T has either a blue vertex which is not a genus 0 leaf, or that T has a genus 0 leaf whose unique edge has even length, and take R to be one such. Then since R is blue, we have $\Pi_T = \mathbb{Z}[\mathcal{L}_T]$ where the set $\mathcal{L}_T$ of Definition 6.14 (which depends on R) consists of the blue vertices different from R whose parent edge is yellow. The pairing on Π_T is given by

$$\langle v_1, v_2\rangle = \begin{cases} 0 & \widehat{v_1} \neq \widehat{v_2}, \\ \delta(v_1 \wedge v_2, \widehat{v_1}) & \text{else}, \end{cases}$$

for $v_1, v_2 \in \mathcal{L}_T$ (see Definition 6.14 for the definitions of $\hat{v}$ and $v_1 \wedge v_2$).

We claim that for any $v_1, v_2 \in \mathcal{L}_T$ we have

$$\langle v_1, v_2\rangle \equiv \begin{cases} 0 \pmod 2 & v_1 \neq v_2, \\ l_p(v) \pmod 2 & v_1 = v = v_2, \end{cases}$$

where for $v \in T$ (not equal to R) $l_p(v)$ is the length of its parent edge. To prove the claim, first take $v_1 \neq v_2 \in \mathcal{L}_T$ and assume that $\widehat{v_1} = \widehat{v_2}$ (otherwise v_1 pairs trivially with v_2 by definition and we are done). Then $v_1 \wedge v_2$ cannot be a leaf and so each (necessarily yellow) edge on the shortest path from $v_1 \wedge v_2$ to $\widehat{v_1}$ has even length. It now follows from Lemma 7.2 that $\delta(v_1 \wedge v_2, \widehat{v_1})$ is an even integer as desired. The case where $v_1 = v_2$ is similar: every edge in the path from v_1 to $\widehat{v_1}$ is a yellow edge not incident to a genus 0 leaf, save possibly for the parent edge of v_1.

Now for $v \in \mathcal{L}_T$, let ϕ_v denote the homomorphism in $\Pi_T^\vee$ dual to v (i.e. sending v to 1 and all other elements of $\mathcal{L}_T$ to 0). Then by the claim, we see that the map $\Pi_T/2\Pi_T \to \Pi_T^\vee/2\Pi_T^\vee$ is given by $v \mapsto l_p(v)\phi_v$. Since the set $\{\phi_v | v \in \mathcal{L}_T\}$ is a basis for $\Pi_T^\vee$ (the dual basis to the standard basis for $\Pi_T = \mathbb{Z}[\mathcal{L}_T]$), the kernel of this map is the $\mathbb{F}_2$-vector space having as basis the elements $v \in \mathcal{L}_T$ for which $l_p(v)$ is even. Now by Lemma 7.2, $v \in \mathcal{L}_T$ can only have $l_p(v)$ odd if it is a genus 0 leaf. In particular, writing O for the set of genus 0 leaves in T whose unique edge has odd length, an $\mathbb{F}_2$-basis for $\ker(\Pi_T/2\Pi_T \to \Pi_T^\vee/2\Pi_T^\vee)$ is given by the set $\mathcal{L}_T \setminus O$. The map sending $v \in \mathcal{L}_T$ to its connected component in T_b is a bijection onto the set of connected components of T_b not containing R. It follows that the isomorphism of the statement in the case $S \neq \emptyset$ holds abstractly.

To additionally obtain the isomorphism as $\operatorname{Aut} T$-modules, recall from the proof of Proposition 6.17 that the canonical isomorphism of $\mathbb{Z}$-lattices $\Pi_T \cong H_1(T, T_b)$ is given by sending $v \in \mathcal{L}_T$ to the unique shortest path $p(v)$ between v and $\hat{v}$. It follows from the argument above that, as $\operatorname{Aut} T$-modules, $\Phi(T)[2]$ is isomorphic to the subgroup of $H_1(T, T_b)/2H_1(T, T_b)$ generated by the set $\{p(v) \mid v \in \mathcal{L}_T \setminus O\}$. One checks that the isomorphism $H_1(T, T_b) \to \tilde{H}_0(T_b)$ coming from the relative homology sequence (Remark 3.24) identifies this subgroup with $\ker\left((\mathbb{Z}/2\mathbb{Z})[S] \xrightarrow{\text{sum}} \mathbb{Z}/2\mathbb{Z}\right)$. Since the map $H_1(T, T_b) \to \tilde{H}_0(T_b)$ is $\operatorname{Aut} T$-equivariant upon passing to quotients by multiplication by 2 on each side (since then we no longer need to consider orientation or signs) we are done.

Suppose now that $S = \emptyset$, so that all blue vertices of T are genus 0 leaves and each of their edges has odd length. Then T necessarily has a yellow vertex and now we take the root R to be one such. Note that now $\mathcal{L}_T$ is precisely the set of genus 0 leaves of T. Now T necessarily has precisely one yellow component, whence Π_T sits in a short exact sequence

$$0 \to \Pi_T \longrightarrow \mathbb{Z}[\mathcal{L}_T] \xrightarrow{\text{sum}} \mathbb{Z} \to 0,$$

the map 'sum' sending $\sum_{v \in \mathcal{L}_T} \lambda_v v$ to the sum of the λ_v. Since Π_T is a free $\mathbb{Z}$-module, the sequence remains exact after applying the functor $\operatorname{Hom}(-, \mathbb{Z})$ (which we denote $(-)^\vee$ for simplicity) and we obtain a commutative diagram with exact rows

$$\begin{array}{ccccccccc} 0 & \longrightarrow & \Pi_T & \longrightarrow & \mathbb{Z}[\mathcal{L}_T] & \longrightarrow & \mathbb{Z} & \longrightarrow & 0 \\ & & \downarrow & & \downarrow & & & & \\ 0 & \longleftarrow & \Pi_T^\vee & \longleftarrow & (\mathbb{Z}[\mathcal{L}_T])^\vee & \longleftarrow & \mathbb{Z}^\vee & \longleftarrow & 0 \end{array}$$

where here the two vertical maps are induced by the pairing (see Definition 6.14). Since each object in the diagram is torsion free, tensoring by $\mathbb{Z}/2\mathbb{Z}$ we obtain a commutative diagram with exact rows

$$\begin{array}{ccccccccc} 0 & \longrightarrow & \Pi_T/2\Pi_T & \longrightarrow & \mathbb{Z}[\mathcal{L}_T]/2\mathbb{Z}[\mathcal{L}_T] & \longrightarrow & \mathbb{Z}/2\mathbb{Z} & \longrightarrow & 0 \\ & & \downarrow & & \downarrow & & & & \\ 0 & \longleftarrow & \Pi_T^\vee/2\Pi_T^\vee & \longleftarrow & (\mathbb{Z}[\mathcal{L}_T])^\vee/2(\mathbb{Z}[\mathcal{L}_T])^\vee & \longleftarrow & \mathbb{Z}^\vee/2\mathbb{Z}^\vee & \longleftarrow & 0. \end{array}$$

The same argument as in the case $S \neq \emptyset$ shows that the rightmost of the two vertical maps sends $v \in \mathcal{L}_T$ to its dual vector ϕ_v (each $l_p(v)$ being odd) and as such is injective. Moreover, the map $\mathbb{Z}^\vee/2\mathbb{Z}^\vee \to (\mathbb{Z}[\mathcal{L}_T])^\vee/2(\mathbb{Z}[\mathcal{L}_T])^\vee$ sends the unique non-trivial element of $\mathbb{Z}^\vee/2\mathbb{Z}^\vee$ to $\sum_{v \in \mathcal{L}_T} \phi_v$. Combining exactness in the middle of the bottom row with the injectivity of the rightmost vertical map shows that $\sum_{v \in \mathcal{L}_T} v$ is the unique non-trivial element of the kernel of the map $\mathbb{Z}[\mathcal{L}_T]/2\mathbb{Z}[\mathcal{L}_T] \to \Pi_T^\vee/2\Pi_T^\vee$ given by composing the rightmost vertical map with the restriction map

$(\mathbb{Z}[\mathcal{L}_T])^\vee/2(\mathbb{Z}[\mathcal{L}_T])^\vee \to \Pi_T^\vee/2\Pi_T^\vee$. Further, the top sequence shows that $\sum_{v\in\mathcal{L}_T} v$ lies in $\Pi_T/2\Pi_T$ if and only if $|\mathcal{L}_T| = \mathrm{rk}\, H_1(T,T_b)+1$ is even. Thus

$$\ker(\Pi_T/2\Pi_T \to \Pi_T^\vee/2\Pi_T^\vee) \cong \begin{cases} 0 & \mathrm{rk}\, H_1(T,T_b) \text{ even}, \\ \mathbb{Z}/2\mathbb{Z} & \mathrm{rk}\, H_1(T,T_b) \text{ odd}, \end{cases}$$

which completes the proof of the theorem (note that we do not need to consider the action of $\mathrm{Aut}\, T$ in this case since the only possible action of any group on $\mathbb{Z}/2\mathbb{Z}$ is trivial). □

COROLLARY 7.13. *Let G be a hyperelliptic graph of genus ≥ 2. Write G_b for the subgraph of G fixed by the hyperelliptic involution and write $G(\mathbb{Z})$ for the set of points on G which are an integer distance from a vertex. Let $\mathcal{W}$ denote the set of connected components of G_b which contain a point of $G(\mathbb{Z})$. Then we have isomorphisms of* $\mathrm{Aut}\, G$*-modules*

$$\Phi(G)[2] \cong \begin{cases} 0 & \mathcal{W}=\emptyset \textit{ and } \mathrm{rk}\, H_1(G) \textit{ even}, \\ \mathbb{Z}/2\mathbb{Z} & \mathcal{W}=\emptyset \textit{ and } \mathrm{rk}\, H_1(G) \textit{ odd}, \\ \ker\Big((\mathbb{Z}/2\mathbb{Z})[\mathcal{W}] \xrightarrow{\text{sum}} \mathbb{Z}/2\mathbb{Z}\Big) & \textit{else}, \end{cases}$$

where 'sum' *denotes the sum of the coefficients map.*

PROOF. Let $T = \underline{T}(G)$ be the BY tree associated to G. Then the quotient map gives a homeomorphism from G_b to T_b. Let Z be a connected component of T_b. Then as yellow vertices of T have only yellow edges, Z necessarily contains a vertex of T. In fact, either Z contains a vertex which is not a genus 0 leaf, or $Z = \{v\}$ for a single genus 0 leaf v. In the first case, the preimage under π of this vertex is a vertex of $\pi^{-1}(Z)$. On the other hand, if $Z=\{v\}$ for a genus 0 leaf v, then $\pi^{-1}(v)$ is not a vertex of G but the midpoint of an ι-anti-invariant edge. In particular, $\pi^{-1}(v) \in G(\mathbb{Z})$ if and only if the parent edge of v has even length. Thus, the set $\mathcal{W}$ corresponds under π to the set of connected components of T_b excluding the genus 0 leaves whose parent edge has odd length. Since the number of components of T_b is equal to $\mathrm{rk}\, H_1(T,T_b)-1$ (see Remark 3.24), the result now follows from Theorem 7.12, along with usual identification of $\mathrm{Aut}\, G$ with $\mathrm{Aut}\, T$ (choices of section here are irrelevant since all signs act trivially on the objects involved). □

8. Classification of semistable types and naming convention

8.1. Types of BY trees (and hyperelliptic graphs/cluster pictures). We propose a naming scheme for cluster pictures, (open) BY trees and (open) hyperelliptic graphs. We define it for BY trees and transport to the other two categories via the one-to-one correspondence.

NOTATION 8.1. Let T be an *open BY tree.* For the edges, we use

$\cdot$	blue edge
$:$	yellow edge
$\cdot_{d/2}$, $:_d$	edge of length d

and for the vertices

U	yellow vertex
I or 0	blue vertex of genus 0
1,2,3,...	blue vertex of genus 1,2,3,...

To define the notation ('Type') of T itself, let e be its open edge, say incident to a vertex v. As a topological space, T decomposes as a disjoint union

$$T = \{v\} \cup \{e\} \cup t_1 \cup \ldots \cup t_k \cup T_1 \cup \ldots \cup T_n.$$

where the t_i are open trees '' (a blue vertex of genus 0 with one open yellow edge, say of length n_i), and the T_i are the other connected components of $T\backslash\{e, v\}$; they are open BY trees. Then we define, inductively,

$$\mathrm{Type}(T) = [e][v]_{n_1,\ldots,n_k}\,\mathrm{Type}(T_1)\cdots\mathrm{Type}(T_n),$$

where $[e]$ is the notation for the edge e, as above, $[v]$ is the notation for the vertex v, as above. To avoid ambiguity, when $n > 0$, unless T is the full tree that we are interested in, we bracket everything after $[e]$ and write

$$\mathrm{Type}(T) = [e]\big([v]_{n_1,\ldots,n_k}\,\mathrm{Type}(T_1)\cdots\mathrm{Type}(T_n)\big).$$

In the non-metric case, the subscripts n_i are placeholders instead of lengths whose purpose is only to record the number of genus 0 leaves[9]. See Example 8.3 below.

NOTATION 8.2. For a *closed BY tree* $\tilde{T}$, recall from Remark 5.15 that there is a canonical open BY tree T with core $\tilde{T}$ obtained by glueing a yellow open edge e_0 to the centre of $\tilde{T}$. We let $\mathrm{Type}(\tilde{T})$ to be the name of T with $[e_0]$ omitted. To emphasize the configurations for which the centre is an edge, we also use an alternative notation

$$\begin{array}{ccc} 0\cdot_m \mathrm{Type}(T_1)\cdot_n \mathrm{Type}(T_2) & \longmapsto & \mathrm{Type}(T_1)\times_{m+n}\mathrm{Type}(T_2) \\ \mathrm{U}:_m \mathrm{Type}(T_1):_n \mathrm{Type}(T_2) & \longmapsto & \mathrm{Type}(T_1)\circ_{m+n}\mathrm{Type}(T_2). \end{array}$$

The symbol $\times$ or $\circ$ can only appear once in the name, so the names of T_1 and T_2 do not have to be bracketed.

EXAMPLE 8.3. Here are a few examples:

T (open)	Type(T)	$\tilde{T}$ (closed)	Type($\tilde{T}$)	genus
	$\cdot 1$		1	1
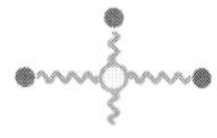	$:\mathrm{U}_{n,m,k}$	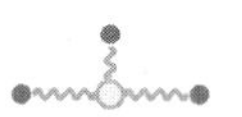	$\mathrm{U}_{n,m,k}$	2
	$\cdot 0:1$		1_n	2
	$:0:(0\cdot \mathrm{I}_n)$		$\mathrm{I}_n \times \mathrm{I}_m$	2

EXAMPLE 8.4. In genus 2 the 7 balanced configurations (Table 9.5) are

$$2 \quad 1_n \quad \mathrm{I}_{n,m} \quad \mathrm{U}_{n,m,k} \quad 1\times 1 \quad 1\times \mathrm{I}_n \quad \mathrm{I}_n\times \mathrm{I}_m$$

EXAMPLE 8.5. The BY tree in Example 3.32 has Type $:0_{5,5}\cdot_1 2:_2(0:_2\mathrm{U}_{6,6})$ and its core Type $0_{5,5}\cdot_1 2:_2(0:_2\mathrm{U}_{6,6})$.

REMARK 8.6. All the main invariants of a BY tree T can be seen from the type name: vertices which are not genus 0 leaves are the capitals I, U, 0, 1, 2, ... in the name. The genus 0 leaves are their subscripts; say there are k of them. Edges not incident to genus 0 leaves are the symbols $\cdot$, $:$, $\times$ and $\circ$. The genus of T is the sum of all (non-subscript) numbers in the type plus

$$\mathrm{rk}\,\Lambda_T = \#\text{colons} + k \text{ - } \#\text{'U's - } \#\text{'o's}$$

and similarly for open BY trees, by ignoring the first symbol if it is ':'.

[9]this agrees with Kodaira and Namikawa-Ueno types for semistable curves of genus 1 and 2

8.2. Automorphisms.

NOTATION 8.7 (Automorphism). Let T be a (possibly open) BY tree, and $E = \{e_1, ..., e_n\} \subset E(T)$ an ordered subset of its edges, preserved by $\operatorname{Aut} T$ (or some subgroup that we care about). By default, we take

$$E = \{\text{all closed edges}\},$$

ordered as follows: if $e = \{v_1, v_2\}$, $e' = \{v_1', v_2'\}$, we check which $v \in \{v_1, v_2, v_1', v_2'\}$ comes last in the name $\operatorname{Type}(T)$ (as a capital letter or its subscript); if it is v_i' then e' comes after e, and vice versa.

We then write σ as a permutation on the indices of the blue edges and $\pm$indices of the yellow edges, where the sign of $\pm\sigma(e)$ is determined by $\epsilon(e)$.

EXAMPLE 8.8. Take a BY tree of genus 2 that corresponds to two nodal genus 0 curves meeting at a point, with two nodes of the same depth:

Type	T	$\Sigma(T)$	$\underline{G}(T)$	$\operatorname{Aut}(T)$
$\mathrm{I}_n \times \mathrm{I}_n$				D_4

Its automorphism group is D_4 (order 8). To write its elements we order the edges as above:

1 2 3

As signed permutations, the elements of $\operatorname{Aut} T$ are

$$\text{id},\ (-1\,1),\ (-3\,3),\ (-1\,1)(-3\,3),$$

$$(13)(-1\,-3),\ (1\,-3)(-1\,3),\ (1\,-3\,-1\,3),\ (1\,3\,-1\,-3)$$

On the corresponding hyperelliptic graph

the element $(-1\,1)$ reflects the left loop in the x-axis, $(-3\,3)$ reflects the right loop, and $(1\;3\;-1\;-3)$ sends the left loop to the right one keeping the orientation and the right one to the right one reversing the orientation. (In this example, it is also reasonable to take $E = \{\text{yellow edges}\}$ instead of all edges.)

8.3. BY trees with an automorphism. For semistable hyperelliptic curves over local fields, it is important to keep track of the action of Frobenius. Therefore we need a naming convention for BY trees with a distinguished automorphism.

NOTATION 8.9 (Type with an automorphism). We incorporate the action of an automorphism on the edges and the signs into the type name, as follows.

Suppose T is an open BY tree, and $\phi \in \operatorname{Aut} T$. As in Notation 8.1, write

$$T = \{v\} \cup \{e\} \cup t_1 \cup \ldots \cup t_k \cup T_1 \cup \ldots \cup T_n.$$

We extend the notation

$$\mathrm{Type}(T) = [e][v]_{n_1,\dots,n_k}\,\mathrm{Type}(T_1)\cdots\mathrm{Type}(T_n).$$

to a notation for $\mathrm{Type}(T,\phi)$ as follows:

The automorphism ϕ permutes the t_i and the T_i, and we rearrange them, if necessary, by ϕ-orbits. In other words, each ϕ-orbit, say of length m, is a block $t_i, \dots, t_{i+m-1}$ or $T_i, \dots, T_{i+m-1}$.

(1) For each ϕ-orbit $t_i, \dots, t_{i+m-1}$ replace commas in $n_i, \dots, n_{i+m-1}$ by $\sim$, a symbol for 'are in the same ϕ-orbit'.
(2) Similarly for each ϕ-orbit $\mathrm{Type}(T_i)\cdots\mathrm{Type}(T_{i+m-1})$, let ϕ^k be the smallest power of ϕ that stabilises T_i. Instead of $\mathrm{Type}(T_j)$ write $\mathrm{Type}(T_j,\phi^k)$, defined inductively, with the first edge symbol '.' or ':' replaced by $\sim$ for $j > i$.
(3) For a closed BY tree (see Notation 8.2), we similarly replace $\times, \circ$ by $\tilde{\times}, \tilde{\circ}$ when the endpoints of the central edge are swapped by ϕ.

We decorate the type with signs as follows. For a vertex v, let ϕ^{k_v} be the smallest power of ϕ that stabilizes v and ϵ_v be the sign of ϕ^{k_v} on its parent edge should it be yellow. For each vertex v that is first in its ϕ-orbit (ordered by appearance in the type):

- if v is yellow such that its parent edge does not lead to a yellow vertex, decorate the symbol for v in the type name with the superscript ϵ_v,
- if v is blue, let $w_1, .., w_s$ be the blue vertices joined to v by a yellow edge leading away from ∞ (ordered by appearance in the type). For each w_i that is first in its ϕ^{k_v}-orbit, decorate the symbol for v in the type name with the superscript ϵ_{w_i}. By convention, these superscripts appear in the same order as the w_i's and are separated by commas.

Finally if the open edge is yellow and incident to a blue vertex, decorate the initial colon with the sign of ϕ on the open edge.

In the case of a closed BY tree $\tilde{T}$ with an automorphism ϕ, define the type $(\tilde{T},\phi)$ to be the type (T,ϕ') with the first dot or colon (and their sign) deleted, where T is obtained from $\tilde{T}$ by glueing a yellow open edge to its centre, and ϕ' extends ϕ. We use an analogous convention as in Notation 8.1 for the cases where the centre is an edge. In these cases, we decorate $\circ$ with the sign of the initial U and we write $\tilde{\circ}$, $\tilde{\times}$ if T_1 and T_2 are swapped by ϕ.

One can check that, in the open or closed case, (T,ϕ) and (T',ϕ') get the same notation if and only if they are isomorphic as pairs, that is there is an isomorphism $\psi : T \to T'$ such that $\psi \circ \phi = \phi' \circ \psi$.

EXAMPLE 8.10 (Elliptic curves). Let T be one of the BY trees

 or

The associated cluster pictures are all possible ones of size 3 (see Table 4.1)

and they correspond to elliptic curves with good and multiplicative reduction. If $\phi \in \mathrm{Aut}\,T$, then $\mathrm{Type}(T,\phi)$ is $\cdot 1$ in the first case, and $\cdot \mathrm{I}_n^+, \cdot \mathrm{I}_n^-$ in the second case, depending on the ϕ-action on the yellow edge. When ϕ is Frobenius, $\cdot \mathrm{I}_n^+$ is 'split multiplicative' and $\cdot \mathrm{I}_n^-$ 'non-split multiplicative' reduction. If one is only interested in elliptic curves and not general curves of genus 1, one could omit the first dot and write the types as 1, I_n^+ and I_n^-.

EXAMPLE 8.11 ($\mathrm{I}_n \times \mathrm{I}_n$). In Example 8.8, for the 5 conjugacy classes of automorphisms $\phi \in D_4 = \operatorname{Aut} T$ the label Type(T, ϕ) is

$$\mathrm{I}_n^+ \times \mathrm{I}_n^+, \qquad \mathrm{I}_n^+ \times \mathrm{I}_n^-, \qquad \mathrm{I}_n^- \times \mathrm{I}_n^-, \qquad \mathrm{I}_n^+ \tilde{\times} \mathrm{I}_n, \qquad \mathrm{I}_n^- \tilde{\times} \mathrm{I}_n.$$

See Table 9.6 for all possible types with an automorphism in genus 2.

EXAMPLE 8.12. The BY tree with automorphism from Example 3.32 has type $:^{+}0^{-,+}_{5\text{-}5} \cdot_1 2{:}_2(0{:}_2\mathrm{U}^{-}_{6,6})$

9. Classification for small genera

Recall that in genus 0,1 and 2 there are, respectively, 1, 2 and 7 'semistable types', that is equivalence classes of hyperelliptic graphs/BY trees/cluster pictures (cf. Theorem 5.1 and Table 3.1 p. 81). It is easy to classify them in any genus g by listing all balanced cluster pictures in $X = \{1, ..., 2g+2\}$, up to S_{2g+2}-conjugacy and then using Theorem 5.1 to obtain the corresponding hyperelliptic graphs and BY trees. The classification in genus 3 is given by the following theorem.

THEOREM 9.1. *There are 32 hyperelliptic graphs of genus 3 up to isomorphism, 32 BY trees of genus 3 up to isomorphism and 32 cluster pictures of genus 3 up to equivalence. The one-to-one correspondence of Theorem* 5.1 *is given in Table* 9.4.

We can similarly classify all open hyperelliptic graphs by listing all possible cluster pictures. This is given in Table 4.1 (p. 91) for genus 0 and 1. For genus 2 the complete classification is the following:

THEOREM 9.2. *There are 45 isomorphism classes of genus 2 open hyperelliptic graphs, open BY trees and cluster pictures. The one-to-one correspondence of Theorem* 4.2 *is given in Table* 9.5.

In the arithmetic setting of Section 1.3 one often wants to know the hyperelliptic graph together with an automorphism. In the context of curves over local fields of odd residue characteristic, these correspond to all possible Frobenius actions on the dual graph of the special fiber of the minimal regular model of semistable genus 2 curves. For elliptic curves, the corresponding types are 1, I_n^+ and I_n^- — good, split multiplicative and non-split multiplicative reduction; see Example 8.10.

Note that by Theorems 5.1 and 5.2, there is a bijection between

- Isomorphism classes of pairs (Σ, ϕ), where Σ is a balanced cluster picture and $\phi \in \operatorname{Aut} \Sigma$ has sign $+1$ on X if X is non-übereven. Here two pairs (Σ, ϕ) and (Σ', ϕ') are isomorphic if there is an isomorphism $\psi : \Sigma \to \Sigma'$ such that $\psi\phi\psi^{-1} = \phi'$.
- Isomorphism classes of pairs (G, ϕ) of hyperelliptic graphs with an automorphism, where two pairs (G, ϕ) and (G', ϕ') are isomorphic if there is an isomorphism $\psi : G \to G'$ such that $\psi\phi\psi^{-1} = \phi'$.

Explicitly, the bijection is given by mapping Σ to the core G of $\underline{G}(\Sigma)$ and ϕ to the restriction of $\underline{G}(\phi)$ to G. This makes it easy to list the types on the level of cluster pictures. The lattice Λ, the ϕ-action on it and the Tamagawa group can also be computed from it. (Definition 3.48).

THEOREM 9.3. *Up to isomorphism, there are 23 genus 2 hyperelliptic graphs together with an automorphism. These are given in Table* 9.6 *together with their corresponding BY trees, balanced cluster pictures and lattices* Λ.

TABLE 9.4. Hyperelliptic graphs, BY trees and balanced cluster pictures in genus 3.

Type name	G	T	Σ
3			
2_n			
$1_{n,m}$			
$\mathrm{I}_{n,m,k}$			
$\mathrm{U}_{n,m,k,l}$			
$2{\cdot}\mathrm{I}_n$			
$2{\cdot}1$			
$1_n{\cdot}\mathrm{I}_m$			
$1_n{\cdot}1$			
$0_{n,m}{\cdot}\mathrm{I}_k$			
$0_{n,m}{\cdot}1$			
$1{\cdot}\mathrm{I}_n{\cdot}\mathrm{I}_m$			
$1{\cdot}1{\cdot}\mathrm{I}_n$			
$1{\cdot}1{\cdot}1$			
$0_n{\cdot}\mathrm{I}_m{\cdot}\mathrm{I}_k$			
$0_n{\cdot}1{\cdot}\mathrm{I}_m$			
$0_n{\cdot}1{\cdot}1$			
$0{\cdot}1\circ 0{\cdot}1$			
$0{\cdot}\mathrm{I}_n\circ 0{\cdot}1$			
$0{\cdot}\mathrm{I}_n\circ 0{\cdot}\mathrm{I}_m$			
$\mathrm{U}_{n,m}\circ 0{\cdot}1$			
$\mathrm{U}_{n,m}\circ 0{\cdot}\mathrm{I}_k$			
$\mathrm{U}_{n,m}\circ \mathrm{U}_{k,l}$			
$\mathrm{I}_n\circ 0{\cdot}1$			
$\mathrm{I}_n\circ 0{\cdot}\mathrm{I}_m$			
$\mathrm{I}_n\circ \mathrm{U}_{m,k}$			
$\mathrm{I}_n\circ \mathrm{I}_m$			
$1\circ 0{\cdot}1$			
$1\circ 0{\cdot}\mathrm{I}_n$			
$1\circ \mathrm{U}_{n,m}$			
$1\circ \mathrm{I}_n$			
$1\circ 1$			

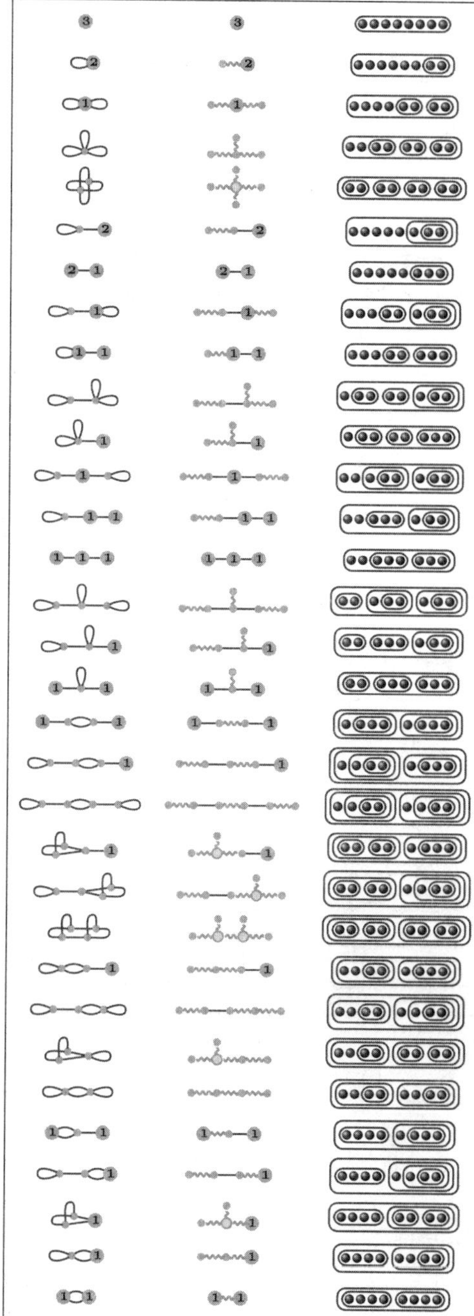

TABLE 9.5. Open hyperelliptic graphs, open BY trees up to isomorphism and cluster pictures in genus 2.

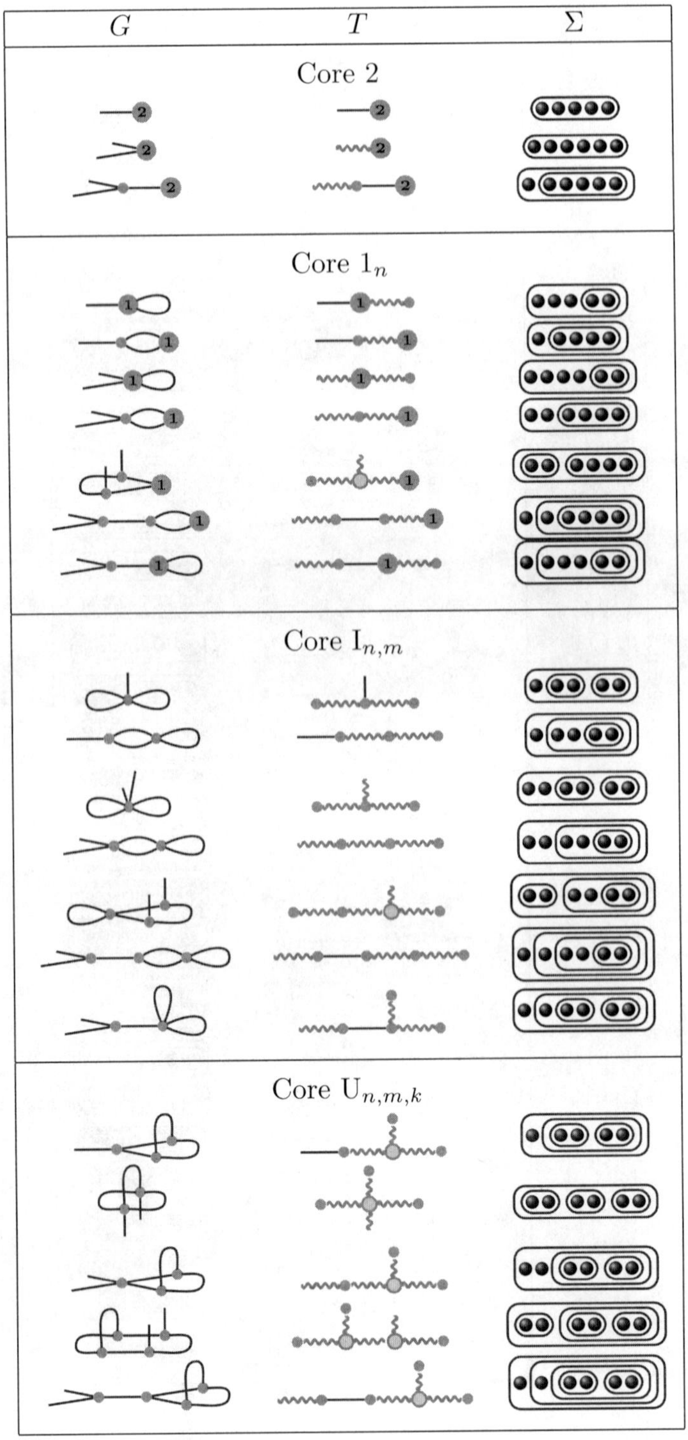

TABLE 9.5. (continued)

G	T	Σ
Core 1×1		
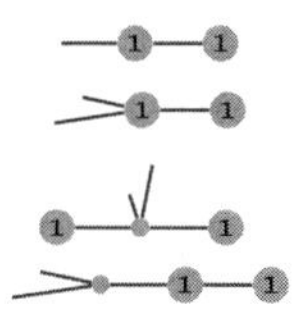	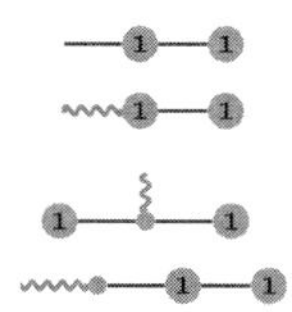	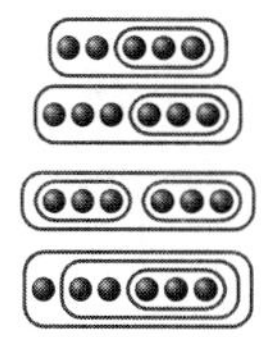
Core $1\times \mathrm{I}_n$		
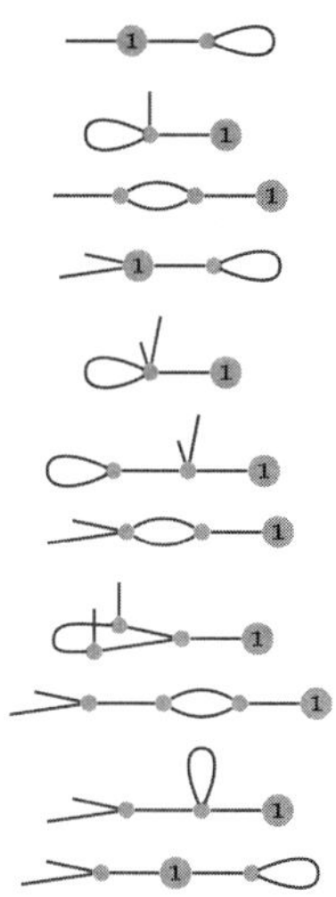	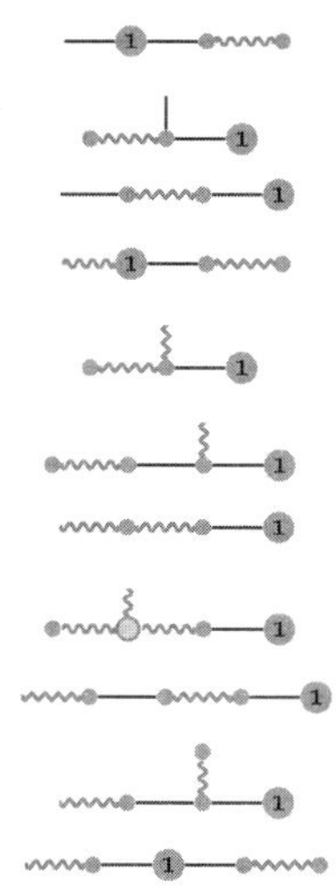	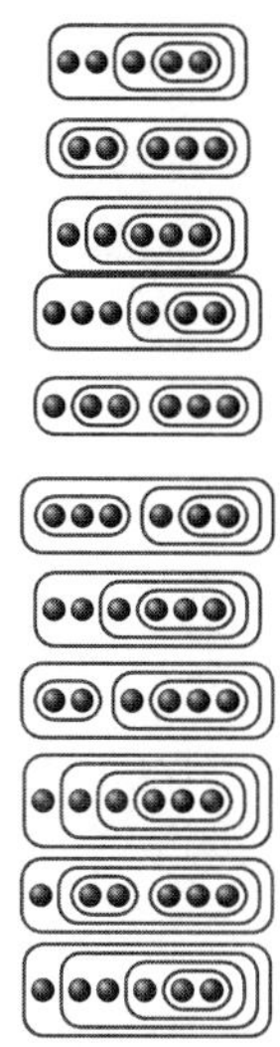
Core $\mathrm{I}_n\times \mathrm{I}_m$		
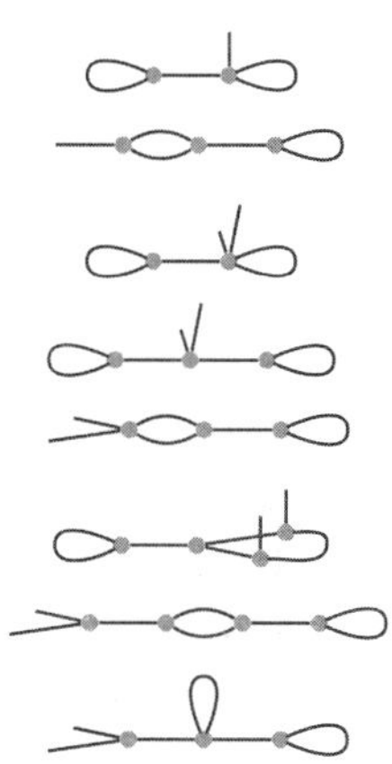	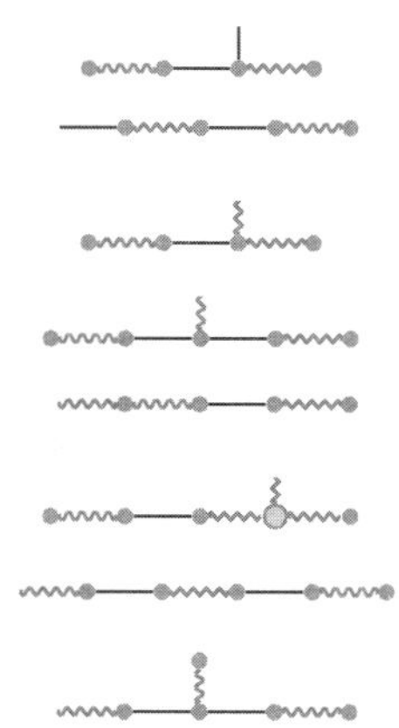	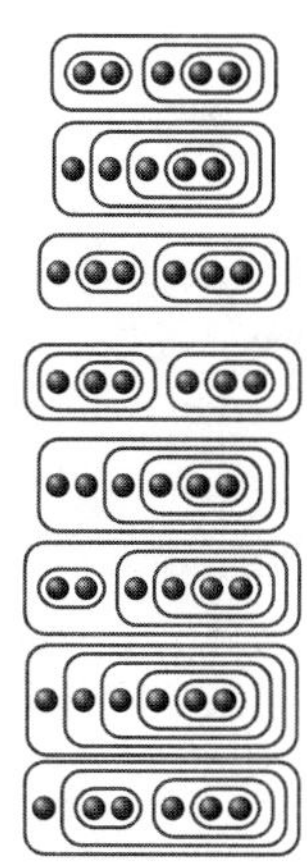

TABLE 9.6. Types with an automorphism in genus 2.

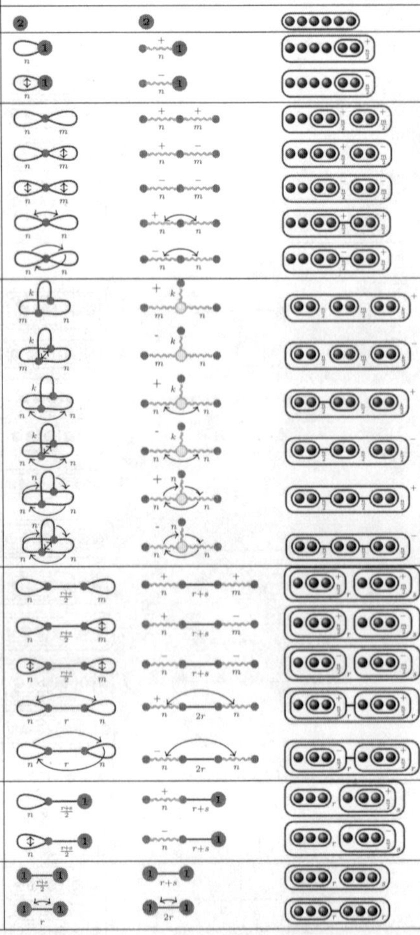

Type with automorphism ϕ	G	T	Σ	pairing $\langle,\rangle$ on Λ	action of ϕ on Λ	$\lvert(\Lambda^\vee/\Lambda)^{\phi=1}\rvert$
2						1
1_n^+				(n)	(1)	n
1_n^-				(n)	(-1)	$\tilde{n}$
$I_{n,m}^{+,+}$				$\begin{pmatrix} n & 0 \\ 0 & m \end{pmatrix}$	$\begin{pmatrix} 1 & 0 \\ 0 & 1 \end{pmatrix}$	nm
$I_{n,m}^{+,-}$				$\begin{pmatrix} n & 0 \\ 0 & m \end{pmatrix}$	$\begin{pmatrix} 1 & 0 \\ 0 & -1 \end{pmatrix}$	$n\tilde{m}$
$I_{n,m}^{-,-}$				$\begin{pmatrix} n & 0 \\ 0 & m \end{pmatrix}$	$\begin{pmatrix} -1 & 0 \\ 0 & -1 \end{pmatrix}$	$\tilde{n}\tilde{m}$
$I_{n\text{-}n}^+$				$\begin{pmatrix} n & 0 \\ 0 & n \end{pmatrix}$	$\begin{pmatrix} 0 & 1 \\ 1 & 0 \end{pmatrix}$	n
$I_{n\text{-}n}^-$				$\begin{pmatrix} n & 0 \\ 0 & n \end{pmatrix}$	$\begin{pmatrix} 0 & -1 \\ 1 & 0 \end{pmatrix}$	$\tilde{n}$
$U_{n,m,k}^+$				$\begin{pmatrix} n+k & k \\ k & m+k \end{pmatrix}$	$\begin{pmatrix} 1 & 0 \\ 0 & 1 \end{pmatrix}$	N
$U_{n,m,k}^-$				$\begin{pmatrix} n+k & k \\ k & m+k \end{pmatrix}$	$\begin{pmatrix} -1 & 0 \\ 0 & -1 \end{pmatrix}$	$\widetilde{N/D}\times\tilde{D}$
$U_{n\text{-}n,k}^+$				$\begin{pmatrix} n+k & k \\ k & n+k \end{pmatrix}$	$\begin{pmatrix} 0 & 1 \\ 1 & 0 \end{pmatrix}$	$n{+}2k$
$U_{n\text{-}n,k}^-$				$\begin{pmatrix} n+k & k \\ k & n+k \end{pmatrix}$	$\begin{pmatrix} 0 & -1 \\ -1 & 0 \end{pmatrix}$	n
$U_{n\text{-}n\text{-}n}^+$				$\begin{pmatrix} 2n & n \\ n & 2n \end{pmatrix}$	$\begin{pmatrix} 0 & 1 \\ -1 & -1 \end{pmatrix}$	3
$U_{n\text{-}n\text{-}n}^-$				$\begin{pmatrix} 2n & n \\ n & 2n \end{pmatrix}$	$\begin{pmatrix} 0 & -1 \\ 1 & 1 \end{pmatrix}$	1
$I_n^+ \times_{\frac{r+s}{2}} I_m^+$				$\begin{pmatrix} n & 0 \\ 0 & m \end{pmatrix}$	$\begin{pmatrix} 1 & 0 \\ 0 & 1 \end{pmatrix}$	nm
$I_n^+ \times_{\frac{r+s}{2}} I_m^-$				$\begin{pmatrix} n & 0 \\ 0 & m \end{pmatrix}$	$\begin{pmatrix} 1 & 0 \\ 0 & -1 \end{pmatrix}$	$n\tilde{m}$
$I_n^- \times_{\frac{r+s}{2}} I_m^-$				$\begin{pmatrix} n & 0 \\ 0 & m \end{pmatrix}$	$\begin{pmatrix} -1 & 0 \\ 0 & -1 \end{pmatrix}$	$\tilde{n}\tilde{m}$
$I_n^+ \tilde{\times}_r I_n$				$\begin{pmatrix} n & 0 \\ 0 & n \end{pmatrix}$	$\begin{pmatrix} 0 & 1 \\ 1 & 0 \end{pmatrix}$	n
$I_n^- \tilde{\times}_r I_n$				$\begin{pmatrix} n & 0 \\ 0 & n \end{pmatrix}$	$\begin{pmatrix} 0 & -1 \\ 1 & 0 \end{pmatrix}$	$\tilde{n}$
$1 \times_{\frac{r+s}{2}} I_n^+$				(n)	(1)	n
$1 \times_{\frac{r+s}{2}} I_n^-$				(n)	(-1)	$\tilde{n}$
$1 \times_{\frac{r+s}{2}} 1$						1
$1 \tilde{\times}_r 1$						1

Notation in the last column: $\tilde{n}=2$ if $2|n$ and $\tilde{n}=1$ if $2\nmid n$; $D=\gcd(m,n,k)$; $N=nm+nk+mk$.
Black arrows in G and T and black lines in Σ indicate the automorphism; $+/-$ in T and Σ indicate the value of ϵ_ϕ.
Numbers indicate lengths of edges in G and T, and distances to the parent clusters in Σ.

Acknowledgements

We would like to thank the EPSRC and the Royal Society for their support, and the Warwick Mathematics Institute where parts of this research were carried out.

References

[1] A. Betts, *On the computation of Tamagawa numbers and Néron component groups of Jacobians of semistable hyperelliptic curves*, arXiv:1808.05479, 2018.

[2] Roland Bacher, Pierre de la Harpe, and Tatiana Nagnibeda, *The lattice of integral flows and the lattice of integral cuts on a finite graph* (English, with English and French summaries), Bull. Soc. Math. France **125** (1997), no. 2, 167–198. MR1478029

[3] Matthew Baker and Serguei Norine, *Harmonic morphisms and hyperelliptic graphs*, Int. Math. Res. Not. IMRN **15** (2009), 2914–2955, DOI 10.1093/imrn/rnp037. MR2525845

[4] Matthew Baker and Serguei Norine, *Riemann-Roch and Abel-Jacobi theory on a finite graph*, Adv. Math. **215** (2007), no. 2, 766–788, DOI 10.1016/j.aim.2007.04.012. MR2355607

[5] Matthew Baker and Joseph Rabinoff, *The skeleton of the Jacobian, the Jacobian of the skeleton, and lifting meromorphic functions from tropical to algebraic curves*, Int. Math. Res. Not. IMRN **16** (2015), 7436–7472, DOI 10.1093/imrn/rnu168. MR3428970

[6] Siegfried Bosch, *Formelle Standardmodelle hyperelliptischer Kurven* (German), Math. Ann. **251** (1980), no. 1, 19–42, DOI 10.1007/BF01420278. MR583822

[7] L. Caporaso, Gonality of algebraic curves and graphs, Algebraic and Complex Geometry, ed. A. Frühbis-Krüger, R. Kloosterman, M. Schütt, Springer, 2014, 77–108.

[8] Melody Chan, *Tropical hyperelliptic curves*, J. Algebraic Combin. **37** (2013), no. 2, 331–359, DOI 10.1007/s10801-012-0369-x. MR3011346

[9] T. Dokchitser, V. Dokchitser, C. Maistret, and A. Morgan, *Arithmetic of hyperelliptic curves over local fields*, arXiv:1808.02936, 2018.

[10] A. Grothendieck, Modèles de Néron et monodromie, SGA 7, Exposé IX, 1972.

[11] Allen Hatcher, *Algebraic topology*, Cambridge University Press, Cambridge, 2002. MR1867354

[12] Ivan Kausz, *A discriminant and an upper bound for ω^2 for hyperelliptic arithmetic surfaces*, Compositio Math. **115** (1999), no. 1, 37–69, DOI 10.1023/A:1000580901251. MR1671741

[13] Shu Kawaguchi and Kazuhiko Yamaki, *Rank of divisors on hyperelliptic curves and graphs under specialization*, Int. Math. Res. Not. IMRN **12** (2015), 4121–4176, DOI 10.1093/imrn/rnu059. MR3356750

[14] K. Kodaira, *On the structure of compact complex analytic surfaces. I*, Amer. J. Math. **86** (1964), 751–798, DOI 10.2307/2373157. MR0187255

[15] C. Maistret, Parity of ranks of Jacobians of hyperelliptic curves of genus 2, PhD Thesis, University of Warwick, 2017.

[16] Yukihiko Namikawa and Kenji Ueno, *The complete classification of fibres in pencils of curves of genus two*, Manuscripta Math. **9** (1973), 143–186, DOI 10.1007/BF01297652. MR0369362

[17] Melanie Matchett Wood, *The distribution of sandpile groups of random graphs*, J. Amer. Math. Soc. **30** (2017), no. 4, 915–958, DOI 10.1090/jams/866. MR3671933

Department of Mathematics, University of Bristol, Bristol BS8 1TW, United Kingdom

Email address: tim.dokchitser@bristol.ac.uk

Department of Mathematics, King's College London, Strand, London WC2R 2LS, United Kingdom

Email address: vladimir.dokchitser@kcl.ac.uk

Department of Mathematics, University of Bristol, Bristol BS8 1TW, United Kingdom

Email address: celine.maistret@bristol.ac.uk

Department of Mathematics, King's College London, Strand, London WC2R 2LS, United Kingdom

Email address: adam.morgan@kcl.ac.uk

Contemporary Mathematics
Volume **724**, 2019
https://doi.org/10.1090/conm/724/14587

Formal deformations of algebraic spaces and generalizations of the motivic Igusa-zeta function

Andrew R. Stout

ABSTRACT. We generalize the notion of the auto-Igusa zeta function to formal deformations of algebraic spaces. By incorporating data from all algebraic transformations of local coordinates, this function can be viewed as a generalization of the traditional motivic Igusa zeta function. Furthermore, we introduce a new series, which we term the canonical auto-Igusa zeta function, whose coefficients are given by the quotient stacks formed from the coefficients of the auto-Igusa zeta function modulo change of coordinates. We indicate the current state of the literature on these generalized Igusa-zeta functions and offer directions for future research.

1. Introduction

Progress has been made recently in understanding certain motivic generating series associated to the germ of a variety. Originally introduced by Schoutens in [**Scha**] and [**Schb**], the auto-Igusa zeta series is a motivic generating series whose coefficients are determined by the algebra automorphisms of the local ring corresponding to a point on a variety. First calculations of this series were completed by Schoutens and secondary calculations were carried out by the author of this paper with the use of computer algebra software in [**Sto17**]. This led to several conjectures concerning the structure of the auto-Igusa zeta function. The author later confirmed that these conjectures are indeed true in [**Sto18**] for plane curve singularities, yet the veracity of these conjectures for broader classes of singularities remains unknown.

In this paper, the author generalizes the auto-Igusa zeta function to stacks. However, we choose to frame almost all of the work in the language of algebraic spaces, as there is a bottleneck for Artin stacks related to the occasional lack of representability of the hom functor for general Artin stacks. The representability results are discussed in §2.

One notable development in this paper concerns the introduction of a new type of moduli stack, which takes into account the canonical group action on the space of endormophism by conjugation of all automorphisms. We call this the moduli stack of Jordan-Normal forms. This space is in many ways more natural than the space

2010 *Mathematics Subject Classification.* Primary 14H20; Secondary 14H50, 14E18.

Support for this project was provided by a PSC-CUNY Award (PSC-Grant Traditional B, # 60784-00 48), jointly funded by the Professional Staff Congress and The City University of New York.

of all endomrophism (from which the auto-Igusa zeta series is formed), yet it is unclear to the author under what conditions this moduli stack is itself an algebraic space or even when it is an Artin stack. Regardless, one may form a generating series from this stack, which we then term the canonical auto-Igusa zeta function.

Another important development in this paper is that it is shown in §6 that both the canonical zeta function and the auto-Igusa zeta function may both be regarded as generalizations of the motivic Igusa zeta function, which has been studied by Denef, Loeser, Cluckers, Nicaise, Mustata, and many others. It is hoped that the motivic Igusa zeta series would offer a key to proving the monodromy conjecture first introduced by Igusa. However, as of yet, the general conjecture remains unproven.

In §7, the author rephrases his work in [**Sto18**] in this new context and offers possible directions for future research. As mentioned at the beginning of this section, the possible rationality of the auto-Igusa zeta function for varieties (and other conjectures) other than plane curves remains mostly elusive to direct proof. It is the hope of the author that the work carried out in this paper will offer progress toward such a proof as well as possible keys to further understanding of local zeta functions in general.

2. Representability results concerning Hom

Let $\mathbf{C}$ be a site. It follows from the Yoneda lemma that the Grothendieck topos $\tau = \mathrm{Sh}(\mathbf{C})$ is a locally cartesian closed monoidal category. In particular, given two objects $\mathcal{X}$ and $\mathcal{Y}$ of τ, we have a presheaf which sends an object U of $\mathbf{C}$ to the set $\mathrm{Hom}_\tau(\mathcal{X} \otimes U, \mathcal{Y} \otimes U)$, and, moreover, this presheaf is in fact a sheaf – i.e., it is an object of τ, which we denote by $\underline{\mathrm{Hom}}_\tau(\mathcal{X}, \mathcal{Y})$. In other words, there is a natural bijection

$$\mathrm{Hom}_\tau(\mathcal{X} \otimes U, \mathcal{Y} \otimes U) \xrightarrow{\sim} \mathrm{Hom}_\tau(U, \underline{\mathrm{Hom}}_\tau(\mathcal{X}, \mathcal{Y})). \tag{2.1}$$

In fact, a similar result holds in higher category theory–i.e., if τ is an $(\infty, 1)$-topos, then τ is a locally cartesian closed monoidal $(\infty, 1)$-category. This follows from the $(\infty, 1)$-Yoneda lemma (cf. Proposition 5.1.3.1 of [**Lur09**]). In particular, if $\mathcal{X}$ and $\mathcal{Y}$ are two stacks over an algebraic space S, then one defines $\underline{\mathrm{Hom}}_S(\mathcal{X}, \mathcal{Y})$ to be the fibered category of groupoids over the site of S-schemes $(\mathbf{Sch}/S)_J$ with a given Grothendieck topology J, which to any obect $U \to S$ of $(\mathbf{Sch}/S)_J$ associates the groupoid of functors from $\mathcal{X} \times_S U$ to $\mathcal{Y} \times_S U$ over U, and, moreover, this fibered category $\underline{\mathrm{Hom}}_S(\mathcal{X}, \mathcal{Y})$ is in fact a stack. It is possible to show that $\underline{\mathrm{Hom}}_S(\mathcal{X}, \mathcal{Y})$ is an Artin stack (or, Deligne-Mumford stack, or algebraic space) depending on certain conditions on $\mathcal{X}$ and $\mathcal{Y}$ (cf. Theorem 1.1 of [**Ols06**]). For example, we have the following result due to Artin.

THEOREM 2.1. *If S is a locally Noetherian algebraic space and if X and Y are algebraic spaces over S with $X \to S$ proper and flat and $Y \to S$ is separated and of finite type, then $\underline{Hom}_S(X, Y)$ is a separated algebraic space locally of finite type over S.*

This result can be found in [**Art69**], and generalizes[1] the following result of Grothendieck. If S is a locally Noetherian scheme and if X and Y are in $\mathbf{Sch}/S$

[1]Essentially, Artin relaxed the condition that $X \to S$ is projective to merely proper, but this came at the cost of moving from the category of schemes to the larger category of algebraic spaces.

with X projective and flat over S and Y quasi-projective over S, then the functor from **Sch**$/S$ to **Sets** defined by sending U to $\mathrm{Hom}_S(X \times_S U, Y \times_S U)$ is represented by a scheme $\underline{\mathrm{Hom}}_S(X, Y)$ which is a separated and locally of finite type over S. This result can be found at the end of page 221-19 of [**Gro62**].

In particular, if X is finite and flat over a locally Noetherian scheme S, then for any quasi-projective object Y in **Sch**$/S$, $\underline{\mathrm{Hom}}_S(X, Y)$ is a scheme over S. This is found in Proposition 5.7 on pages 221-27 of [**Gro62**], and it usually goes by the name Weil restriction or restriction of scalars. If $Y \to S$ is separated (respectively, affine, finite type), then $\underline{\mathrm{Hom}}_S(X, Y) \to S$ is separated (respectively, affine, finite type) over S (cf. Section 7 on pages 221-27 and 221-28 of [**Gro62**]).

REMARK 2.2. In fact, regarding either Artin's result or Grothendieck's result, the assumption that S is locally Noetherian can be relaxed if we assume that $X \to S$ and $Y \to S$ are locally of finite presentation. Note that a finite type morphism with a locally Noetherian target is locally of finite presentation.

3. Projective systems of Hilbert spaces

Let $X \to S$ be a proper and flat morphism of algebraic spaces with S locally Noetherian. There is a functor $\mathrm{Hilb}_{X/S} : \mathbf{Sch}/S \to \mathbf{Sets}$ defined by

(3.1)
$$\mathrm{Hilb}_{X/S}(T) = \{[j : Z \hookrightarrow X \times_S T] \mid Z \to T \text{ proper and flat, and } j \text{ a closed immersion}\},$$

where $[j : Z \hookrightarrow X \times_S T]$ denotes the isomorphism class of j. This is in fact represented by a separated algebraic space locally of finite type over S (cf. Section 6 of [**Art69**]) which we denote by $\underline{\mathrm{Hilb}}_{X/S}$. The natural transformation $\iota : \mathrm{Hom}_S(X, Y) \hookrightarrow \mathrm{Hilb}_{(X \times_S Y)/S}$ defined by

$$\begin{aligned} \iota(T) : \mathrm{Hom}_S(X, Y)(T) &\hookrightarrow \mathrm{Hilb}_{(X \times_S Y)/S}(T) \\ [f : X \times_S T \to Y \times_S T] &\mapsto [\Gamma_f \hookrightarrow X \times_S \times Y \times_S T] \end{aligned} \tag{3.2}$$

where Γ_f denotes the graph of f induces an open immersion of algebraic spaces $\underline{\mathrm{Hom}}_S(X, Y) \hookrightarrow \underline{\mathrm{Hilb}}_{(X \times_S Y)/S}$ provided $X \to S$ is proper and flat and $Y \to S$ is separated and of finite type.

Now, if $f : S'' \to S'$ is an S-morphism of algebraic spaces, then there is a natural pullback f^* given by the natural transformation $f^* : \mathrm{Hilb}_{S'/S} \to \mathrm{Hilb}_{S''/S}$ defined by

$$\begin{aligned} f^*(T) : \mathrm{Hilb}_{S'/S}(T) &\to \mathrm{Hilb}_{S''/S}(T) \\ Z &\mapsto Z \times_{S' \times_S T} (S'' \times_S T). \end{aligned} \tag{3.3}$$

Given two S-morphisms $f : X' \to X$ and $g : Y' \to Y$ of algebraic spaces with $X' \to S$ and $X \to S$ proper and flat and with $Y' \to S$ and $Y \to S$ separated and finite type, we have the canonical morphism $(f, g) : X' \times_S Y' \to X \times_S Y$ and the corresponding pullback morphism of Hilbert spaces

$$(f, g)^* : \underline{\mathrm{Hilb}}_{X \times_S Y/S} \to \underline{\mathrm{Hilb}}_{X' \times_S Y'/S}. \tag{3.4}$$

If, in addition, $g : Y' \hookrightarrow Y$ is a closed immersion, then the restriction of $(f, g)^*$ to $\underline{\mathrm{Hom}}(X, Y)$ induces a well-defined morphism of algebraic spaces

$$(f, g)^* : \underline{\mathrm{Hom}}_S(X, Y) \to \underline{\mathrm{Hom}}_S(X', Y'). \tag{3.5}$$

Note that $\underline{\mathrm{Hom}}_S(X,Y)$ is contravariant in the first argument and covariant in the second; however, the morphism $(f,g)^*$ sends a morphism $h : X \to Y$ to $h \circ f : X' \to Y$ and then restricts the range of $h \circ f$ to the closed subset Y' of Y.

An injective system $\{P_i\}_{i\in I}$ of algebraic spaces over S with affine transition maps and with $P_i \to S$ proper and flat gives rise to a projective system $\{\underline{\mathrm{Hilb}}_{P_i/S}\}_{i\in I}$ of algebraic spaces with affine transition maps, and it is therefore the case that $\varprojlim \underline{\mathrm{Hilb}}_{P_i/S}$ exists in the category of algebraic spaces over S. In particular, if $\{Y_i\}_{i\in I}$ is an injective system whose transition maps are closed immersions with $Y_i \to S$ separated and finite type and if $\{X_i\}_{i\in I}$ is any other injective system with $X_i \to S$ proper and flat, then $\varprojlim \underline{\mathrm{Hom}}_S(X_i, Y_i) \hookrightarrow \varprojlim \underline{\mathrm{Hilb}}_{X_i\times_S Y_i/S}$ is an open immersion of algebraic spaces over S.

Furthermore, if $\{X_i\}_{i\in I}$ is an injective system of algebraic spaces over S whose transitions maps are closed immersions and where each $X_i \to S$ is proper and flat then there is a projective system of open immersions

$$\underline{\mathrm{Aut}}_S(X_i) \hookrightarrow \underline{\mathrm{End}}_S(X_i) \tag{3.6}$$

where $\underline{\mathrm{Aut}}(X_i)$ is the algebraic space which represents the subfunctor of $\mathrm{Hom}_S(X_i, X_i)$ which associates to $T \to S$ the set of all automorphisms of $X_i \times_S T$. Here, we write $\underline{\mathrm{End}}_S(X_i)$ in place of $\underline{\mathrm{Hom}}_S(X_i, X_i)$. Note that $\underline{\mathrm{End}}_S(X_i)$ is a monoid object and $\underline{\mathrm{Aut}}_S(X_i)$ is a group object in the category of algebraic spaces over S. Since the transition maps are closed immersions, the inclusion

$$\varprojlim \underline{\mathrm{Aut}}_S(X_i) \hookrightarrow \varprojlim \underline{\mathrm{End}}_S(X_i) \tag{3.7}$$

is an open immersion of algebraic spaces over S.

REMARK 3.1. As in Remark 2.2, we may relax the condition that S is locally Noetherian in this section by assuming that all morphisms are locally of finite presentation.

4. Finite free algebras

LEMMA 4.1. *Let $X \to S$ be a finite and flat morphism of algebraic spaces with S locally Noetherian. Then, $X \to S$ is finite locally free. In particular, $X \to S$ is faithfully flat.*

PROOF. Note that flat is local on the source and target in the étale topology (cf. Remark 5.4.13 of [**Ols16**]). Therefore, it is enough to check when S is affine by Lemma 45.3 of [**Sta17**]. Since $X \to S$ is finite and hence also an affine morphism, X is affine. Hence, we reduce to the case where $(A, \mathfrak{m})$ is a Noetherian local ring with residue field $k = A/\mathfrak{m}$ and a flat morphism $A \to B$, which induces a surjective morphism $A^n \to B$ of A-modules (by finiteness). Therefore, we have a short exact sequence

$$0 \to I \to A^n \to B \to 0$$

where I is finitely generated (since A is Noetherian). Tensoring this short exact sequence by k gives the short exact sequence

$$0 \to I \otimes_A k \to k^n \to B/\mathfrak{m}B \to 0$$

since B is flat over A. Since $B/\mathfrak{m}B$ is a free k-module, we may choose n above (i.e., generators $b_1, \ldots, b_n$ of B over A) so that $B/\mathfrak{m}B$ is isomorphic to k^n, which implies that $I/\mathfrak{m}I = I \otimes_A k = 0$. Therefore, by Nakayama's Lemma (cf, Part (2) of Lemma 10.19.1 of [**Sta17**]), $I = 0$, and B is a free A-module. □

REMARK 4.2. The Noetherian condition on S in Lemma 4.1 may be relaxed. In fact, the morphism $X \to S$ of algebraic spaces is a finite, flat, and locally of finite presentation if and only if it is a finite locally free morphism. This follows from 1.4.7 of [**GD67**], Lemma 45.3 [**Sta17**], and the fact that flatness and locally of finite presentation are both local on the source and target in the étale topology (cf. Remark 5.4.13 of [**Ols16**]).

Assume that A is a Noetherian local ring with B finite and flat over A, and whence free. Let $A^n \xrightarrow{\sim} B$ be a presentation of B as a free A-algebra obtained by sending the standard basis elements e_i of A^n to the generating elements b_i of B for $i = 1, \ldots, n$. Consider the polynomial ring $R = A[x_1, \ldots, x_n]$ and the surjective homomorphism $\phi : R \to B$ defined by $\phi(x_i) = b_i$ for $i = 1, \ldots, n$. Then, by Hilbert's Basis Theorem, R is Noetherian and therefore the kernel $\mathrm{Ker}(\phi)$ is finitely generated – i.e., $\mathrm{Ker}(\phi) = (f_1, \ldots, f_s)$. Moreover, there is a canonical isomorphism $R/\mathrm{Ker}(\phi) \xrightarrow{\sim} B$ in the category of A-algebras. Therefore, an A-algebra endomorphism γ of B is given by a choice $\gamma(x_i) = P_i$ with P_i any element of $B = A[x_1, \ldots, x_n]/(f_1, \ldots, f_s)$ for each $i = 1, \ldots, n$ such that $f_j(P_1, \ldots, P_n) = 0$ in B for $j = 1, \ldots, s$.

LEMMA 4.3. *Let $S = Spec(A)$ with A is a reduced Noetherian local ring and let $X = Spec(B)$ with $B = A[x_1, \ldots, x_d]/(x_1, \ldots, x_d)^n$ for $n \in \mathbb{N}$. Set $r = d \cdot (\ell - 1)$, where ℓ is the (generic) rank of B over A. Then, $X \to S$ is finite and flat, and we have the following isomorphisms*

$$(4.1)\qquad (\underline{End}_S(X))^{\mathrm{red}} \cong \mathbb{A}^r_S \quad \text{and} \quad (\underline{Aut}_S(X))^{\mathrm{red}} \cong GL_{d,S} \times_S \mathbb{A}^{r-d^2}_S.$$

PROOF. Consider the A-algebra endomorphism of $A[x_1, \ldots, x_d]/(x_1, \ldots, x_d)^n$ defined by $x_i \mapsto P_i$, where we define $P_i = \sum_{|j|<n} a_{i,j} x^j$ where j is a multi-index (i.e., $j = (j_1, \ldots, j_d)$, $x^j = \prod_{s=1}^d x_s^{j_s}$, and $|j| = \sum_{s=1}^d j_s < n$) and where $a_{i,j} \in A$ for $i = 1, \ldots, d$. Then, the equations defining the space of S-endomorphisms of X are given by

$$0 = P_i^n = (a_{i,0} + Q_i)^n, \quad \forall i = 1, \ldots, d$$

where 0 is treated as the multi-index $(0, \ldots, 0)$ and $Q_i \in (x_1, \ldots, x_d)/(x_1, \ldots, x_d)^n$. This implies that $0 = (a_{i,0})^n$ for all $i = 1, \ldots, d$. Thus, in the reduction, $0 = a_{i,0}$ for all $i = 1, \ldots, d$ since A is reduced. Clearly, $Q^n = 0$ for all $Q \in (x_1, \ldots, x_d)/(x_1, \ldots, x_d)^n$ from which the first isomorphism follows.

Note that those endomorphisms above which also lie in $(\underline{\mathrm{Aut}}_S(X))^{\mathrm{red}}$ are given by $x_i \mapsto \sum_{j=1}^d a_{i,j} x_j + T_i$ with $T_i \in (x_1, \ldots, x_d)^2/(x_1, \ldots, x_d)^n$ and where the $d \times d$-matrix $M = (a_{i,j})$ is invertible over A – i.e., $\det(M)$ is a unit of A. The second isomorphism then follows from this fact. □

REMARK 4.4. In the case where $d = 1$, we obtain

$$(4.2)\quad (\underline{\mathrm{End}}_S(X))^{\mathrm{red}} \cong \mathbb{A}^{n-1}_S \quad \text{and} \quad (\underline{\mathrm{Aut}}_S(X))^{\mathrm{red}} \cong \mathbb{G}_{m,S} \times_S \mathbb{A}^{n-2}_S \text{ for } n > 1.$$

One simple case where Lemma 4.1 may be applied is when $S = \mathrm{Spec}(A)$ with $(A, \mathfrak{m})$ an Artinian local ring. This is because if B is a finitely generated A-module with A an Artinian local ring, then B will have finite length over the residue field $k = A/\mathfrak{m}$ and hence B will also be an Artinian local ring. In particular, B is automatically free over A. Thus, we may form the full subcategory **Fat**$/A$ of **Sch**$/A$

defined by all finite maps $X \to \operatorname{Spec}(A)$. We call[2] this category the *category of fat points over A*. Usually, we assume $A = k$ is a field (and, in this case, Lemma 4.3 will apply), and often, we will further assume that it has characteristic zero and that it is algebraically closed.

In general, if X is proper and flat over an algebraic space S, then the group action

$$\underline{\mathrm{End}}_S(X)^{\mathrm{red}} \times_S \underline{\mathrm{Aut}}_S(X)^{\mathrm{red}} \to \underline{\mathrm{End}}_S(X)^{\mathrm{red}} \tag{4.3}$$

given by conjugation is well-defined in the category of algebraic spaces over S. The resulting quotient stack

$$\mathcal{M}_{\mathrm{JN}}(X) := [\underline{\mathrm{End}}_S(X)^{\mathrm{red}}/\underline{\mathrm{Aut}}_S(X)^{\mathrm{red}}] \tag{4.4}$$

is termed *the moduli stack of Jordan norm forms of X over S*, and it is of general interest. In the case of Lemma 4.3, assuming further that A is an algebraically closed field k, the points of $\mathcal{M}_{\mathrm{JN}}(X)$ correspond to $d \times d$ Jordan normal forms over k.

5. Auto-arc spaces of formal deformations

DEFINITION 5.1. Let S be an algebraic space and let I be a directed countable set. Let $\{X_i\}_{i\in I}$ be an injective system of objects of $\mathbf{Fat}/S$ such that all transition maps $X_i \hookrightarrow X_j$ are closed immersions. We say such a system is *an admissible system of fat points over S* if

1) the structure morphism $j : X_0 \to S$ is surjective étale and
2) the underlying topological spaces of X_0 and X_i are homeomorphic ($\forall i \in I$).

DEFINITION 5.2. We say that a system $f_i : Y_i \hookrightarrow Y_{i+1}$ of closed S-immersions is *a formal deformation over an algebraic space S* if there is an admissible system of fat points $\{X_i\}_{i\in I}$ over S such that for all i in I, there are flat morphisms $\varphi_i : Y_i \to X_i$ which make the following diagram commutative:

$$\begin{array}{ccccccc}
Y_0 & \overset{f_0}{\hookrightarrow} & Y_1 & \overset{f_1}{\hookrightarrow} & Y_2 & \overset{f_2}{\hookrightarrow} & \cdots \\
\downarrow{\scriptstyle \varphi_0} & & \downarrow{\scriptstyle \varphi_1} & & \downarrow{\scriptstyle \varphi_2} & & \\
X_0 & \hookrightarrow & X_1 & \hookrightarrow & X_2 & \hookrightarrow & \cdots \\
\downarrow{\scriptstyle j} & & & & & & \\
S & & & & & &
\end{array}$$

and where φ_i induces an isomorphism

$$Y_{i-1} \cong Y_i \times_{X_i} X_{i-1}. \tag{5.1}$$

Given an admissible system $\{X_i\}_{i\in I}$ over S and a morphism $Y \to S$ of algebraic spaces, we have the trivial formal deformation given by $Y_i := Y \times_S X_i$. More generally, given an ind-object $\mathcal{Y}$ in the category of algebraic spaces over S – i.e., the filtered colimit $\mathcal{Y} = \varinjlim_{i\in I} Y_i$ where the transition maps $Y_i \hookrightarrow Y_{i+1}$ are closed immersions of algebraic spaces – and, an admissible system $\{X_i\}_{i\in I}$ with $\mathcal{X} = \varinjlim_{i\in I} X_i$, we let $\mathtt{Def}_{\mathcal{X}}$ denote the fibered category of formal deformations over S

[2] In general, if S is an algebraic space, we call the full subcategory $\mathbf{C}$ of $\mathbf{Sch}/S$ whose objects are finite, flat, and locally finitely presented over S *the category of fat points over S*, and we denote this category by $\mathbf{Fat}/S$.

with respect to the admissible system $\{X_i\}$. Thus, in particular, if $Y \to S$ is a morphism of algebraic space then the category $\mathtt{Def}_{\mathcal{X}}(Y)$ has at least one object.

An object $\mathcal{Y} = \varinjlim_{i\in I} Y_i$ of $\mathtt{Def}_{\mathcal{X}}$ gives rise to a projective system $\{\mathcal{A}_i(\mathcal{Y})\}_{i\in I}$ of algebraic spaces over S defined by

$$\mathcal{A}_i(\mathcal{Y}) := \underline{\mathrm{Hom}}_S(X_i, Y_i)^{\mathrm{red}}. \tag{5.2}$$

We call $\mathcal{A}_i(\mathcal{Y})$ *the truncated auto-arc space of* $\mathcal{Y}$ *at level* i. Moreover, we may form the projective limit

$$\mathcal{A}(\mathcal{Y}) = \varprojlim_{i\in I} \mathcal{A}_i(\mathcal{Y}), \tag{5.3}$$

which is termed *the infinite auto-arc space of* $\mathcal{Y}$. If $\mathcal{Y}$ is the trivial deformation of $Y \to S$, then

$$\mathcal{A}_i(\mathcal{Y}) = (\underline{\mathrm{Hom}}_S(X_i, Y)^{\mathrm{red}} \times_{S^{\mathrm{red}}} \underline{\mathrm{End}}_S(X_i)^{\mathrm{red}})^{\mathrm{red}} \tag{5.4}$$

In general, the group scheme $\underline{\mathrm{Aut}}_S(X_i)^{\mathrm{red}}$ acts on $\mathcal{A}_i(\mathcal{Y})$ by conjugation for any formal deformation $\mathcal{Y}$, and this give rise to the quotient stack

$$\mathcal{M}_i(\mathcal{Y}) := [\mathcal{A}_i(\mathcal{Y})/\underline{\mathrm{Aut}}_S(X_i)^{\mathrm{red}}]. \tag{5.5}$$

In particular, when $\mathcal{Y}$ is the trivial deformation, then

$$\mathcal{M}_i(\mathcal{Y}) \cong \mathcal{A}_i(Y) \times_{S^{\mathrm{red}}} \mathcal{M}_{JN}(X_i), \tag{5.6}$$

where $\mathcal{M}_{JN}(X_i)$ is the moduli stack of Jordan normal forms introduced at the end of section 4. The projective limit of quotient stacks

$$\mathcal{M}(\mathcal{Y}) = \varprojlim_{i\in I} \mathcal{M}_i(\mathcal{Y}) \tag{5.7}$$

is of general interest.

6. A generalization of the motivic Igusa zeta function

DEFINITION 6.1. Let S be an algebraic space. Let $\mathbf{Alg}/S$ denote the category of finitely presented algebraic spaces over S. The *Grothendieck ring of algebraic spaces over* S, denoted by $\mathrm{K}_0(\mathbf{Alg}/S)$, is the ring formed by introducing relations on the free abelian group of isomorphism classes $\langle X\rangle$ of objects X of $\mathbf{Alg}/S$:

1. $\langle X\rangle = \langle X \setminus Y\rangle + \langle Y\rangle$ whenever $Y \hookrightarrow X$ is a locally closed S-immersion.
2. $\langle Z\rangle = \langle X \times_S \mathbb{A}^n_S\rangle$ whenever $Z \to X$ is a vector bundle of constant rank n.

We denote the class of an algebraic space X in $K_0(\mathbf{Alg}/S)$ by $[X]$. The multiplicative structure is defined by $[X]\cdot[Y] := [X \times_S Y]$.

REMARK 6.2. Note that when $S = \mathrm{Spec}(k)$ with k a field, then $\mathrm{K}_0(\mathbf{Alg}/S)$ is isomorphic to the Grothendieck ring of algebraic varieties $\mathrm{K}_0(\mathbf{Var}/S)$. This isomorphism is induced by the natural inclusion of sets $\mathbf{Var}/S \hookrightarrow \mathbf{Alg}/S$. Moreover, relations of the form given in 2 above are superfluous in this case (cf., Section 1 of [**Eke09**]).

We define $\mathbb{L} := [\mathbb{A}^1_S]$ and call it the *Leftshetz motive* over S. We may invert this element to obtain the *localized Grothendieck ring* $\mathcal{G}_S := \mathrm{K}_0(\mathbf{Alg}/S)[\mathbb{L}^{-1}]$ *of algebraic spaces over* S. We say that an element of the power series ring $\mathcal{G}_S[[t]]$ is a *motivic generating series of algebraic spaces over* S *in one variable*, or, more briefly, we will call it a motivic generating series.

DEFINITION 6.3. Let $\mathcal{X} = \varinjlim_{i\in I} X_i$ be given by an admissible system of fat points $\{X\}_{i\in I}$ over S. Let $\mathcal{Y} = \varinjlim_{i\in I} Y_i$ be an object of $\mathtt{Def}_X$ with Y_i an object of $\mathbf{Alg}/S$ for all $i \in I$. We define the *auto-Igusa zeta function of* $\mathcal{Y}$ to be the motivic generating series given by

$$\zeta_{\mathcal{Y}}(t) = \sum_{i\in I} [\mathcal{A}_i(\mathcal{Y})]\mathbb{L}^{-n_i} t^i \tag{6.1}$$

where $n_i = \dim_S(Y_i)(\mathrm{rank}_S(X_i) - 1) + e_i(\mathrm{rank}_S(X_i) - 1)$ where e_i is the whole number which makes the coefficient $[\mathcal{A}_i(\mathcal{Y})]\mathbb{L}^{-n_i}$ dimensionless[3].

Let $\mathcal{Y}$ be as in Definition 6.3 and assume further that $\mathcal{M}_i(\mathcal{Y})$ is an algebraic space for all $i \in I$, we may form the motivic generating function

$$\eta_{\mathcal{Y}}(t) = \sum_{i\in I} [\mathcal{M}_i(\mathcal{Y})]\mathbb{L}^{-m_i} t^i \tag{6.2}$$

where $m_i = \dim_S(Y_i)(\mathrm{rank}_S(X_i) - 1) + e_i$ where e_i is chosen to be the whole number which makes the coefficient $[\mathcal{M}_i(\mathcal{Y})]\mathbb{L}^{-m_i}$ dimensionless. We call $\eta_{\mathcal{Y}}(t)$ the *canonical auto-Igusa zeta function of* $\mathcal{Y}$.

EXAMPLE 6.4. Let $\mathcal{Y}$ be the trivial formal deformation of a morphism $Y \to S$ where $S = \mathrm{Spec}(A)$ with A reduced with respect to the admissible system $\{X_i\}_{i\in I}$ with $X_i = \mathrm{Spec}(A[t]/t^{i+1})$. Then,

$$\mathcal{A}_i(\mathcal{Y}) \cong \mathcal{L}_i(Y) \times_S \mathbb{A}_S^i \tag{6.3}$$

where $\mathcal{L}_i(Y)$ denotes the classical jet space of Y over S. Thus,

$$\zeta_{\mathcal{Y}}(t) = \sum_{i\in I} [\mathcal{L}_i(Y)]\mathbb{L}^{-d\cdot i} t^i \tag{6.4}$$

where $d = \dim_S(Y)$. This is the classical motivic Igusa zeta function of Y over S. Moreover, in this case, the canonical auto-Igusa zeta function is also the classical motivic Igusa-zeta function–i.e., in this case, we have

$$\eta_{\mathcal{Y}}(t) = \zeta_{\mathcal{Y}}(t). \tag{6.5}$$

EXAMPLE 6.5. Let $S = \mathrm{Spec}(k)$ with k an algebraically closed field of characteristic zero. Let $\mathcal{Y}$ be the trivial deformation of the point $\mathrm{Spec}(k) \to X_0 = S$, and let $X_i = \mathrm{Spec}(\mathcal{O}_{C,O}/\mathfrak{m}_O^{i+1})$ where C is the cuspidal cubic defined by $y^2 = x^3$ over k and O is the origin given by the singular point $(0, 0)$. Then,

$$\mathcal{A}_i(\mathcal{Y}) \cong \mathcal{L}_{2(i-3)}(C) \times_S \mathbb{A}_S^7 \tag{6.6}$$

for all $i > 3$ (cf. Theorem 6.1 of [**Sto17**]). Moreover, in Section 7 of loc. cit., we use formula 6.6 to explicitly calculate the auto-Igusa zeta function to obtain

$$\zeta_{\mathcal{Y}}(t) = \mathbb{L}^{-1} + \mathbb{L}t + \mathbb{L}^2t^2 + \frac{(\mathbb{L}^7 - \mathbb{L}^6)t^3 + \mathbb{L}^7t^4 + \mathbb{L}^7t^7}{(1 - \mathbb{L}t^3)(1 - t)}. \tag{6.7}$$

[3]It is anticipated that $e_i = \mathrm{rank}_S(\Omega_{X_0/S})$ for all $i \in I$.

7. Motivic rationality for plane curve singularities

We will need to refer to the following Set-up below.

SET-UP 7.1. *Let $S = Spec(k)$ where k is an algebraically closed field of characteristic zero. Let C be an algebraic curve on a smooth surface and let p be a point on C. We form the admissible injective system of fat points $\{X_i\}_{i\in\mathbb{N}}$ over S by defining X_i to be the formal neighborhood $Spec(\mathcal{O}_{C,p}/\mathfrak{m}_p^{i+1})$ of the point p on C. Let $\mathcal{Y} = \varinjlim_{i\in\mathbb{N}} Y_i$ be any object of $\mathtt{Def}_{\mathcal{X}}$ such that $\varphi_i : Y_i \to X_i$ is smooth with Y_i of pure dimension d_i for all $i \in \mathbb{N}$.*

The main rationality result the author has obtained is the following.

THEOREM 7.2. *Let $\mathcal{Y}$ be a formal deformation as in Set-up 7.1. Then,*

$$\zeta_{\mathcal{Y}}(t) = p(t) \cdot \prod_{i=1}^{n} (1 - \mathbb{L}^{a_i} t^{b_i})^{-1}, \tag{7.1}$$

where $p(t) \in \mathcal{G}_k[t]$ and where $a_i \in \mathbb{Z}$ and $b_i \in \mathbb{N}$.

PROOF. By assumption, $\varphi_i : Y_i \to X_i$ is smooth, and, by shrinking Y_i if necessary, φ_i factors through an étale morphism[4] $\psi : Y_i \to \mathbb{A}^{d_i}_{X_i}$. Therefore,

$$\underline{\mathrm{Hom}}_S(X_i, Y_i) \cong \underline{\mathrm{Hom}}_S(X_i, \mathbb{A}^{d_i}_{X_i}) \times_{\mathbb{A}^{d_i}_{X_i}} Y_i \cong \underline{\mathrm{End}}_S(X_i) \times_S \mathbb{A}_S^{d_i(\mathrm{rank}_S(X_i)-1)} \times_S Y_i \tag{7.2}$$

Since $(Y_i)^{\mathrm{red}} = (Y_0)^{\mathrm{red}}$, we have

$$[\mathcal{A}_i(\mathcal{Y})]\mathbb{L}^{-d_i(\mathrm{rank}_S(X_i)-1)-e_i(\mathrm{rank}_S(X_i)-1)} = [Y_0][\underline{\mathrm{End}}_S(X_i)]\mathbb{L}^{-e_i(\mathrm{rank}_S(X_i)-1)} \tag{7.3}$$

Multiplying t^i on both sides and summing over i in $\mathcal{G}_k[[t]]$, we have

$$\zeta_{\mathcal{Y}}(t) = [Y_0] \sum_{i=0}^{\infty} [\underline{\mathrm{End}}_S(X_i)]\mathbb{L}^{-e_i(\mathrm{rank}_S(X_i)-1)} t^i. \tag{7.4}$$

Note that then that $\zeta_{\mathcal{Y}}(t) = [Y_0]\zeta_{C,p}(t)$, where $\zeta_{C,p}(t)$ is the auto-Igusa zeta of the algebraic germ (C, p) defined in [**Sto18**]. In loc. cit., we show that this series is of the form displayed in formula 7.1. □

Part of the argument in [**Sto18**] relies on rationality results of [**DL98**] and [**DL99**] and their later generalizations in [**CL08**] and [**CL15**]. In fact, the first part of this argument holds whenever the morphisms $\varphi_i : Y_i \to X_i$ are smooth with $\{X_i\}_{i\in I}$ any admissible injective system of fat points over any algebraic space S – i.e., regardless of whether or not X_i is subscheme of a plane curve C, formula 7.4 holds provided $\varphi_i : Y_i \to X_i$ is smooth for all $i \in I$. However, as of yet, rationality is only proven for the case when $\mathcal{Y}$ is as in Set-up 7.1. Thus, there are several directions for future results:

QUESTION 7.3. Let $\mathcal{Y}$ be a formal deformation such that $\varphi_i : Y_i \to X_i$ are smooth for all $i \in I$ with $\{X_i\}_{i\in I}$ any admissible system of fat points over an algebraic space S. Under what conditions on the admissible system $\{X_i\}_{i\in I}$ and on S will $\zeta_{\mathcal{Y}}(t)$ be rational – i.e., when will $\zeta_{\mathcal{Y}}(t)$ have a similar form as in formula 7.1?

---

[4]The existence of an étale morphism ψ follows from the proof of Theorem 8.3.3, page 182 of [**Ols16**] whose argument is taken from the proof Theorem 8.1 of [**LMB00**]. In fact, one may generalize by letting X_i can be merely a Deligne-Mumford Stack.

QUESTION 7.4. Assuming that the appropriate conditions are placed on $\{X_i\}_{i\in I}$ and S through investigations of the type discussed in Question 7.3, which formal deformations $\mathcal{Y}$ over $\{X_i\}_{i\in I}$ will have a zeta function $\zeta_{\mathcal{Y}}(t)$ of the form given in formula 7.1?

QUESTION 7.5. Note that formula 6.5 gives a relationship between $\eta_{\mathcal{Y}}(t)$ and $\zeta_{\mathcal{Y}}(t)$ in the most basic case (i.e., they are both equal to the classical motivic Igusa-zeta function). In general, what is the relationship between $\eta_{\mathcal{Y}}(t)$ and $\zeta_{\mathcal{Y}}(t)$? In particular, is there a way to transfer results vis-a-vis Question 7.3 and Question 7.4 to corresponding statments about $\eta_{\mathcal{Y}}(t)$? In general, if $\zeta_{\mathcal{Y}}(t)$ is rational, must $\eta_{\mathcal{Y}}(t)$ also be rational?

QUESTION 7.6. In the case of Set-up 7.1, there are cases where explicit computations of the auto Igusa-zeta function are possible. Are similar computations possible for the canonical Igusa-zeta function $\eta_{\mathcal{Y}}(t)$?

References

[Art69] M. Artin, *Algebraization of formal moduli. I*, Global Analysis (Papers in Honor of K. Kodaira), Univ. Tokyo Press, Tokyo, 1969, pp. 21–71. MR0260746

[CL08] Raf Cluckers and François Loeser, *Constructible motivic functions and motivic integration*, Invent. Math. **173** (2008), no. 1, 23–121, DOI 10.1007/s00222-008-0114-1. MR2403394

[CL15] Raf Cluckers and François Loeser, *Motivic integration in all residue field characteristics for Henselian discretely valued fields of characteristic zero*, J. Reine Angew. Math. **701** (2015), 1–31, DOI 10.1515/crelle-2013-0025. MR3331725

[DL98] Jan Denef and François Loeser, *Motivic Igusa zeta functions*, J. Algebraic Geom. **7** (1998), no. 3, 505–537. MR1618144

[DL99] Jan Denef and François Loeser, *Germs of arcs on singular algebraic varieties and motivic integration*, Invent. Math. **135** (1999), no. 1, 201–232, DOI 10.1007/s002220050284. MR1664700

[Eke09] Torsten Ekedahl, *The grothendieck group of algebraic stacks*, `arXiv:0903.3143v2`, 2009.

[GD67] Alexander Grothendieck and Jean Dieudonné, *Éléments de géométrie algébrique IV*, Publications Mathématiques, vol. 20, 24, 28, 32, Institute des Hautes Études Scientifiques., 1964-1967.

[Gro62] Alexander Grothendieck, *Fondements de la géométrie algébrique. [Extraits du Séminaire Bourbaki, 1957–1962.]* (French), Secrétariat mathématique, Paris, 1962. MR0146040

[LMB00] Gérard Laumon and Laurent Moret-Bailly, *Champs algébriques* (French), Ergebnisse der Mathematik und ihrer Grenzgebiete. 3. Folge. A Series of Modern Surveys in Mathematics [Results in Mathematics and Related Areas. 3rd Series. A Series of Modern Surveys in Mathematics], vol. 39, Springer-Verlag, Berlin, 2000. MR1771927

[Lur09] Jacob Lurie, *Higher topos theory*, Annals of Mathematics Studies, vol. 170, Princeton University Press, Princeton, NJ, 2009. MR2522659

[Ols06] Martin C. Olsson, Hom*-stacks and restriction of scalars*, Duke Math. J. **134** (2006), no. 1, 139–164, DOI 10.1215/S0012-7094-06-13414-2. MR2239345

[Ols16] Martin C. Olsson, *Algebraic spaces and stacks*, American Mathematical Society, 2016.

[Scha] H. Schoutens, *Schemic Grothendieck Rings I*, `websupport1.citytech.cuny.edu/faculty/hschoutens/PDF/SchemicGrothendieckRingPartII.pdf`, accessed on 4/01/2017.

[Schb] H. Schoutens, *Schemic Grothendieck Rings II*, `websupport1.citytech.cuny.edu/faculty/hschoutens/PDF/SchemicGrothendieckRingPartII.pdf`, accessed on 4/01/2017.

[Sta17] The Stacks Project Authors, *Stacks Project*, `http://stacks.math.columbia.edu`, 2017.

[Sto17] Andrew R. Stout, *On the auto Igusa-zeta function of an algebraic curve*. part 1, J. Symbolic Comput. **79** (2017), no. part 1, 156–185, DOI 10.1016/j.jsc.2016.08.011. MR3550360

[Sto18] Andrew R. Stout, *The auto Igusa-zeta function of a plane curve singularity is rational*, Proc. Amer. Math. Soc. **In Press.** (2018).

Department of Mathematics, Borough of Manhattan Community College, CUNY, New York, New York 10007

Current address: 199 Chambers Street, New York, New York 10007

Email address: astout@bmcc.cuny.edu

Contemporary Mathematics
Volume **724**, 2019
https://doi.org/10.1090/conm/724/14588

Computing heights on weighted projective spaces

Jorgo Mandili and Tony Shaska

Dedicated to Mehmet Likaj, on the occasion of his 70th birthday

Abstract. In this note we extend the concept height on projective spaces to that of weighted height on weighted projective spaces and show how such a height can be computed. We prove some of the basic properties of the weighted height and show how it can be used to study hyperelliptic curves over $\mathbb{Q}$. Some examples are provided from the weighted moduli space of binary sextics and octavics.

1. Introduction

Let $\mathbb{WP}^n_w(K)$ be the weighted projective space of weights $\mathfrak{w} = (w_0, \dots, w_n)$ over a field K of characteristic zero. Is there a way to measure the "size" of points in $\mathbb{WP}^n_w(K)$ similar to the height on the projective space $\mathbb{P}^n(K)$? The answer comes from **[9]** where the concept of the height is defined for weighted projective spaces. How does one compute the height of a point in $\mathbb{WP}^n_w(K)$? Moreover, how do we get a tuple $\mathbf{x} = (x_0, \dots x_n)$ such that it is the minimal representative for the point $\mathfrak{p} = [x_0, \cdots, x_n] \in \mathbb{WP}^n(K)$? In this short paper we explore these questions.

The motivation for considering the above question comes from the theory of hyperelliptic or superelliptic curves. The isomorphism classes of a genus $g \geq 2$ hyperelliptic curve $C: \; y^2 z^{2g} = f(x, z)$ correspond to the tuple of generators of the ring of invariants $S(2, 2g+2)$ of binary forms evaluated at the binary form $f(x, z)$. Such ring of invariants is a weighted projective space. Hence, determining a canonical minimal tuple for any point in $\mathbb{WP}^n_w(K)$ would give a one to one correspondence between the isomorphism classes of curves and such minimal tuples. We illustrate briefly with the genus 2 curves.

In **[6]** we created a database of isomorphism classes of genus 2 curves defined over $\mathbb{Q}$. Every such isomorphism class was identified uniquely by a set of absolute invariants (i_1, i_2, i_3); see **[6]** for details. These invariants are defined in terms of the Igusa invariants J_2, J_4, J_6, J_{10}. Why not identify the curve with the tuple (J_2, J_4, J_6, J_{10}) instead of (i_1, i_2, i_3)? If we do so then we have determine how to pick the smaller size tuple for any point $\mathfrak{p} = [J_2, J_4, J_6, J_{10}]$ and how to do this in a canonical way. The goal of this paper is to address such issues for any weighted projective space.

2010 *Mathematics Subject Classification.* Primary 14H10, 14H45.

In this paper we define a *normalization* of points $\mathfrak{p} \in \mathbb{WP}^n_w(\mathbb{Q})$ which is the representing tuple of $\mathfrak{p}$ with smallest coefficients. We show that this normalization is unique up to multiplication by a primitive d-th root of unity, where $d = \gcd(w_0, \ldots, w_n)$ and is unique when $\mathbb{WP}^n_w(\mathbb{Q})$ is well-formed. The height of a point $\mathfrak{p} \in \mathbb{WP}^n_{\mathfrak{w}}(\mathbb{Q})$ is the weighted absolute value of coordinates of $\mathfrak{p}$, when $\mathfrak{p}$ is normalized.

We also define the *absolutely normalized* tuples which is a normalization over the algebraic closure $\bar{\mathbb{Q}}$. This is a normalization by multiplying by scalars which are allowed to be in $\bar{\mathbb{Q}}$. The height of an absolutely normalized tuple is called an *absolute height* for analogy with the terminology in [**30**]. In other words the *absolute height* of a point $\mathfrak{p} \in \mathbb{WP}^n_{\mathfrak{w}}(\mathbb{Q})$ is the the weighted absolute value of coordinates of $\mathfrak{p}$, when $\mathfrak{p}$ is absolutely normalized.

The paper is organized as follows. In Section 2 we give a brief introduction to weighted projective spaces $\mathbb{WP}^n_w(K)$. A standard reference here is [**16**]. We consider both *well-formed* and not *well-formed* weighted projective spaces. For any point $\mathbf{x} = (x_0, \ldots, x_n) \in (\mathbb{Z})^{n+1} \setminus \{0\}$, we define the *weighted greatest common divisor* $wgcd(\mathbf{x})$ as the product of all primes $p \in \mathbb{Z}$ such that for all $i = 0, \ldots, n$, we have $p^i \mid x_i$. We will call a point $\mathfrak{p} \in \mathbb{WP}^n_{\mathfrak{w}}(\mathbb{Q})$ **normalized** if it has $wgcd(\mathfrak{p}) = 1$.

In Section 3 we follow [**9**] and define the *weighted projective height* on $\mathbb{WP}^n_w(\mathbb{Q})$ and show that this is well-defined. We prove a version of the Northcott's theorem for the weighted projective height and determine for what conditions on integers $w_o, \ldots w_n$ the normalized tuple is unique. Analogously we extend the definitions and results over the algebraic closure. We show how to determine all twists of a given point $\mathfrak{p} \in \mathbb{WP}^n_{\mathfrak{w}}(\mathbb{Q})$ of height $h \leq \mathfrak{h}(\mathfrak{p})$. When the set of weights is $\mathfrak{w} = (1, \ldots, 1)$, then the weighted projective space is simply the projective space $\mathbb{P}^n(K)$ and our weighted moduli height becomes the usual height on $\mathbb{P}^n(K)$ as defined in [**11**].

The notion of weighted height and absolute weighted height is used in [**5**], to study the weighted moduli space of binary sextics and in [**7**] to study the weighted moduli space of binary octavics. Both cases lead to creating databases of genus 2 or genus 3 hyperelliptic curves with small absolute moduli height. For connections of weighted projective spaces and the algebraic curves or other topics on databases of hyperelliptic curves the reader can check [**6**]. We give some examples for genus 2 curves and genus 3 hyperelliptic curves, which were the main motivation behind this paper. It remains to be seen if there are any explicit relations between the weighted moduli height, moduli height, and height as in [**30**].

The concept of weighted height in weighted projective spaces, surprisingly seems unexplored before. The only reference we could find was the unpublished report in [**14**] which defines the function *Size* similarly to our height with different motivations. Our goal in writing this short note was to simply provide a brief introduction to heights in weighted projective spaces. We assume the reader is familiar with the concept of height in projective spaces as in [**11**] and [**20**]. A more detailed study of heights on weighted projective spaces over any number field is given in [**9**].

Notation The algebraic closure of a field K is denoted by $\bar{K}$. For an algebraically number field K we denote by $\mathcal{O}_K$ its ring of integers and by M_K the set of all absolute values in K. A point $\mathfrak{p} \in \mathbb{WP}^n_{\mathfrak{w}}(K)$ is denoted by $\mathfrak{p} = [x_0 : x_1 : \cdots : x_n]$ and the tuple of coordinates $\mathbf{x} = (x_0, x_1, \ldots, x_n)$. By a "curve" we always mean the isomorphism class of a smooth, irreducible curve.

2. Weight projective spaces

Let K be a field of characteristic zero and $(q_0, \dots, q_n) \in \mathbb{Z}^{n+1}$ a fixed tuple of positive integers called **weights**. Consider the action of $K^\star = K \setminus \{0\}$ on $\mathbb{A}^{n+1}(K)$ as follows

$$\lambda \star (x_0, \dots, x_n) = (\lambda^{q_0} x_0, \dots, \lambda^{q_n} x_n)$$

for $\lambda \in K^*$. The quotient of this action is called a **weighted projective space** and denoted by $\mathbb{WP}^n_{(q_0,\dots,q_n)}(K)$. It is the projective variety $Proj\,(K[x_0, ..., x_n])$ associated to the graded ring $K[x_0, \dots, x_n]$ where the variable x_i has degree q_i for $i = 0, \dots, n$.

We denote greatest common divisor of $q_0, \dots, q_n$ by $\gcd(q_0, \dots, q_n)$. The space $\mathbb{WP}^n_w$ is called **well-formed** if

$$\gcd(q_0, \dots, \hat{q}_i, \dots, q_n) = 1, \quad \text{for each } \ i = 0, \dots, n.$$

While most of the papers on weighted projective spaces are on well-formed spaces, we do not assume that here. We will denote a point $\mathfrak{p} \in \mathbb{WP}^n_w(K)$ by $\mathfrak{p} = [x_0 : x_1 : \dots : x_n]$.

Let K be a number field and $\mathcal{O}_K$ its ring of integers. The group action $K^\star$ on $\mathbb{A}^{n+1}(K)$ induces a group action of $\mathcal{O}_K$ on $\mathbb{A}^{n+1}(K)$. By $Orb(\mathfrak{p})$ we denote the $\mathcal{O}_K$-orbit in $\mathbb{A}^{n+1}(\mathcal{O}_K)$ which contains $\mathfrak{p}$. For any point $\mathfrak{p} = [x_0 : \dots : x_n] \in \mathbb{WP}^n_w(K)$ we can assume, without loss of generality, that $\mathfrak{p} = [x_0 : \dots : x_n] \in \mathbb{WP}^n_w(\mathcal{O}_K)$. The height for weighted projective spaces will be defined in the next section.

For the rest of this section we assume $K = \mathbb{Q}$. For the tuple $\mathbf{x} = (x_0, \dots, x_n) \in \mathbb{Z}^{n+1}$ we define the **weighted greatest common divisor** with respect to the absolute value $|\cdot|_v$, denoted by $wgcd_v(\mathbf{x})$,

$$wgcd_v(\mathbf{x}) := \prod_{\substack{d^{q_i} \mid x_i \\ d \in \mathbb{Z}}} |d|_v$$

as the product of all divisors $d \in \mathbb{Z}$ such that for all $i = 0, \dots, n$, we have $d^i \mid x_i$. In other words, the *weighted greatest common divisor with respect to the set of weights* $\mathfrak{w}$ is the largest integer d such that

$$d^{q_i} \mid x_i, \quad \text{for all } i = 0, \dots, n.$$

We will call a point $\mathfrak{p} \in \mathbb{WP}^n_{\mathfrak{w}}(\mathbb{Q})$ **normalized** if $wgcd(\mathfrak{p}) = 1$.

DEFINITION 1. We will call a point $\mathfrak{p} \in \mathbb{WP}^n_{\mathfrak{w}}(\mathbb{Q})$ a **normalized point** if the weighted greatest common divisor of its coordinates is 1.

LEMMA 1. *Let* $\mathfrak{w} = (q_0, \dots, q_n)$ *be a set of weights and* $d = \gcd(q_0, \dots, q_n)$. *For any point* $\mathfrak{p} \in \mathbb{WP}^n_{\mathfrak{w}}(\mathbb{Q})$, *the point*

$$\mathfrak{q} = \frac{1}{wgcd(\mathfrak{p})} \star \mathfrak{p}$$

is normalized. Moreover, this normalization is unique up to a multiplication by a d-root of unity.

PROOF. Let $\mathfrak{p} = [x_0 : \dots, x_n] \in \mathbb{WP}^n_w(\mathbb{Q})$ and $\mathfrak{p}_1 = [\alpha_0 : \dots : \alpha_n]$ and $\mathfrak{p}_2 = [\beta_0 : \dots : \beta_n]$ two different normalizations of $\mathfrak{p}$. Then exists non-zero $\lambda_1, \lambda_2 \in \mathbb{Q}$ such that

$$\mathfrak{p} = \lambda_1 \star \mathfrak{p}_1 = \lambda_2 \star \mathfrak{p}_2,$$

or in other words

$$(x_0, \dots, x_n) = (\lambda_1^{q_0}\alpha_0, \dots, \lambda_1^{q_i}\alpha_i, \dots) = (\lambda_2^{q_0}\beta_0, \dots, \lambda_2^{q_i}\beta_i, \dots).$$

Thus,

$$(\alpha_0, \dots, \alpha_i, \dots, \alpha_n) = (r^{q_0}\beta_0, \dots, r^{q_i}\beta_i, \dots, r^{q_n}\beta_n).$$

for $r = \frac{\lambda_2}{\lambda_1} \in K$. Thus, $r^{q_i} = 1$ for all $i = 0, \dots, n$. Therefore, $r^d = 1$. This completes the proof. □

Thus we have the following:

COROLLARY 1. *For any point* $\mathfrak{p} = [x_0 : \cdots : \mathbf{x}_n] \in \mathbb{WP}^n_{\mathfrak{w}}(\mathbb{Q})$, *if the greatest common divisors of all the weights is 1, then the normalization of* $\mathfrak{p}$ *is unique.*

Here is an example which illustrates the Lemma.

EXAMPLE 1. *Let* $\mathfrak{p} = [x_0, x_1, x_2, x_3] \in \mathbb{WP}^3_{(2,4,6,10)}(\mathbb{Q})$ *be a normalized point. Hence,*

$$wgcd(x_0, x_1, x_2, x_3) = 1.$$

Since $d = \gcd(2, 4, 6, 10) = 2$, *then we can take* r *such that* $r^2 = 1$. *Hence,* $r = \pm 1$. *Therefore, the point*

$$(-1) \star \mathfrak{p} = [-x_0 : x_1 : -x_2 : -x_3]$$

is also be normalized.

However, if $\mathfrak{p} = [x_0, x_1, x_2, x_3] \in \mathbb{WP}^3_{(1,2,3,5)}(\mathbb{Q})$ *is normalized then it is unique, unless some of the coordinates are zero. For example the points* $[0, 1, 0, 0]$ *and* $[0, -1, 0, 0]$ *are equivalent and both normalized.*

Next we give two examples, which were the main motivation behind this note.

EXAMPLE 2 (Weighted projective space of binary sextics). *The ring of invariants of binary sextics is generated by the* **basic arithmetic invariants**, *or as they sometimes called,* **Igusa invariants** (J_2, J_4, J_6, J_{10}) *as defined in* **[21]**. *Two genus 2 curves* $\mathcal{X}$ *and* $\mathcal{X}'$ *are isomorphic if and only if there exists* $\lambda \in K^*$ *such that*

$$J_{2i}(\mathcal{X}) = \lambda^{2i} J_{2i}(\mathcal{X}'), \quad for \quad i = 1, 2, 3, 5.$$

We take the set of weights $\mathfrak{w} = (2, 4, 6, 10)$ *and considered the weighted projective space* $\mathbb{WP}_{(2,4,6,10)}(\mathbb{Q})$. *Thus, the invariants of a sextic define a point in a weighted projective space* $[J_2 : J_4 : J_6 : J_{10}] \in \mathbb{WP}_{\mathfrak{w}}(\mathbb{Q})$ *and every genus 2 curve correspond to a point in* $\mathbb{WP}^3_{\mathfrak{w}}(\mathbb{Q}) \setminus \{J_{10} \neq 0\}$. *There is a bijection between*

$$\phi: \ \mathbb{WP}^3_{(2,4,6,10)} \setminus \{J_{10} \neq 0\} \to \mathcal{M}_2,$$

with ϕ *provided explicitly in* [**25**, Theorem 1].

Using the notion of a normalized point as above we have the following:

COROLLARY 2. *Normalized points in* $\mathbb{WP}^3_{(2,4,6,10)}(\mathbb{Q})$ *occur in pairs. In other words, for every normalized point* $\mathfrak{p} = [J_2, J_4, J_6, J_{10}]$, *there is another normalized point* $\mathfrak{p}' = [-J_2, J_4, -J_6, -J_{10}]$ *equivalent to* $\mathfrak{p}$. *Moreover,* $\mathfrak{p}$ *and* $\mathfrak{p}'$ *are isomorphic over the Gaussian integers.*

PROOF. Let $\mathcal{X}$ be a genus 2 curve with equation $y^2 = f(x)$ and $[J_2, J_4, J_6, J_{10}]$ its corresponding invariants. The transformation $x \mapsto \sqrt{-1} \cdot x$ with give a curve $\mathcal{X}'$ with invariants $[-J_2 : J_4 : -J_6 : -J_{10}]$ and the same weighted moduli height.

If two weighted moduli points have the same minimal absolute height, then they differ up to a multiplication by a unit. Hence,

$$[J_2' : J_4' : J_6' : J_{10}'] = [d^2 \cdot J_2 : d^4 \cdot J_4 : d^6 \cdot J_6 : d^{10} \cdot J_{10}]$$

such that d^2 is a unit. Then, $d^2 = \pm 1$. Hence, $d = \sqrt{-1}$. □

So unfortunately for any genus 2 curve we have two corresponding normalized points $[\pm J_2, J_4, \pm J_6, \pm J_{10}]$. In **[5]** this problem is solve by taking always the point $[|J_2|, J_4, \pm J_6, \pm J_{10}]$ or by considering the space $\mathbb{WP}^3_{(1,2,3,5)}(\mathbb{Q})$ instead.

EXAMPLE 3 (Weighted projective space of binary octavics). *Every irreducible, smooth, hyperelliptic genus 3 curve has equation $y^2 z^6 = f(x,z)$, where $f(x,z)$ is a binary octavic with non-zero discriminant. The ring of invariants of binary octavics is generated by invariants $J_2, \dots, J_8$, which satisfy an algebraic equation as in* **[29**, Thm. 6**]**. *Two genus 3 hyperelliptic curves $\mathcal{X}$ and $\mathcal{X}'$ are isomorphic over a field K if and only if there exists some $\lambda \in k \setminus \{0\}$ such that*

$$J_i(\mathcal{X}) = \lambda^i J_i(\mathcal{X}'), \ \textit{for} \ \ i = 2, \dots, 7.$$

There is another invariant J_{14} given in terms of $J_2, \dots J_7$ which is the discriminant of the binary octavic.

Hence, there is a bijection between the hyperelliptic locus in the moduli space of genus 3 curves and the weighted projective space $\mathbb{WP}^5_{(2,3,4,5,6,7)}(K) \setminus \{J_{14} \neq 0\}$. Since $d = \gcd(2,3,4,5,6,7) = 1$ then we have:

COROLLARY 3. *For every genus 3 hyperelliptic curve $\mathcal{X}$, defined over a field K, the corresponding normalized point*

$$\mathfrak{p} = [J_2 : J_3 : J_4 : J_5 : J_6 : J_7] \in \mathbb{WP}^5_{(2,3,4,5,6,7)}(K)$$

is unique.

EXAMPLE 4. *Consider the curve $y^2 = x^8 - 1$. The moduli point in $\mathbb{WP}^5_{\mathfrak{w}}(\mathbb{Q})$ is*

$$\mathfrak{p} = \left[-2^3 \cdot 5 \cdot 7, 0, 2^{10} \cdot 7^4, 0, 2^{15} \cdot 7^6, 0, -2^{19} \cdot 5 \cdot 7^8\right]$$

Then, wgcd$(\mathbf{x}) = 2$. Hence, the point $\mathfrak{p}$ normalized becomes

$$\frac{1}{2} \star \mathfrak{p} = \left[-2 \cdot 5 \cdot 7, 0, 2^6 \cdot 7^4, 0, 2^9 \cdot 7^6, 0, -2^{11} \cdot 5 \cdot 7^8\right].$$

In **[7]** we use such normalized points to create a database of genus 3 hyperelliptic curves defined over $\mathbb{Q}$.

2.1. Absolutely normalized points. For any point $\mathfrak{p} = [x_0 : \cdots : x_n] \in \mathbb{WP}^n_{\mathfrak{w}}(\mathbb{Q})$ we may assume that $x_i \in \mathbb{Z}$ for $i = 0, \dots, n$ and define

$$\overline{wgcd}(\mathfrak{p}) = \prod_{\lambda \in \bar{\mathbb{Q}},\, \lambda^{q_i} | x_i} |\lambda|$$

as the product of all $\lambda \in \bar{\mathbb{Q}}$, such that for all $i = 0, \dots, n$, $\lambda^i \in \mathbb{Z}$ and $\lambda^i | x_i$. In other words, the *absolute weighted greatest common divisor* of a tuple $\mathbf{x} = (x_0, \dots, x_n)$ with respect to the set of weights $\mathfrak{w} = (q_0, \dots, q_n)$ is the largest real number d such that

$$d^{q_i} \in \mathbb{Z} \quad \text{and} \quad d^{q_i} \mid x_i, \quad \text{for all } i = 0, \dots n.$$

A point $\mathfrak{p} = [x_0 : \cdots : x_n] \in \mathbb{WP}^n_{\mathfrak{w}}(\mathbb{Q})$ is called **absolutely normalized** or **normalized over** $\bar{\mathbb{Q}}$ if $\overline{wgcd}(\mathfrak{p}) = 1$.

Definition 2. A point $\mathfrak{p} = [x_0 : \cdots : x_n] \in \mathbb{WP}^n_{\mathfrak{w}}(\mathbb{Q})$ is called **absolutely normalized** or **normalized over the algebraic closure** if $\overline{wgcd}(\mathfrak{p}) = 1$.

Lemma 2. *For any point $\mathfrak{p} = [x_0 : \cdots : x_n] \in \mathbb{WP}^n_{\mathfrak{w}}(\mathbb{Q})$ its normalization over the algebraic closure*

$$\bar{\mathfrak{p}} = \frac{1}{\overline{wgcd}(\mathfrak{p})} \star \mathfrak{p}$$

is unique up to a multiplication by a d-th root of unity.

Proof. Let $\mathfrak{p} = [x_0 : \ldots, x_n] \in \mathbb{WP}^n_w(\mathbb{Q})$ and $\mathfrak{p}_1 = [\alpha_0 : \cdots : \alpha_n]$ and $\mathfrak{p}_2 = [\beta_0 : \cdots : \beta_n]$ two different normalizations of $\mathfrak{p}$ over $\overline{\mathbb{Q}}$. Then exists non-zero $\lambda_1, \lambda_2 \in \overline{\mathbb{Q}}$ such that

$$\mathfrak{p} = \lambda_1 \star \mathfrak{p}_1 = \lambda_2 \star \mathfrak{p}_2,$$

or in other words

$$(x_0, \ldots, x_n) = (\lambda_1^{q_0}\alpha_0, \ldots, \lambda_1^{q_i}\alpha_i, \ldots) = (\lambda_2^{q_0}\beta_0, \ldots, \lambda_2^{q_i}\beta_i, \ldots).$$

Thus,

$$(\alpha_0, \ldots, \alpha_i, \ldots, \alpha_n) = (r^{q_0}\beta_0, \ldots, r^{q_i}\beta_i, \ldots, r^{q_n}\beta_n).$$

for $r = \frac{\lambda_2}{\lambda_1} \in \overline{\mathbb{Q}}$. Thus, $r^{q_i} = 1$ for all $i = 0, \ldots, n$. Therefore, $r^d = 1$. This completes the proof. □

Two points $\mathfrak{p}$ and $\mathfrak{q}$ in $\mathbb{WP}^n_{\mathfrak{w}}(\mathbb{Q})$ are called **twists** of each other if they are equivalent in $\mathbb{WP}^n_{\mathfrak{w}}(\overline{\mathbb{Q}})$ but $Orb_{\mathbb{Q}}(\mathfrak{p})$ is not the same as $Orb_{\mathbb{Q}}(\mathfrak{q})$. Hence, we have the following.

Lemma 3. *Let $\mathfrak{p}$ and $\mathfrak{p}'$ be normalized points in $\mathbb{WP}^n_{\mathfrak{w}}(\mathbb{Q})$. Then $\mathfrak{p}$ and $\mathfrak{p}'$ are twists of each other if and only if there exists $\lambda \in \overline{\mathbb{Q}}^\star$ such that $\lambda \star \mathfrak{p} = \mathfrak{p}'$.*

Next we see another example from genus 2 curves.

Example 5. *Let $\mathcal{X}$ be the genus two curve with equation $y^2 = x^6 - 1$ and J_2, J_4, J_6, and J_{10} its Igusa invariants. Then the isomorphism class of $\mathcal{X}$ is determined by the point $\mathfrak{p} = [240, 1620, 119880, 46656] \in \mathbb{WP}^3_{(2,4,6,10)}(\mathbb{Q})$. Thus,*

$$\mathfrak{p} = [240, 1620, 119880, 46656] = [2^4 \cdot 3 \cdot 5;\ 2^2 \cdot 3^4 \cdot 5;\ 2^3 \cdot 3^4 \cdot 5 \cdot 37;\ 2^6 \cdot 3^6].$$

Therefore,

$$wgcd(240, 1620, 119880, 46656) = 1$$
$$\overline{wgcd}(240, 1620, 119880, 46656) = \sqrt{6}.$$

Hence, $\mathfrak{p}$ is normalized but not absolutely normalized. The point $\mathfrak{p}$ has twists,

$$\mathfrak{p}_1 = \frac{1}{\sqrt{2}} \star \mathfrak{p}_1 = [120, 405, 14985, 1458] = [2^3 \cdot 3 \cdot 5 \ :\ 3^4 \cdot 5 \ :\ 3^4 \cdot 5 \cdot 37 \ :\ 2 \cdot 3^6],$$

$$\mathfrak{p}_2 = \frac{1}{\sqrt{3}} \star \mathfrak{p}_1 = [80, 180, 4440, 192] = [2^4 \cdot 5 \ :\ 2^2 \cdot 3^2 \cdot 5 \ :\ 2^3 \cdot 3 \cdot 5 \cdot 37 \ :\ 2^6 \cdot 3],$$

and the absolutely normalized point of $\mathfrak{p}$ which is

$$\bar{\mathfrak{p}} = \frac{1}{\sqrt{6}} \star \mathfrak{p}_1 = [40, 45, 555, 6] = [2^3 \cdot 5,\ 3^2 \cdot 5,\ 3 \cdot 5 \cdot 37,\ 2 \cdot 3]$$

Notice that $\bar{\mathfrak{p}}$ has only one twist

$$\bar{\mathfrak{p}}' = [-2^3 \cdot 5,\ 3^2 \cdot 5,\ -3 \cdot 5 \cdot 37,\ -2 \cdot 3]$$

which is also normalized.

We can do better even with the genus 3 curve from Example 4.

EXAMPLE 6. *The normalized moduli point in* $\mathbb{WP}_{\mathfrak{w}}^5(\mathbb{Q})$ *of the curve* $y^2 = x^8 - 1$ *is*

$$\frac{1}{2} \star \mathfrak{p} = \left[-2 \cdot 5 \cdot 7, 0, 2^6 \cdot 7^4, 0, 2^9 \cdot 7^6, 0, -2^{11} \cdot 5 \cdot 7^8\right].$$

Then, $\overline{wgcd}(\mathfrak{p}) = \frac{\mathfrak{i}}{\sqrt{14}}$, *for* $\mathfrak{i}^2 = -1$. *Then its absolutely normalized form is*

$$\bar{\mathfrak{p}} = \left[5, 0, 2^4 \cdot 7^2, 0, 2^6 \cdot 7^3, 0, -2^7 \cdot 5 \cdot 7^4\right].$$

□

In the next section we will introduce some measure of the magnitude of points in weighted moduli spaces $\mathbb{WP}_{\mathfrak{w}}^n(K)$ and show that the process of normalization and absolute normalization lead us to the representation of points in $\mathbb{WP}_{\mathfrak{w}}^n(K)$ with smallest possible coordinates.

3. Heights on the weighted projective spaces

Let K be an algebraic number field and $[K : \mathbb{Q}] = n$ and its ring of integers $\mathcal{O}_K$. With M_K we denote the set of all absolute values in K. For $v \in M_K$, the **local degree at** v, denoted n_v is $n_v = [K_v : \mathbb{Q}_v]$, where $K_v, \mathbb{Q}_v$ are the completions with respect to v. As above $\mathbb{WP}^n(K)$ is the projective space with weights $w = (q_0, \dots, q_n)$, and $\mathfrak{p} \in \mathbb{WP}^n(K)$ a point with coordinates $\mathfrak{p} = [x_0, \dots, x_n]$, for $x_i \in K$. The **multiplicative height** of $\mathfrak{p}$ is defined as follows

$$\mathfrak{h}_K(\mathfrak{p}) := \prod_{v \in M_K} \max\left\{ |x_0|_v^{n_v/q_0}, \dots, |x_n|_v^{n_v/q_n} \right\}$$

Let $\mathfrak{p} = [x_0, \dots, x_n] \in \mathbb{WP}^n(\mathbb{Q})$ with weights $w = (q_0, \dots, q_n)$. It is clear that $\mathfrak{p}$ will have a representative $[y_0, \dots, y_n]$ such that $y_i \in \mathbb{Z}$ for all i and $wgcd(y_0, \dots, y_n) = 1$. With such representative for the coordinates of $\mathfrak{p}$, the non-Archimedean absolute values give no contribution to the height, and we obtain

$$\mathfrak{h}_{\mathbb{Q}}(\mathfrak{p}) = \max_{0 \le j \le n} \left\{ |x_j|_\infty^{1/q_j} \right\}$$

So for a tuple $\mathbf{x} = (x_0 : \dots : x_n)$ the height of the corresponding point $\mathfrak{p} = [\mathbf{x}]$ is

$$\mathfrak{h}(\mathfrak{p}) = \frac{1}{wgcd(\mathbf{x})} \max\left\{|x_0|^{1/q_0}, \dots, |x_n|^{1/q_n}\right\}.$$

We combine some of the properties of $\mathfrak{h}(\mathfrak{p})$ in the following:

PROPOSITION 1. *Then the following are true:*

i) The function $\mathfrak{h} : \mathbb{WP}_{\mathfrak{w}}^n(\mathbb{Q}) \to \mathbb{R}$ *is well-defined.*

ii) A normalized point $\mathfrak{p} = [x_0 : \dots : x_n] \in \mathbb{WP}_{\mathfrak{w}}^n(\mathbb{Q})$ *is the point with smallest coordinates in its orbit* $Orb(\mathfrak{p})$.

iii) For any constant $c > 0$ *there are only finitely many points* $\mathfrak{p} \in \mathbb{WP}_w^n(\mathbb{Q})$ *such that* $\mathfrak{h}(\mathfrak{p}) \le c$.

PROOF. i) It is enough to show that two normalizations of the same point $\mathfrak{p} \in \mathbb{WP}_{\mathfrak{w}}^n(\mathbb{Q})$ have the same height. Let $\mathfrak{p}$ and $\mathfrak{q}$ be such normalizations. Then from Lemma 1 we have $\mathfrak{p} = r \star \mathfrak{q}$, where $r^d = 1$. Thus,

$$\mathfrak{h}(\mathfrak{p}) = \mathfrak{h}(r \star \mathfrak{q}) = |r| \cdot \mathfrak{h}(\mathfrak{q}) = \mathfrak{h}(\mathfrak{q}).$$

ii) This is obvious from the definition.

iii) Let $\mathfrak{p} \in \mathbb{WP}^n_{\mathfrak{w}}(\mathbb{Q})$. It is enough to count only normalized points $\mathfrak{p} = [x_0 : \cdots : x_n] \in \mathbb{WP}^n_w(\mathbb{Z})$ such that $\mathfrak{h}(\mathfrak{p}) \leq c$. For every coordinate x_i there are only finitely values in $\mathbb{Z}$ such that $|x_i|_v^{1/q_i} \leq c$. Hence, the result holds.

□

Part iii) of the above is the analogue of the Northcott's theorem in projective spaces.

REMARK 1. *If the set of weights* $\mathfrak{w} = (1, \dots 1)$ *then* $\mathbb{WP}^n_{\mathfrak{w}}(\mathbb{Q})$ *is simply the projective space* $\mathbb{P}^n(\mathbb{Q})$ *and the height* $\mathfrak{h}(\mathfrak{p})$ *correspond to the height of a projective point as defined in* **[30]**.

Let's see an example how to compute the height of a point.

EXAMPLE 7. *Let* $\mathfrak{p} = (2^2, 2 \cdot 3^4, 2^6 \cdot 3, 2^{10} \cdot 5^{10}) \in \mathbb{WP}^3_{(2,4,6,10)}(\mathbb{Q})$. *Notice that* $\mathfrak{p}$ *is normalized, which implies that*

$$\mathfrak{h}(\mathfrak{p}) = \max \left\{ 2, 2^{1/4} \cdot 3, 3^{1/6}, 2 \cdot 5 \right\} = 10$$

However, the point $\mathfrak{q} = (2^2, 2^4 {\cdot} 3^4, 2^6 {\cdot} 3, 2^{10} {\cdot} 5^{10}) \in \mathbb{WP}^3_{(2,4,6,10)}(\mathbb{Q})$ *can be normalized to* $(1, 3^4, 3, 5^{10})$ *which has height*

$$\mathfrak{h}(\mathfrak{q}) = \max \left\{ 1, 3, 3^{1/6}, 5 \right\} = 5.$$

A proof for the following will be provided in **[9]**.

LEMMA 4. *Let* $\mathfrak{p} \in \mathbb{WP}^n(K)$ *with weights* $w = (q_0, \dots, q_n)$ *and* L/K *be a finite extension. Then,*

$$\mathfrak{h}_L(P) = \mathfrak{h}_K(P)^{[L:K]}.$$

3.1. Absolute heights. We can define the height on $\mathbb{WP}^n(\overline{\mathbb{Q}})$. The height of a point on $\mathbb{WP}^n(\overline{\mathbb{Q}})$ is called the **weighted absolute (multiplicative) height** and is the function

$$\begin{aligned} \tilde{\mathfrak{h}} : \mathbb{WP}^n(\bar{\mathbb{Q}}) &\to [1, \infty) \\ \tilde{\mathfrak{h}}(\mathfrak{p}) &= \mathfrak{h}_K(P)^{1/[K:\mathbb{Q}]}, \end{aligned}$$

where $\mathfrak{p} \in \mathbb{WP}^n(K)$, for any K. Then, the absolute weight height is given by

$$\tilde{\mathfrak{h}}_{\mathbb{Q}}(\mathfrak{p}) = \frac{1}{wgcd(\mathfrak{p})} \max \left\{ |x_0|^{1/q_0}, \dots, |x_n|^{1/q_n} \right\} \tag{1}$$

Let's see an example which compares the height of a point with the absolute height.

EXAMPLE 8. *Let* $\mathfrak{p} = [0 : 2 : 0 : 0] \in \mathbb{WP}^3_{(2,4,6,10)}(\mathbb{Q})$. *Then* $\mathfrak{p}$ *is normalized and therefore* $\mathfrak{h}(\mathfrak{p}) = 2$. *However, it absolute normalization is* $\mathfrak{q} = \frac{1}{2^{1/4}} \star \mathfrak{p} = [0 : 0 : 1 : 0]$. *Hence,* $\tilde{\mathfrak{h}}(\mathfrak{p}) = 1$.

REMARK 2. *As a consequence of the above results it is possible to "sort" the points in* $\mathbb{WP}^n_{\mathfrak{w}}(\bar{K})$ *according to the absolute height and even determine all the twists for each point when the weighted projective space is not well-formed. This is used in* **[5]** *to create a database of genus 2 curves and similarly in* **[7]** *for genus 3 hyperelliptic curves.*

The **weighted absolute height** of $\mathfrak{p} = [\mathbf{x}] \in \mathbb{WP}^n_{\mathfrak{w}}(K)$, where $\mathbf{x} = (x_0 : \cdots : x_n)$, for any number field K, is

$$(2) \qquad \tilde{\mathfrak{h}}_K(\mathfrak{p}) = \frac{1}{\overline{wgcd}(\mathbf{x})} \prod_{v \in M_K} \max\left\{ |x_0|^{1/q_0}, \ldots, |x_n|^{1/q_n} \right\}$$

The concept of weighted absolute height correspond to that of absolute height in **[30]**. In **[30]** a curve with minimum absolute height has an equation with the smallest possible coefficients. In this paper, the absolute height says that there is a representative tuple of $\mathfrak{p} \in \mathbb{WP}^n_{\mathfrak{w}}(K)$ with smallest magnitude of coordinates.

Then we have the following:

PROPOSITION 2. *Let K be a number field and $\mathcal{O}_K$ its ring of integers. Then the following are true:*

i) The absolute height function $\tilde{\mathfrak{h}}_K : \mathbb{WP}^n_{\mathfrak{w}}(K) \to \mathbb{R}$ is well-defined.

ii) $\tilde{\mathfrak{h}}(\mathfrak{p})$ is the minimum of heights of all twists of $\mathfrak{p}$.

iii) For any constant $c > 0$ there are only finitely many points $\mathfrak{p} \in \mathbb{WP}^n_w(K)$ such that $\tilde{\mathfrak{h}}(\mathfrak{p}) \leq c$.

PROOF. Part ii) and iii) are obvious. We prove part i). We have to show that two different normalizations over the algebraic closure have the same absolute height. Let $\mathfrak{p}$ and $\mathfrak{q}$ be such normalizations. Then from Lemma 2 we have $\mathfrak{p} = r \star \mathfrak{q}$, where $r^d = 1$. Thus,

$$\tilde{\mathfrak{h}}(\mathfrak{p}) = \tilde{\mathfrak{h}}(r \star \mathfrak{q}) = |r| \cdot \tilde{\mathfrak{h}}(\mathfrak{q}) = \mathfrak{h}(\mathfrak{q}).$$

This completes the proof. □

For more details we direct the reader to **[9]**. Let's revisit again our example from genus 2 curves.

EXAMPLE 9. *Let $\mathcal{X}$ be the genus two curve with equation $y^2 = x^6 - 1$ and moduli point $\mathfrak{p} = [240, 1620, 119880, 46656] \in \mathbb{WP}^3_{(2,4,6,10)}(\mathbb{Q})$. We showed that $\mathfrak{p}$ is normalized and therefore has height $\mathfrak{h}(\mathfrak{p}) = 4\sqrt{15}$. Its absolute normalization is*

$$\bar{\mathfrak{p}} = [40, 45, 555, 6] = [2^3 \cdot 5,\ 3^2 \cdot 5,\ 3 \cdot 5 \cdot 37,\ 2 \cdot 3]$$

Hence, the absolute height is $\tilde{\mathfrak{h}}(\mathfrak{p}) = 2\sqrt{10}$.

4. Computing the weighted height

Given a point $\mathfrak{p} \in \mathbb{WP}^n(K)$, how easy is it to compute its weighted height $\mathfrak{h}_K(\mathfrak{p})$? From the previous section this would be equivalent to computing the weighted greatest common divisor $wgcd(\mathbf{x})$ for a point $\mathbf{x} \in \mathcal{O}_K^{n+1}$, such that $\mathfrak{p} = [\mathbf{x}]$. There are issues to be resolved when computing over $\mathcal{O}_K$, so for the purposes of this paper we continue to assume $K = \mathbb{Q}$.

Computing $wgcd(\mathbf{x})$ is equivalent with factoring every coordinate over $\mathbb{Z}$. Hence, this approach is not very effective for points with large coordinates. Hence, the main part of concern for any algorithm of computing the weighted height of a point $\mathfrak{p} \in \mathbb{WP}^n(\mathbb{Q})$ is the *normalization* of a point in $\mathbb{WP}^n(\mathbb{Q})$. We have implemented this algorithm in SageMath and it has been used in **[5]** and **[7]** to create databases of binary sextics and binary octavics of small weighted height. It works well for small heights $\mathfrak{h}$. Recall that the size of the coordinates for any $\mathfrak{p} \in \mathbb{WP}^3_{2,4,6,10}(\mathbb{Q})$ and $\mathfrak{h}(\mathfrak{p}) \leq c$ is $\leq c^{10}$. In general, for a point $\mathfrak{p} \in \mathbb{WP}^n$ with maximal weight among coordinates w, the worst bound is c^w.

One of the main problems of arithmetic related to such heights is the following. Consider $f(x, y) \in K[x, y]$ a binary form of degree $d \geq 2$. In classical mathematics determining conditions on the coefficients of $f(x, y)$ such that $f(x, y)$ has minimal discriminant has been well studied, but only understood for small n (i.e., $n = 2, 3$). However, the discriminant is only one of the invariants of the degree d binary forms, so one of the coordinates of the corresponding point $\mathfrak{p}$ in the weighted projective space $\mathbb{WP}_w(K)$. The complete problem would be restated as:

Problem 1. *Determine conditions on the coefficients of the binary form $f(x, y)$ such that the corresponding point $\mathfrak{p}$ in the weighted projective space $\mathbb{WP}_w(K)$ is absolutely normalized.*

This seems out of reach for any degree $n > 3$. However, our algorithm suggested above would work well over $\mathbb{Q}$ in finding a binary form g equivalent to f over $\mathbb{Q}$ with such that the corresponding point in $\mathbb{WP}^n(\mathbb{Q})$ is normalized. For more on this problem see [**8**].

In [**8**] we suggest an algorithm to compute equations of hyperelliptic or superelliptic curves which correspond to a normalized point in the weighted projective space. This algorithm is an extension of Tate's algorithm of elliptic curves [**31**] and methods suggested in [**28**].

There is another problem that comes from the analogy with the discriminant. A classical problem as determining the number of curves with bounded discriminant (or a good bound for such number of curves), becomes now the problem of determining a good bound for the number of curves with bounded weighted moduli point. It is unclear of any such good bounds, since now we don't want to estimate the number of tuples with bounded weighted height, but the number of equivalence classes of such tuples. Number that must be significantly less than the number of tuples. Some heuristically data for the space $\mathbb{WP}^3_{2,4,6,10}(\mathbb{Q})$ of binary sextics is given in [**5**].

Heights on weighted projective spaces, surprisingly have not been explored before. There is an unpublished preprint by A. Deng (1998) with the intention of counting the rational points in weighted projective spaces; see [**14**]. This is the first article where the concept of the height is defined in weighted projective spaces. A full account of heights in weighted projective spaces and their properties is intended in [**9**].

References

[1] C. Bagiński, J. J. Etayo, G. Gromadzki, and E. Martínez, *On real forms of a Belyi action of the alternating groups*, Albanian J. Math. **10** (2016), no. 1, 3–10. MR3490049

[2] Edoardo Ballico, *Symmetric tensor rank and the identification of a point using linear spans of an embedded variety*, Albanian J. Math. **8** (2014), no. 1, 9–21. MR3270073

[3] Mauro Beltrametti and Lorenzo Robbiano, *Introduction to the theory of weighted projective spaces*, Exposition. Math. **4** (1986), no. 2, 111–162. MR879909

[4] Marie José Bertin and Odile Lecacheux, *Automorphisms of certain Niemeier lattices and elliptic fibrations*, Albanian J. Math. **11** (2017), no. 1, 13–33. MR3724575

[5] L. Beshaj and S. Guest, *Some remarks of the weighted moduli space of binary sextics*, Algebraic curves and their applications, 2018.

[6] L. Beshaj, R. Hidalgo, S. Kruk, A. Malmendier, S. Quispe, and T. Shaska, *Rational points in the moduli space of genus two*, Higher genus curves in mathematical physics and arithmetic geometry, Contemp. Math., vol. 703, Amer. Math. Soc., Providence, RI, 2018, pp. 83–115, DOI 10.1090/conm/703/14132. MR3782461

[7] L. Beshaj and M. Polak, *On hyperelliptic curves of genus 3*, Algebraic curves and their applications, 2018.

[8] L. Beshaj and T. Shaska, *Equations of curves with minimal invariants*, submitted (2018).

[9] L. Beshaj, J. Gutierrez, and T. Shaska, *Heights on weighted projective spaces*, 2018. submitted.

[10] Lubjana Beshaj and Fred Thompson, *Equations for superelliptic curves over their minimal field of definition*, Albanian J. Math. **8** (2014), no. 1, 3–8. MR3253208

[11] Enrico Bombieri and Walter Gubler, *Heights in Diophantine geometry*, New Mathematical Monographs, vol. 4, Cambridge University Press, Cambridge, 2006. MR2216774

[12] S. Allen Broughton, *Quasi-platonic $PSL_2(q)$-actions on closed Riemann surfaces*, Albanian J. Math. **9** (2015), no. 1, 31–61. MR3438844

[13] Harris B. Daniels, *Siegel functions, modular curves, and Serre's uniformity problem*, Albanian J. Math. **9** (2015), no. 1, 3–29. MR3338631

[14] An-Wen Deng, *Rational points on weighted projective spaces* (1998), available at `math/9812082`.

[15] Anand Deopurkar, Maksym Fedorchuk, and David Swinarski, *Gröbner techniques and ribbons*, Albanian J. Math. **8** (2014), no. 1, 55–70. MR3293901

[16] Igor Dolgachev, *Weighted projective varieties*, Group actions and vector fields (Vancouver, B.C., 1981), Lecture Notes in Math., vol. 956, Springer, Berlin, 1982, pp. 34–71, DOI 10.1007/BFb0101508. MR704986

[17] G. Frey and T. Shaska, *Curves, Jacobians, and cryptography*, Algebraic curves and their applications, 2018, pp. 295–350.

[18] T. Alden Gassert, *A note on the monogeneity of power maps*, Albanian J. Math. **11** (2017), no. 1, 3–12. MR3659215

[19] Ruben Hidalgo and Tony Shaska, *On the field of moduli of superelliptic curves*, Higher genus curves in mathematical physics and arithmetic geometry, Contemp. Math., vol. 703, Amer. Math. Soc., Providence, RI, 2018, pp. 47–62, DOI 10.1090/conm/703/14130. MR3782459

[20] Marc Hindry and Joseph H. Silverman, *Diophantine geometry*, Graduate Texts in Mathematics, vol. 201, Springer-Verlag, New York, 2000. An introduction. MR1745599

[21] Jun-ichi Igusa, *Arithmetic variety of moduli for genus two*, Ann. of Math. (2) **72** (1960), 612–649, DOI 10.2307/1970233. MR0114819

[22] Jun-Ichi Igusa, *On the ring of modular forms of degree two over* **Z**, Amer. J. Math. **101** (1979), no. 1, 149–183. MR527830

[23] Qing Liu, *Modèles entiers des courbes hyperelliptiques sur un corps de valuation discrète* (French, with English summary), Trans. Amer. Math. Soc. **348** (1996), no. 11, 4577–4610, DOI 10.1090/S0002-9947-96-01684-4. MR1363944

[24] A. Malmendier and T. Shaska, *The Satake sextic in F-theory*, J. Geom. Phys. **120** (2017), 290–305, DOI 10.1016/j.geomphys.2017.06.010. MR3712162

[25] Andreas Malmendier and Tony Shaska, *A universal genus-two curve from Siegel modular forms*, SIGMA Symmetry Integrability Geom. Methods Appl. **13** (2017), Paper No. 089, 17, DOI 10.3842/SIGMA.2017.089. MR3731039

[26] Rezart Muço, Nejme Pjero, Ervin Ruci, and Eustrat Zhupa, *Classifying families of superelliptic curves*, Albanian J. Math. **8** (2014), no. 1, 23–35. MR3270074

[27] Nguyen Tien Quang, Che Thi Kim Phung, and Ngo Sy Tung, *Abelian crossed modules and strict Picard categories*, Albanian J. Math. **7** (2013), no. 1, 37–48. MR3069882

[28] Rachel Shaska, *Equations of curves with minimal discriminant* (201407), available at `1407.7064`.

[29] T. Shaska, *Some remarks on the hyperelliptic moduli of genus 3*, Comm. Algebra **42** (2014), no. 9, 4110–4130, DOI 10.1080/00927872.2013.791305. MR3200084

[30] T. Shaska and L. Beshaj, *Heights on algebraic curves*, Advances on superelliptic curves and their applications, NATO Sci. Peace Secur. Ser. D Inf. Commun. Secur., vol. 41, IOS, Amsterdam, 2015, pp. 137–175. MR3525576

[31] J. Tate, *Algorithm for determining the type of a singular fiber in an elliptic pencil*, Modular functions of one variable, IV (Proc. Internat. Summer School, Univ. Antwerp, Antwerp, 1972), Springer, Berlin, 1975, pp. 33–52. Lecture Notes in Math., Vol. 476. MR0393039

Department of Mathematics, Research Institute of Sciences and Technology, Vlorë, Albania
Email address: mandili@risat.org

Department of Mathematics and Statistics, Oakland University, Rochester, Michigan 48309
Email address: shaska@oakland.edu

Contemporary Mathematics
Volume **724**, 2019
https://doi.org/10.1090/conm/724/14589

On hyperelliptic curves of genus 3

L. Beshaj and M. Polak

ABSTRACT. We study the moduli space of genus 3 hyperelliptic curves via the weighted projective space of binary octavics. This enables us to create a database of all genus 3 hyperelliptic curves defined over $\mathbb{Q}$, of weighted moduli height $\mathfrak{h} \leq 1.5$.

1. Introduction

Genus 3 curves have been interesting objects in classical mathematics for many reasons. They are the lowest genus curves which are not hyperelliptic. Their generic equation is a ternary quartic given explicitly as in [**31**]. The lowest genus curve, defined over a field of characteristic zero, which reaches the maximum size for the automorphism group is a genus three curve, namely the well celebrated Klein's quartic which has automorphism group of order 168.

The moduli space of genus three curves $\mathcal{M}_3$ is a coarse moduli space of dimension 6. The hyperelliptic locus $\mathcal{H}_3$ in $\mathcal{M}_3$ has dimension 5. It is precisely this fact that makes genus 3 curves very attractive in certain cryptographical applications; see [**14**] for details. The existence of isogenies between genus 3 hyperelliptic Jacobians and genus 3 non-hyperelliptic Jacobians are the focus of much research lately; see [**10**] where fibrations of such surfaces are studied via the Satake polynomial as in [**22**].

This paper started with the idea of creating a database of genus 3 hyperelliptic curves similar to the database for genus 2 curves in [**5**]. There are a couple of approaches to this. In [**29**] and [**7**] we created such database for genus 2 curves with small coefficients by making use of the notion of height in the projective space. Such techniques could be used for any genus $g > 2$ hyperelliptic curves providing that one can handle invariants of binary forms. However, a new important point of view has become available after Shaska in [**23**] introduced the concept of *weighted moduli height* for weighted moduli spaces. This makes it possible to created a database with small invariants instead of small coefficients.

There are several benefits in using the weighted projective moduli space instead of the regular moduli space. First, using the weighted moduli space we avoid special cases that we have to address during the compactification of $\mathcal{H}_3$. Another benefit is computational: one has to store powers of invariants when dealing with $\mathcal{H}_3$, but this is not necessary when dealing with the weighted moduli space. The case of

2010 *Mathematics Subject Classification.* Primary 54C40, 14E20; Secondary 46E25, 20C20.
Key words and phrases. Invariants, binary forms, genus 3, algebraic curves.

$g = 2$ treated in **[4]** illustrates how things become much easier when using the weighted moduli space instead of the regular moduli space.

The main idea of this paper is to use the approach from **[4]** for binary sextics and to study the weighted moduli space of binary octavics, including their twists and the weighted moduli height. We will follow the definition of the weighted moduli height as in **[23]**. There are a few things that are slightly different for binary octavics from the binary sextics, as we will see in the coming sections.

It is important to notice that the set of generators of the ring of invariants of binary octavics (or the so-called *Shioda invariants*), namely J_2, J_3, J_4, J_5, J_6, J_7, J_8 satisfy five syzygies among them **[33]** which were determined by Shaska in **[28]** as a single degree 5 equation. Hence, a tuple $(J_2, \dots, J_8)$ doesn't necessary correspond to a genus 3 hyperelliptic curve. There are two conditions that need to be checked here. First, if such tuple satisfies the equation $F(J_2, \dots, J_8) = 0$ given in **[28]** and second if the corresponding discriminant J_{14} is non-zero.

This paper is organized as follows. In section two we define the invariants of the binary octavics in terms of transvections and in terms of root differences, see **[28, 33]** for more details. As we will see, two genus three curves $\mathcal{X}$ and $\mathcal{X}'$ are isomorphic over an algebraically closed field k if and only if there exists some $\lambda \in k \setminus \{0\}$ such that

$$J_i(\mathcal{X}) = \lambda^i J_i(\mathcal{X}'), \; for \;\; i = 2, \dots, 7,$$

and the invariants $J_2, \dots, J_8$ satisfy **[28**, Eq. (14)**]**. Hence, isomorphic curves correspond to elements of the projective space $\mathbb{P}_{(2,3,4,5,6,7,8)}$ such that the coordinates satisfy the two conditions mentioned above. Therefore, enumerating points on the moduli space can be done by enumerating points on a variety defined inside the weighted projective space. Hence, we can built a database of isomorphism classes of genus three hyperelliptic curves using $[J_2, \dots, J_8]$ as a point in the weighted projective space $\mathbb{WP}^6_w(\mathbb{Q})$, with weight $w = (2, \dots, 8)$, satisfying **[28**, Eq. (14)**]**.

In section three, we describe the weighted moduli space of binary octavics following closely the notation in **[23]**. Then we define the weighted moduli height which will be used in the later section to construct the database. Using this height for $\mathbb{WP}^6_w(\mathbb{Q})$, $w = (2, \dots, 8)$, we are able to determine a unique tuple $(J_2, \dots, J_8)$ for a point $\mathfrak{p} \in \mathbb{WP}^6_w(\mathbb{Q})$ and compile a database of genus three hyperelliptic curves.

In the last section we describe in details how we can build a database of genus three hyperelliptic curves.

2. Preliminaries

Let k be an algebraically closed field. A binary form of degree d is a homogeneous polynomial $f(X, Y)$ of degree d in two variables over k. Let V_d be the k-vector space of binary forms of degree d. The group $GL_2(k)$ of invertible 2×2 matrices over k acts on V_d by coordinate change. Many problems in algebra involve properties of binary forms which are invariant under these coordinate changes. In particular, any hyperelliptic genus g curve over k has a projective equation of the form $Z^2Y^{2g} = f(X, Y)$, where f is a binary form of degree $d = 2g + 2$ and non-zero discriminant. Two such curves are isomorphic if and only if the corresponding binary forms are conjugate under $GL_2(k)$. Therefore the moduli space $\mathcal{H}_g$ of hyperelliptic genus g curves is the affine variety whose coordinate ring is the ring of

$GL_2(k)$-invariants in the coordinate ring of the set of elements of V_d with non-zero discriminant. Throughout this section char $(k) \neq 2, 3, 5, 7$.

2.1. Covariants and invariants of binary octavics. We will use the symbolic method of classical theory to construct covariants of binary octavics. Our standard reference for this section is [**33**] or [**28**]. Let f denote a binary octavic then $f(X,Y)$ is given by the formula:

$$f(X,Y) = \sum_{i=0}^{8} a_i X^i Y^{8-i}.$$

For any two binary forms f, g of degree of degree n and m respectively the level r transvection is

$$(f,g)^r = \frac{(m-r)!(n-r)!}{n!m!} \sum_{k=0}^{r} (-1)^k \binom{r}{k} \cdot \frac{\partial^r f}{\partial X^{r-k} \partial Y^k} \cdot \frac{\partial^r g}{\partial X^r \partial Y^{r-k}}$$

As in Eq. (8) in [**28**] we define the following covariants:

$$\begin{aligned} & g = (f,f)^4, \quad k = (f,f)^6, \quad h = (k,k)^2, \quad m = (f,k)^4, \\ & n = (f,h)^4, \quad p = (g,k)^4, \quad q = (g,h)^4. \end{aligned} \tag{1}$$

Then, the following

$$\begin{aligned} & \mathcal{J}_2 = (f,f)^8, && \mathcal{J}_3 = (f,g)^8, && \mathcal{J}_4 = (k,k)^4, && \mathcal{J}_5 = (m,k)^4, \\ & \mathcal{J}_6 = (k,h)^4, && \mathcal{J}_7 = (m,h)^4, && \mathcal{J}_8 = (p,h)^4, && \mathcal{J}_9 = (n,h)^4, \\ & \mathcal{J}_{10} = (q,h)^4 \end{aligned} \tag{2}$$

are $SL_2(k)$- invariants. There is another invariant $\mathcal{J}_{14}$ which is the discriminant of the binary form $f(X,Y)$. For the purposes of this paper we will denote $\Delta = \mathcal{J}_{14}$.

THEOREM 1. *The graded ring $\mathcal{R}_8$ of invariants of binary octavics is generated by elements $\mathcal{J}_2, \dots, \mathcal{J}_{10}$.*

See [**33**, Thm. 3] for a computational proof and [**28**, Thm. 5] for a proof via the Reynold's operator. Moreover, the ring of invariants $\mathcal{R}_8$ of binary octavics is finitely generated as a module over $k[\mathcal{J}_2, \dots, \mathcal{J}_7]$; see [**33**] or [**28**].

In this paper we will follow closely the computations on [**28**]. For reasons that will be explained in the following section we prefer that the definition of $[\mathcal{J}_2 : \mathcal{J}_4 : \dots : \mathcal{J}_8]$ constitutes an integer tuple. Thus, we modify the definitions as follows.

$$\begin{aligned} & J_2 = 2^2 \cdot 5 \cdot 7 \cdot (f,f)^8, && J_3 = \frac{1}{3} \cdot 2^4 \cdot 5^2 \cdot 7^3 \cdot (f,g)^8, \\ & J_4 = 2^9 \cdot 3 \cdot 7^4 \cdot (k,k)^4, && J_5 = 2^9 \cdot 5 \cdot 7^5 \cdot (m,k)^4, \\ & J_6 = 2^{14} \cdot 3^2 \cdot 7^6 \cdot (k,h)^4, && J_7 = 2^{14} \cdot 3 \cdot 5 \cdot 7^7 \cdot (m,h)^4, \\ & J_8 = 2^{17} \cdot 3 \cdot 5^2 \cdot 7^9 \cdot (p,h)^4, && J_9 = 2^{19} \cdot 3^2 \cdot 5 \cdot 7^9 \cdot (n,h)^4, \\ & J_{10} = 2^{22} \cdot 3^2 \cdot 5^2 \cdot 7^{11} (q,h)^4 \end{aligned} \tag{3}$$

In [**28**, Thm. 6] it was shown that invariants $J_2, \dots, J_8$ satisfy the following equation

$$J_8^5 + \frac{I_8}{3^4 \cdot 5^3} J_8^4 + 2 \cdot \frac{I_{16}}{3^8 \cdot 5^6} J_8^3 + \frac{I_{24}}{2 \cdot 3^{12} \cdot 5^6} J_8^2 + \frac{I_{32}}{3^{16} \cdot 5^{10}} J_8 + \frac{I_{40}}{2^2 \cdot 3^{20} \cdot 5^{12}} = 0, \tag{4}$$

where $I_8, I_{16}, I_{24}, I_{32}$ and I_{40} are displayed in [**28**, Appendix].

Thus, in our efforts to create a list of genus 3 hyperelliptic curves we will create a list of ordered tuples $(J_2, \dots, J_8)$ such that its coordinates satisfy Eq. (4). When do two such tuples give the same curve? We have the following:

PROPOSITION 1. *Two genus 3 hyperelliptic curves $\mathcal{X}$ and $\mathcal{X}'$ are isomorphic over k if and only if there exists some $\lambda \in k \setminus \{0\}$ such that*

$$J_i(\mathcal{X}) = \lambda^i J_i(\mathcal{X}'), \ \textit{for} \ \ i = 2, \dots, 7,$$

and $J_2, \dots, J_8$ satisfy the [**28**, Eq. (14)]. *Moreover, the isomorphism is given by $f = f^M$ where $M \in GL_2(k)$ and $\lambda = (\det M)^4$.*

See [**28**, Thm. 7] for details of the proof. In the next section we will explain how the tuples $(J_2, \dots, J_8)$ are the points in the weighted projective space.

There is no known rational set of generators for the graded ring $\mathbb{R}_8$. In [**28**] Shaska has defined a set of absolute invariants $t_1, \dots, t_6$ which seem to work very well for curves of relatively small naive height, but as expected they are not defined everywhere.

2.2. Invariants from root differences. In 1986 Tsuyumine, while studying the Siegel modular forms of degree 3, gave a proof that the graded ring $S(2,8)$ of invariants of binary octavics is generated by invariants expressed in terms of root differences. Let $\xi_1, ..., \xi_8$ denote the roots of $\sum \binom{8}{i} a_i x^{8-i}$ and (ij) denotes the roots difference $\xi_i - \xi_j$. We define the following as in [**34**, pg. 772].

$$\begin{aligned}
I_2 &= \sum (13)(14)(23)(24)(57)(58)(67)(68), \\
I_3 &= \sum (12)^2(34)^2(56)^2(78)^2(13)(24)(57)(68), \\
I_4 &= \sum (12)^4(345,678)^2, \\
I_5 &= \sum (12)^4(345,678)^2(15)(26)(37)(48), \\
I_6 &= \sum (1234,5678)^2, \qquad (5)\\
I_7 &= \sum (1234,5678)^2(15)(26)(37)(48), \\
I_8 &= \sum (1234,5678)^2(15)(16)(25)(26)(37)(38)(47)(48), \\
I_9 &= \sum (1234,5678)^2(15)(16)(17)(26)(27)(28)(35)(36)(38)(45)(46)(48), \\
I_{10} &= \sum (1234,5678)^2(15)^2(26)^2(37)^2(48)^2(16)(17)(25)(28)(35)(38)(46)(47),
\end{aligned}$$

where $(1234,5678)$ denotes $(12)(13)(14)(23)(24)(34) \times (56)(57)(58)(67)(68)(78)$ and $(345,678)$ denotes $(34)(35)(45)(67)(68)(78)$. Then we have the following.

PROPOSITION 2 (Tsuyumine). *The graded ring $\mathcal{R}_8$ of invariants of binary octavics is generated by $I_2, I_3, I_4, I_5, I_6, I_7, I_8, I_9, I_{10}$.*

In [**34**] all degree 3 Siegel modular forms are expressed in terms of invariants $I_2, \dots, I_{10}$ and in terms of thetanulls. In [**26**] we display all thetanulls for genus 3 hyperelliptic curves. Moreover, the thetanull constraints that define the loci for each automorphism group are calculated.

As far as we are aware, the explicit formulas converting $I_2, \dots, I_{10}$ to $J_2, \dots, J_{10}$ and vice-versa have not appeared in print.

PROBLEM 1. *Express $I_2, \dots, I_{10}$ in terms of $J_2, \dots, J_{10}$ and vice-versa*

One can express the above sets of invariants in terms of the Siegel modular forms, but that is outside the scope of this paper. In the next section, we will describe the weighted projective space and define the weighted moduli height.

3. Weighted moduli space of binary octavics

Let k be a field of characteristic zero and $q_0, \dots, q_n$ a fixed tuple of positive integers called **weights**. Consider the action of $k^\star$ on $\mathbb{A}^{n+1}$ as follows

$$\lambda \star (x_0, \dots, x_n) = (\lambda^{q_0} x_0, \dots, \lambda^{q_n} x_n)$$

for $\lambda \in k^*$. The quotient of this action is called a **weighted projective space** and denoted by $\mathbb{P}^n_{(q_0,\dots,q_n)}$. It is the projective variety $Proj\,(k[x_0, ..., x_n])$ associated to the graded ring $k[x_0, \dots, x_n]$ where the variable x_i has degree q_i for $i = 0, \dots, n$. Next we focus our attention to the weighted projective space of binary octavics.

3.1. The weighted moduli space of binary octavics. From above we know that the invariant ring of binary octavics is generated by invariants $J_2, \dots, J_8$. Since J_i, for $i = 2, \dots, 7$ are all homogenous polynomials of degree i, we take the set of weights $w = (2, 3, 4, 5, 6, 7, 8)$ and considered the weighted projective space $\mathbb{WP}^6_w(k)$.

Invariants of octavics define a point in a weighted projective space $\mathfrak{p} = [J_2 : J_3 : J_4 : J_5 : J_6 : J_7 : J_8] \in \mathbb{WP}^6_w$. However, not every point in $\mathbb{WP}^6_w$ correspond to a genus 3 hyperelliptic curve.

PROPOSITION 3. *Let $\mathfrak{p} \in \mathbb{WP}^6_w$, where $w = (2, 3, 4, 5, 6, 7, 8)$. Then $\mathfrak{p}$ correspond to the isomorphism class of a genus 3 hyperelliptic curve if and only if $\mathfrak{p} \in \mathbb{WP}^6_w(k) \setminus \{\Delta = 0\}$ and its coordinates satisfy Eq. (4).*

Hence, in our goal of creating a list of all points in $\mathbb{WP}^6_w$ which correspond to genus 3 hyperelliptic curves we will have to through away all the points left out by the Prop. 3. But first, let's determine a way of ordering such points in $\mathbb{WP}^6_w$.

3.2. The height of $\mathbb{WP}^6_w$. Let $K \subset k$ be a number field and $\mathcal{O}_K$ its ring of integers. For a point $\mathfrak{p}$ such that $\mathfrak{p} = [J_2 : J_3 : J_4 : J_5 : J_6 : J_7 : J_8] \in \mathbb{WP}^6_w(\mathcal{O}_K)$, we say that $\mathfrak{p}$ is defined over $\mathcal{O}_K$. We call the tuple $\mathfrak{p}$ a **minimal tuple** or a **normalized weighted moduli point** if there is no prime $p \in \mathcal{O}_K$ such that $p^i | J_i$, for $i = 2, 3, 4, 5, 6, 7, 8$. Two minimal tuples $\mathfrak{p}$ and $\mathfrak{p}'$ are called **twists** of each other if there exists $\lambda \in k$ such that

$$\lambda \star \mathfrak{p} = \mathfrak{p}'.$$

Let $\mathfrak{p} = [J_2 : J_3 : J_4 : J_5 : J_6 : J_7 : J_8]$ be any tuple in $\mathbb{WP}^6_w(K)$. We define the **weighted moduli height** of $\mathfrak{p}$ (or simply **the height**) to be

$$\mathfrak{h}(\mathfrak{p}) = \frac{\max\{|J_i|^{\frac{1}{i}}\}}{\prod\limits_{p^i | J_i} p}, \tag{6}$$

where the product is taken over all the primes $p \in \mathcal{O}_K$. In **[23]** was proved that this is a well-defined height in a weighted projective space. The weighted moduli height of a minimal tuple is simply $\mathfrak{h}(\mathfrak{p}) = \max\{|J_i|^{\frac{1}{i}}\}$. It is easy to verify that:

LEMMA 1. *Let $\mathfrak{p} = [J_2, \dots, J_7, J_8] \in \mathbb{WP}^6_w$. Then the following hold:*
i) $\mathfrak{p}$ is a minimal tuple if and only if it has minimal height.
ii) If $\mathfrak{p}$ is a minimal tuple and it has a twist with minimal height, then there exists a square free integer d such that $d^i | J_i$ for $i = 2, \dots, 8$.

A point $\mathfrak{p} = [J_2, \dots, J_8] \in \mathbb{WP}^6_w$ is called an **absolute minimal tuple** if it has the smallest height among all the twists.

LEMMA 2. *A minimal tuple $\mathfrak{p} = [J_2, \dots, J_8] \in \mathbb{WP}^6_w(\mathbb{Q})$ is an absolute minimum tuple if and only if there is no $\lambda \in \mathbb{C} \setminus \{0\}$ such that $\lambda^i \in \mathbb{Z}$ and $\lambda^i | J_i$, for all $i = 2, \dots, 8$ such that $J_i \neq 0$.*

PROOF. Let $\mathfrak{p}$ be an absolute minimal tuple. Suppose that there exist $\lambda \in \mathbb{C} \setminus \{0\}$ such that $\lambda^i \in \mathbb{Z}$ and $\lambda^i | J_i$, for all $i = 2, \dots, 8$ such that $J_i \neq 0$. Hence,

$$\mathfrak{p}' = \left[\frac{J_2}{\lambda^2}, \dots, \frac{J_8}{\lambda^8}\right] = [J_2', \dots, J_8'] \in \mathbb{WP}^6_w(\mathbb{Z})$$

is a twist of $\mathfrak{p}$ such that $J_i = \lambda^i J_i'$, for all $i = 2, \dots, 8$. Therefore, $J_i \geq J_i'$ for all $i = 2, \dots, 8$ and

$$\mathfrak{h}(\mathfrak{p}) = \max\{|J_i|^{\frac{1}{i}}\} \geq \max\{|J_i'|^{\frac{1}{i}}\} = \mathfrak{h}(\mathfrak{p}'),$$

which contradicts the assumption.

Conversely, let $\mathfrak{p}$ be a minimal tuple such that there is no $\lambda \in \mathbb{C} \setminus \{0\}$ such that $\lambda^i \in \mathbb{Z}$ and $\lambda^i | J_i$, for all $i = 2, \dots, 8$. Hence $\mathfrak{p}$ cannot be written as $\left[\lambda^2 J_2', \dots, \lambda^8 J_8'\right]$, where $J_i' \in \mathbb{Z}$ for all $i = 2, \dots, 8$. Hence, there is no twist of $\mathfrak{p}$ in $\in \mathbb{WP}^6_w(\mathbb{Q})$. $\square$

REMARK 1. *Absolute minimal tuples may be not unique over $\overline{\mathbb{Q}}$. Consider minimal tuples: $[-1, -1, 0, -1, 0, 0, 0] \in \mathbb{WP}^6_w(\mathbb{Q})$ and $[-1, 1, 0, 1, 0, 0, 0] \in \mathbb{WP}^6_w(\mathbb{Q})$. For both tuples $\mathfrak{h}([-1, -1, 0, -1, 0, 0, 0]) = 1$ and $\mathfrak{h}([-1, 1, 0, 1, 0, 0, 0]) = 1$. The first tuple is a twist of the second one, with $\lambda = i$: $i \star [-1, -1, 0, -1, 0, 0, 0] = [-1, 1, 0, 1, 0, 0, 0]$.*

For any absolute minimal tuple $\mathfrak{p}$ there may exist other absolute minimal tuples $(-1) \star \mathfrak{p}$ or $i \star \mathfrak{p}$. For any isomorphic class of hyperelliptic curves of genus 3 we want to store only one absolute minimal tuple. For this reason we set a convention that defines which absolute minimal tuples shall be stored.

- **i:** For any two corresponding absolute minimal tuples we choose the one with more positive invariants J_i.
- **ii:** For any two corresponding absolute minimal tuples with the same number of positive invariants J_i we choose the one with greater value $\sum_{i=2}^{8} i \cdot J_i$.

We will call an absolute minimal tuple that satisfies this convention a **normalized absolute minimal tuple**.

PROPOSITION 4. *For every point $\mathfrak{q} \in \mathbb{WP}^6_w(\mathbb{Q})$ there is a unique absolute minimal tuple.*

PROOF. Let $\mathfrak{p}$ and $\mathfrak{p}'$ be absolute minimal tuples corresponding to $\mathfrak{q} \in \mathbb{WP}^6_w(\mathbb{Q})$. Then there are $\lambda_1, \lambda_2 \in \mathbb{C}$ such that

$$\mathfrak{q} = \lambda_1 \star \mathfrak{p} = \lambda_2 \star \mathfrak{p}',$$

and there is no prime p dividing all coordinates. Hence,

$$J_i = \lambda^i J_i', \quad i = 2, \dots, 8,$$

and λ is not divisible by any prime p. Hence, λ is a unit in $\mathbb{Z}$ and therefore $\lambda = \pm 1$. If $\lambda = 1$ we have uniqueness, otherwise $\lambda = -1$. But now the tuple $[\lambda^2 J_2, \lambda^3 j_3, \dots, \lambda^8 J_8]$ is not an integer tuple any longer and therefore not in $\mathbb{WP}^6_w(\mathbb{Q})$.

□

The above result makes it possible to create a database with all points in the moduli space $\mathcal{H}_3$ with field of moduli $\mathbb{Q}$ and bounded height. We describe next how this is done, but first let's see a couple of examples.

EXAMPLE 1. *Let $\mathcal{X}$ be the curve with equation*

$$y^2 = x^7 - 1.$$

Then the moduli point is

$$\mathfrak{p} = [0, 0, 0, 0, 0, -395300640, 0]$$

Notice that multiplying this tuple by $\lambda = \frac{1}{(-395300640)^{1/7}}$ gives us the absolute minimal tuple

$$\mathfrak{p} = [0, 0, 0, 0, 0, 1, 0].$$

Let us consider another example.

EXAMPLE 2. *Let $\mathcal{X}$ be the curve with equation*

$$y^2 = x^8 - 1.$$

Then the moduli point is

$$\mathfrak{p} = [-2^3 \cdot 5 \cdot 7, 0, 2^{10} \cdot 7^4, 0, 2^{15} \cdot 7^6, 0, -2^{19} \cdot 5 \cdot 7^8]$$

Notice that multiplying this tuple by $\lambda = \frac{1}{2}$ we get a twist

$$\mathfrak{p} = [-2 \cdot 5 \cdot 7, 0, 2^6 \cdot 7^4, 0, 2^9 \cdot 7^6, 0, -2^{11} \cdot 5 \cdot 7^8].$$

However, we can further reduce its height by multiplying by $\lambda = I\frac{1}{\sqrt{2}}$ and we get

$$\mathfrak{p} = [5 \cdot 7, 0, 2^4 \cdot 7^4, 0, 2^6 \cdot 7^6, 0, -2^7 \cdot 5 \cdot 7^8].$$

Furthermore, by multiplying with $\lambda = \frac{1}{\sqrt{7}}$ we get

$$\mathfrak{p} = [5, 0, 2^4 \cdot 7^2, 0, 2^6 \cdot 7^3, 0, -2^7 \cdot 5 \cdot 7^4].$$

which is an absolute minimum tuple.

In the next section we show how to create a database of genus 3 hyperelliptic curves which will be ordered by the moduli height.

4. A database of genus 3 hyperelliptic curves

In this part we give a quick computational view on how to create a database of genus 3 hyperelliptic curves of weighted moduli height $\leq \mathfrak{h}$, where $\mathfrak{h}$ is a positive integer. We briefly describe each step of our algorithm below.

(1) Since we want all the points of height $\leq \mathfrak{h}$ that means that we want all ordered tuples $(x_2, \dots, x_8) \in \mathbb{Z}^7$ such that $|x_i| \leq \mathfrak{h}^i$ (tuple with all zeros is not considered). Let the list of all such tuples be denoted by A. There are $\prod_{i=2}^{8}(2\mathfrak{h}^i + 1) - 1$ possible tuples in this list. But, not all tuples in A are valid tuples for us. Valid tuples are only the ones that satisfy Eq. (4) and $J_{14} \neq 0$.

(2) Next, for each tuple in A we check if they satisfy the Eq. (4):

$$J_8^5 + \frac{I_8}{3^4 \cdot 5^3} J_8^4 + 2 \cdot \frac{I_{16}}{3^8 \cdot 5^6} J_8^3 + \frac{I_{24}}{2 \cdot 3^{12} \cdot 5^6} J_8^2 + \frac{I_{32}}{3^{16} \cdot 5^{10}} J_8 + \frac{I_{40}}{2^2 \cdot 3^{20} \cdot 5^{12}} = 0,$$

We throw away all tuples which do not satisfy this equation. Let the new set of tuples be denoted by B. Elements of the set B are points in a weighted projective space $\mathbb{WP}_w^6(\mathbb{Q})$.

(3) For each tuple in B we calculate the discriminant $\Delta = J_{14}$. We throw away all tuples with $\Delta = 0$ and denote the new set by C.

(4) Next, for each tuple in the set C we compute the corresponding minimal tuple. This step is the most expensive one since it requires factorization of each integer coordinate of the tuple into prime factors. Let the new set of tuples be denoted by D.

(5) The last step is to compute a normalized absolute minimal tuple for each element in set D.

In step (2) we compute invariants $I_8, I_{16}, I_{24}, I_{32}$ and I_{40} by formulas in **[28]**. In step (3) we compute Δ as function of $J_2, J_3, \dots, J_{10}$.

First, we considered tuples of $\mathfrak{h} = 1$.

(1) A list of all ordered minimal tuples $[x_2, \dots, x_8] \in \mathbb{Z}^7$, such that $|x_i| \leq 1$ has 2186 elements.

(2) There are only 34 tuples ($\#B = 34$) that satisfy the equation:

$$J_8^5 + a_4 J_8^4 + a_3 J_8^3 + a_2 J_8^2 + a_1 J_8 + a_0 = 0.$$

(3) There are 24 tuples with $\Delta \neq 0$ ($\#C = 24$).

(4) Since all elements in C have $\mathfrak{h} = 1$, $D = C$.

(5) For each minimal tuple in D we compute the corresponding normalized absolute minimal tuples and store them in a database (see Table 4).

Notice that round 70% of minimal tuples of height $\mathfrak{h} = 1$ in a weighted projective space $\mathbb{WP}_w^6(\mathbb{Q})$ correspond to hyperelliptic curves of genus 3. But, only $\approx 35\%$ of minimal tuples of height $\mathfrak{h} = 1$ in a weighted projective space $\mathbb{WP}_w^6(\mathbb{Q})$ are normalized absolute minimal tuples that correspond to hyperelliptic curves of $g = 3$.

We considered tuples with $\mathfrak{h} \leq 1.5$.

(1) A list of all ordered minimal tuples $[x_2, \dots, x_8] \in \mathbb{Z}^7$, such that $|x_i| \leq (1.5)^i$ has 237092624 elements.

(2) There are only 748 tuples ($\#B = 748$) that satisfy the equation:

$$J_8^5 + a_4 J_8^4 + a_3 J_8^3 + a_2 J_8^2 + a_1 J_8 + a_0 = 0.$$

(3) There are 544 tuples with $\Delta \neq 0$ ($\#C = 544$).

TABLE 1. Absolute minimal tuples of height 1 (set C)

#	$\mathfrak{p} = [J_2 : \dots : J_8]$	#	$\mathfrak{p} = [J_2 : \dots : J_8]$	#	$\mathfrak{p} = [J_2 : \dots : J_8]$
1	-1 -1 0 -1 0 0 0	2	-1 -1 0 0 0 0 0	3	-1 -1 0 1 0 0 0
4	-1 0 0 -1 0 0 0	5	-1 0 0 0 0 0 0	6	-1 0 0 1 0 0 0
7	-1 1 0 -1 0 0 0	8	-1 1 0 0 0 0 0	9	-1 1 0 1 0 0 0
10	0 -1 0 -1 0 0 0	11	0 -1 0 1 0 0 0	12	0 0 0 0 0 -1 0
13	0 0 0 0 0 1 0	14	0 1 0 -1 0 0 0	15	0 1 0 1 0 0 0
16	1 -1 0 -1 0 0 0	17	1 -1 0 0 0 0 0	18	1 -1 0 1 0 0 0
19	1 0 0 -1 0 0 0	20	1 0 0 0 0 0 0	21	1 0 0 1 0 0 0
22	1 1 0 -1 0 0 0	23	1 1 0 0 0 0 0	24	1 1 0 1 0 0 0

TABLE 2. Normalized absolute minimal tuples of height 1

#	$\mathfrak{p} = [J_2 : \dots : J_8]$	#	$\mathfrak{p} = [J_2 : \dots : J_8]$	#	$\mathfrak{p} = [J_2 : \dots : J_8]$
1	-1 -1 0 1 0 0 0	2	-1 0 0 1 0 0 0	3	-1 1 0 0 0 0 0
4	-1 1 0 1 0 0 0	5	0 -1 0 1 0 0 0	6	0 0 0 0 0 1 0
7	0 1 0 1 0 0 0	8	1 -1 0 1 0 0 0	9	1 0 0 0 0 0 0
10	1 0 0 1 0 0 0	11	1 1 0 0 0 0 0	12	1 1 0 1 0 0 0

(4) All elements in C are minimal tuples so $D = C$.

(5) For each minimal tuple in D we compute the corresponding normalized absolute minimal tuple and store them in a database (see Table 5). There is $246+12 = 258$ normalized absolute minimal tuples of hight $0 < \mathfrak{h} \leq 1.5$.

Notice that round 73.(73)% of minimal tuples of height $1 \leq \mathfrak{h} \leq 1.5$ in a weighted projective space $\mathbb{WP}_w^6(\mathbb{Q})$ correspond to hyperelliptic curves of genus 3. But, only $\approx$ 34% of minimal tuples of hight $1 \leq \mathfrak{h} \leq 1.5$ in a weighted projective space $\mathbb{WP}_w^6(\mathbb{Q})$ are normalized absolute minimal tuples that correspond to hyperelliptic curves of genus 3.

5. Concluding remarks

The arithmetic of genus 3 moduli space is not as well understood as in the case of genus 2 curves. Part of it is because the existence of non-hyperelliptic curves but also because we do not know as much about the space of binary octavics as we do about binary sextics. Thus creating such databases for genus 3 curves is an interesting problem.

In upcoming projects we intent to create such tables for $\mathfrak{h} \leq 4$ and see what percentage of $g = 3$ curves are hyperelliptic and see how this compares to the $g = 2$ curves. Also, in an analogue way with [**4**], an interesting thing to consider would be to check how the weighted moduli height of genus 3 curves is compared to their naive height and see if this lower bound is sharp for any of those curves.

Acknowledgment

We want to thank Prof. Shaska for sharing his Maple package on genus 3 hyperelliptic curves, including the implementation in Maple of the equation [**28**, Eq. (14)].

TABLE 3. Normalized absolute minimal tuples of height $1 < \mathfrak{h} \leq 1.5$

#	$\mathfrak{p} = [J_2 : \dots : J_8]$	#	$\mathfrak{p} = [J_2 : \dots : J_8]$	#	$\mathfrak{p} = [J_2 : \dots : J_8]$
1	-2 -3 0 1 0 0 0	2	-2 -3 0 2 0 0 0	3	-2 -3 0 3 0 0 0
4	-2 -3 0 4 0 0 0	5	-2 -3 0 5 0 0 0	6	-2 -3 0 6 0 0 0
7	-2 -3 0 7 0 0 0	8	-2 -2 0 1 0 0 0	9	-2 -2 0 2 0 0 0
10	-2 -2 0 3 0 0 0	11	-2 -2 0 4 0 0 0	12	-2 -2 0 5 0 0 0
13	-2 -2 0 6 0 0 0	14	-2 -2 0 7 0 0 0	15	-2 -1 0 1 0 0 0
16	-2 -1 0 2 0 0 0	17	-2 -1 0 3 0 0 0	18	-2 -1 0 4 0 0 0
19	-2 -1 0 5 0 0 0	20	-2 -1 0 6 0 0 0	21	-2 -1 0 7 0 0 0
22	-2 0 0 1 0 0 0	23	-2 0 0 2 0 0 0	24	-2 0 0 3 0 0 0
25	-2 0 0 4 0 0 0	26	-2 0 0 5 0 0 0	27	-2 0 0 6 0 0 0
28	-2 0 0 7 0 0 0	29	-2 1 0 0 0 0 0	30	-2 1 0 1 0 0 0
31	-2 1 0 2 0 0 0	32	-2 1 0 3 0 0 0	33	-2 1 0 4 0 0 0
34	-2 1 0 5 0 0 0	35	-2 1 0 6 0 0 0	36	-2 1 0 7 0 0 0
37	-2 2 0 0 0 0 0	38	-2 2 0 1 0 0 0	39	-2 2 0 2 0 0 0
40	-2 2 0 3 0 0 0	41	-2 2 0 4 0 0 0	42	-2 2 0 5 0 0 0
43	-2 2 0 6 0 0 0	44	-2 2 0 7 0 0 0	45	-2 3 0 0 0 0 0
46	-2 3 0 1 0 0 0	47	-2 3 0 2 0 0 0	48	-2 3 0 3 0 0 0
49	-2 3 0 4 0 0 0	50	-2 3 0 5 0 0 0	51	-2 3 0 6 0 0 0
52	-2 3 0 7 0 0 0	53	-1 -3 0 1 0 0 0	54	-1 -3 0 2 0 0 0
55	-1 -3 0 3 0 0 0	56	-1 -3 0 4 0 0 0	57	-1 -3 0 5 0 0 0
58	-1 -3 0 6 0 0 0	59	-1 -3 0 7 0 0 0	60	-1 -2 0 1 0 0 0
61	-1 -2 0 2 0 0 0	62	-1 -2 0 3 0 0 0	63	-1 -2 0 4 0 0 0
64	-1 -2 0 5 0 0 0	65	-1 -2 0 6 0 0 0	66	-1 -2 0 7 0 0 0
67	-1 -1 0 2 0 0 0	68	-1 -1 0 3 0 0 0	69	-1 -1 0 4 0 0 0
70	-1 -1 0 5 0 0 0	71	-1 -1 0 6 0 0 0	72	-1 -1 0 7 0 0 0
73	-1 0 0 2 0 0 0	74	-1 0 0 3 0 0 0	75	-1 0 0 4 0 0 0
76	-1 0 0 5 0 0 0	77	-1 0 0 6 0 0 0	78	-1 0 0 7 0 0 0
79	-1 1 0 2 0 0 0	80	-1 1 0 3 0 0 0	81	-1 1 0 4 0 0 0
82	-1 1 0 5 0 0 0	83	-1 1 0 6 0 0 0	84	-1 1 0 7 0 0 0
85	-1 2 0 0 0 0 0	86	-1 2 0 1 0 0 0	87	-1 2 0 2 0 0 0
88	-1 2 0 3 0 0 0	89	-1 2 0 4 0 0 0	90	-1 2 0 5 0 0 0
91	-1 2 0 6 0 0 0	92	-1 2 0 7 0 0 0	93	-1 3 0 0 0 0 0
94	-1 3 0 1 0 0 0	95	-1 3 0 2 0 0 0	96	-1 3 0 3 0 0 0
97	-1 3 0 4 0 0 0	98	-1 3 0 5 0 0 0	99	-1 3 0 6 0 0 0
100	-1 3 0 7 0 0 0	101	0 -3 0 1 0 0 0	102	0 -3 0 2 0 0 0
103	0 -3 0 3 0 0 0	104	0 -3 0 4 0 0 0	105	0 -3 0 5 0 0 0
106	0 -3 0 6 0 0 0	107	0 -3 0 7 0 0 0	108	0 -2 0 1 0 0 0
109	0 -2 0 2 0 0 0	110	0 -2 0 3 0 0 0	111	0 -2 0 4 0 0 0
112	0 -2 0 5 0 0 0	113	0 -2 0 6 0 0 0	114	0 -2 0 7 0 0 0
115	0 -1 0 2 0 0 0	116	0 -1 0 3 0 0 0	117	0 -1 0 4 0 0 0
118	0 -1 0 5 0 0 0	119	0 -1 0 6 0 0 0	120	0 -1 0 7 0 0 0
121	0 0 -3 0 -10 0 -5	122	0 0 -3 0 10 0 -5	123	0 0 3 0 -8 0 -5
124	0 0 3 0 -2 0 -5	125	0 0 3 0 2 0 -5	126	0 0 3 0 8 0 -5
127	0 1 0 2 0 0 0	128	0 1 0 3 0 0 0	129	0 1 0 4 0 0 0
130	0 1 0 5 0 0 0	131	0 1 0 6 0 0 0	132	0 1 0 7 0 0 0
133	0 2 0 1 0 0 0	134	0 2 0 2 0 0 0	135	0 2 0 3 0 0 0
136	0 2 0 4 0 0 0	137	0 2 0 5 0 0 0	138	0 2 0 6 0 0 0
139	0 2 0 7 0 0 0	140	0 3 0 1 0 0 0	141	0 3 0 2 0 0 0

TABLE 4. Normalized absolute minimal tuples of height $1 < \mathfrak{h} \leq 1.5$

#	$\mathfrak{p} = [J_2 : \ldots : J_8]$	#	$\mathfrak{p} = [J_2 : \ldots : J_8]$	#	$\mathfrak{p} = [J_2 : \ldots : J_8]$
142	0 3 0 3 0 0 0	143	0 3 0 4 0 0 0	144	0 3 0 5 0 0 0
145	0 3 0 6 0 0 0	146	0 3 0 7 0 0 0	147	1 -3 0 1 0 0 0
148	1 -3 0 2 0 0 0	149	1 -3 0 3 0 0 0	150	1 -3 0 4 0 0 0
151	1 -3 0 5 0 0 0	152	1 -3 0 6 0 0 0	153	1 -3 0 7 0 0 0
154	1 -2 0 1 0 0 0	155	1 -2 0 2 0 0 0	156	1 -2 0 3 0 0 0
157	1 -2 0 4 0 0 0	158	1 -2 0 5 0 0 0	159	1 -2 0 6 0 0 0
160	1 -2 0 7 0 0 0	161	1 -1 0 2 0 0 0	162	1 -1 0 3 0 0 0
163	1 -1 0 4 0 0 0	164	1 -1 0 5 0 0 0	165	1 -1 0 6 0 0 0
166	1 -1 0 7 0 0 0	167	1 0 0 2 0 0 0	168	1 0 0 3 0 0 0
169	1 0 0 4 0 0 0	170	1 0 0 5 0 0 0	171	1 0 0 6 0 0 0
172	1 0 0 7 0 0 0	173	1 1 0 2 0 0 0	174	1 1 0 3 0 0 0
175	1 1 0 4 0 0 0	176	1 1 0 5 0 0 0	177	1 1 0 6 0 0 0
178	1 1 0 7 0 0 0	179	1 2 0 0 0 0 0	180	1 2 0 1 0 0 0
181	1 2 0 2 0 0 0	182	1 2 0 3 0 0 0	183	1 2 0 4 0 0 0
184	1 2 0 5 0 0 0	185	1 2 0 6 0 0 0	186	1 2 0 7 0 0 0
187	1 3 0 0 0 0 0	188	1 3 0 1 0 0 0	189	1 3 0 2 0 0 0
190	1 3 0 3 0 0 0	191	1 3 0 4 0 0 0	192	1 3 0 5 0 0 0
193	1 3 0 6 0 0 0	194	1 3 0 7 0 0 0	195	2 -3 0 1 0 0 0
196	2 -3 0 2 0 0 0	197	2 -3 0 3 0 0 0	198	2 -3 0 4 0 0 0
199	2 -3 0 5 0 0 0	200	2 -3 0 6 0 0 0	201	2 -3 0 7 0 0 0
202	2 -2 0 1 0 0 0	203	2 -2 0 2 0 0 0	204	2 -2 0 3 0 0 0
205	2 -2 0 4 0 0 0	206	2 -2 0 5 0 0 0	207	2 -2 0 6 0 0 0
208	2 -2 0 7 0 0 0	209	2 -1 0 1 0 0 0	210	2 -1 0 2 0 0 0
211	2 -1 0 3 0 0 0	212	2 -1 0 4 0 0 0	213	2 -1 0 5 0 0 0
214	2 -1 0 6 0 0 0	215	2 -1 0 7 0 0 0	216	2 0 0 1 0 0 0
217	2 0 0 2 0 0 0	218	2 0 0 3 0 0 0	219	2 0 0 4 0 0 0
220	2 0 0 5 0 0 0	221	2 0 0 6 0 0 0	222	2 0 0 7 0 0 0
223	2 1 0 0 0 0 0	224	2 1 0 1 0 0 0	225	2 1 0 2 0 0 0
226	2 1 0 3 0 0 0	227	2 1 0 4 0 0 0	228	2 1 0 5 0 0 0
229	2 1 0 6 0 0 0	230	2 1 0 7 0 0 0	231	2 2 0 0 0 0 0
232	2 2 0 1 0 0 0	233	2 2 0 2 0 0 0	234	2 2 0 3 0 0 0
235	2 2 0 4 0 0 0	236	2 2 0 5 0 0 0	237	2 2 0 6 0 0 0
238	2 2 0 7 0 0 0	239	2 3 0 0 0 0 0	240	2 3 0 1 0 0 0
241	2 3 0 2 0 0 0	242	2 3 0 3 0 0 0	243	2 3 0 4 0 0 0
244	2 3 0 5 0 0 0	245	2 3 0 6 0 0 0	246	2 3 0 7 0 0 0

References

[1] C. Bagiński, J. J. Etayo, G. Gromadzki, and E. Martínez, *On real forms of a Belyi action of the alternating groups*, Albanian J. Math. **10** (2016), no. 1, 3–10. MR3490049

[2] Edoardo Ballico, *Symmetric tensor rank and the identification of a point using linear spans of an embedded variety*, Albanian J. Math. **8** (2014), no. 1, 9–21. MR3270073

[3] Marie José Bertin and Odile Lecacheux, *Automorphisms of certain Niemeier lattices and elliptic fibrations*, Albanian J. Math. **11** (2017), no. 1, 13–33. MR3724575

[4] L. Beshaj and S. Guest, *The weighted moduli space of binary sextics*, Algebraic curves and their applications, Contemp. Math., vol. 724, Amer. Math. Soc., Providence, RI, 2019, pp. 33–44.

[5] L. Beshaj, R. Hidalgo, S. Kruk, A. Malmendier, S. Quispe, and T. Shaska, *Rational points in the moduli space of genus two*, Higher genus curves in mathematical physics and arithmetic geometry, Contemp. Math., vol. 703, Amer. Math. Soc., Providence, RI, 2018, pp. 83–115, DOI 10.1090/conm/703/14132. MR3782461

[6] L. Beshaj, T. Shaska, and C. Shor, *On Jacobians of curves with superelliptic components*, Riemann and Klein surfaces, automorphisms, symmetries and moduli spaces, Contemp. Math., vol. 629, Amer. Math. Soc., Providence, RI, 2014, pp. 1–14, DOI 10.1090/conm/629/12557. MR3289629

[7] Lubjana Beshaj, *Reduction theory of binary forms*, Advances on superelliptic curves and their applications, NATO Sci. Peace Secur. Ser. D Inf. Commun. Secur., vol. 41, IOS, Amsterdam, 2015, pp. 84–116. MR3525574

[8] Lubjana Beshaj and Fred Thompson, *Equations for superelliptic curves over their minimal field of definition*, Albanian J. Math. **8** (2014), no. 1, 3–8. MR3253208

[9] S. Allen Broughton, *Quasi-platonic $PSL_2(q)$-actions on closed Riemann surfaces*, Albanian J. Math. **9** (2015), no. 1, 31–61. MR3438844

[10] Adrian Clingher, Andreas Malmendier, and Tony Shaska, *Six line configurations and string dualities* (2018), available at `1806.07460`.

[11] Harris B. Daniels, *Siegel functions, modular curves, and Serre's uniformity problem*, Albanian J. Math. **9** (2015), no. 1, 3–29. MR3338631

[12] Barry H. Dayton, *Algorithms for real numerical varieties with application to parameterizing quadratic surface intersection curves*, Albanian J. Math. **7** (2013), no. 2, 61–75. MR3109478

[13] Anand Deopurkar, Maksym Fedorchuk, and David Swinarski, *Gröbner techniques and ribbons*, Albanian J. Math. **8** (2014), no. 1, 55–70. MR3293901

[14] Gerhard Frey and Tony Shaska, *Curves, jacobians, and cryptography*, Algebraic curves and their applications, 2018, pp. 3–72.

[15] Von Gall, *Das vollständige Formensystem einer binären Form achter Ordnung* (German), Math. Ann. **17** (1880), no. 1, 31–51, DOI 10.1007/BF01444117. MR1510048

[16] Von Gall, *Ueber das vollständige System einer binären Form achter Ordnung* (German), Math. Ann. **17** (1880), no. 1, 139–152, DOI 10.1007/BF01444131. MR1510062

[17] Paul Gordan, *Beweis, dass jede Covariante und Invariante einer binären Form eine ganze Function mit numerischen Coefficienten einer endlichen Anzahl solcher Formen ist* (German), J. Reine Angew. Math. **69** (1868), 323–354, DOI 10.1515/crll.1868.69.323. MR1579424

[18] Jaime Gutierrez, D. Sevilla, and T. Shaska, *Hyperelliptic curves of genus 3 with prescribed automorphism group*, Computational aspects of algebraic curves, 2005, pp. 109–123. MR2182037

[19] Ruben Hidalgo and Tony Shaska, *On the field of moduli of superelliptic curves*, Higher genus curves in mathematical physics and arithmetic geometry, Contemp. Math., vol. 703, Amer. Math. Soc., Providence, RI, 2018, pp. 47–62, DOI 10.1090/conm/703/14130. MR3782459

[20] David Kalaj and Ken-ichi Sakan, *Quasiconformal harmonic mappings onto a convex domain revisited*, Albanian J. Math. **7** (2013), no. 2, 51–60. MR3109477

[21] P. Katsylo, *Rationality of the moduli variety of curves of genus* 3, Comment. Math. Helv. **71** (1996), no. 4, 507–524, DOI 10.1007/BF02566434. MR1420508

[22] A. Malmendier and T. Shaska, *The Satake sextic in F-theory*, J. Geom. Phys. **120** (2017), 290–305, DOI 10.1016/j.geomphys.2017.06.010. MR3712162

[23] J. Mandili and T. Shaska, *Computing heights on weighted projective spaces*, submitted (201801), 1–10, available at `1801.06250`.

[24] Rezart Muço, Nejme Pjero, Ervin Ruci, and Eustrat Zhupa, *Classifying families of superelliptic curves*, Albanian J. Math. **8** (2014), no. 1, 23–35. MR3270074

[25] Monika Polak and Eustrat Zhupa, *Graph based linear error correcting codes*, Albanian J. Math. **10** (2016), no. 1, 37–45. MR3589800

[26] E. Previato, T. Shaska, and G. S. Wijesiri, *Thetanulls of cyclic curves of small genus*, Albanian J. Math. **1** (2007), no. 4, 253–270. MR2367218

[27] Nguyen Tien Quang, Che Thi Kim Phung, and Ngo Sy Tung, *Abelian crossed modules and strict Picard categories*, Albanian J. Math. **7** (2013), no. 1, 37–48. MR3069882

[28] T. Shaska, *Some remarks on the hyperelliptic moduli of genus 3*, Comm. Algebra **42** (2014), no. 9, 4110–4130, DOI 10.1080/00927872.2013.791305. MR3200084

[29] T. Shaska and L. Beshaj, *Heights on algebraic curves*, Advances on superelliptic curves and their applications, NATO Sci. Peace Secur. Ser. D Inf. Commun. Secur., vol. 41, IOS, Amsterdam, 2015, pp. 137–175. MR3525576

[30] T. Shaska and F. Thompson, *Bielliptic curves of genus 3 in the hyperelliptic moduli*, Appl. Algebra Engrg. Comm. Comput. **24** (2013), no. 5, 387–412, DOI 10.1007/s00200-013-0209-9. MR3118614

[31] Tanush Shaska and Jennifer L. Thompson, *On the generic curve of genus 3*, Affine algebraic geometry, Contemp. Math., vol. 369, Amer. Math. Soc., Providence, RI, 2005, pp. 233–243, DOI 10.1090/conm/369/06814. MR2126664

[32] Tony Shaska and Caleb M. Shor, *2-Weierstrass points of genus 3 hyperelliptic curves with extra involutions*, Comm. Algebra **45** (2017), no. 5, 1879–1892, DOI 10.1080/00927872.2016.1226861. MR3582832

[33] Tetsuji Shioda, *On the graded ring of invariants of binary octavics*, Amer. J. Math. **89** (1967), 1022–1046, DOI 10.2307/2373415. MR0220738

[34] Shigeaki Tsuyumine, *On Siegel modular forms of degree three*, Amer. J. Math. **108** (1986), no. 4, 755–862, DOI 10.2307/2374517. MR853217

[35] Elke Wolf, *Eigenvalues of composition combined with differentiation*, Albanian J. Math. **10** (2016), no. 1, 11–19. MR3589798

[36] Mohamed Zaatra, *A characterization of some Laguerre-Hahn orthogonal polynomials of class one*, Albanian J. Math. **10** (2016), no. 1, 21–36. MR3589799

Department of Mathematical Sciences, United States Military Academy, West Point, New York 10996

Email address: lubjana.beshaj@westpoint.edu

Department of Computer Science, Rochester Institute of Technology, Rochester, New York 14623

Email address: mkp@cs.rit.edu

Contemporary Mathematics
Volume **724**, 2019
https://doi.org/10.1090/conm/724/14590

On automorphisms of algebraic curves

A. Broughton, T. Shaska, and A. Wootton

ABSTRACT. An irreducible, algebraic curve $\mathcal{X}_g$ of genus $g \geq 2$ defined over an algebraically closed field k of characteristic $\operatorname{ch} k = p \geq 0$, has finite automorphism group $\operatorname{Aut}(\mathcal{X}_g)$. In this paper we describe methods of determining the list of groups $\operatorname{Aut}(\mathcal{X}_g)$ for a fixed $g \geq 2$. Moreover, equations of the corresponding families of curves are given when possible.

1. Introduction

Algebraic curves are some of the most studied mathematical objects. One of the first questions asked about algebraic curves was if they had any symmetries (automorphisms) and if so how many of them? It was noticed early in XIX-century that if $g = 0, 1$, then the curve could have infinitely many automorphisms. However, if $g \geq 2$ then the set of automorphisms is finite, this was first proved by Schwartz.

Given a genus $g \geq 2$ curve $\mathcal{X}_g$ defined over a perfect field k, char $k = p \geq 0$, let $F := k(\mathcal{X}_g)$ be its function field and denote by $\operatorname{Aut}(\mathcal{X}_g) := \operatorname{Aut}(F/k)$. Historically, the main questions addressed when it comes to automorphisms of curves have been the following.

i) What is the order $|\operatorname{Aut}(\mathcal{X}_g)|$?
ii) What is the list of groups $G = \operatorname{Aut}(\mathcal{X}_g)$, for a fixed g?
iii) For a given $G = \operatorname{Aut}(\mathcal{X}_g)$, can one determine an equation for $\mathcal{X}_g$?

The answer to each of the above questions becomes simpler if we assume that the curve $\mathcal{X}_g$ is a smooth, irreducible algebraic curve defined over an algebraically closed field k, which will be the focus of this paper.

The answer to question i) is well-known and has been long established. When $p = 0$ then $|\operatorname{Aut}(\mathcal{X}_g)| \leq 84(g-1)$ (the so called Hurwitz bound) and when $p > 0$ then $|\operatorname{Aut}(\mathcal{X}_g)| \leq 16g^4$. In the case $p = 0$ the bound is sharp and curves which achieve this bound are called Hurwitz curves. The first example of a Hurwitz curve occurs in genus $g = 3$ and is the celebrated Klein's quartic with group $\operatorname{PSL}(2, 7)$. The next Hurwitz curve occurs for $g = 7$. It was first discovered by Fricke and later by Macbeath [**46**]. The next two Hurwitz curves occur for $g = 14, 17$. If $p > 0$ the bound is naturally higher, even though cases with $|G| > 8g^3$ are very special though well-known (cf. Thm. 18).

The answer to question ii) is known for $p = 0$ for small genus due to recent advances in computational group theory. There is a huge amount of literature for

2010 *Mathematics Subject Classification.* Primary 14H10, 14H45.

$p = 0$ from the analytic point of view via Fuchsian groups and Riemann surfaces. In [**14**] Breuer computed all possible signatures of Riemann surfaces up to genus $g = 48$. The restriction on the genus was simply a restriction on the GAP's SmallGroupLibrary which included all groups of order ≤ 1000. This bound has been extended recently to all groups of order < 2000 and therefore Breuer's result can be easily extended. In [**49**] the authors, using previous results of Singerman [**67**], Ries [**53**], et al., and their moduli dimension gave a method of how to pick from all Breuer's signatures those signatures which occur as full automorphism groups. Moreover, all the signatures of groups of size $\geq 4(g-1)$, also called "large groups", and $g \leq 10$ are given in [**49**]. Hence, for $p = 0$ we can explicitly determine the list of all groups $G = \mathrm{Aut}(\mathcal{X}_g)$ for any reasonably chosen $g \geq 2$.

In the case of char $k = p > 0$ we are not aware of any comparable results. However, the list of groups for $p > 2g + 1$ is the same as for $p = 0$. Hence, for any given genus g the fields that need to be checked are those with characteristic $2 \leq p \leq 2g + 1$. Thm. 18 shows that there are very few families of curves with $|G| \geq 8g^3$. They are all of moduli dimension 0 and are superelliptic curves or cyclic curves. (A superelliptic curve (or cyclic curve) has an equation of the form $y^n = f(x)$ with some restrictions. See Section 5 for background literature and Section 2.2 for a definition in terms of the automorphism group). So for all practical purposes, if we want to find a list of groups G for $p > 0$, then $|G| < 8g^3$ is the practical bound.

As previously remarked, the special cases of Thm. 18 are well known falling into the family of superelliptic curves (or cyclic curves). Such curves are the best understood families of curves and interesting from many points of view; see [**8**], [**52**] for more details. In [**54**] all groups which occur as automorphism groups $\mathrm{Aut}(\mathcal{X}_g)$ over any field k of char $k = p \neq 2$ are determined.

The well-known hyperelliptic curves are a special case of superelliptic curves. They are the best understood among all families of algebraic curves and were the first examples used to understand Jacobians, theta-functions, etc in algebraic geometry. In [**61**], [**59**], [**19**], [**13**], [**63**] one can find complete details of groups that occur as automorphism groups of hyperelliptic curves, the corresponding locus in the moduli space for each such group, and invariants parameterizing some of these spaces.

The answer to question iii) is not yet completely answered. The only families of curves for which we fully understand how to write down equations and even determine such equations over a field of moduli are the superelliptic curves. Such equations are given in [**55**] for all superelliptic curves over a field of characteristic $p \neq 2$. In [**33**] conditions are given when such equations are defined over a field of moduli. It is still an open problem, even for characteristic $p = 0$, to determine such equations for all curves.

The main goal of this paper is to give a comprehensive survey of the main results of the topic and more importantly to provide complete results for small genus g or for known families of curves. We prefer the algebraic approach which makes it possible to give a more unified approach for any $p \geq 0$, but we also briefly describe the analytic approach via Riemann surfaces, Fuchsian groups, etc. We give a very brief review of definitions of the main results and refer the reader to more complete works in this topic such as [**70**], [**35**].

Throughout the paper we highlight what can be accomplished with superelliptic curves in terms of Weierstrass points, automorphism groups, and equations of curves. We give precise lists of groups for such curves in all characteristics.

Notation: By a **curve** $\mathcal{X}_g$ we always mean a smooth, irreducible, projective curve of genus $g \geq 2$ defined over a perfect field k. $\Sigma_{\mathcal{X}}(k)$ will denote the set of k-points of $\mathcal{X}$. By $F := k(\mathcal{X}_g)$ we denote its function field. For any function field F, a place in F is denoted by $\mathfrak{p}$ and the set of places of F by $\mathcal{P}_F$ (see Section 2.2 for definition of place). A finite field will be denoted by $\mathbb{F}_q$ and the characteristic of a field k by char $k = p$.

For a curve $\mathcal{X}_g$ defined over k, we denote by $\mathrm{Aut}_k(\mathcal{X}_g)$ the automorphism group of $\mathcal{X}_g$. Since we will only consider such groups for curves defined over algebraically closed fields, then $\mathrm{Aut}(\mathcal{X}_g)$ will be used instead. The cyclic group of order n will be denoted by C_n, the dihedral group of order $2n$ will be D_n and the symmetric and alternating groups on n letters by S_n and A_n respectively. In many of our results the GAP identity of a group will be used. A group G with GAP identity (m, n) means that $|G| = m$ and that the group is the n-th in the list of SmallGroup library in GAP.

2. Algebraic curves and their function fields

We assume that the reader is familiar with the basic definitions of field extensions. This part is intended more to settle the notation used in the rest of the paper than as an introduction to algebraic curves. Throughout k is a perfect field. For more details the reader is encouraged to see [**70**] or [**35**] among other places.

Let us establish some notation and basic facts about algebraic curves and their function fields.

2.1. Algebraic curves. The following definitions are easily extended to any algebraic variety, but we will stick with curves. Let k be a perfect field and $\mathcal{X}$ an algebraic curve defined over k. Then there is a homogeneous ideal $I_{\mathcal{X}} \subset k[X_0, X_1, \dots, X_n]$ defining $\mathcal{X}$, and the curve $\mathcal{X}$ is irreducible if and only if $I_{\mathcal{X}}$ is a *prime* ideal in $k[X_0, X_1, \dots, X_n]$. The (homogenous) coordinate ring of $\mathcal{X}$ is $\Gamma_h(\mathcal{X}) := k[X_0, X_1, \dots, X_n]/I_{\mathcal{X}}$, which is an integral domain. The function field of $\mathcal{X}$ is the quotient field of $\Gamma_h(\mathcal{X})$ and denoted by $k(\mathcal{X})$. Since $\mathcal{X}$ is an algebraic variety of dimension one, then $k(\mathcal{X})$ is an algebraic function field of one variable.

Let $P = (a_0, a_1, \dots, a_n) \in \mathcal{X}$. The ring

$$\mathcal{O}_P(\mathcal{X}) = \{f \in k(\mathcal{X}) \mid f \text{ is defined at } P\} \subset k(\mathcal{X})$$

is a local ring with maximal ideal

$$M_P(\mathcal{X}) = \{f \in \mathcal{O}_P(\mathcal{X}) \mid f(P) = 0\}.$$

The point $P \in \mathcal{X}$ is a *non-singular point* if the local ring $\mathcal{O}_P(\mathcal{X})$ is a discrete valuation ring. There is a 1-1 correspondence between points $P \in \mathcal{X}$ and the places of $k(\mathcal{X})/k$, given by $P \mapsto M_P(\mathcal{X})$. This correspondence makes it possible to translate definitions from algebraic function fields to algebraic curves and vice-versa.

2.2. Algebraic extensions of function fields. An algebraic function field F/k of one variable over k is a finite algebraic extension of $k(x)$ for some $x \in F$ which is transcendental over k. A **place** $\mathfrak{p}$ of the function field F/k is the maximal ideal for some valuation ring $\mathcal{O}$ of F/k. We will denote by $\mathcal{P}_F$ the set of all places of F/k. Equivalently $\Sigma_{\mathcal{X}}(k)$ will denote the set of k-points of $\mathcal{X}$.

An algebraic function field F'/k' is called an algebraic extension of F/k if F' is an algebraic extension of F and $k \subset k'$.

A place $\mathfrak{p}' \in \mathcal{P}_{F'}$ is said to **lie over** $\mathfrak{p} \in \mathcal{P}_F$ if $\mathfrak{p} \subset \mathfrak{p}'$. We write $\mathfrak{p}'|\mathfrak{p}$. In this case there exists an integer $e \geq 1$ such that $v_{\mathfrak{p}'}(x) = e \cdot v_{\mathfrak{p}}(x)$, for all $x \in F$. This integer is denoted by $e(\mathfrak{p}'|\mathfrak{p}) := e$ and is called the **ramification index** of $\mathfrak{p}'$ over $\mathfrak{p}$. We say that $\mathfrak{p}'|\mathfrak{p}$ is *ramified* when $e(\mathfrak{p}'|\mathfrak{p}) > 1$ and otherwise *unramified.*

For any place $\mathfrak{p} \in \mathcal{P}_F$ denote by $F_{\mathfrak{p}} := \mathcal{O}/\mathfrak{p}$. The integer $f(\mathfrak{p}'|\mathfrak{p}) := [F'_{\mathfrak{p}'} : F_{\mathfrak{p}}]$ is called the *relative degree* of $\mathfrak{p}'|\mathfrak{p}$.

THEOREM 1 (Fundamental equality). *Let F'/k' be a finite extension of F/k and $\mathfrak{p}$ a place of F/k. Let $\mathfrak{p}_1, \dots, \mathfrak{p}_m$ be all the places in F'/k' lying over $\mathfrak{p}$ and $e_i := e(\mathfrak{p}_i|\mathfrak{p})$ and $f_i := f(\mathfrak{p}_i|\mathfrak{p})$ the relative degree of $\mathfrak{p}_i|\mathfrak{p}$. Then*

$$\sum_{i=1}^{m} e_i f_i = [F' : F].$$

For a place $\mathfrak{p} \in \mathcal{P}_F$ let $\mathcal{O}'_{\mathfrak{p}}$ be the integral closure of $\mathcal{O}_{\mathfrak{p}}$ in F'. The complementary module over $\mathcal{O}_{\mathfrak{p}}$ is given by $t \cdot \mathcal{O}'_{\mathfrak{p}}$. Then for $\mathfrak{p}'|\mathfrak{p}$ we define the **different exponent** of $\mathfrak{p}'$ over $\mathfrak{p}$ as

$$d(\mathfrak{p}'|\mathfrak{p}) := -v_{\mathfrak{p}'}(t).$$

By [**70**, Prop. 3.4.2] $d(\mathfrak{p}'|\mathfrak{p})$ is well-defined and $d(\mathfrak{p}'|\mathfrak{p}) \geq 0$. Moreover, $d(\mathfrak{p}'|\mathfrak{p}) = 0$ for almost all $\mathfrak{p} \in \mathcal{P}_F$. The **different divisor** is defined as

$$\operatorname{Diff}(F'/F) := \sum_{\mathfrak{p} \in \mathcal{P}_F} \sum_{\mathfrak{p}'|\mathfrak{p}} d(\mathfrak{p}'|\mathfrak{p}) \cdot \mathfrak{p}'.$$

The following well-known formula for Riemann surfaces can now be generalized for function fields as follows.

THEOREM 2 (Hurwitz genus formula). *Let F/k be an algebraic function field of genus g and F'/F a finite separable extension. Let k' denote the constant field of F' and g' the genus of F'/k'. Then,*

$$2(g'-1) = \frac{[F' : F]}{[k' : k]}(2g-2) + \deg \operatorname{Diff}(F'/F) \tag{1}$$

For a proof see [**70**, Thm. 3.4.13]. A special case of the above is the following:

COROLLARY 1. *Let F/k be a function field of genus g and $x \in F \setminus k$ such that $F/k(x)$ is separable. Then,*

$$2g - 2 = -2[F : k(x)] + \deg \operatorname{Diff}(F/k(x))$$

The ramification index and the different exponent are closely related, as made precise by the Dedekind theorem.

THEOREM 3 (Dedekind different theorem). *For all $\mathfrak{p}'|\mathfrak{p}$ we have:*
i) $d(\mathfrak{p}'|\mathfrak{p}) \geq e(\mathfrak{p}'|\mathfrak{p}) - 1$.
ii) $d(\mathfrak{p}'|\mathfrak{p}) = e(\mathfrak{p}'|\mathfrak{p}) - 1$ *if and only if* $e(\mathfrak{p}'|\mathfrak{p})$ *is not divisible by the* char k.

An extension $\mathfrak{p}'|\mathfrak{p}$ is said to be *tamely* ramified if $e(\mathfrak{p}'|\mathfrak{p}) > 1$ and char k does not divide $e(\mathfrak{p}'|\mathfrak{p})$. If $e(\mathfrak{p}'|\mathfrak{p}) > 1$ and char k does divide $e(\mathfrak{p}'|\mathfrak{p})$ we say that $\mathfrak{p}'|\mathfrak{p}$ is *wildly* ramified.

The extension F'/F is called *ramified* if there is at least one place $\mathfrak{p} \in \mathcal{P}_F$ which is ramified in F'/F. The extension F'/F is called *tame* if there is no place $\mathfrak{p} \in \mathcal{P}_F$ which is wildly ramified in F'/F.

LEMMA 1. *Let F'/F be a finite separable extension of algebraic function fields. Then*

a) *$\mathfrak{p}'|\mathfrak{p}$ is ramified if and only if $\mathfrak{p}' \leq \mathrm{Diff}(F'/F)$. Moreover, if $\mathfrak{p}'/\mathfrak{p}$ is ramified then:*
   - *i) $d(\mathfrak{p}'|\mathfrak{p}) = e(\mathfrak{p}'|\mathfrak{p}) - 1$ if and only if $\mathfrak{p}'|\mathfrak{p}$ is tamely ramified*
   - *ii) $d(\mathfrak{p}'|\mathfrak{p}) > e(\mathfrak{p}'|\mathfrak{p}) - 1$ if and only if $\mathfrak{p}'|\mathfrak{p}$ is wildly ramified*

b) *Almost all places $\mathfrak{p} \in \mathcal{P}_F$ are unramified in F'/F.*

From now on we will use the term "curve" and its function field interchangeably, depending on the context. It is more convenient to talk about function fields than curves in most cases.

2.3. Divisors and the Riemann-Roch theorem. For a given curve $\mathcal{X}$ defined over k, a divisor D is called the formal finite sum

$$D = \sum_{\mathfrak{p} \in \Sigma_{\mathcal{X}}(k)} z_{\mathfrak{p}} P.$$

The set of all divisors of $\mathcal{X}$ is denoted by $\mathrm{Div}_{\mathcal{X}}(k)$.

2.3.1. *Riemann-Roch spaces.* Define a partial ordering of elements in $\mathrm{Div}_{\mathcal{X}}(k)$ as follows; D is *effective* ($D \geq 0$) if $z_{\mathfrak{p}} \geq 0$ for every $\mathfrak{p}$, and $D_1 \geq D_2$ if $D_1 - D_2 \geq 0$. The **Riemann-Roch space** associated to D is

$$\mathcal{L}(D) = \{f \in k(\mathcal{X})^* \text{ with } (f) \geq -D\} \cup \{0\}.$$

So the elements $x \in \mathcal{L}(D)$ are defined by the property that $w_{\mathfrak{p}}(x) \geq -z_{\mathfrak{p}}$ for all $\mathfrak{p} \in \Sigma_{\mathcal{X}}(k)$. Basic properties of valuations imply immediately that $\mathcal{L}(D)$ is a vector space over k. This vector space has positive dimension if and only if there is a function $f \in k(\mathcal{X})^*$ with $D + (f) \geq 0$, or equivalently, $D \sim D_1$ with $D_1 \geq 0$.

Here are some immediately obtained facts: $\mathcal{L}(0) = k$ and if $\deg(D) < 0$ then $\mathcal{L}(D) = \{0\}$. If $\deg(D) = 0$ then either D is a principal divisor or $\mathcal{L}(D) = \{0\}$. The following result is easy to prove but fundamental.

PROPOSITION 1. *Let $D = D_1 - D_2$ with $D_i \geq 0$. Then*

$$\dim(\mathcal{L}(D)) \leq \deg(D_1) + 1.$$

We remark that for $D \sim D'$ we have $\mathcal{L}(D) \sim \mathcal{L}(D')$. In particular $\mathcal{L}(D)$ is a finite-dimensional k-vector space. We follow traditional conventions and denote the dimension of $\mathcal{L}(D)$ by

$$\ell(D) := \dim_k(\mathcal{L}(D)). \tag{2}$$

Computing $\ell(D)$ is a fundamental problem which is solved by the Riemann-Roch Theorem. A first estimate is a generalization of the proposition above:

For all divisors D we have the inequality

$$\ell(D) \leq \deg(D) + 1.$$

For a proof one can assume that $\ell(D) > 0$ and so $D \sim D' > 0$. The important fact is that one can estimate the interval given by the inequality.

THEOREM 4 (**Riemann**). *For given curve $\mathcal{X}$ there is a minimal number $g_{\mathcal{X}} \in \mathbb{N} \cup \{0\}$ such that for all $D \in \mathrm{Div}_{\mathcal{X}}$ we have*

$$\ell(D) \geq \deg(D) + 1 - g_{\mathcal{X}}.$$

For a proof see [**70**, Proposition 1.4.14]. Therefore,

$$g_{\mathcal{X}} = \max\{\deg D - \ell(D) + 1;\ D \in \mathrm{Div}_{\mathcal{X}}(k)\}$$

exists and is a non-negative integer independent of D. The integer $g_{\mathcal{X}}$ is called the **genus** of $\mathcal{X}$.

The genus does not change under constant field extensions because we have assumed that k is perfect. This can be wrong in general if the constant field of $\mathcal{X}$ has inseparable algebraic extensions. There is a corollary of the theorem.

COROLLARY 2. *There is a number $n_{\mathcal{X}}$ such that for $\deg(D) > n_{\mathcal{X}}$ we get equality*

$$\ell(D) = \deg(D) + 1 - g_{\mathcal{X}}.$$

Thm. 4 together with its corollary is the "Riemann part" of the Theorem of Riemann-Roch for curves. To determine $n_{\mathcal{X}}$ and to get more information about the inequality for small degrees one needs canonical divisors.

2.3.2. *Canonical divisors.* Let $k(\mathcal{X})$ be the function field of a curve $\mathcal{X}$ defined over k. To every $f \in k(\mathcal{X})$ we attach a symbol df, the *differential* of f lying in a $k(\mathcal{X})$-vector space $\Omega(k(\mathcal{X}))$ generated by the symbols df modulo the following relations:

For $f, g \in k(\mathcal{X})$ and $\lambda \in k$ we have:

i) $d(\lambda f + g) = \lambda df + dg$
ii) $d(f \cdot g) = f dg + g df$.

The relation between derivations and differentials is given by the

DEFINITION 5 (Chain rule). *Let x be as above and $f \in k(\mathcal{X})$. Then $df = (\partial f/\partial x)dx$.*

The $k(\mathcal{X})$-vector space of differentials $\Omega(k(\mathcal{X}))$ has dimension 1 and it is generated by dx for any $x \in k(\mathcal{X})$ for which $k(\mathcal{X})/k(x)$ is finite and separable. We use a well known fact from the theory of function fields F in one variable.

Let $\mathfrak{p}$ be a place of F, i.e. an equivalence class of discrete rank one valuations of F trivial on k). Then there exist a function $t_{\mathfrak{p}} \in F$ with $w_{\mathfrak{p}}(t_{\mathbb{P}}) = 1$ and $F/k(t_{\mathfrak{p}})$ separable.

We apply this to $F = k(\mathcal{X})$. For all $\mathfrak{p} \in \Sigma_{\mathcal{X}}(k)$ we choose a function $t_{\mathfrak{p}}$ as above. For a differential $0 \neq \omega \in \Omega(k(\mathcal{X}))$ we get $\omega = f_{\mathfrak{p}} \cdot dt_{\mathfrak{p}}$. The divisor (ω) is given by

$$(\omega) := \sum_{\mathfrak{p} \in \Sigma_{\mathfrak{p}}} w_{\mathfrak{p}}(f_{\mathfrak{p}}) \cdot \mathfrak{p}$$

and is a called a **canonical divisor** of $\mathcal{X}$. The chain rule implies that this definition is independent of the choices, and the relation to differentials yields that (ω) is a divisor. Since $\Omega(k(\mathcal{X}))$ is one-dimensional over $k(\mathcal{X})$ it follows that the set of canonical divisors of $\mathcal{X}$ form a divisor class $\mathcal{K}_{\mathcal{X}} \in \mathrm{Pic}_{\mathcal{X}}(k)$ called the **canonical class** of $\mathcal{X}$. We are now ready to formulate the Riemann-Roch Theorem.

THEOREM 6 (**Riemann-Roch theorem**). *Let* (W) *be a canonical divisor of* $\mathcal{X}$. *For all* $D \in \mathrm{Div}_{\mathcal{X}}(k)$ *we have*

$$\ell(D) = \deg(D) + 1 - g_{\mathcal{X}} + \ell(W - D).$$

For a proof see [**70**, Section 1.5].

A differential ω is *holomorphic* if (ω) is an effective divisor. The set of holomorphic differentials is a k-vector space denoted by $\Omega^0_{\mathcal{X}}$ which is equal to $\mathcal{L}(W)$. If we take $D = 0$ respectively $D = W$ in the theorem of Riemann-Roch we get the following:

COROLLARY 3. $\Omega^0_{\mathcal{X}}$ *is a* $g_{\mathcal{X}}$*-dimensional* k*-vector space and* $\deg(W) = 2g_{\mathcal{X}} - 2$.

For our applications there are two further important consequences of the Riemann-Roch theorem.

COROLLARY 4. *The following are true:*

(1) *If* $\deg(D) > 2g_{\mathcal{X}} - 2$ *then* $\ell(D) = \deg(D) + 1 - g_{\mathcal{X}}$.
(2) *In every divisor class of degree* g *there is a positive divisor.*

PROOF. Take D with $\deg(D) \geq 2g_{\mathcal{X}} - 1$. Then $\deg(W - D) \leq -1$ and therefore $\ell(W - D) = 0$. Take D with $\deg(D) = g_{\mathcal{X}}$. Then $\ell(D) = 1 + \ell(W - D) \geq 1$ and so there is a positive divisor in the class of D. □

2.4. Function fields and branched covers. We continue to assume that $\mathcal{X} = \mathcal{X}_g$ is a smooth, irreducible, curve of genus $g = g_{\mathcal{X}}$ over an algebraically closed field k, with function field $k(\mathcal{X})$. Correspondingly, given a field K of transcendence degree 1 over k, then $K \backsimeq k(\mathcal{X}')$ for some curve $\mathcal{X}'$. As k is algebraically closed, each place of $k(\mathcal{X})/k$ may be identified with a geometric point P on a smooth model of $\mathcal{X}$. For $P \in \mathcal{X}$, let $\mathcal{O}_P(\mathcal{X})$ and $M_P(\mathcal{X})$ be as previously defined. Since $\mathcal{X}$ is smooth at P and one dimensional, there is a local parameter $z \in M_P(\mathcal{X})$ such that $M_P(\mathcal{X}) = z\mathcal{O}_P(\mathcal{X})$. We may write every $f \in k(\mathcal{X})$ as $f = z^e v$, $v \in \mathcal{O}^*_P(\mathcal{X})$. The number $e = e_P(f) = \nu_P(f)$ is the valuation, also called the **order of vanishing**.

We are particularly interested in the relationship between the non-constant morphisms $\pi : \mathcal{X} \to \mathcal{Y}$ of curves, which we shall call **branched coverings**, and the function fields $k(\mathcal{X})$ and $k(\mathcal{Y})$. Let $\pi : \mathcal{X} \to \mathcal{Y}$ be a branched covering. Then the induced map $\pi^* : k(\mathcal{Y}) \to k(\mathcal{X})$, $f \to f \circ \pi$ is an embedding of fields, realizing $k(\mathcal{X})$ as a finite degree extension of $k(\mathcal{Y})$. Conversely, if L is a finite extension of $k(\mathcal{Y})$ then there is an $\mathcal{X}$ and a morphism $\pi : \mathcal{X} \to \mathcal{Y}$ such that the extension is induced via π^*. If $n = [k(\mathcal{X}) : k(\mathcal{Y})]$ is the degree of the extension, and the extension is separable, then for all but finitely many points $Q \in \mathcal{Y}$, $\pi^{-1}(Q)$ has n points. The correspondence $\pi \leftrightarrow \pi^*$ is contravariant.

We can give a more precise statement of description of "$|\pi^{-1}(Q)| = n$ generically" using the notion of ramification degree. Given $\pi : \mathcal{X} \to \mathcal{Y}$, let $P \in \mathcal{X}, Q = \pi(P)$, and $z \in M_P(\mathcal{X})$, $w \in M_Q(\mathcal{Y})$ be local parameters. The function $\pi^*(w) \in M_P(\mathcal{X})$, so we can define the **ramification degree** of π at P to be $e_\pi(P) = e_P(\pi^*(w))$. The integer $e_\pi(P) \geq 1$ and we say that π is ramified at P if $e_\pi(P) > 1$. Note that $e_\pi(P) > 1$ if and only if $d\pi_P = 0$. We have the following proposition.

PROPOSITION 2. *Let* $\pi : \mathcal{X} \to \mathcal{Y}$ *be a branched covering of degree* n. *Then for* $Q \in \mathcal{Y}$ *we have*

$$n = \sum_{\pi(P)=Q} e_\pi(P) \tag{3}$$

If $\pi : \mathcal{X} \to \mathcal{Y}$ is induced by a G-action and the field extension is separable, then for each $Q \in \mathcal{Y}$ number of points lying over Q is $|G| / |G_P|$ for any $P \in \pi^{-1}(Q)$. It follows then that

$$n = e_\pi(P) \frac{|G|}{|G_P|}, \tag{4}$$

where $e_\pi(P) = |G_P|$ (See equation (5)).

Equation (3) is simply Thm. 1.

EXAMPLE 1. *Let $k = \overline{\mathbb{F}}_p$, and consider the Frobenius morphism $\pi : \mathbb{P}^1(k) \to \mathbb{P}^1(k)$, given by $x \to x^p$. The map is an injective but not invertible morphism. The induced maps on fields $k(y) \to k(x)$ is given by $y = x^p$, and has degree p. By direct calculation it can be shown that $e_\pi(x) = p$ for all $x \in \mathbb{P}^1(k)$, or alternatively $d\pi = 0$. Thus equation (3) is satisfied. This strange behavior is linked to the fact that the extension $k(y) = k(x^p) \to k(x)$ is purely inseparable.*

From now on we will consider only separable extensions.

A branched cover $\pi : \mathcal{X} \to \mathcal{Y}$ is called **tamely ramified** if all branching orders $e_\pi(P)$, $P \in \mathcal{X}$ are relatively prime to the characteristic p. In characteristic $p = 0$ all branched covers are considered to be tamely ramified. A cover is **wildly ramified** if it is not tamely ramified.

2.5. Automorphism groups, G-actions, stabilizers. Let $\mathcal{X}_g$ be an irreducible and non-singular algebraic curve defined over k of genus $g \geq 2$. We denote its function field by $F := k(\mathcal{X}_g)$. The automorphism group of $\mathcal{X}_g$ is the group $G := \mathrm{Aut}(F/k)$ (i.e., all field automorphisms of F fixing k). We will denote it by $\mathrm{Aut}_k(\mathcal{X}_g)$. When k is algebraically closed then we will simply use $\mathrm{Aut}(\mathcal{X}_g)$. The rest of this paper will focus on determining $\mathrm{Aut}(\mathcal{X}_g)$, for any given $g \geq 2$.

We say that a finite group G acts (birationally, conformally) on $\mathcal{X}$ if there is a monomorphism $\epsilon : G \to \mathrm{Aut}(\mathcal{X})$. There is an induced action on function fields given by $\epsilon^* : h \to \left(h^{-1}\right)^*$. Assuming G is finite, then the field of invariant functions $k(\mathcal{X})^G$ is a subfield such that $k(\mathcal{X})$ is an extension of $k(\mathcal{X})^G$ of degree $|G|$. The subfield $k(\mathcal{X})^G$ corresponds to some $k(\mathcal{Y})$ and there is a morphism $\pi_G : \mathcal{X} \to \mathcal{Y}$. We denote $\mathcal{Y}$ by $\mathcal{X}/G$. It can be shown that G acts transitively on the fibers of π_G, so that $\mathcal{Y}$ is an orbit space, as a set, so that degree of π_G is $|G|$. The set of **branch points** or **branch locus** is denoted $\mathcal{B}_G$. The covering $\pi_G : \mathcal{X}^\circ \to \mathcal{Y}^\circ$, where $\mathcal{Y}^\circ = \mathcal{Y} - \mathcal{B}_G$ and $\mathcal{X}^\circ = \mathcal{X} - \pi_G^{-1}(\mathcal{B}_G)$ is an unramified Galois covering of affine curves.

A first difference that we see in positive characteristic is the structure and action of stabilizers. Given G acting on $\mathcal{X}$ and $P \in \mathcal{X}$, the **stabilizer** at P is defined by

$$G_P = \{g \in G : gP = P\}, \tag{5}$$

also known as the decomposition group. In the characteristic 0 case G_P is cyclic, and acts faithfully on the tangent space $T_P(\mathcal{X})$. This may fail in the $p > 0$ case.

The action of stabilizers on $M_P(\mathcal{X})/M_P^{s+1}(\mathcal{X})$ is faithful for some s. Indeed, if x is a local parameter at P, and g^* is the identity on $M_P(\mathcal{X})/M_P^{s+1}(\mathcal{X})$ then $g^*x - x \in M_P^{s+1}(\mathcal{X})$ for all s. It follows that $g^*x - x = 0$ as $\bigcap\limits_{s\geq 0} M_P^{s+1}(\mathcal{X}) = \{0\}$. Now let $f \in \mathcal{O}_P(\mathcal{X})$ be arbitrary. By considering a Taylor series expansion in x

at P we see that $g^*f = f$. It follows that g^* is trivial on all of $k(\mathcal{X})$ and hence g is trivial on all of $\mathcal{X}$. From the format of the matrix for g^* it follows that the action on $GL_k\left(M_P(\mathcal{X})/M_P^2(\mathcal{X})\right) \backsimeq k^*$ is trivial on elements of order p, so that the map $G_P \to k^*$ has cyclic image and the kernel is a p-group. It then follows that $G_P \backsimeq C_m \ltimes H$ where H is a p-group and that $e_\pi(P) = m \times q$ where q is some p-power. For $g \in G_P$ we call the image of $(g^{-1})^*$ in k^* the **rotation number** of g at P.

2.6. Cyclic n-gonal curves. A cyclic n-gonal curve has a cyclic $G = C_n$ action for which the genus of $\mathcal{X}/G$ is zero. A defining equation is of the form $y^n = f(x)$, where $f(x)$ is a rational function. The curve can be put in a canonical form:

$$y^n = (x-a_1)^{n_1}\cdots(x-a_t)^{n_t},$$

where $n, n_1, \ldots, n_t \in \mathbb{Z}^+$, $a_1, \ldots, a_t \in k$ satisfy:

(1) $a_1, \ldots, a_t$ are distinct,
(2) $0 < n_i < n$,
(3) n divides $n_1 + \cdots + n_t$
(4) $\gcd(n_1, \ldots, n_t) = 1$.

The conditions 1 and 2 simplify the model and eliminate degeneracies, condition 3 ensures that curve is not ramified over ∞, and condition 4 ensures that curve is irreducible. The curve needs to be normalized to make it smooth. There is a G-action, where $G = U_n = \{u \in k : u^n = 1\}$, given by $(x, y) \to (x, uy)$. We must assume the p does not divide n otherwise the field extension will be inseparable. The quotient map is $(x, y) \to x$, and is branched over $a_1, \ldots, a_t$. The local equation is $y^n = b(x)(x-a_i)^{n_i}$, where $b(a_i) \neq 0$. There are $d_i = \gcd(n, n_i)$ branches of the curve meeting at $(a_i, 0)$, and in the normalization the ramification degree is $m_i = n/d_i$. For a discussion of the $p = 0$ case, which extends in many ways to the $p > 0$ case see [**18**],[**16**]. In the complex case, the vector $(n_1, \ldots, n_t)$ is a generating vector (see Section 6.1) for $G = C_n$ by the conditions $2, 3, 4$. In the $p > 0$ case $(n_1, \ldots, n_t)$ is a classifier for the C_n action and may be thought of as a generating vector for Galois covers of $\mathcal{Y} = \mathbb{P}^1(k)$, ramified exactly over $\mathcal{B} = \{a_1, \ldots, a_t\}$, with Galois group $G = C_n$. It plays the role of generating vector, though it is not constructed topologically. We will look at such curves in more detail in Section 5.

3. Weierstrass Gap theorem and Weierstrass points

We assume that the reader is familiar with basic definitions on divisors on curves. For a short introduction see [**27**] in this volume or [**70**].

Let P be a point on $\mathcal{X}_g$ and consider the vector spaces $\mathcal{L}(nP)$ for $n = 0, 1, \ldots, 2g-1$. These vector spaces contain functions with poles only at P up to a specific order. This leads to a chain of inclusions

$$\mathcal{L}(0) \subseteq \mathcal{L}(P) \subseteq \mathcal{L}(2P) \subseteq \cdots \subseteq \mathcal{L}((2g-1)P)$$

with a corresponding non-decreasing sequence of dimensions

$$\ell(0) \leq \ell(P) \leq \ell(2P) \leq \cdots \leq \ell((2g-1)P).$$

The following proposition shows that the dimension goes up by at most 1 in each step.

PROPOSITION 3. *For any $n > 0$,*

$$\ell((n-1)P) \leq \ell(nP) \leq \ell((n-1)P) + 1.$$

PROOF. It suffices to show $\ell(nP) \leq \ell((n-1)P) + 1$. To do this, suppose $f_1, f_2 \in \mathcal{L}(nP) \setminus \mathcal{L}((n-1)P)$. Since f_1 and f_2 have the same pole order at P, using the series expansions of f_1 and f_2 with a local coordinate, one can find a linear combination of f_1 and f_2 to eliminate their leading terms. That is, there are constants $c_1, c_2 \in k$ such that $c_1 f_1 + c_2 f_2$ has a strictly smaller pole order at P, so $c_1 f_1 + c_2 f_2 \in \mathcal{L}((n-1)P)$. Then f_2 is in the vector space generated by a basis of $\mathcal{L}((n-1)P)$ along with f_1. Since this is true for any two functions f_1, f_2, we conclude $\ell(nP) \leq \ell((n-1)P) + 1$, as desired. □

For any integer $n > 0$, we call n a **Weierstrass gap number of** P if $\ell(nP) = \ell((n-1)P)$; that is, if there is no function $f \in k(\mathcal{X}_g)^\times$ such that $(f)_\infty = nP$. Weierstrass stated and proved the Gap theorem, or **Lückensatz**, on gap numbers in the 19th century, likely in the 1860s.

THEOREM 7 (The Weierstrass Gap theorem). *For any point P, there are exactly g gap numbers $\alpha_i(P)$ with*

$$1 = \alpha_1(P) < \alpha_2(P) < \cdots < \alpha_g(P) \leq 2g - 1.$$

This theorem is a special case of the Noether Gap theorem, which we state and prove below.

The set of gap numbers, denoted by G_P, forms the **Weierstrass gap sequence** for P. The non-gap numbers form a semi-group under addition since they correspond to pole orders of functions. For any curve the gap sequence is the same for all points with finitely many exceptional points. All curves in characteristic 0 and most curves in positive characteristic have the **classical gap sequence** $\{1, 2, \ldots, g\}$ for the generic gap sequence. In this section we discuss only curves with a classical gap sequence. See [**57**] for examples of curves with a non-classical gap sequence.

DEFINITION 8 (Weierstrass point). *If the gap sequence at P is anything other than $\{1, 2, \ldots, g\}$, then P is called a* ***Weierstrass point****.*

Equivalently, P is a Weierstrass point if $\ell(gP) > 1$; that is, if there is a function f with $(f)_\infty = mP$ for some m with $1 < m \leq g$. The following was proved by F. K. Schmidt (1939).

THEOREM 9 (Schmidt [**57**]). *Every algebraic curve of genus $g \geq 2$ has at least one Weierstrass point.*

The notion of gaps can be generalized, which we briefly describe. Let $P_1, P_2, \ldots,$ be a sequence of (not necessarily distinct) points on $\mathcal{X}_g$. Let $D_0 = 0$ and, for $n \geq 1$, let $D_n = D_{n-1} + P_n$. One constructs a similar sequence of vector spaces

$$\mathcal{L}(D_0) \subseteq \mathcal{L}(D_1) \subseteq \mathcal{L}(D_2) \subseteq \cdots \subseteq \mathcal{L}(D_n) \subseteq \cdots$$

with a corresponding non-decreasing sequence of dimensions

$$\ell(D_0) \leq \ell(D_1) \leq \ell(D_2) \leq \cdots \leq \ell(D_n) \leq \cdots.$$

If $\ell(D_n) = \ell(D_{n-1})$, then n is a **Noether gap number** of the sequence $P_1, P_2, \ldots$.

THEOREM 10 (The Noether Gap theorem). *For any sequence $P_1, P_2, \ldots,$ there are exactly g Noether gap numbers n_i with*

$$1 = n_1 < n_2 < \cdots < n_g \leq 2g - 1.$$

PROOF. In analog with Prop. 3, one can show the dimension goes up by at most 1 in each step; that is,

$$\ell(D_{n-1}) \leq \ell(D_n) \leq \ell(D_{n-1}) + 1$$

for all $n > 0$. First, note that the Riemann-Roch theorem is an equality for $n > 2g-1$, so the dimension goes up by 1 in each step, so there are no gap numbers greater than $2g-1$.

Now, consider the chain $\mathcal{L}(D_0) \subseteq \cdots \subseteq \mathcal{L}(D_{2g-1})$. By Riemann-Roch, $\ell(D_0) = 1$ and $\ell(D_{2g-1}) = g$, so in this chain of vector spaces, the dimension must increase by 1 exactly $g-1$ times in $2g-1$ steps. Thus, for $n \in \{1, 2, \ldots, 2g-1\}$, there are g values of n such that $\ell(D_n) = \ell(D_{n-1})$. These g values are the Noether gap numbers. □

For a complete treatment of Weierstrass points and their weights for $p = 0$ see [**65**].

3.1. Weierstrass points via holomorphic differentials. Continuing with a point P on a curve $\mathcal{X}_g$, recall that n is a gap number precisely when $\ell(nP) = \ell((n-1)P)$. By Riemann-Roch, this occurs exactly when

$$\ell(K - (n-1)P) - \ell(K - nP) = 1$$

for a canonical divisor K, which is the divisor associated to some differential dx. Thus there is $f \in k(\mathcal{X}_g)^\times$ such that $(f) + K - (n-1)P \geq 0$ and $(f) + K - nP \not\geq 0$, which implies that $\text{ord}_P(f \cdot dx) = n-1$. Since $(f) + K \geq (n-1)P \geq 0$ (for $n \geq 1$), n is a gap number of P exactly when there is a holomorphic differential $f \cdot dx$ such that $\text{ord}_P(f \cdot dx) = n-1$.

For $H^0(\mathcal{X}_g, \Omega^1)$ the space of holomorphic differentials on $\mathcal{X}_g$, by Riemann-Roch, the dimension of $H^0(\mathcal{X}_g, \Omega^1)$ is g. Let $\{\psi_i\}$, for $i = 1, \ldots, g$, be a basis, chosen in such a way that

$$\text{ord}_P(\psi_1) < \text{ord}_P(\psi_2) < \cdots < \text{ord}_P(\psi_g).$$

Let $n_i = \text{ord}_P(\psi_i) + 1$. The **1-gap sequence at** P is $\{n_1, n_2, \ldots, n_g\}$. We then have the following equivalent definition of a Weierstrass point.

DEFINITION 11 (Weierstrass point). *If the 1-gap sequence at P is anything other than $\{1, 2, \ldots, g\}$, then P is a Weierstrass point.*

With this formulation, we see P is a Weierstrass point exactly when there is a holomorphic differential $f \cdot dx$ with $\text{ord}_P(f \cdot dx) \geq g$.

DEFINITION 12 (Weierstrass weight). *The* ***Weierstrass weight*** *of a point P is*

$$w(P) = \sum_{i=1}^{g} (n_i - i).$$

In particular, P is a Weierstrass point if and only if $w(P) > 0$.

3.2. Bounds for weights of Weierstrass points. Suppose $\mathcal{X}_g$ is a curve of genus $g \geq 1$, $P \in \mathcal{X}_g$, and consider the 1-gap sequence of P $\{n_1, n_2, \ldots, n_g\}$. We will refer to the non-gap sequence of P as the complement of this set within the set $\{1, 2, \ldots, 2g\}$. That is, the non-gap sequence is the sequence $\{\alpha_1, \ldots, \alpha_g\}$ where

$$1 < \alpha_1 < \cdots < \alpha_g = 2g.$$

PROPOSITION 4. *For each integer j with $0 < j < g$, $\alpha_j + \alpha_{g-j} \geq 2g$.*

PROPOSITION 5. *For $P \in \mathcal{X}_g$, $w(P) \leq g(g-1)/2$, with equality if and only if P is a branch point on a hyperelliptic curve $\mathcal{X}_g$.*

COROLLARY 5. *For a curve of genus $g \geq 2$, there are between $2g+2$ and $g^3 - g$ Weierstrass points. The lower bound of $2g+2$ occurs only in the hyperelliptic case.*

PROOF. The total weight of the Weierstrass points is $g^3 - g$. In Prop. 5, we see that the maximum weight of a point is $g(g-1)/2$, which occurs in the hyperelliptic case. Thus, there must be at least $\dfrac{g^3 - g}{g(g-1)/2} = 2g + 2$ Weierstrass points. On the other hand, the minimum weight of a point is 1, so there are at most $g^3 - g$ Weierstrass points. □

THEOREM 13 (Weierstrass normal form). *Let $\mathcal{X}$ be an irreducible curve of genus $g \geq 1$ defined over k. For a place $\mathfrak{p} \in k(\mathcal{X})$ let m be the first non-gap at $\mathfrak{p}$ and n be the least non-gap which is prime to m. Then*

i) $\mathcal{X}$ has affine equation

$$f(x,y) = y^m + u_1(x)y^{m-1} + \cdots + u_{m-1}(x)y + u_m(x), \tag{6}$$

where $u_i(x) \in k[x]$, $\deg u_m = n$ and $\deg u_i(x) < in/m$, for $i = 1, \ldots, m-1$.

ii) $p = (x,y)$ is a generic point of $\mathcal{X}$ for $\operatorname{div}(x)_\infty = m\mathfrak{p}$ and $\operatorname{div}(y)_\infty = n\mathfrak{p}$.

iii) the branch of $\mathcal{X}$ associated to $\mathfrak{p}$ is the unique branch of $\mathcal{X}$ with centre at y_∞.

The following is a well-known result.

LEMMA 2. *Let $\mathcal{X}$ be a genus g curve in Weierstrass normal form as in Eq. (6). Then*

i) $g \leq \frac{1}{2}(n-1)(m-1)$

ii) If no point of $\mathcal{X}$ other than y_∞ is singular, then $g = \frac{1}{2}(n-1)(m-1)$

For more details about Weierstrass points and their weights see [**64**], [**65**] where Weierstrass points of superelliptic curves are studied.

4. Automorphisms of curves

Let $\mathcal{X}$ be an irreducible and non-singular algebraic curve defined over k. We denote its function field by $F := k(\mathcal{X})$. The automorphism group of $\mathcal{X}$ is the group $G := \operatorname{Aut}(F/k)$ (i.e., all field automorphisms of F fixing k).

4.1. The action of k-automorphisms on places. G acts on the places of F/k. Since there is a 1-1 correspondence between places of F/k and points of $\mathcal{X}$, this action naturally extends to the points of $\mathcal{X}$. For $\alpha \in G$ and $P \in \mathcal{X}$, we denote its image under α by P^α. In a natural way we extend this G-action to $\operatorname{Div}_k(\mathcal{X})$. Let $D \in \operatorname{Div}_k(\mathcal{X})$, say $D = \sum n_P \cdot P$. Then

$$D^\alpha = \sum n_p \cdot P^\alpha.$$

LEMMA 3. *G acts on the set $\mathcal{W}$ of Weierstrass points.*

PROOF. The set $\mathcal{W}$ of Weierstrass points do not depend on the choice of the local coordinate and so it is invariant under any $\sigma \in \operatorname{Aut}(\mathcal{X}_g)$. □

Hence, in order to determine the automorphism group we can just study the action of the group on the set of Weierstrass point of the curve. Then we have the following.

PROPOSITION 6. *Let $\alpha \in \mathrm{Aut}(\mathcal{X})$ be a non-identity element. Then α has at most $2g+2$ fixed places.*

PROOF. Let α be a non-trivial element of $\mathrm{Aut}(F/k)$. Since α is not the identity, there is some place $\mathfrak{p} \in \mathcal{P}_F$ not fixed by α. Take $g+1$ distinct places $\mathfrak{p}_1, \ldots, \mathfrak{p}_{g+1}$ in $\mathcal{P}_F$ such that $D = \mathfrak{p}_1 + \cdots + \mathfrak{p}_{g+1}$ and D^α share no place. By [**35**, Thm. 6.82] there is $z \in F \setminus k$ such that $\mathrm{div}(z)_\infty = D$. Then consider $w = z - \alpha(z)$. Since z and $\alpha(z)$ have different poles then $w \neq 0$. Hence, w has exactly $2g+2$ poles. Then w has exactly $2g+2$ zeroes. But every fixed place of α is a zero of w. Hence α has at most $2g+2$ fixed places. □

Let $\mathcal{W}$ be the set of Weierstrass points. From Cor. 5 we know that $\mathcal{W}$ is finite. Since for every $\alpha \in \mathrm{Aut}(\mathcal{X})$, from Lem. 3 we have $\alpha(\mathcal{W}) = \mathcal{W}$. Then we have the following; see [**35**, Thm. 11.24] for the proof.

THEOREM 14. *Let $\mathcal{X}$ be a genus $g \geq 2$ irreducible, non-hyperelliptic curve defined over k such that* char $k = p$ *and $\alpha \in \mathrm{Aut}(\mathcal{X})$. If $p = 0$ or $p > 2g-2$ then α has finite order.*

Then we have the following; see [**35**, Lemma 11.25].

LEMMA 4. *If $p = 0$ and $g \geq 2$ then every automorphism is finite.*

In the case of $p = 0$ Hurwitz [**37**] showed $|\alpha| \leq 10(g-1)$. In 1895, Wiman improved this bound to be $|\alpha| \leq 2(2g+1)$ and showed this is best possible. If $|\alpha|$ is a prime then $|\alpha| \leq 2g+1$. Homma [**36**] (1980) shows that this bound is achieved for a prime $q \neq p$ if and only if the curve is birationally equivalent to

$$y^{m-s}(y-1)^s = x^q, \quad for \quad 1 \leq s < m \leq g+1.$$

If $p > 0$ then we have the following; see [**35**, Thm. 11.34].

THEOREM 15. *Let $\mathcal{X}$ be a genus $g \geq 2$, irreducible curve defined over k, with* char $k = p > 0$ *and $\alpha \in \mathrm{Aut}(\mathcal{X})$ which fixes a place $\mathfrak{p} \in \mathcal{P}_F$. Then the order of α is bounded by*

$$|\alpha| \leq 2p(g+1)(2g+1)^2.$$

4.2. Finiteness of $\mathrm{Aut}(\mathcal{X})$**.** The main difference for $g = 0, 1$ and $g \geq 2$ is that for $g \geq 2$ the automorphism group is a finite group. This result was proved first by Schmid (1938).

THEOREM 16 (Schmid [**56**]). *Let $\mathcal{X}$ be a genus $g \geq 2$, irreducible curve defined over a field k,* char $k = p \geq 0$. *Then* $\mathrm{Aut}(\mathcal{X})$ *is finite.*

4.2.1. *Characteristic $p = 0$.* As an immediate consequence of the Hurwitz theorem for char $k = p = 0$ we have that

$$|\mathrm{Aut}(\mathcal{X})| \leq 84(g-1).$$

Curves which obtain this bound are called *Hurwitz curves.* Klein's quartic is the only Hurwitz curve of genus $g \leq 3$. Fricke showed that the next Hurwitz group occurs for $g = 7$ and has order 504. Its group is $SL(2, 8)$, and an equation for it was computed by Macbeath [**46**] in 1965. Further Hurwitz curves occur for $g = 14$ and $g = 17$ (and for no other values of $g \leq 19$).

For a fixed $g \geq 2$ denote by $N(g)$ the maximum of the $|\mathrm{Aut}(\mathcal{X}_g)|$. Accola [**1**] and Maclachlan [**47**] independently show that $N(g) \geq 8(g+1)$ and this bound is sharp for infinitely many g's. If g is divisible by 3 then $N(g) \geq 8(g+3)$.

The following terminology is rather standard. We say $G \leq \mathrm{Aut}(\mathcal{X}_g)$ is a **large automorphism group** in genus g if

$$|G| \; > \; 4(g-1).$$

Then the quotient of $\mathcal{X}_g$ by G is a curve of genus 0, and the number of points of this quotient ramified in $\mathcal{X}_g$ is 3 or 4 (see [**14**], Lemma 3.18, or [**26**], pages 258-260). Singerman [**68**] (1974) shows that Riemann surfaces with large cyclic, Abelian, or Hurwitz groups are symmetric (admit an anti-conformal involution). Kulkarni [**40**](1997) classifies Riemann surfaces admitting large cyclic automorphism groups and works out full automorphism groups of these surfaces. Matsuno [**50**](1999) investigates the Galois covering of the projective line from compact Riemann surfaces with large automorphism groups.

4.2.2. *Characteristic $p > 0$.* In the case of positive characteristic the bound is higher due to possibly wild ramifications. The following was proved by Stichtenoth (1973) by extending previous results of P. Roquette and others.

THEOREM 17 (Stichtenoth [**69**]). *Let $\mathcal{X}$ be a genus $g \geq 2$, irreducible curve defined over a field k,* char $k = p > 0$. *Then*

$$|\,\mathrm{Aut}(\mathcal{X})| < 16 \cdot g^4,$$

unless $\mathcal{X}$ is the curve with equation

$$y^{p^n} + y = x^{p^{n+1}},$$

in which case it has genus $g = \frac{1}{2}p^n(p^n-1)$ and $|\,\mathrm{Aut}(\mathcal{X})| = p^{3n}(p^{3n}+1)(p^{2n}-1)$.

Hence we have a bound for curves of genus $g \geq 2$ even in characteristic $p > 0$. It turns out that all curves with large groups of automorphisms are special curves. So getting "better" bounds while such curves are left out has always been interesting. There is a huge amount of literature on this topic due to the interest of such bounds in coding theory. Perhaps the following theorem, which is due to Henn [**32**], provides a better bound if the following four families of curves are left out. As Henn points out in a footnote, this result may be sharpened to show that the order of $\mathrm{Aut}(\mathcal{X})$ is less than $3 \cdot (2g)^{5/2}$ except when $k(\mathcal{X})$ belongs to one of five types of function fields.

THEOREM 18 (Henn [**32**]). *Let $\mathcal{X}$ be an irreducible curve of genus $g \geq 2$. If $|G| \geq 8g^3$, then $\mathcal{X}$ is isomorphic to one of the following:*

i) The hyperelliptic curve

$$y^2 + y + x^{2^k+1} = 0,$$

defined over a field of characteristic $p = 2$. In this case the genus is $g = 2^{k-1}$ and $|G| = 2^{2k+1}(2^k+1)$.

ii) The hyperelliptic curve

$$y^2 = x^q - x,$$

defined over a field of characteristic $p > 2$ such that q is a power of p. In this case $g = \frac{1}{2}(q-1)$ and the reduced group $\bar{G}$ is isomorphic to $\mathrm{PSL}(2,q)$ *or* $\mathrm{PGL}(2,q)$.

iii) The Hermitian curve

$$y^q + y = x^{q+1},$$

defined over a field of characteristic $p \geq 2$ such that q is a power of p. In this case $g = \frac{1}{2}(q^2-q)$ and G is isomorphic to $PSU(3,q)$ or $PGU(3,q)$.

iv) The curve

$$y^q + y = x^{q_0}(x^q + x),$$

for $p = 2$, $q_0 = 2^r$, *and* $q = 2q_0^2$. *In this case,* $g = q_0(q-1)$ *and* $G \cong Sz(q)$.

4.2.3. *Bounds for* $p = 0$. Next we will consider some special curves, namely superelliptic or cyclic curves. Most of the groups of automorphisms for a fixed $g \geq 2$ correspond to such curves and they have large automorphism groups. A more general interesting problem which we do not consider here is the following.

PROBLEM 1. *Let* $\mathcal{X}$ *be an irreducible, smooth algebraic curve of genus* $g \geq 2$ *which is not superelliptic. Determine a bound for* $|\operatorname{Aut}(\mathcal{X})|$.

5. Superelliptic curves

There is a lot of literature published on superelliptic curves in the last two decades. We mostly follow the terminology from [**9**], [**51**], or [**34**].

Let k be a field of characteristic p and $\mathcal{X}_g$ be a genus $g \geq 2$ cyclic curve given by the equation $y^n = f(x)$ for some $f \in k[x]$. Then $k(x)$ is degree n genus zero subfield of $K = k(\mathcal{X})$. Let $G = \operatorname{Aut}(K/k)$. We are going to assume that $C_n := \operatorname{Gal}(K/k(x)) = \langle w \rangle$, with $w^n = 1$ and which satisfies $\langle w \rangle \lhd G$. (This is assumed in the literature cited above, and is the definition we take in Section 6.4.) Then, the reduced group $\bar{G} := G/C_n$ acts on $\mathbb{P}^1(k)$, and so satisfies $\bar{G} \leq \operatorname{PGL}(2,k)$. Hence $\bar{G}$ is isomorphic to one of the following: C_m, D_m, A_4, S_4, A_5, *semidirect product of an elementary Abelian group with cyclic group*, $\operatorname{PSL}(2,q)$ and $\operatorname{PGL}(2,q)$, see [**72**].

The group $\bar{G}$ acts on $k(x)$ naturally. The fixed field is a genus 0 field, say $k(z)$. Thus z is a degree $|\bar{G}|$ rational function in x, say $z = \phi(x)$, yielding a Galois branched covering $\phi : \mathbb{P}^1 \to \mathbb{P}^1$.

Let $\phi_0 : \mathcal{X}_g \to \mathbb{P}^1$ be the cover which corresponds to the degree n extension $K/k(x)$. Then $\Phi := \phi \circ \phi_0$ has monodromy group $G := \operatorname{Aut}(\mathcal{X}_g)$. From basic covering theory, the group G is embedded in the group S_l where $l = \deg \Phi$. There is an r-tuple $\overline{\sigma} := (\sigma_1, \dots, \sigma_r)$, where $\sigma_i \in S_l$ such that $\sigma_1, \dots, \sigma_r$ generate G and $\sigma_1 \cdots \sigma_r = 1$. The signature of Φ is an r-tuple of conjugacy classes $\mathbf{C} := (C_1, \dots, C_r)$ in S_l such that C_i is the conjugacy class of σ_i. We use the notation n to denote the conjugacy class of permutations which is cycle of length n. Using the signature of $\phi : \mathbb{P}^1 \to \mathbb{P}^1$ one finds out the signature of $\Phi : \mathcal{X}_g \to \mathbb{P}^1$ for any given $g \geq 2$ and G.

Let E be the fixed field of G, the Hurwitz genus formula states that

$$2(g_K - 1) = 2(g_E - 1)|G| + \deg(\operatorname{Diff}_{K/E}) \tag{7}$$

with g_K and g_E the genera of K and E respectively and $\operatorname{Diff}_{K/E}$ the different of K/E. Let $\overline{\mathfrak{p}}_1, \overline{\mathfrak{p}}_2, \dots, \overline{\mathfrak{p}}_r$ be ramified primes of E. If we set $d_i = \deg(\overline{\mathfrak{p}}_i)$, let e_i be the ramification index of the $\overline{\mathfrak{p}}_i$ and β_i be the exponent of $\overline{\mathfrak{p}}_i$ in $\operatorname{Diff}_{K/E}$. Hence, the above equation may be written as

$$2(g_K - 1) = 2(g_E - 1)|G| + |G| \sum_{i=1}^{r} \frac{\beta_i}{e_i} d_i \tag{8}$$

If $\overline{\mathfrak{p}}_i$ is tamely ramified then $\beta_i = e_i - 1$ or if $\overline{\mathfrak{p}}_i$ is wildly ramified then $\beta_i = e_i^* q_i + q_i - 2$ with $e_i = e_i^* q_i$, e_i^* relatively prime to p, q_i a power of p and $e_i^* | q_i - 1$. For fixed G, $\mathbf{C}$ the family of covers $\Phi : \mathcal{X}_g \to \mathbb{P}^1$ is a Hurwitz space $\mathcal{H}(G, \mathbf{C})$.

The space $\mathcal{H}(G, \mathbf{C})$ is an irreducible algebraic variety of dimension $\delta(G, \mathbf{C})$. Using equation (8) and signature $\mathbf{C}$ one can find out the dimension for each G.

Next we want to determine the cover $z = \phi(x) : \mathbb{P}^1 \to \mathbb{P}^1$ for all characteristics. Notice that the case of char $k = 0$ is already worked out in [**59**] and for char $k = p \neq 2$ is done in [**54**] which we are following here.

We define a semidirect product of elementary Abelian group with cyclic group as follows.

$$K_m := \langle \{\sigma_a, t | \, a \in U_m\} \rangle$$

where $t(x) = \xi^2 x$, $\sigma_a(x) = x + a$, for each $a \in U_m$,

$$U_m := \{a \in k \,|\, (a \prod_{j=0}^{\frac{p^t-1}{m}-1} (a^m - b_j)) = 0\}$$

$b_j \in \mathbb{F}_q^*$, $m|p^t - 1$ and ξ is a primitive $2m$-th root of unity. Obviously U_m is a subgroup of the additive group of k.

LEMMA 5. *Let k be an algebraically closed field of characteristic p, $\bar{G}$ be a finite subgroup of* $\mathrm{PGL}(2, k)$ *acting on the field $k(x)$. Then, $\bar{G}$ is isomorphic to one of the following groups C_m, D_m, A_4, S_4, A_5, $U = C_p^t$, K_m,* $\mathrm{PSL}(2, q)$ *and* $\mathrm{PGL}(2, q)$, *where $q = p^f$ and $(m, p) = 1$. Moreover, the fixed subfield $k(x)^{\bar{G}} = k(z)$ is given by Table 1, where $\alpha = \frac{q(q-1)}{2}$, $\beta = \frac{q+1}{2}$. The subgroup H_t is a subgroup of the additive group of k with $|H_t| = p^t$ and $b_j \in k^*$.*

Next we want to determine groups which occur as automorphism group G of genus $g \geq 2$ cyclic curves, their signatures and the dimension of the corresponding locus. We know that $\bar{G} := G/G_0$, where $G_0 := \mathrm{Gal}(k(x, y)/k(x))$ and $\bar{G}$ is isomorphic to C_m, D_m, A_4, S_4, A_5, U, K_m, $\mathrm{PSL}(2, q)$, $\mathrm{PGL}(2, q)$. By considering the lifting of ramified points in each $\bar{G}$, we divide each $\bar{G}$ into subcases. We determine the signature of each subcase by looking at the behavior of lifting and ramification of $\bar{G}$. Using the signature and Eq. (8) we calculate δ for each case. We list all possible automorphism groups G as separate theorems for each $\bar{G}$.

We assume that $5 < p \leq 2g + 1$. The case $p > 2g + 1$ is same as $p = 0$; see [**59**]

THEOREM 19. *Let $g \geq 2$ be a fixed integer, $\mathcal{X}$ a genus g cyclic curve, $G =$* $\mathrm{Aut}(\mathcal{X})$ *and $C_n \triangleleft G$ such that $g(\mathcal{X}^{C_n}) = 0$. The signature of the cover $\Phi : \mathcal{X} \to \mathcal{X}^G$ and dimension δ is given in* [**54**, Table 2], *where $m = |\,\mathrm{PSL}(2, q)|$ for cases 38-41 and $m = |\,\mathrm{PGL}(2, q)|$ for cases 42-45.*

There are 45 signatures from the above theorem (not all occur in every genus $g \geq 2$). The following theorem gives us all possible automorphism groups of genus $g \geq 2$ cyclic curves defined over the finite field of characteristic p.

THEOREM 20. *Let $\mathcal{X}_g$ be a genus $g \geq 2$ irreducible cyclic curve defined over an algebraically closed field k,* char $k = p \neq 2$. *Assume the cyclic group $C_n \triangleleft G =$* $\mathrm{Aut}(\mathcal{X}_g)$ *and let $\bar{G} = \mathrm{Aut}(\mathcal{X}_g)/C_n$ be its reduced automorphism group.*

(1) *If $\bar{G} \cong C_m$ then $G \cong C_{mn}$ or*

$$\langle r, \sigma | \, r^n = 1, \sigma^m = 1, \sigma r \sigma^{-1} = r^l \rangle$$

where (l,n)=1 and $l^m \equiv 1 \pmod n$.

TABLE 1. Rational functions corresponding to each $\bar{G}$.

Case	$\bar{G}$	z	*Ramification*
1	C_m, $(m,p)=1$	x^m	(m,m)
2	D_{2m}, $(m,p)=1$	$x^m+\frac{1}{x^m}$	$(2,2,m)$
3	A_4, $p\neq 2,3$	$\frac{x^{12}-33x^8-33x^4+1}{x^2(x^4-1)^2}$	$(2,3,3)$
4	S_4, $p\neq 2,3$	$\frac{(x^8+14x^4+1)^3}{108(x(x^4-1))^4}$	$(2,3,4)$
5	A_5, $p\neq 2,3,5$	$\frac{(-x^{20}+228x^{15}-494x^{10}-228x^5-1)^3}{(x(x^{10}+11x^5-1))^5}$	$(2,3,5)$
	A_5, $p=3$	$\frac{(x^{10}-1)^6}{(x(x^{10}+2ix^5+1))^5}$	$(6,5)$
6	U	$\prod_{a\in H_t}(x+a)$	(p^t)
7	K_m	$(x\prod_{j=0}^{\frac{p^t-1}{m}-1}(x^m-b_j))^m$	(mp^t,m)
8	$\mathrm{PSL}(2,q)$, $p\neq 2$	$\frac{((x^q-x)^{q-1}+1)^{\frac{q+1}{2}}}{(x^q-x)^{\frac{q(q-1)}{2}}}$	(α,β)
9	$\mathrm{PGL}(2,q)$	$\frac{((x^q-x)^{q-1}+1)^{q+1}}{(x^q-x)^{q(q-1)}}$	$(2\alpha,2\beta)$

(2) *If* $\bar{G}\cong D_{2m}$ *then* $G\cong D_{2m}\times C_n$ *or*

$$G_5=\langle r,\sigma,t\,|\,r^n=1,\sigma^2=r,t^2=1,(\sigma t)^m=1,\sigma r\sigma^{-1}=r,trt^{-1}=r^{n-1}\rangle$$
$$G_6=D_{2mn}$$
$$G_7=\langle r,\sigma,t\,|\,r^n=1,\sigma^2=r,t^2=r^{n-1},(\sigma t)^m=1,\sigma r\sigma^{-1}=r,trt^{-1}=r\rangle$$
$$G_8=\langle r,\sigma,t\,|\,r^n=1,\sigma^2=r,t^2=1,(\sigma t)^m=r^{\frac{n}{2}},\sigma r\sigma^{-1}=r,trt^{-1}=r^{n-1}\rangle$$
$$G_9=\langle r,\sigma,t\,|\,r^n=1,\sigma^2=r,t^2=r^{n-1},(\sigma t)^m=r^{\frac{n}{2}},\sigma r\sigma^{-1}=r,trt^{-1}=r\rangle$$

(3) *If* $\bar{G}\cong A_4$ *and* $p\neq 3$ *then* $G\cong A_4\times C_n$ *or*

$$G'_{10}=\langle r,\sigma,t\,|\,r^n=1,\sigma^2=1,t^3=1,(\sigma t)^3=1,\sigma r\sigma^{-1}=r,trt^{-1}=r^l\rangle$$
$$G'_{12}=\langle r,\sigma,t\,|\,r^n=1,\sigma^2=1,t^3=r^{\frac{n}{3}},(\sigma t)^3=r^{\frac{n}{3}},\sigma r\sigma^{-1}=r,trt^{-1}=r^l\rangle$$

where $(l,n)=1$ *and* $l^3\equiv 1$ *(mod* n*) or*

$$\langle r,\sigma,t\,|\,r^n=1,\sigma^2=r^{\frac{n}{2}},t^3=r^{\frac{n}{2}},(\sigma t)^5=r^{\frac{n}{2}},\sigma r\sigma^{-1}=r,trt^{-1}=r\rangle$$

or

$$G_{10} = \langle r, \sigma, t \,|\, r^n = 1, \sigma^2 = 1, t^3 = 1, (\sigma t)^3 = 1, \sigma r \sigma^{-1} = r, trt^{-1} = r^k \rangle$$

$$G_{13} = \langle r, \sigma, t \,|\, r^n = 1, \sigma^2 = r^{\frac{n}{2}}, t^3 = 1, (\sigma t)^3 = 1, \sigma r \sigma^{-1} = r, trt^{-1} = r^k \rangle$$

where $(k, n) = 1$ *and* $k^3 \equiv 1 \pmod n$.

(4) *If* $\bar{G} \cong S_4$ *and* $p \neq 3$ *then* $G \cong S_4 \times C_n$ *or*

$$G_{16} = \langle r, \sigma, t \,|\, r^n = 1, \sigma^2 = 1, t^3 = 1, (\sigma t)^4 = 1, \sigma r \sigma^{-1} = r^l, trt^{-1} = r \rangle$$

$$G_{18} = \langle r, \sigma, t \,|\, r^n = 1, \sigma^2 = 1, t^3 = 1, (\sigma t)^4 = r^{\frac{n}{2}}, \sigma r \sigma^{-1} = r^l, trt^{-1} = r \rangle$$

$$G_{20} = \langle r, \sigma, t \,|\, r^n = 1, \sigma^2 = r^{\frac{n}{2}}, t^3 = 1, (\sigma t)^4 = 1, \sigma r \sigma^{-1} = r^l, trt^{-1} = r \rangle$$

$$G_{22} = \langle r, \sigma, t \,|\, r^n = 1, \sigma^2 = r^{\frac{n}{2}}, t^3 = 1, (\sigma t)^4 = r^{\frac{n}{2}}, \sigma r \sigma^{-1} = r^l, trt^{-1} = r \rangle$$

where $(l, n) = 1$ *and* $l^2 \equiv 1 \pmod n$.

(5) *If* $\bar{G} \cong A_5$ *and* $p \neq 5$ *then* $G \cong A_5 \times C_n$ *or*

$$\langle r, \sigma, t \,|\, r^n = 1, \sigma^2 = r^{\frac{n}{2}}, t^3 = r^{\frac{n}{2}}, (\sigma t)^5 = r^{\frac{n}{2}}, \sigma r \sigma^{-1} = r, trt^{-1} = r \rangle$$

(6) *If* $\bar{G} \cong U$ *then* $G \cong U \times C_n$ *or*

$$< r, \sigma_1, \sigma_2, ..., \sigma_t | r^n = \sigma_1^p = \sigma_2^p = ... = \sigma_t^p = 1, \\ \sigma_i \sigma_j = \sigma_j \sigma_i, \sigma_i r \sigma_i^{-1} = r^l, 1 \le i, j \le t >$$

where $(l, n) = 1$ *and* $l^p \equiv 1 \pmod n$.

(7) *If* $\bar{G} \cong K_m$ *then* $G \cong$

$$< r, \sigma_1, ..., \sigma_t, v | r^n = \sigma_1^p = ... = \sigma_t^p = v^m = 1, \sigma_i \sigma_j = \sigma_j \sigma_i, \\ vrv^{-1} = r, \sigma_i r \sigma_i^{-1} = r^l, \sigma_i v \sigma_i^{-1} = v^k, 1 \le i, j \le t >$$

where $(l, n) = 1$ *and* $l^p \equiv 1 \pmod n$, $(k, m) = 1$ *and* $k^p \equiv 1 \pmod m$ *or*

$$\langle r, \sigma_1, ..., \sigma_t | r^{nm} = \sigma_1^p = ... = \sigma_t^p = 1, \sigma_i \sigma_j = \sigma_j \sigma_i, \sigma_i r \sigma_i^{-1} = r^l, 1 \le i, j \le t \rangle$$

where $(l, nm) = 1$ *and* $l^p \equiv 1 \pmod{nm}$.

(8) *If* $\bar{G} \cong \mathrm{PSL}(2, q)$ *then* $G \cong \mathrm{PSL}(2, q) \times C_n$ *or* $SL_2(3)$.

(9) *If* $\bar{G} \cong \mathrm{PGL}(2, q)$ *then* $G \cong \mathrm{PGL}(2, q) \times C_n$.

In [**55**] the corresponding equations are given for each case. In [**33**] for each group, it is discussed if the corresponding curve is defined over its field of moduli.

Applying Thm. 20 we obtain the automorphism groups of all genus 3 superelliptic curves defined over algebraically closed field of characteristic $p \neq 2$. Below we list the GAP group ID's of each of those groups.

LEMMA 6. *Let $\mathcal{X}_g$ be a genus 3 superelliptic curve defined over a field of characteristic $p \neq 2$. Then the automorphism groups of $\mathcal{X}_g$ is one of the following.*

i): $p = 3$: (2,1), (4,2), (3,1), (4,1), (8,2), (8,3), (7,1), (14,2), (6,2), (8,1), (8,5), (16,11), (16,10), (32,9), (30,2), (16,7), (16,8), (6,2).

ii): $p = 5$: (2,1), (4,2), (3,1), (4,1), (8,2), (8,3), (7,1), (21,1), (14,2), (6,2), (12,2), (9,1), (8,1), (8,5), (16,11), (16,10), (32,9), (42,3), (12,4), (16,7), (24,5), (18,3), (16,8), (48,33), (48,48).

iii): $p = 7$: (2,1), (4,2), (3,1), (4,1), (8,2), (8,3), (7,1), (21,1), (6,2), (12,2), (9,1), (8,1), (8,5), (16,11), (16,10), (32,9), (30,2), (42,3), (12,4), (16,7), (24,5), (18,3), (16,8), (48,33), (48,48).

iv): $p = 0$ *or* $p \geq 11$: $(2,1)$, $(4,2)$, $(3,1)$, $(4,1)$, $(8,2)$, $(8,3)$, $(7,1)$, $(21,1)$, $(14,2)$, $(6,2)$, $(12,2)$, $(9,1)$, $(8,1)$, $(8,5)$, $(16,11)$, $(16,10)$, $(32,9)$, $(30,2)$, $(42,3)$, $(12,4)$, $(16,7)$, $(24,5)$, $(18,3)$, $(16,8)$, $(48,33)$, $(48,48)$.

Recall that the list for $p = 0$ is the same as for $p > 7$.

Again applying Thm. 20, we obtain the possible automorphism groups of genus 4 cyclic curves defined over an algebraically closed field of characteristic 0,3,5,7 and bigger than 7. We list the GAP group ID of these groups in following theorem.

LEMMA 7. *Let $\mathcal{X}_g$ be a genus 4 cyclic curve defined over a field of characteristic p. Then the automorphism group of $\mathcal{X}_g$ is one of the following.*

i): $p = 3$: $(2,1)$, $(4,2)$, $(3,1)$, $(6,2)$, $(5,1)$, $(10,2)$, $(20,1)$, $(9,1)$, $(18,2)$, $(15,1)$, $(4,1)$, $(20,4)$, $(8,3)$, $(40,8)$, $(12,5)$, $(16,7)$, $(20,5)$, $(32,19)$, $(24,10)$, $(8,4)$, $(9,2)$, $(18,5)$.

ii): $p = 5$: $(2,1)$, $(4,2)$, $(3,1)$, $(6,2)$, $(9,2)$, $(5,1)$, $(10,2)$, $(20,1)$, $(9,1)$, $(27,4)$, $(18,2)$, $(4,1)$, $(18,3)$, $(8,3)$, $(12,5)$, $(36,12)$, $(54,4)$, $(16,7)$, $(20,5)$, $(32,19)$, $(24,10)$, $(8,4)$, $(60,9)$, $(36,11)$, $(24,3)$, $(72,42)$, $(10,2)$, $(18,5)$.

iii): $p = 7$: $(2,1)$, $(4,2)$, $(3,1)$, $(6,2)$, $(9,2)$, $(5,1)$, $(10,2)$, $(20,1)$, $(9,1)$, $(27,4)$, $(18,2)$, $(15,1)$, $(4,1)$, $(20,4)$, $(18,3)$, $(8,3)$, $(40,8)$, $(12,5)$, $(36,12)$, $(54,4)$, $(16,7)$, $(20,5)$, $(32,19)$, $(24,10)$, $(8,4)$, $(60,9)$, $(36,11)$, $(24,3)$, $(72,42)$.

iv): $p = 0$ *or* $p \geq 11$: $(2,1)$, $(4,2)$, $(3,1)$, $(6,2)$, $(9,2)$, $(5,1)$, $(10,2)$, $(20,1)$, $(9,1)$, $(27,4)$, $(18,2)$, $(15,1)$, $(4,1)$, $(20,4)$, $(18,3)$, $(8,3)$, $(40,8)$, $(12,5)$, $(36,12)$, $(54,4)$, $(16,7)$, $(20,5)$, $(32,19)$, $(24,10)$, $(8,4)$, $(60,9)$, $(36,11)$, $(24,3)$, $(72,42)$.

The above two lemmas are a simple search going through all the cases of the theorem, but they illustrate the idea that for cyclic (superelliptic curves) all isomorphism classes of curves can be written out (including a parametric equation in each case).

There is one case missing from all the results of this section, namely $p = 2$. Next we will try to explain that case.

A hyperelliptic curve over an algebraically closed field k of characteristic 2 admits an Artin-Schreier model

$$\mathcal{X} : y^2 + y = R(x)$$

where $R(x)$ is a rational function with no poles of even order. An isomorphism to another curve $\mathcal{X}' : y^2 + y = Q(x)$ induces an automorphism of the projective x-line. One can then determine possible normal forms for $R(x)$, work out which fractional linear transformations preserve each form and see how these lift to $\mathcal{X}$ and interact with the hyperelliptic involution.

Now the details. Let $\mathcal{X}'$ be a hyperelliptic curve of genus g over an algebraically closed field k of characteristic 2. We use an Artin-Schreier model $y'^2 + y' = R'(x)$. As a consequence of Hasse's theory we can find a rational function $Q(x) \in k(x)$ such that the rational function $R'(x) + Q(x) + Q(x)^2$ has no poles of even order. Let $R(x) = R'(x) + Q(x) + Q(x)^2$ and $y = y' + Q(x)$ to get a curve $\mathcal{X}$ in normalized form $y^2 + y = R(x)$. Then y is unique up to transformations of the form $y \longmapsto y + B(x)$, where $B(x)$ is a rational function of x.

Now, take two hyperelliptic curves, $\mathcal{X} : y^2 + y = R(x)$ and $\mathcal{X}' : y^2 + y = Q(x)$. Then, given an isomorphism $\phi : \mathcal{X} \longmapsto \mathcal{X}'$ and the finite morphisms $f_1 : \mathcal{X} \longmapsto \mathbb{P}^1$, and $f_2 : \mathcal{X}' \longmapsto \mathbb{P}^1$ of degree 2, there exists a unique automorphism σ of $\mathbb{P}^1$ such that $f_2 \circ \phi = \sigma \circ f_1$. Any isomorphism between these curves has the form

$$(x, y) \longmapsto (\sigma(x), y + S(x)) = \left(\frac{ax + b}{cx + d}, y + S(x) \right)$$

for some $S(x) \in k(x)$. Hence, these curves are isomorphic if and only if

$$Q(x) = R\left(\frac{ax + b}{cx + d} \right) + S(x) + S(x)^2.$$

Let $\mathrm{div(R)}_\infty = \Sigma \mathrm{n_a(a)}$ be the polar divisor of $R(x)$ on the projective line $\mathbb{P}^1$. $\mathcal{X}$ is ramified at each a and if P_a is the unique point of $\mathcal{X}$ over a then the curve $y^2 + y = R(x)$ has the different

$$\mathrm{Diff}(\mathcal{X}/\mathbb{P}^1) = \Sigma(n_a + 1)P_a,$$

where the n_a are odd ([**70**], Prop III.7.8)

$$2g - 2 = -2[F : k(x)] + \deg(\mathrm{Diff}(\mathcal{X}/\mathbb{P}^1)) \implies \deg(\mathrm{Diff}(\mathcal{X}/\mathbb{P}^1)) = 2g + 2$$

The ramification types determine the isomorphism classes of the hyperelliptic curves. The solutions of the equation $\Sigma(n_a + 1) = 2g + 2$ in the unknown odd positive integers give us the following ramification types: $(1, 1, 1, 1)$, $(3, 1, 1)$, $(3, 3)$, $(5, 1)$, (7) for genus $g = 3$. Therefore we get the following normal forms.

$$y^2 + y = \begin{cases} \alpha_1 x + \alpha_2 x^{-1} + \alpha_3 (x-1)^{-1} + \alpha_4 (x - \lambda)^{-1} \\ x^3 + \alpha x + \beta x^{-1} + \gamma (x-1)^{-1} \\ x^3 + \alpha x + \beta x^{-3} + \gamma x^{-1} \\ x^5 + \alpha x^3 + \beta x^{-1} \\ x^7 + \alpha x^5 + \beta x^3 \end{cases} \tag{9}$$

For genus $g = 4$ we have $(1, 1, 1, 1, 1)$, $(3, 1, 1, 1)$, $(3, 3, 1)$, $(5, 1, 1)$, $(5, 3)$, $(7, 1)$, (9) and therefore we get the following normal forms for genus 3 and 4 respectively.

$$y^2 + y = \begin{cases} \alpha_1 x + \alpha_2 x^{-1} + \alpha_3 (x-1)^{-1} + \alpha_4 (x - \lambda)^{-1} + \alpha_5 (x - \mu)^{-1} \\ x^3 + \alpha x + \beta_1 x^{-1} + \beta_2 (x-1)^{-1} + \beta_3 (x - \lambda)^{-1} \\ x^3 + \alpha x + \beta x^{-3} + \gamma x^{-1} + \sigma (x-1)^{-1} \\ x^5 + \alpha x^3 + \beta x^{-1} + \gamma (x-1)^{-1} \\ x^5 + \alpha x^3 + \beta x^{-3} + \gamma x^{-1} \\ x^7 + \alpha x^5 + \beta x^3 + \gamma x^{-1} \\ x^9 + \alpha_1 x^7 + \alpha_2 x^5 + \alpha_3 x^3 \end{cases} \tag{10}$$

Using a case by case analysis in [**23**] it is proved that

THEOREM 21. *Let C be a genus $g \geq 2$ hyperelliptic curve defined over a field k such that* char $k = 2$. *Then*

i) if $g = 3$ then $\mathrm{Aut}(\mathcal{X})$ *is isomorphic to* C_2, C_4, V_4, C_2^3, C_6, C_{14}, D_{12}

ii) if $g = 4$ then $\mathrm{Aut}(\mathcal{X})$ *is isomorphic to* C_2, V_4, C_4, C_2^3, C_6, C_{18}, D_{20}.

The corresponding equations are given in each case. The higher genus cases can be determined in similar way.

REMARK 1. *It seems as the above methods can be extended to determine complete lists of automorphism groups of any superelliptic curve of genus $g \geq 2$ and any $2 \leq p \leq 2g+1$. We are not aware if such lists are determined for $p = 2, 3, 5$ for large g.*

We still have to consider the case $p = 0$ and equivalently $p > 2g + 1$. However, here we can use the theory of compact Riemann surfaces and Fuchsian groups to give a complete answer to the question of determining the list of automorphism groups for any $g \geq 2$.

REMARK 2. *Isomorphism classes of superelliptic curves are determined by invariants of binary forms. For superelliptic curves with extra automorphisms other invariants are introduced in* [**29**], [**3**]. *Finding equations of such superelliptic curves over a field of moduli is interesting on its own right; see* [**33**].

6. Automorphism groups of compact Riemann surfaces

To contrast with the case $p > 0$, and to provide some historical footing, we now turn our attention to the classical case: compact Riemann surfaces. Due to a contemporary adaptation of Riemann's Existence Theorem, see Thm. 23, and its amenability to computational group theory, in modern work one of the most utilized tools for classification of automorphism groups of compact Riemann surfaces is also one of the more classical ones: uniformization and Fuchsian groups. In this section, we provide a short exposition outlining this traditional approach, describe some of the current results and how, with a little further direction, this method can be used to help determine full automorphism groups.

6.1. Fuchsian groups and signatures. The Uniformization Theorem states that any compact Riemann surface $\mathcal{X}$ of genus $g \geq 2$ is conformally equivalent to a quotient of its universal cover, the upper half plane $\mathbb{H}$, by a torsion free discrete subgroup Λ of $\mathrm{Aut}(\mathbb{H}) = \mathrm{PSL}(2, \mathbb{R})$. The group Λ is isomorphic to Π_g, the fundamental group of $\mathcal{X}$, and is called a *surface group* for $\mathcal{X}$. For a given surface $\mathcal{X}$, surface groups are unique up to conjugacy in $\mathrm{PSL}(2, \mathbb{R})$ meaning two compact Riemann surfaces $\mathcal{X}$ and $\mathcal{X}'$ are conformally equivalent if and only if surface groups for $\mathcal{X}$ and $\mathcal{X}'$ are conjugate in $\mathrm{PSL}(2, \mathbb{R})$.

Now if $\phi\colon \mathcal{X} \to \mathcal{X}$ is an automorphism of $\mathcal{X}$, then it can be lifted to an automorphism ϕ_Λ of $\mathbb{H}$ which normalizes Λ. In particular, if G is a group of automorphisms of $\mathcal{X}$, then G can be lifted to a discrete subgroup Γ of $\mathrm{PSL}(2, \mathbb{R})$, called a *Fuchsian group*, containing Λ with index $|G|$ and which normalizes Λ. We call Γ the Fuchsian group corresponding to G, and if Λ has been fixed, we call G the automorphism group corresponding to Γ.

Conversely, if Γ is a Fuchsian group and Λ is a normal subgroup of Γ which is isomorphic to Π_g, then there is a natural action of the quotient group Γ/Λ on the quotient surface $\mathbb{H}/\Lambda$ which is a surface of genus g.

These observations illustrate the primary basic tool for determining group actions on compact Riemann surfaces of genus $g \geq 2$: for a given genus, determine all Fuchsian groups (up to isomorphism) for which there exists a normal subgroup isomorphic to Π_g. To explain this process in more detail, we need some additional preliminary results.

For a cocompact Fuchsian group Γ, the quotient surface $\mathbb{H}/\Gamma$ is a compact Riemann surface and the quotient map $\pi_\Gamma\colon \mathbb{H} \to \mathbb{H}/\Gamma$ is a holomorphic, possibly branched, map. We define the *signature* of Γ to be the tuple $(g_\Gamma; m_1, m_2, \ldots, m_r)$ where g_Γ is the genus of $\mathbb{H}/\Gamma$ and the quotient map π_Γ branches over r points with ramification indices m_i for $1 \le i \le r$. We call g_Γ the **orbit genus** of Γ and the numbers $m_1, \ldots, m_r$ the **periods** of Γ. The signature of Γ provides information regarding a presentation for Γ:

THEOREM 22. *If Γ is a Fuchsian group with signature $(g_\Gamma; m_1, \ldots, m_r)$ then there exist group elements $\alpha_1, \beta_1, \ldots, \alpha_{g_\Gamma}, \beta_{g_\Gamma}, \gamma_1, \ldots \gamma_r \in \mathrm{PSL}(2, \mathbb{R})$, such that;*

(1) $\Gamma = \langle \alpha_1, \beta_1, \ldots, \alpha_{g_\Gamma}, \beta_{g_\Gamma}, \gamma_1, \ldots \gamma_r \rangle$.
(2) *Defining relations for Γ are*

$$\gamma_1^{m_1}, \gamma_2^{m_2}, \ldots, \gamma_r^{m_r}, \prod_{i=1}^{g_\Gamma} [\alpha_i, \beta_i] \prod_{j=1}^{r} \gamma_j.$$

(3) *Each elliptic element (the elements of finite order) lies in a unique conjugate of $\langle \gamma_i \rangle$ for suitable i. Furthermore, the cyclic groups $\langle \gamma_i \rangle$ are self-normalizing in Γ.*
(4) *Each elliptic element of Γ has a unique fixed point in $\mathbb{H}$. All other elements (the hyperbolic elements) act fixed point freely on $\mathbb{H}$.*

We call a set of elements of Γ satisfying Thm. 22 **canonical generators** for Γ. Notice that if Γ is a surface group for a surface of genus g, since it is torsion free, it must have signature $(g; -)$.

Now, if G acts conformally on $\mathcal{X}$, Λ is surface group for $\mathcal{X}$, and Γ is the Fuchsian group corresponding to G with signature $(g_\Gamma; m_1, \ldots, m_r)$, then there exists an epimorphism, called a **surface kernel epimorphism**, $\rho\colon \Gamma \to G$ with kernel Λ. This epimorphism can be neatly summarized in the context of finite groups by a **generating vector**. Specifically, if $\alpha_1, \beta_1, \ldots, \alpha_{g_\Gamma}, \beta_{g_\Gamma}, \gamma_1, \ldots \gamma_r \in \mathrm{PSL}(2, \mathbb{R})$ are canonical generators for Γ, then we get a $(2g_\Gamma + r)$-tuple of elements from G, $(a_1, \ldots a_{g_\Gamma}, b_1, \ldots b_{g_\Gamma}, c_1, \ldots c_r)$ where $\rho(\alpha_i) = a_i$, $\rho(\beta_i) = b_i$ and $\rho(\gamma_i) = c_i$ called a $(g_\Gamma; m_1, \ldots, m_r)$-generating vector for G. Moreover, since ρ is an epimorphism with torsion free kernel, we have:

- $\displaystyle\prod_{i=1}^{g_\Gamma} [a_i, b_i] \prod_{j=1}^{r} c_j = 1$ (the identity)
- $\mathrm{O}(c_i) = m_i$ where O denotes element order

Moreover, provided a set of canonical generators of Γ have been fixed, there is a one to one correspondence between the set of $(g_\Gamma; m_1, \ldots, m_r)$ - generating vectors of G and Epi(Γ, G), epimorphisms $\Gamma \to G$ preserving the orders of the γ_j.

Now suppose that Γ is a Fuchsian group with signature $(g_\Gamma; m_1, \ldots, m_r)$ and suppose that Λ is a normal surface subgroup of Γ for a surface of genus g. Letting $G = \Gamma/\Lambda$ and identifying the orbit spaces $\mathbb{H}/\Gamma$ and $\mathcal{X}/G$ we get the tower of covers and quotient maps given in Fig. 1.

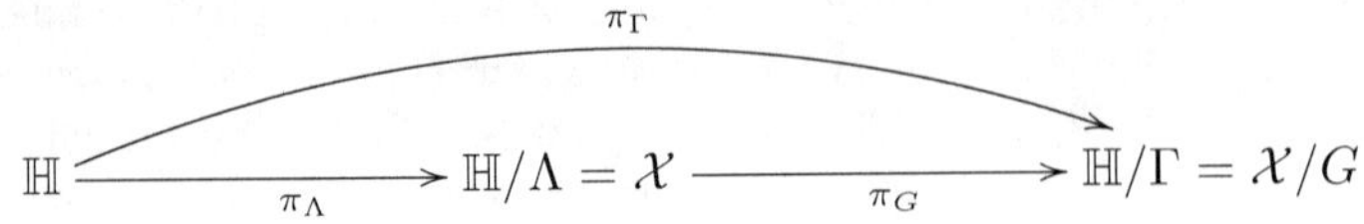

FIGURE 1. Holomorphic quotient maps and surface identifications.

Since the universal covering map π_Λ is unramified, it follows that the quotient map $\pi_G\colon \mathcal{X} \to \mathcal{X}/G$ is branched over the same points as $\pi_\Gamma\colon \mathbb{H} \to \mathbb{H}/\Gamma$ with the same ramification indices. In particular, $\mathcal{X}/G$ has genus g_Γ and π_G is a degree $|G|$ map branched over r points with ramification indices $m_1, \dots m_r$. Consequently, we often say that G acts or has the same signature as Γ.

Now since the map π_G is between compact Riemann surfaces, the Riemann-Hurwitz formula holds, giving:

$$g - 1 = |G|(g_\Gamma - 1) + \frac{|G|}{2}\sum_{j=1}^{r}\left(1 - \frac{1}{m_j}\right).$$

Combining our observations, we get the following modern adaption of Riemann's existence theorem which provides necessary and sufficient conditions for the existence of the action of a group G on a compact Riemann surface of genus g with signature $(g_\Gamma; m_1, \dots, m_r)$:

THEOREM 23. *A finite group G acts on a compact Riemann surface S of genus $g \geq 2$ with signature $(g_\Gamma; m_1, \dots, m_r)$ if and only if:*

(1) *the Riemann–Hurwitz formula is satisfied:*

$$g - 1 = |G|(g_\Gamma - 1) + \frac{|G|}{2}\sum_{j=1}^{r}\left(1 - \frac{1}{m_j}\right).$$

(2) *there exists an $(g_\Gamma; m_1, \dots, m_r)$-generating vector for G.*

REMARK 3. *Let $\mathcal{Y}$ denote the orbit surface $\mathcal{X}/G = \mathbb{H}/\Gamma$ and $\pi_G : \mathcal{X} \to \mathcal{Y}$ the quotient morphism. The set of branch points over which π_G is ramified is denoted $\mathcal{B} = \mathcal{B}_G$. The covering $\pi_G : \mathcal{X}^\circ \to \mathcal{Y}^\circ$, where $\mathcal{Y}^\circ = \mathcal{Y} - \mathcal{B}_G$ and $\mathcal{X}^\circ = \mathcal{X} - \pi_G^{-1}(\mathcal{B}_G)$ is an unramified Galois covering of affine curves. The fundamental group $\pi_1(\mathcal{Y}^\circ)$ has a presentation*

$$\pi_1(\mathcal{Y}^\circ) = \langle \alpha_1, \beta_1, \dots, \alpha_{g_\Gamma}, \beta_{g_\Gamma}, \gamma_1, \dots \gamma_r : \prod_{i=1}^{g_\Gamma}[\alpha_i, \beta_i]\prod_{j=1}^{r}\gamma_j = 1\rangle.$$

Any Galois cover $\mathcal{X} \to \mathcal{Y}$, ramified exactly over $\mathcal{B}_G$, with Galois group G, is defined by an epimorphism $\eta : \pi_1(\mathcal{Y}^\circ) \to G$, and, hence, defined by a generating vector, where we do not impose specific orders on the c_j. The generating vectors classify (not uniquely) the finite covers of $\mathcal{Y}$, branched over $\mathcal{B}$. There are only finitely many covers for a given pair $(\mathcal{Y}, \mathcal{B})$ and group order $|G|$. This surface construction approach allows us adopt a similar, though unwieldy, approach for $p > 0$ using the étale fundamental group $\pi_1^{et}(\mathcal{Y}^\circ)$.

6.2. Translating the problem into finite group theory. The importance of Thm. 23 is that it translates the problem of determining group actions of compact Riemann surfaces from a problem about infinite discrete groups into a problem about finite groups through the introduction of generating vectors. In particular, it makes the problem amenable to computational group theory and accordingly, classification results have significantly improved over the last few decades.

There are many different approaches to classifying automorphism groups, but perhaps the most common approach is to do so by genus; that is, fix a genus g and then find all possible automorphism groups that can act on a surface of that genus and the signatures with which they act. The basic approach to this form of classification for a fixed genus g is as follows:

(1) Find all possible signatures for each possible group order which satisfy the Riemann-Hurwitz formula.
(2) For each group order and signature, run over all groups of that possible order to either construct a generating vector, or show that no such generating vector exists.

We illustrate with a couple of examples.

EXAMPLE 2. *Suppose that $g = 2$ and consider the group order $|G| = 2$. Then the Riemann-Hurwitz formula gives:*

$$2 - 1 = 1 = 2(g_\Gamma - 1) + \frac{2}{2}\sum_{j=1}^{r}\left(1 - \frac{1}{m_j}\right).$$

Since $|G| = 2$, and the m_i's are all element orders of G, it follows that $m_i = 2$ for all i, so simplifying, we get

$$3 = 2g_\Gamma + \frac{r}{2}.$$

Solving, we get $g_\Gamma = 0$ and $r = 6$ or $g_\Gamma = 1$ and $r = 2$, so the signatures $(0; 2, 2, 2, 2, 2, 2)$ and $(1; 2, 2)$. Now the only group G of order 2 is cyclic, so let x be a generator of a group G of order 2. Then (x, x, x, x, x, x) is a $(0; 2, 2, 2, 2, 2, 2)$ and (x, x, x, x) is a $(1; 2, 2)$-generating vector for G, and hence G acts on a surface of genus 2 with signatures $(0; 2, 2, 2, 2, 2, 2)$ and $(1; 2, 2)$ but no other signatures.

As illustrated in Example 2, for a fixed group order and genus g, after simplification we obtain a linear equality in $k + 1$ variables where k is the number of distinct element orders in G. However, since all the variables in the equality are non-negative integers, since we have fixed g and the group order $|G|$, there are just finitely many solutions and hence the problem can be solved fairly easily by computer (or even by hand). Of course, complete classification relies on running over all possible group orders in a given genus, but from the Hurwitz bound, we know there are only finitely many possibilities. In particular, the problem of classification can be solved completely computationally through an algorithm similar to the following:

ALGORITHM 1. *For $2 \leq n \leq 84(g - 1)$, we do the following:*

(1) *Solve for all signatures satisfying*

$$g - 1 = n(g_\Gamma - 1) + \frac{n}{2}\sum_{j=1}^{r}\left(1 - \frac{1}{m_j}\right)$$

where the m_j are divisors of n.

(2) *For each signature $(g_\Gamma; m_1, \ldots m_r)$ found in (a), we do the following:*
   (a) *For each group G of order n with elements of orders $m_1, \ldots, m_r$, construct all vectors of elements of G of length $2g_\Gamma + r$ where the first $2g_\Gamma$ elements are any elements of G, and the $2g_\Gamma + i$th element has order m_i.*
   (b) *For each vector $(a_1, b_1, \ldots, a_{g_\Gamma}, b_{g_\Gamma}, c_1, \ldots, c_r)$, test the relation*

$$\prod_{i=1}^{g_\Gamma}[a_i, b_i]\prod_{j=1}^{r} c_j = 1.$$

If there exists a vector for which this relation holds, we have constructed a generating vector and hence an action of G exists with this signature. If no such vector satisfies this relation, G does not act with this signature.

Using essentially Algorithm 1, one can determine all possible signatures for a given genus $g \geq 2$. There is a huge amount of literature on this; see [**49**] and [**35**] for a complete list of references. As previously mentioned, some of the most comprehensive results are from [**14**] where Breuer was able to determine all possible signatures of genus up to 48 using GAP.

6.3. Full automorphism groups.

6.3.1. *Candidates for non-maximal automorphism groups and Singerman's list.* For each group and signature in most of the classification data available, all subgroups are also in the list. This raises the question how to pick out those groups that occur as the **full automorphism group** of some surface of genus g. One way to answer this question is to translate the problem back into a problem about Fuchsian groups. Specifically, suppose that G acts as a group of automorphisms on $\mathcal{X}$ with signature $(g_\Gamma; m_1, \dots m_r)$ and that for the surface group Λ, Γ is the Fuchsian group corresponding to G. Then G is not the full automorphism group of $\mathcal{X}$ if and only if there exists a Fuchsian overgroup Γ_1 of Γ which also contains Λ as a normal subgroup. Therefore, we need to understand the subgroup and overgroup structure of Fuchsian groups based on signature, and in particular when, for a given Fuchsian group Γ, there is an overgroup Γ_1. This exact problem was solved by Singerman in [**67**], and we summarize his results.

THEOREM 24. *If Γ is a Fuchsian group whose signature does not appear in the second column of Table 2, then Γ is isomorphic to a finitely-maximal Fuchsian group, that is, a group that is not contained with finite index in any other Fuchsian group. If the signature of Γ does appear in the second column of Table 2, then Γ is a subgroup of a Fuchsian group Γ_1 with signature from the third column with finite index given in the last column.*

The importance of Thm. 24 is that it tells us that unless G acts with one of the signatures in the second column of Table 2, then there always exists a Fuchsian group with that signature that is not contained with finite index in any other Fuchsian group Γ_1. In particular, the surface $\mathcal{X} = \Gamma/\mathrm{Ker}(\rho)$ where ρ is the surface kernel epimorphism from ρ to G given by the generating vector of G has G as its full group of automorphisms.

Alternatively, there is a *moduli dimension* argument in [**49**] which determines which signatures give full automorphism groups. Moreover, from methods in [**49**] for any fixed $g \geq 2$ one can determine completely inclusion among the subloci of the moduli space $\mathcal{M}_g$ for all the groups. See for example such diagrams for $g = 3, 4$ in [**8**]. It is noted in [**8**] that the majority of cases come from (cyclic) superelliptic curves, and as previously discussed, such curves are well understood.

6.3.2. *Conditions for when a group is not a full automorphism group.* Now, just because a signature does appear in Singerman's list does not necessarily mean that a corresponding automorphism group is always contained in some larger automorphism group – just that it might. By considering containments of the Fuchsian groups given in Singerman's list however, it is possible to determine necessary and sufficient conditions in terms of generating vectors or surface kernel epimorphisms

TABLE 2. Singerman's list.

Case	Signature Γ	Signature Γ_1	$[\Gamma_1 : \Gamma]$
N1	$(2; -)$	$(0; 2, 2, 2, 2, 2, 2)$	2
N2	$(1; t, t)$	$(0; 2, 2, 2, 2, t)$	2
N3	$(1; t)$	$(0; 2, 2, 2, 2t)$	2
N4	$(0; t, t, t, t)$, $t \geq 3$	$(0; 2, 2, 2, t)$	4
N5	$(0; t, t, u, u)$, $t + u \geq 5$	$(0; 2, 2, t, u)$	2
N6	$(0; t, t, t)$, $t \geq 4$	$(0; 3, 3, t)$	3
N7	$(0; t, t, t)$, $t \geq 4$	$(0; 2, 3, 2t)$	6
N8	$(0; t, t, u)$, $t \geq 3$,$t + u \geq 7$	$(0; 2, t, 2u)$	2
T1	$(0; 7, 7, 7)$	$(0; 2, 3, 7)$	24
T2	$(0; 2, 7, 7)$	$(0; 2, 3, 7)$	9
T3	$(0; 3, 3, 7)$	$(0; 2, 3, 7)$	8
T4	$(0; 4, 8, 8)$	$(0; 2, 3, 8)$	12
T5	$(0; 3, 8, 8)$	$(0; 2, 3, 8)$	10
T6	$(0; 9, 9, 9)$	$(0; 2, 3, 9)$	12
T7	$(0; 4, 4, 5)$	$(0; 2, 4, 5)$	6
T8	$(0; n, 4n, 4n)$, $n \geq 2$	$(0; 2, 3, 4n)$	6
T9	$(0; n, 2n, 2n)$, $n \geq 3$	$(0; 2, 4, 2n)$	4
T10	$(0; 3, n, 3n)$, $n \geq 3$	$(0; 2, 3, 3n)$	4
T11	$(0; 2, n, 2n)$, $n \geq 4$	$(0; 2, 3, 2n)$	3

for when a group does extend to some larger group. This was the primary goal of [**20**] which we summarize in the following two theorems.

THEOREM 25. *Let G be a finite group acting with a non-maximal and non-triangular Fuchsian signature on a compact Riemann surface $\mathcal{X}$ of genus g.*

(1) *Suppose G acts with signature $(2; -)$ and has corresponding $(2; -)$-generating vector (a_1, b_1, a_2, b_2). Then G is contained in some larger group of automorphisms with corresponding signature $(0; 2, 2, 2, 2, 2, 2)$ if and only if the assignment $a_1 \to a_1^{-1}$, $b_1 \to a_1 b_1^{-1} a_1^{-1}$, $a_2 \to (b_1^{-1} a_2 b_2) a_2^{-1} (b_1^{-1} a_2 b_2)^{-1}$ and $b_2 \to (b_1^{-1} a_2) b_2^{-1} (b_1^{-1} a_2)^{-1}$ is an automorphism of G.*

(2) *Suppose G acts with signature $(1; t, t)$ and has corresponding $(1; t, t)$-generating vector $(a_1, b_1, c_1, ([a_1, b_1]x)^{-1})$. Then G is contained in some larger group of automorphisms with corresponding signature $(0; 2, 2, 2, 2, t)$ if and only if the assignment $a_1 \to a_1^{-1}$, $b_1 \to b_1^{-1}$ and $c_1 \to (a_1 b_1)^{-1} c_1^{-1} (ba)$ is an automorphism of G.*

(3) *Suppose* G *acts with signature* $(1;t)$ *and has corresponding* $(1,t)$*-generating vector* $(a_1, b_1, [a_1,b_1]^{-1})$. *Then* G *is contained in some larger group of automorphisms with corresponding signature* $(0;2,2,2,2t)$ *if and only if the assignment* $a_1 \to a_1^{-1}$, $b_1 \to b_1^{-1}$ *is an of* G.

(4) *Suppose* G *acts with signature* $(0;t,t,u,u)$, *where* $t+u>5$ *and has corresponding* $(0;t,t,u,u)$*-generating vector* (c_1,c_2,c_3,c_4). *Then* G *is contained in some larger group of automorphisms with corresponding signature* $(0;2,2,t,u)$ *if and only if the assignment* $c_1 \to c_2$, $c_2 \to c_1$, $c_3 \to c_1^{-1}c_4c_1$ *and* $c_4 \to c_2c_3c_2^{-1}$ *is an automorphism of* G.

THEOREM 26. *Let* G *be a finite group acting on a compact Riemann surface* $\mathcal{X}$ *of genus* g *with a triangular signature* $(0;m_1,m_2,m_3)$ *with corresponding* $(0;m_1,m_2,m_3)$*-generating vector* (c_1,c_2,c_3). *Then* G *is not the full automorphism group of* $\mathcal{X}$ *if and only if at least one of the following conditions is satisfied (up to permutation of the periods* m_1, m_2, m_3*).*

(1) G *acts with signature* $(0;t,t,t)$ *where* $t>4$, *and the assignment* $c_1 \to c_2$, $c_2 \to c_3$ *and* $c_3 \to c_1$ *induces an automorphism of* G.

(2) G *acts with signature* $(0;t,t,u)$ *where* $t>3$ *and* $t+u>7$, *and the assignment* $c_1 \to c_2$, $c_2 \to c_1$, $c_3 \to c_2c_3c_2^{-1}$ *induces an automorphism of* G.

(3) G *acts with signature* $(0;2,7,7)$, *the conjugates of* $c_2c_3^{-1}c_2c_1c_3^3$ *generate a normal subgroup* K *of index* 56 *in* G, *and* G *is extendable to a group* G' *containing* G *as a subgroup of index* 9 *such that* G' *is generated by* c_1 *and an element* α *which normalizes* K *and satisfies* $\alpha^3=1$, $(c_1\alpha)^7=1$, $c_2=(\alpha c_1\alpha)^{-1}c_1\alpha(\alpha c_1\alpha)$ *and* $c_3=(\alpha c_1\alpha^{-1})c_1\alpha(\alpha c_1\alpha^{-1})$.

(4) G *acts with signature* $(0;3,3,7)$, *the conjugates of* $c_2c_1c_3^2$ *generate a normal subgroup* K *of index* 21 *in* G, *and* G *is extendable to a group* G' *containing* G *as a subgroup of index* 8 *such that* G' *is generated by* c_3 *and an element* α *which normalizes* K *and satisfies* $\alpha^3=1$, $(\alpha c_3)^2=1$, $c_1=\alpha c_3^{-2}\alpha c_3^2\alpha^{-1}$ *and* $c_2=\alpha^{-1}c_3^2\alpha c_3^{-2}\alpha$.

(5) G *acts with signature* $(0;3,8,8)$, *conjugates of* $c_2^2c_1c_3^2$ *and* $c_3^{-1}c_2c_1^{-1}c_2^{-1}c_1c_2^{-1}$ *generate a normal subgroup* K *of index* 72 *in* G, *and* G *is extendable to a group* G' *containing* G *as a subgroup of index* 10 *such that* G' *is generated by* c_3 *and an element* α *which normalizes* K *and satisfies* $\alpha^3=1$, $(\alpha c_3)^2=1$, $c_1=\alpha c_3^{-2}\alpha c_3^2\alpha^{-1}$ *and* $c_2=\alpha^{-1}c_3^2\alpha^{-1}c_3\alpha c_3^{-2}\alpha$.

(6) G *acts with signature* $(0;4,4,5)$, *the conjugates of* $c_1^{-1}c_2^{-1}c_3^2$ *generate a normal subgroup* K *of index* 20 *in* G, *and* G *is extendable to a group* G' *containing* G *as a subgroup of index* 6 *such that* G' *is generated by* c_3 *and an element* α *which normalises* K *and satisfies* $\alpha^4=1$, $(\alpha c_3)^2=1$, $c_1=\alpha^2c_3\alpha c_3^{-1}\alpha^2$ *and* $c_2=\alpha^{-1}c_3\alpha c_3^{-1}\alpha$

(7) G *acts with signature* $(0;3,n,3n)$ *where* $n>3$, *the conjugates of* c_2 *generate a normal subgroup* K *of index* 3 *in* G, *and* G *is extendable to a group* G' *containing* G *as a subgroup of index* 4 *such that* G' *is generated by* c_3 *and an element* α *which normalizes* K *and satisfies* $\alpha^2=1$, $(c_3\alpha)^3=1$, $c_1=\alpha c_3(c_3\alpha)^{-1}c_3^{-1}\alpha$ *and* $c_2=\alpha c_3^3\alpha$

(8) G *acts with signature* $(0;2,n,2n)$ *where* $n>4$, *the conjugates of* c_2 *generate a normal subgroup* K *of index* 2 *in* G, *and* G *is extendable to a*

group G' containing G as a subgroup of index 3 such that G' is generated by c_3 and an element α which normalizes K and satisfies $\alpha^3 = 1$, $c_1 = \alpha(\alpha c_3)\alpha^{-1}$ and $c_2 = \alpha^{-1}c_3^2\alpha$.

We illustrate with an example.

EXAMPLE 3. *The vector (x, x, x^3) is a $(0; 5, 5, 5)$-generating vector for the cyclic group $C_5 = \langle x \rangle$, so applying the Riemann-Hurwitz formula, we get a C_5 action on a surface $\mathcal{X}$ of genus $g = 2$ with signature $(0; 5, 5, 5)$. Letting $(c_1, c_2, c_3) = (x, x, x^3)$, we see that the assignment $c_1 \to c_2$, $c_2 \to c_1$ and $c_3 \to c_2c_3c_1^{-1}$ induces an automorphism of G (the trivial automorphism), so it follows by (2) of Thm. 26 that C_5 is not the full automorphism group of $\mathcal{X}$.*

6.3.3. *Finding full automorphism groups using counting methods.* Thm. 25 and Thm. 26 provide explicit ways to determine whether or not a given group is the full automorphism group of a compact Riemann surface by looking at the corresponding generating vectors and possible the overgroup structures. In certain cases, we can avoid the actual explicit computation of a generating vector to determine full automorphism groups by using a different method that relies on counting epimorphisms instead. An advantage to this alternate method is that the process is somewhat iterative and uses existing lists such as Breuers, [**14**]. The main disadvantage to this alternate method is that it does not work in all cases, and in some cases, we still need to return to generating vectors.

A Fuchsian group Γ with a signature of the form $(0; m_1, m_2, m_3)$ is unique up to conjugation in $\mathrm{PSL}(2, \mathbb{R})$. It follows that if we fix a Fuchsian group Γ with signature $(0; m_1, m_2, m_3)$ and $\mathcal{X}$ is any surface on which a finite group G acts with signature $(0; m_1, m_2, m_3)$, then there is a surface group Λ for $\mathcal{X}$ which is normal in Γ with $\Gamma/\Lambda = G$. In particular, the number of distinct surfaces with such a G-action will be equal to the number of non-$\mathrm{PSL}(2, \mathbb{R})$-conjugate torsion free normal subgroups of Γ with quotient G. Of course, in general it is difficult to determine whether two subgroups of $\mathrm{PSL}(2, \mathbb{R})$ are conjugate, but for triangle groups we have the following proved in [**28**]:

THEOREM 27. *If the $\mathrm{PSL}(2, \mathbb{R})$-conjugate surface groups Λ and Λ' are both normal subgroups of the triangle group Γ, then $\Lambda' = \alpha\Lambda\alpha^{-1}$ for some $\alpha \in N(\Gamma)$ or $\alpha \in N(N(\Lambda))$ (where $N(.)$ denotes normalizer).*

In particular, two normal surface subgroups of a triangle group Γ will be $\mathrm{PSL}(2, \mathbb{R})$-conjugate if and only if they are conjugate within some triangle group Γ_1 containing Γ, and we know all possible such pairs of triangle groups from Singerman's list. Thus, for a given triangle group Γ with signature $(0; m_1, m_2, m_3)$ and finite group G, we can count the number of distinct surfaces (up to conformal equivalence) on which G acts with signature $(0; m_1, m_2, m_3)$ by doing the following:

(1) Count the number of epimorphisms from Γ onto G.
(2) Divide this number by $|\mathrm{Aut}(G)|$, the size of the automorphism group of G. This will give the number of distinct surface subgroups of Γ with quotient G.
(3) For each triangle group Γ_1 containing Γ, sort the surface subgroups of Γ with quotient G into Γ_1-conjugacy classes.

We discuss these steps in a little more detail, and explain how they can help with the problem of determining full automorphism groups.

First, at least currently, there is no straightforward way to calculate the number of epimorphisms from Γ onto G without constructing generating vectors. Fortunately, there is a way to use the characters of G to count torsion free homomorphisms. Specifically the following is a consequence of the more general main result from [**38**]:

THEOREM 28. *Suppose that Γ has signature $(0; m_1, m_2, m_3)$ and denote by L_i for $1 \leq i \leq 3$ the union of all conjugacy classes of G of elements of orders m_i respectively. The number of homomorphisms from Δ to G with torsion free kernel, denoted $|\mathrm{Hom}(\Delta, G)|$, is given by:*

$$\frac{1}{|G|}\sum_{\chi}\left\{\chi(1)^{-1}[\sum_{x_1\in L_1}\chi(x_1)][\sum_{x_2\in L_2}\chi(x_2)][\sum_{x_3\in L_3}\chi(x_3)]\right\} \tag{11}$$

where χ runs over the irreducible characters of G.

Of course, we need to be a little careful as some of the homomorphisms counted in Thm. 28 may not be surjective onto G, but rather define an epimorphism onto a strict subgroup H of G. However, if this is the case, the Riemann-Hurwitz formula ensures a corresponding action of H on some surface of smaller genus. In particular, if we are working iteratively through a list, then we already know the number of epimorphisms of Γ onto H. Thus to refine the number of homomorphisms to epimorphisms given in Thm. 28, we simply subtract off the number of epimorphisms onto the subgroups of G which we have determined already earlier in our list.

For step (3), for many triangle groups, there are only a few containments, so it is often fairly straightforward to reason when different groups have to be conjugate. We illustrate with an example.

EXAMPLE 4. *The cyclic group C_5 acts with signature $(0; 5, 5, 5)$ on a surface of genus 2. Using Thm. 28, the number of homomorphisms from Γ with signature $(0; 5, 5, 5)$ onto C_5 is 12. Since there are no non-trivial normal subgroups of C_5, all of these must be epimorphisms, and thus there are $12/|\mathrm{Aut}(C_5)| = 12/4 = 3$ distinct surface subgroups of Γ with quotient C_5. From Breuer's lists, we know there is no group action on a surface of genus 2 with signature $(0; 3, 3, 5)$. In particular, none of these surface subgroups are normal in the Fuchsian overgroup Γ_1 with signature $(0; 3, 3, 5)$. Since Γ is normal in Γ_1, the three surface subgroups must all be conjugate within Γ_1 (since $[\Gamma_1 : \Gamma] = 3$). In particular, there is exactly one surface of genus 2 up to conformal equivalence on which the group C_5 acts with signature $(0; 5, 5, 5)$.*

We use this information to determine full automorphism groups in the following way. Suppose we know a particular containment of Fuchsian triangle groups $\Gamma \leq \Gamma_1$, both of which appear in a list such as Breuer's with corresponding finite groups G and G_1. Then if we know that G_1 contains a subgroup isomorphic to G, it is possible that G is not the full automorphism group of a given surface on which G acts. In order to decide this, we can count the number of distinct surfaces whose surface subgroup is contained in Γ normally with quotient G, and compare that to the number of distinct surfaces whose surface subgroup is contained in Γ_1 normally with quotient G_1 – if there are more subgroups inside Γ, then we know for sure there exists at least one surface $\mathcal{X}$ for which G acts but does not extend to G_1. Of course, this doesn't mean it doesn't extend to other actions given by other Fuchsian group containments, but we can check each containment individually (and many are unique anyway).

Things are trickier if there are the same number or less than the same number of normal surface subgroups of Γ with quotient G than normal surface subgroups of Γ_1 with quotient G_1. The difficulty lies in the fact that it is not always guaranteed if G_1 acts on $\mathcal{X}$ then the subgroup G corresponds to Γ – it could correspond to a completely different Fuchsian group which may not even be a triangle group. Fortunately, it is easy to determine signatures of Fuchsian subgroups of a given index using the following result of Singerman:

THEOREM 29. *Let Γ_1 have signature $(g_{\Gamma_1}; m_1, \dots, m_r)$. Then Γ_1 contains a subgroup Γ of finite index with signature $(g_\Gamma; n_{1,1}, n_{1,2}, \dots, n_{1,\theta_1}, \dots n_{r,\theta_r})$ if and only if*

(1) *There exists a finite permutation group G transitive on $[\Gamma_1 : \Gamma]$ points and an epimorphism $\Phi\colon \Gamma_1 \to G$ such that the permutation $\Phi(\zeta_j)$ has precisely θ_j cycles of length less then m_j, the lengths of these cycles being*

$$m_j/n_{j,1}, \dots m_j/n_{j,\theta_j}.$$

(2) $[\Gamma_1 : \Gamma] = (2g_\Gamma - 2 + \sum_{j=1}^{r} \sum_{i=1}^{\theta_j} (1 - \frac{1}{n_{i,\theta_j}}))/(2g_{\Gamma_1} - 2 + \sum_{i=1}^{r} (1 - \frac{1}{m_i}))$.

Now, if Γ is the unique subgroup of a given index, then each surface subgroup of Γ_1 with quotient G_1 will also be a surface subgroup of Γ with quotient G. In particular, if there are the same number of surface subgroups, then all such surfaces have G_1 as their full automorphism group and not G. If Γ is not unique the problem becomes much more difficult, and we need to start looking at generating vectors for G_1 corresponding to each of the different surfaces to determine the signature of the subgroups corresponding to G using Thm. 29. Rather than present complete details, we illustrate with an explicit example below, and then use these techniques to explain how to classify full automorphism groups of superelliptic surfaces in the next section.

EXAMPLE 5. *In Example 4, we saw that there is exactly one surface of genus 2 up to conformal equivalence on which the group C_5 acts with signature $(0; 5, 5, 5)$. In Breuer's list for genus 2, there is also an action by the cyclic group C_{10} with signature $(0; 2, 5, 10)$. Using Thm. 28, the number of homomorphisms from Γ_1 with signature $(0; 2, 5, 10)$ onto C_{10} is 4. Since there are no non-trivial normal subgroups of C_{10} containing an element of order 10, all of these must be epimorphisms, and thus there is $4/|\operatorname{Aut}(C_{10})| = 4/4 = 1$ distinct surface subgroup of Γ_1 with quotient C_{10}. Now, using Thm. 29, it is easy to see that Γ_1 has a unique subgroup of index 2, this subgroup having signature $(0; 5, 5, 5)$. In particular, it follows that corresponding unique surface of genus 2 on which C_5 acts with signature $(0; 5, 5, 5)$ has a larger automorphism group – at least C_{10}. In particular, since this is the unique surface on which C_5 acts with this signature, it cannot possibly be the full automorphism group of a genus 2 surface.*

6.4. Superelliptic curves revisited. In order to illustrate the methods we have outlined for finding full automorphism groups, we conclude our discussion on compact Riemann surfaces by describing how to use the techniques we have described to find the full automorphism group of all superelliptic surfaces of level n

(henceforth just superelliptic surfaces). By *superelliptic surface*, we mean a compact Riemann surface $\mathcal{X}$ with a cyclic group of automorphisms H of order n, called a *superelliptic group*, with the property that $\mathcal{X}/H$ has genus 0 and that every branch point of the map $\pi_H \colon \mathcal{X} \to \mathcal{X}/H$ has order n. Such surfaces are natural generalizations of hyperelliptic surfaces, where $n = 2$ whose full automorphism groups were completely classified in [**19**]. It can be shown that, assuming the above definition of superelliptic, the curve has an equation of the form $y^n = f(x)$ with some restrictions on the factors of $f(x)$. See the references at the beginning Section 5.

6.5. Preliminaries on superelliptic surfaces. Before we start our analysis, we shall first introduce some basic terminology, notation and facts about superelliptic surfaces from the point of view of Fuchsian groups. Henceforth, for our analysis, $\mathcal{X}$ will denote a superelliptic surface and H a superelliptic group with generator τ. Now since $\mathcal{X}/H$ has genus 0 and every branch point of the map $\pi_H \colon \mathcal{X} \to \mathcal{X}/H$ has order n, it follows that H has signature $(0; n, \ldots, n)$.

Now, if $\mathrm{Aut}(\mathcal{X}) > H$, then using the notation we have introduced, after appropriate identifications, we have the tower of groups and epimorphisms illustrated in Fig. 2 and corresponding to this, the tower of surfaces and holomorphic maps between them illustrated in Fig. 3.

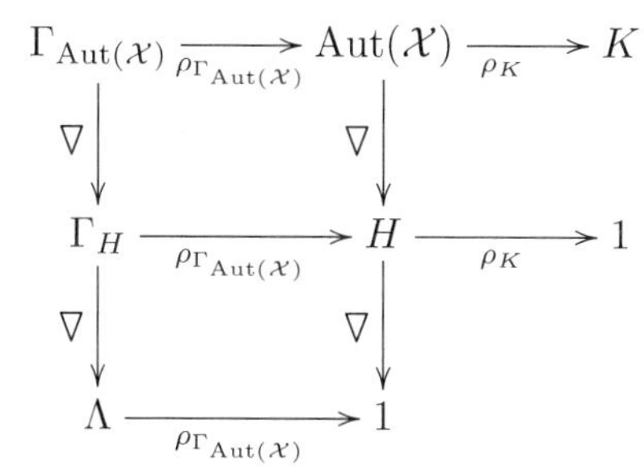

FIGURE 2. Groups and quotients.

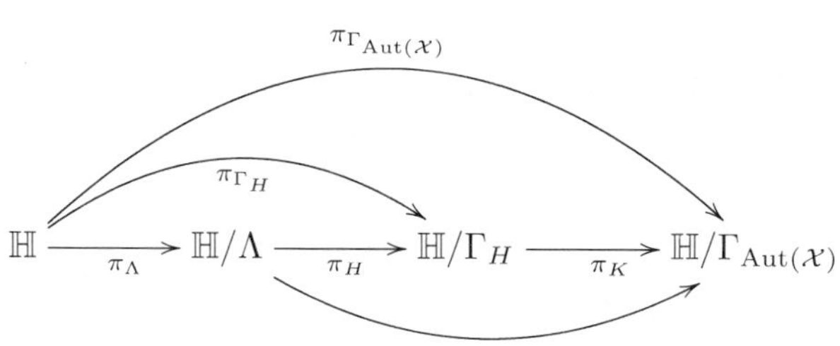

FIGURE 3. Maps and quotient spaces.

Observe that the map π_K is a finite Galois map with covering group K from the Riemann sphere to itself, and all such maps are well known. We summarize the possibilities for K and the ramification data of the map π_K in Table 3. The importance of these observations is that we can use it to help find $\mathrm{Aut}(\mathcal{X})$ and the possible signatures of $\Gamma_{\mathrm{Aut}(\mathcal{X})}$. Specifically, since H is normal in $\mathrm{Aut}(\mathcal{X})$ with quotient K, $\mathrm{Aut}(\mathcal{X})$ will be a group satisfying the short exact sequence

$$1 \to H \to \mathrm{Aut}(\mathcal{X}) \to K \to 1$$

and all such groups are relatively easy to find. For our purposes, we can take the more general list of such groups in any characteristic given in Thm. 20 and refine this list to the admissible K for characteristic $p = 0$.

TABLE 3. Groups of automorphisms of the Riemann sphere and branching data.

Group	Branching Data
C_n	(n, n)
D_n	$(2, 2, n)$
A_4	$(2, 3, 3)$
S_4	$(2, 3, 4)$
A_5	$(2, 3, 5)$

Now the signature of $\Gamma_{\mathrm{Aut}(\mathcal{X})}$ depends only on which branch points of the map π_H are also ramification points of the map π_K. We summarize (see for example Proposition 3 in [**76**] for details).

THEOREM 30. *The signature of* $\Gamma_{\mathrm{Aut}(\mathcal{X})}$ *takes one of the following forms:*

(1) *If* $K = C_m$ *and* $b_1, b_2 \in \mathcal{X}/\mathrm{Aut}(\mathcal{X})$ *are the branch points of* π_K, *the signature of* $\Gamma_{\mathrm{Aut}(\mathcal{X})}$ *is* $(0; a_1m, a_2m, n, \dots, n)$ *where* $a_i = n$ *if* $\pi_K^{-1}(b_i)$ *contains a branch point of the map* π_H *and* $a_i = 1$ *else.*
(2) *If* $K \neq C_m$ *and* (d_1, d_2, d_3) *is the branching data of the quotient map* π_K *with corresponding branch points* $b_1, b_2, b_3 \in \mathcal{X}/\mathrm{Aut}(\mathcal{X})$ *respectively, the signature of* $\Gamma_{\mathrm{Aut}(\mathcal{X})}$ *is* $(0; a_1d_1, a_2d_2, a_3d_3, n, \dots, n)$ *where* $a_i = n$ *if* $\pi_K^{-1}(b_i)$ *contains a branch point of the map* π_H *and* $a_i = 1$ *else.*

6.6. Finding full automorphism groups of superelliptic surfaces. With the necessary terminology and notation introduced, we can now formalize the problem and explain how to solve it. First, we recall the problem: for a valid group signature pair $(G,\mathcal{S})$ for a superelliptic surface, we want to know if there exists a superelliptic surface $\mathcal{X}$ on which G acts with signature $\mathcal{S}$ as the full automorphism group of $\mathcal{X}$, or conversely, if for every such superelliptic surface, G is always contained in some larger group.

Two immediate observations. First, clearly if the signature $\mathcal{S}$ does not appear in Singerman's list, then there always exists a superelliptic surface $\mathcal{X}$ on which G acts with signature $\mathcal{S}$ as the full automorphism group of $\mathcal{X}$. Therefore, to solve this problem, we just need to consider possible group signature pairs for which the signature appears in Singerman's list. Second, since we are assuming $H < G$, the orbit genus of the signature of Γ_G must be 0, so we can eliminate cases N1 through N4 from Singerman's list as possibilities for $\mathcal{S}$.

Next, we can eliminate further possibilities using the fact that our surface is superelliptic. Specifically, if an extension of G' of G exists then H would also be central in G' and would therefore be isomorphic to one of the groups appearing in Thm. 20 (with appropriate K) with corresponding quotient $K' = G'/H$ being an extension of $K = G/H$ with corresponding ramification data from Table 3. In particular, when $K = S_4$ or A_5, there are no possible extensions, so in either of these cases, G acts as the full automorphism group. There are also restrictions to which groups a given K can extend to for the three remaining cases, A_4, D_m and C_m which we shall use in our analysis below.

At this point, we have enough information to tabulate all possibilities for the signatures which might arise as signatures for non-maximal groups of automorphisms of superelliptic surfaces. We shall tabulate these signatures organizing them by isomorphism classes of K. For a given signature of Γ_G in each such table, we also include the signature for any Fuchsian group Δ with $\Delta \geq \Gamma_G$ with signature coming from Singerman's list together with K', the corresponding extension of K which is needed to test for the maximality of G. The arguments for why we can eliminate each group and signature is a little different for each of the possible $K's$, so we break up our arguments accordingly.

Suppose first that $K = C_m$ and suppose that K' is some extension corresponding to G', an extension of G. Now, it is not possible for $K = C_{ms}$ since using Thm. 29, the number of branch points of Γ_G would be strictly bigger than that of $\Gamma_{G'}$, and so the signatures would not appear in Singerman's list. Next, it is always possible for $K' = D_m$ with corresponding signature $(0; 2, 2, m)$. Moreover, if K is contained in some larger dihedral group, it will necessarily also be contained in D_m, so to check for maximality, we just need to check extension to D_m. For $K' = S_4$ or A_5, each cyclic subgroup C_m is also a subgroup of D_m within K', so in particular, to check for maximality, we just need to check extension to D_m. Finally, for $K = A_4$, any C_2 is contained in D_2, so to check for maximality, we just need to check extension to D_2. For $C_3 \leq A_4$ however, there are no intermediate subgroups, so to check for maximality, we need to check extension to A_4.

TABLE 4. Potential non-maximal signatures with $K = C_m$.

K	Signature of Γ_G	K'	Signature of Δ	Further Conditions
C_m	(m, m, n, n)	D_m	$(0; 2, 2, m, n)$	
C_m	(mn, mn, n, n)	D_m	$(0; 2, 2, mn, n)$	
C_m	(m, m, n)	D_m	$(0, 2, m, 2n)$	
C_3	$(3, n, 3n)$	A_4	$(0; 2, 3, 3n)$	$m = 3$
C_m	(mn, mn, n)	D_m	$(0; 2, mn, 2n)$	

Now suppose that $K = D_m$ and suppose that K' is some extension corresponding to G', an extension of G. Using Thm. 29 to find the corresponding signature of the surface group for Λ in Γ_G and $\Gamma_{G'}$, we see that if $K' = D_{ms}$, then either $s = 2$ or $s = 4$. In particular, in each case, it will necessarily be contained in D_{2m}, so to check for maximality, we just need to check extension to D_{2m}. The only dihedral subgroup of $K' = A_4$ is D_2 with $m = 2$, and in this case we need to check for extension to A_4. For $K' = S_4$, there are three different dihedral subgroups – D_2, D_3 and D_4. If a given D_2 extends to S_4, then it also extends to D_4, so to check maximality, we only need to check maximality in D_4. The group D_3 has index 4 in S_4. Neither of the signatures $T9$ or $T10$ with index 4 are valid signatures for D_3 and S_4 actions, so in this case we do not need to check maximality. Finally, the group D_4 has index 3 in S_4. Of the two signatures in Singerman's list with index 3, only $T11$ is valid for D_4 and S_4 with $(0; 2, 2n, 4n)$ and $(0; 2, 3, 4n)$ respectively, so this is the only case where we need to check maximality of D_4 in S_4.

TABLE 5. Potential non-maximal signatures.

K	Signature of Γ_G	K'	Signature of Δ	Further Conditions
D_m	$(0; 2n, 2n, n, n)$	D_{2m}	$(0; 2, 2, n, 2n)$	$m = n$
D_m	$(0; 2n, 2n, m)$	D_{2m}	$(0; 2, 2n, 2m)$	
D_m	$(0; 2n, 2n, mn)$	D_{2m}	$(0; 2, 2n, 2nm)$	
D_2	$(0; 2n, 2n, 2n)$	A_4	$(0; 2n, 3, 3)$	$m = 2$
D_4	$(0; 2, 2n, 4n)$	S_4	$(0; 2, 3, 4n)$	$m = 4$

Lastly, now suppose that $K = A_4$ and suppose that K' is some extension corresponding to G', an extension of G. The only possibilities for K' are S_4 and A_5, so we look at each of these cases individually. If $K' = S_4$, then A_4 has index 2 in S_4, and in this case there are multiple signatures from Singerman's list of the form $N8$ which we need to check to determine maximality. When $K' = A_5$, then $K = A_4$ is non-normal and has index 6 in K'. This leaves $T7$ and $T8$ as the only possible signature pairs, but neither of these work for the inclusion of A_4 in A_5. We summarize in Table 6.

TABLE 6. Potential non-maximal signatures.

K	Signature of Γ_G	K'	Signature of Δ	Further Conditions
A_4	$(0; 2, 3n, 3n)$	S_4	$(0; 2, 3n, 4)$	
A_4	$(0; 2n, 3, 3)$	S_4	$(0; 2, 3, 4n)$	
A_4	$(0; 2n, 3n, 3n)$	S_4	$(0; 2, 3n, 4n)$	

At this point, we now have enough information that to finish the problem we can either apply Thm. 25 and Thm. 26 to the different possible generating vectors for each group and signature pair, or alternatively apply the counting methods developed in Section 6.3.3. Rather than go through every individual case, we illustrate with an explicit example. The other signature pairs given in Table 4, Table 5 and Table 6 yield results with similar congruence conditions given in Example 6.

EXAMPLE 6. *Consider the signature* $(0; mn, mn, n)$. *Since the signature has periods of order* mn, *the only possible quotient group in this case is* C_{nm}. *If* $C_{nm} = \langle x \rangle$, *then after appropriate automorphism of* C_{nm}, *any* $(0; mn, mn, n)$*-generating vector of* C_{nm} *will have the form* (x, x^a, x^{nm-a-1}) *where* $(a, nm) = 1$ *and* $(nm, nm - a - 1) = m$.

Applying (2) *of Thm. 25, this group extends to a group with signature* $(0; 2, mn, 2n)$ *if and only if the maps* $x \to x^a$ *and* $x^a \to x$ *induce an automorphism of* C_{nm} *(note that the condition on the third generator is trivially satisfied since the group is Abelian). However, this happens only if* $x^{a^2} = x$ *i.e.* $a^2 \equiv 1 \mod nm$. *In particular, unless* $a^2 \equiv 1 \mod nm$ *for every* a *satisfying the two different congruences, then there exists at least one surface on which* C_{nm} *acts as the full automorphism group with signature* $(0; mn, mn, n)$.

We note that there are examples from Example 6 for which the group is never maximal, such as when $n = 4$ and $m = 10$, and when there exist maximal actions such as $m = n = 8$.

References

[1] Robert D. M. Accola, *On the number of automorphisms of a closed Riemann surface*, Trans. Amer. Math. Soc. **131** (1968), 398–408. MR0222281

[2] Robert D. M. Accola, *Two theorems on Riemann surfaces with noncyclic automorphism groups.*, Proc. Amer. Math. Soc. **25** (1970), 598–602, DOI 10.2307/2036651. MR0259105

[3] Jannis A. Antoniadis and Aristides Kontogeorgis, *On cyclic covers of the projective line*, Manuscripta Math. **121** (2006), no. 1, 105–130, DOI 10.1007/s00229-006-0028-4. MR2258533

[4] C. Bagiński, J. J. Etayo, G. Gromadzki, and E. Martínez, *On real forms of a Belyi action of the alternating groups*, Albanian J. Math. **10** (2016), no. 1, 3–10. MR3490049

[5] Walter L. Baily Jr., *On the automorphism group of a generic curve of genus* > 2, J. Math. Kyoto Univ. **1** (1961/1962), no. 101–108; correction, 325, DOI 10.1215/kjm/1250525063. MR0142552

[6] Marie José Bertin and Odile Lecacheux, *Automorphisms of certain Niemeier lattices and elliptic fibrations*, Albanian J. Math. **11** (2017), no. 1, 13–33. MR3724575

[7] Hans Ulrich Besche and Bettina Eick, *The groups of order at most 1000 except 512 and 768*, J. Symbolic Comput. **27** (1999), no. 4, 405–413, DOI 10.1006/jsco.1998.0259. MR1681347

[8] L. Beshaj, T. Shaska, and E. Zhupa, *The case for superelliptic curves*, Advances on superelliptic curves and their applications, NATO Sci. Peace Secur. Ser. D Inf. Commun. Secur., vol. 41, IOS, Amsterdam, 2015, pp. 1–14. MR3525570

[9] Lubjana Beshaj, Valmira Hoxha, and Tony Shaska, *On superelliptic curves of level n and their quotients, I*, Albanian J. Math. **5** (2011), no. 3, 115–137. MR2846162

[10] Lubjana Beshaj and Fred Thompson, *Equations for superelliptic curves over their minimal field of definition*, Albanian J. Math. **8** (2014), no. 1, 3–8. MR3253208

[11] O. V. Bogopolʹskiĭ, *Classifying the actions of finite groups on orientable surfaces of genus* 4 *[translation of proceedings of the institute of mathematics, 30 (Russian), 48–69, Izdat. Ross. Akad. Nauk, Sibirsk. Otdel., Inst. Mat., Novosibirsk, 1996]*, Siberian Adv. Math. **7** (1997), no. 4, 9–38. Siberian Advances in Mathematics. MR1604157

[12] Rolf Brandt, *Uber die automorphismengruppen von algebraischen funktionenkorpern*, Ph.D. Thesis, 1988.

[13] Rolf Brandt and Henning Stichtenoth, *Die Automorphismengruppen hyperelliptischer Kurven* (German, with English summary), Manuscripta Math. **55** (1986), no. 1, 83–92, DOI 10.1007/BF01168614. MR828412

[14] Thomas Breuer, *Characters and automorphism groups of compact Riemann surfaces*, London Mathematical Society Lecture Note Series, vol. 280, Cambridge University Press, Cambridge, 2000. MR1796706

[15] S. Allen Broughton, *Classifying finite group actions on surfaces of low genus*, J. Pure Appl. Algebra **69** (1991), no. 3, 233–270, DOI 10.1016/0022-4049(91)90021-S. MR1090743

[16] S. Allen Broughton, *Superelliptic surfaces as p-gonal surfaces*, Riemann and Klein surfaces, automorphisms, symmetries and moduli spaces, Contemp. Math., vol. 629, Amer. Math. Soc., Providence, RI, 2014, pp. 15–28, DOI 10.1090/conm/629/12570. MR3289630

[17] S. Allen Broughton, *Quasi-platonic $PSL_2(q)$-actions on closed Riemann surfaces*, Albanian J. Math. **9** (2015), no. 1, 31–61. MR3438844

[18] S. Allen Broughton and Aaron Wootton, *Exceptional automorphisms of (generalized) super elliptic surfaces*, Riemann and Klein surfaces, automorphisms, symmetries and moduli spaces, Contemp. Math., vol. 629, Amer. Math. Soc., Providence, RI, 2014, pp. 29–42, DOI 10.1090/conm/629/12573. MR3289631

[19] E. Bujalance, J. M. Gamboa, and G. Gromadzki, *The full automorphism groups of hyperelliptic Riemann surfaces*, Manuscripta Math. **79** (1993), no. 3-4, 267–282, DOI 10.1007/BF02568345. MR1223022

[20] Emilio Bujalance, F. J. Cirre, and Marston Conder, *On extendability of group actions on compact Riemann surfaces*, Trans. Amer. Math. Soc. **355** (2003), no. 4, 1537–1557, DOI 10.1090/S0002-9947-02-03184-7. MR1946404

[21] Marston Conder, *Hurwitz groups: a brief survey*, Bull. Amer. Math. Soc. (N.S.) **23** (1990), no. 2, 359–370, DOI 10.1090/S0273-0979-1990-15933-6. MR1041434

[22] Harris B. Daniels, *Siegel functions, modular curves, and Serre's uniformity problem*, Albanian J. Math. **9** (2015), no. 1, 3–29. MR3338631

[23] Yasin Demirbas, *Hyperelliptic curves of genus 3 and 4 in characteristic 2*, Computational aspects of algebraic curves, Lecture Notes Ser. Comput., vol. 13, World Sci. Publ., Hackensack, NJ, 2005, pp. 145–162, DOI 10.1142/9789812701640_0011. MR2182039

[24] A. Duma and C. Tanasi, *Groups of automorphisms of a family of compact Riemann surfaces of genus four* (Italian), Ricerche Mat. **34** (1985), no. 2, 377–390. MR870833

[25] Andrei Duma, *Automorphisms of compact Riemann surfaces* (Italian), Confer. Sem. Mat. Univ. Bari **207** (1985), ii+46 pp. (1986). MR853404

[26] H. M. Farkas and I. Kra, *Riemann surfaces*, Second, Graduate Texts in Mathematics, vol. 71, Springer-Verlag, New York, 1992. MR1139765

[27] Gerhard Frey and Tony Shaska, *Curves, Jacobians, and cryptography*, Algebraic curves and their applications, 2018, pp. 280–346.

[28] Ernesto Girondo and Jürgen Wolfart, *Conjugators of Fuchsian groups and quasiplatonic surfaces*, Q. J. Math. **56** (2005), no. 4, 525–540, DOI 10.1093/qmath/hah054. MR2182464

[29] J. Gutierrez and T. Shaska, *Hyperelliptic curves with extra involutions*, LMS J. Comput. Math. **8** (2005), 102–115, DOI 10.1112/S1461157000000917. MR2135032

[30] Jaime Gutierrez, D. Sevilla, and T. Shaska, *Hyperelliptic curves of genus 3 with prescribed automorphism group*, Computational aspects of algebraic curves, 2005, pp. 109–123. MR2182037

[31] W. J. Harvey, *Cyclic groups of automorphisms of a compact Riemann surface*, Quart. J. Math. Oxford Ser. (2) **17** (1966), 86–97. MR0201629

[32] Hans-Wolfgang Henn, *Funktionenkörper mit grosser Automorphismengruppe* (German), J. Reine Angew. Math. **302** (1978), 96–115, DOI 10.1515/crll.1978.302.96. MR511696

[33] Ruben Hidalgo and Tony Shaska, *On the field of moduli of superelliptic curves*, Higher genus curves in mathematical physics and arithmetic geometry, Contemp. Math., vol. 703, Amer. Math. Soc., Providence, RI, 2018, pp. 47–62, DOI 10.1090/conm/703/14130. MR3782459

[34] Ruben A. Hidalgo, Saúl Quispe, and Tony Shaska, *On generalized superelliptic riemann surfaces* (201609), available at `1609.09576`.

[35] J. W. P. Hirschfeld, G. Korchmáros, and F. Torres, *Algebraic curves over a finite field*, Princeton Series in Applied Mathematics, Princeton University Press, Princeton, NJ, 2008. MR2386879

[36] Masaaki Homma, *Automorphisms of prime order of curves*, Manuscripta Math. **33** (1980/81), no. 1, 99–109, DOI 10.1007/BF01298341. MR596381

[37] A. Hurwitz, *Ueber algebraische Gebilde mit eindeutigen Transformationen in sich* (German), Math. Ann. **41** (1892), no. 3, 403–442, DOI 10.1007/BF01443420. MR1510753

[38] Gareth A. Jones, *Enumeration of homomorphisms and surface-coverings*, Quart. J. Math. Oxford Ser. (2) **46** (1995), no. 184, 485–507, DOI 10.1093/qmath/46.4.485. MR1366619

[39] W. T. Kiley, *Automorphism groups on compact Riemann surfaces*, Trans. Amer. Math. Soc. **150** (1970), 557–563. MR0264058

[40] Ravi S. Kulkarni, *Riemann surfaces admitting large automorphism groups*, Extremal Riemann surfaces (San Francisco, CA, 1995), Contemp. Math., vol. 201, Amer. Math. Soc., Providence, RI, 1997, pp. 63–79, DOI 10.1090/conm/201/02610. MR1429195

[41] Akikazu Kuribayashi and Keizo Hayakawa, *On the Hurwitz group* (Japanese, with English summary), Bull. Fac. Sci. Engrg. Chuo Univ. Ser. I Math. **37** (1994), 1–23 (1995). MR1334195

[42] Akikazu Kuribayashi and Hideyuki Kimura, *Automorphism groups of compact Riemann surfaces of genus five*, J. Algebra **134** (1990), no. 1, 80–103, DOI 10.1016/0021-8693(90)90212-7. MR1068416

[43] Akikazu Kuribayashi, Akeo Kobayashi, and Shoji Ohmori, *Automorphism groups of compact Riemann surfaces*, Rev. Roumaine Math. Pures Appl. **40** (1995), no. 1, 65–75. Travaux de la Conférence Internationale d'Analyse Complexe et du 7e Séminaire Roumano-Finlandais (1993). MR1406121

[44] Joseph Lehner and Morris Newman, *On Riemann surfaces with maximal automorphism groups*, Glasgow Math. J. **8** (1967), 102–112, DOI 10.1017/S001708950000015X. MR0220924

[45] A. M. Macbeath, *On a theorem of Hurwitz*, Proc. Glasgow Math. Assoc. **5** (1961), 90–96 (1961). MR0146724

[46] A. M. Macbeath, *On a curve of genus* 7, Proc. London Math. Soc. (3) **15** (1965), 527–542, DOI 10.1112/plms/s3-15.1.527. MR0177342

[47] C. Maclachlan, *Abelian groups of automorphisms of compact Riemann surfaces*, Proc. London Math. Soc. (3) **15** (1965), 699–712, DOI 10.1112/plms/s3-15.1.699. MR0179348

[48] C. Maclachlan, *A bound for the number of automorphisms of a compact Riemann surface.*, J. London Math. Soc. **44** (1969), 265–272, DOI 10.1112/jlms/s1-44.1.265. MR0236378

[49] K. Magaard, T. Shaska, S. Shpectorov, and H. Völklein, *The locus of curves with prescribed automorphism group*, Sūrikaisekikenkyūsho Kōkyūroku **1267** (2002), 112–141. Communications in arithmetic fundamental groups (Kyoto, 1999/2001). MR1954371

[50] Takanori Matsuno, *Compact Riemann surfaces with large automorphism groups*, J. Math. Soc. Japan **51** (1999), no. 2, 309–329, DOI 10.2969/jmsj/05120309. MR1674751

[51] Rezart Muço, Nejme Pjero, Ervin Ruci, and Eustrat Zhupa, *Classifying families of superelliptic curves*, Albanian J. Math. **8** (2014), no. 1, 23–35. MR3270074

[52] E. Previato, T. Shaska, and G. S. Wijesiri, *Thetanulls of cyclic curves of small genus*, Albanian J. Math. **1** (2007), no. 4, 253–270. MR2367218

[53] John F. X. Ries, *Subvarieties of moduli space determined by finite groups acting on surfaces*, Trans. Amer. Math. Soc. **335** (1993), no. 1, 385–406, DOI 10.2307/2154274. MR1097170

[54] R. Sanjeewa, *Automorphism groups of cyclic curves defined over finite fields of any characteristics*, Albanian J. Math. **3** (2009), no. 4, 131–160. MR2578064

[55] R. Sanjeewa and T. Shaska, *Determining equations of families of cyclic curves*, Albanian J. Math. **2** (2008), no. 3, 199–213. MR2492096

[56] Hermann Ludwig Schmid, *Über die Automorphismen eines algebraischen Funktionenkörpers von Primzahlcharakteristik* (German), J. Reine Angew. Math. **179** (1938), 5–15, DOI 10.1515/crll.1938.179.5. MR1581581

[57] Friedrich Karl Schmidt, *Zur arithmetischen Theorie der algebraischen Funktionen. II. Allgemeine Theorie der Weierstraßpunkte* (German), Math. Z. **45** (1939), no. 1, 75–96, DOI 10.1007/BF01580274. MR1545805

[58] David Sevilla and Tanush Shaska, *Hyperelliptic curves with reduced automorphism group A_5*, Appl. Algebra Engrg. Comm. Comput. **18** (2007), no. 1-2, 3–20, DOI 10.1007/s00200-006-0030-9. MR2280308

[59] T. Shaska, *Subvarieties of the hyperelliptic moduli determined by group actions*, Serdica Math. J. **32** (2006), no. 4, 355–374. MR2287373

[60] T. Shaska and F. Thompson, *Bielliptic curves of genus 3 in the hyperelliptic moduli*, Appl. Algebra Engrg. Comm. Comput. **24** (2013), no. 5, 387–412, DOI 10.1007/s00200-013-0209-9. MR3118614

[61] Tanush Shaska, *Determining the automorphism group of a hyperelliptic curve*, Proceedings of the 2003 International Symposium on Symbolic and Algebraic Computation, ACM, New York, 2003, pp. 248–254, DOI 10.1145/860854.860904. MR2035219

[62] Tanush Shaska, *Some special families of hyperelliptic curves*, J. Algebra Appl. **3** (2004), no. 1, 75–89, DOI 10.1142/S0219498804000745. MR2047637

[63] Tanush Shaska and Helmut Völklein, *Elliptic subfields and automorphisms of genus 2 function fields*, Algebra, arithmetic and geometry with applications (West Lafayette, IN, 2000), Springer, Berlin, 2004, pp. 703–723. MR2037120

[64] Tony Shaska and Caleb M. Shor, *2-Weierstrass points of genus 3 hyperelliptic curves with extra involutions*, Comm. Algebra **45** (2017), no. 5, 1879–1892, DOI 10.1080/00927872.2016.1226861. MR3582832

[65] C. Shor and T. Shaska, *Weierstrass points of superelliptic curves*, Advances on superelliptic curves and their applications, NATO Sci. Peace Secur. Ser. D Inf. Commun. Secur., vol. 41, IOS, Amsterdam, 2015, pp. 15–46. MR3525571

[66] David Singerman, *Subgroups of Fuschian groups and finite permutation groups*, Bull. London Math. Soc. **2** (1970), 319–323, DOI 10.1112/blms/2.3.319. MR0281805

[67] David Singerman, *Finitely maximal Fuchsian groups*, J. London Math. Soc. (2) **6** (1972), 29–38. MR0322165

[68] David Singerman, *Symmetries of Riemann surfaces with large automorphism group*, Math. Ann. **210** (1974), 17–32. MR0361059

[69] Henning Stichtenoth, *Über die Automorphismengruppe eines algebraischen Funktionenkörpers von Primzahlcharakteristik. I. Eine Abschätzung der Ordnung der Automorphismengruppe* (German), Arch. Math. (Basel) **24** (1973), 527–544, DOI 10.1007/BF01228251. MR0337980

[70] Henning Stichtenoth, *Algebraic function fields and codes*, 2nd ed., Graduate Texts in Mathematics, vol. 254, Springer-Verlag, Berlin, 2009. MR2464941

[71] Nobumasa Takigawa, *Compact Riemann surfaces with an automorphism group of order p*, Math. Rep. Toyama Univ. **4** (1981), 115–127. MR627968

[72] Robert C. Valentini and Manohar L. Madan, *A hauptsatz of L. E. Dickson and Artin-Schreier extensions*, J. Reine Angew. Math. **318** (1980), 156–177. MR579390

[73] Robert A. Wilson, *The Monster is a Hurwitz group*, J. Group Theory **4** (2001), no. 4, 367–374, DOI 10.1515/jgth.2001.027. MR1859175

[74] A. Wootton, *Counting Belyĭ surfaces with many automorphisms*, Proceedings of the 10th International ACA Conference, 2004, pp. 74–80.

[75] A. Wootton, *The full automorphism group of a cyclic p-gonal surface*, J. Algebra **312** (2007), no. 1, 377–396, DOI 10.1016/j.jalgebra.2007.01.018. MR2320463

[76] Aaron Wootton, *Non-normal Belyĭ p-gonal surfaces*, Computational aspects of algebraic curves, 2005, pp. 95–108. MR2182036

Department of Mathematics (Emeritus), Rose-Hulman Institute of Technology, Terre Haute, Indianna 47803

Email address: `brought@rose-hulman.edu`

Department of Mathematics and Statistics, Oakland University, Rochester, Michigan 48309

Email address: `shaska@oakland.edu`

Department of Mathematics, University of Portland, Portland, Oregon 97203

Email address: `wootton@up.edu`

Contemporary Mathematics
Volume **724**, 2019
https://doi.org/10.1090/conm/724/14591

On the algebraic classification of subgroups of hyperbolic planar crystallographic groups

Ismael Cortázar and Antonio F. Costa

Abstract. A planar Euclidean or hyperbolic crystallographic group Δ is a subgroup of the group of isometries of the Euclidean plane $\mathbb{E}^2$, respectively the hyperbolic plane $\mathbb{H}^2$, with compact orbit space. These groups are classified algebraically by a symbol called signature and an equivalence relation defined on the set of signatures. In 1990 A.H.M. Hoare gave an algorithm to obtain the signature of a finite index subgroup of a planar crystallographic group. Recently the authors completed the algorithm of Hoare and implemented it on a computer system.

In the signature of the hyperbolic groups there is a sign $+$ or $-$, in the case of $+$ sign the cyclic order on some integers in signatures are essential to determine the isomorphism class. In this article we show examples where such cyclic order is necessary to determine if two subgroups of a given hyperbolic group are isomorphic. Finally we announce the implementation of the algorithm to compute the signature of subgroups of hyperbolic crystallographic groups on the computer system for group theory GAP.

1. Introduction

A planar Euclidean or hyperbolic crystallographic group Δ is a subgroup of the group of (orientation preserving and reversing) isometries of either the Euclidean plane $\mathbb{E}^2$ or the hyperbolic plane $\mathbb{H}^2$, such that the action of Δ is discontinuous and the orbit space with the induced topology is compact. For instance in the Euclidean case the planar crystallographic groups are the well known wallpaper groups. The hyperbolic crystallographic groups are not so popular outside mathematical contexts. These groups are the symmetry groups of hyperbolic tessellations and then of some of the works of the famous graphic artist M.C. Escher as the Woodcut "Circle Limit IV".

A first step in the study of planar crystallographic groups is their algebraic classification, i.e. to obtain the isomorphism classes of such groups. In this article we shall be concerned with this first step. The second step is the classification up to conjugation in the group of isometries and this is related to the study of Teichmüller and Moduli spaces ([**17**]).

2010 *Mathematics Subject Classification.* Primary 20H10, 20H15, 30F10.

The second author was partially supported by the project MTM2014-55812.

The isomorphism classes of planar crystallographic groups are obtained using a symbol or signature, composed mainly by integers and with a particular equivalence relation.

Let Δ be an Euclidean or hyperbolic planar crystallographic group. The signature of Δ is a symbol :

$$(g;\sigma;[m_1,...,m_r];\{(n_{11},...,n_{1s_1}),...,(n_{k1},...,n_{ks_k})\})$$

where $g\geq 0$, $m_i\geq 0$ and $n_{ij}\geq 0$ are integers and σ is a sign $+$ or $-$ (see [**16, 21**]). The numbers m_i are called proper periods and the ordered sets of integers $(n_{11},...,n_{1s_1})$ are called period-cycles.

Now we shall describe the equivalence relation between signatures. Two signatures:

$$(g;\sigma;[m_1,...,m_r];\{(n_{11},...,n_{1s_1}),...,(n_{k1},...,n_{ks_k})\})$$

$$(g';\sigma';[m'_1,...,m'_{r'}];\{(n'_{11},...,n'_{1s_1}),...,(n'_{k'1},...,n'_{k's_{k'}})\})$$

are equivalent if:

1. $g=g'$; $\sigma=\sigma'$; $r=r'$; $\{m_1,...,m_r\}=\{m'_1,...,m'_{r'}\}$ (the order is not important); $k=k'$; $s_i=s'_{\delta(i)}$ where δ is a permutation of $\{1,...,k\}$;

2. Let α_i be the cyclic permutation $(1,...,s_i)$, $i=1,...,k$. There exists integers $l_1,...,l_k$ such that:

if $\sigma=\sigma'=+$, either:

- for all i, $(n_{i1},...,n_{is_i})=(n'_{\delta(i)\theta_i(1)},...,n'_{\delta(i)\theta_i(s_i)})$, where $\theta_i=\alpha_i^{l_i}$, $0\leq l_i\leq s_i$,

or

- for all i, $(n_{is_i},...,n_{i1})=(n'_{\delta(i)\theta_i(1)},...,n'_{\delta(i)\theta_i(s_i)})$, where $\theta_i=\alpha_i^{l_i}$, $0\leq l_i\leq s_i$;

if $\sigma=\sigma'=-$, for each i, $i=1,...,k$, either:

- $(n_{i1},...,n_{is_i})=(n'_{\delta(i)\theta_i(1)},...,n'_{\delta(i)\theta_i(s_i)})$, $\theta_i=\alpha_i^{l_i}$, $0\leq l_i\leq s_i$, or
- $(n_{is_i},...,n_{i1})=(n'_{\delta(i)\theta_i(1)},...,n'_{\delta(i)\theta_i(s_i)})$, $\theta_i=\alpha_i^{l_i}$, $0\leq l_i\leq s_i$.

In other words in the orientable case ($\sigma=+$) corresponding pair of period-cycles are all paired in the same cyclic way, all directly or all inversely. In the non-orientable case, some are paired directly and some inversely.

The algebraic classification theorem of Wilkie and Macbeath (see [**16, 21**]) establishes that two planar crystallographic groups are isomorphic if and only if the corresponding signatures are equivalent.

This classification is related with the theory of orbifolds. In fact the signature describes the topological type of the 2-orbifold that appears as quotient of the Euclidean or hyperbolic plane under the action of the crystallographic group. Let Δ be an Euclidean or hyperbolic crystallographic group and $\mathbb{U}^2=\mathbb{E}^2$ or $\mathbb{H}^2$ depending on Δ. The integer g is the genus of the surface $\mathbb{U}^2/\Delta$, $\sigma=+$ or $-$ is the orientability character of such surface, $[m_1,...,m_r]$ is the set of branch indices (periods) of the covering $\mathbb{U}^2\to\mathbb{U}^2/\Delta$ with values in interior points of $\mathbb{U}^2/\Delta$, one period for each branch value, and the ordered sets (period-cycles) of branched indices: $(n_{11},...,n_{1s_1}),...,(n_{k1},...,n_{ks_k})$, correspond to branched values in the k boundary components of $\mathbb{U}^2/\Delta$.

For the case of Euclidean groups we have the 17 types of wall-paper groups. In the Table 1 we show the 17 signatures of crystallographic groups with the corresponding universal notation.

TABLE 1

p1	cm	pm	pg
$(1;+;[-];\{-\})$	$(1;-;[-];\{(-)\})$	$(0;+;\{(-),(-)\})$	$(2;-;\{-\})$
p2	cmm		
$(0;+;[2,2,2,2])$	$(0;+;[2];\{(2,2)\})$		
	pmm	pmg	pgg
	$(0;+;[-];\{(2,2,2,2)\})$	$(0;+;[2,2];\{(-)\})$	$(1;-;[2,2];\{-\})$
p4	p4m	p4g	
$(0;+;[2,4,4];\{-\})$	$(0;+;[-];\{(2,4,4)\})$	$(0;+;[4];\{(2)\})$	
p3	p31m	p3m1	
$(0;+;[3,3,3];\{-\})$	$(0;+;[-];\{(3,3,3)\})$	$(0;+;[3];\{(3)\})$	
p6	p6m		
$(0;+;[2,3,6];\{-\})$	$(0;+;[-];\{(2,3,6)\})$		

Each planar crystallographic group admits a canonical presentation. The geometrical type of the generators and the word expressions of relations of a canonical presentation is given by the signature (see Section 2).

If Γ is a finite index subgroup of a crystallographic group Δ then Γ is also a crystallographic group. From a presentation of Δ and the action of the generators of such presentation on the cosets Δ/Γ, an algebraic method (Reidemeister-Schreier method), provides a non-canonical presentation of Γ. But to obtain the signature of Γ from a non-canonical presentation is not an easy task.

If Δ contains only orientation preserving isometries, there is a direct method due to Singerman [**18**] to obtain the signature of a subgroup Γ of Δ. If Δ is a generic crystallographic group this method was extended by Hoare in [**11**] (with some improvements in [**7**], see also [**12**] and [**13**]). The Hoare method uses a canonical presentation $\mathcal{P}$ of Δ and the coset graph $\mathcal{H}(\Delta,\Gamma,\mathcal{P})$ with vertices the right cosets of Γ in Δ and edges labelled with the generators of $\mathcal{P}$. Some examples of computations using this method are in [**19**], [**7**] and [**6**].

The computation of signatures of subgroups of crystallographic groups is a natural problem from the crystallographic point of view and it has many mathematical applications. Let us mention some ones: finite regular colourations of tessellations, automorphisms of Riemann and Klein surfaces ([**4**]), branch loci of moduli spaces ([**8**] and [**2**]), topological types of real algebraic curves ([**3, 5, 8, 9**]), study of regular maps and hypermaps on surfaces ([**14**]), Belyi curves and Dessins d'enfants theory of Grothendieck ([**15, 20**]).

In case of Euclidean groups the computation of signatures of subgroups is very easy and reduces to know some geometrical properties of the isometries of the subgroup. For instance if we have a subgroup of a Euclidean crystallographic group with signature $(0;+;[-];\{(2,3,6)\})$ ($p6m$) and we know that there are no rotations of order 6 and all the rotations of order 3 are on axes of reflections of the subgroup, we may directly conclude that the subgroup has signature $(0;+;[-];\{(3,3,3)\})$ ($p31m$). In the hyperbolic setting the situation is very involved. For instance, there exist two non-isomorphic subgroups of the same crystallographic group whose usual crystallographic characteristics coincide: the orders of rotations, the conjugacy classes of reflections, the position of the fixed points of rotations on the axis of reflections, etc. In section 4 we present examples of non-isomorphic subgroups of a crystallographic group whose signatures just differ in the direction of the cyclic order of two period cycles. Hence the determination of this direction is a crucial problem.

In [7] we made a program on MatLab system to compute the signature of subgroups of planar crystallographic groups. Now we announce a package in the free software GAP (Groups Algorithms and Programming) for the Hoare method with the modifications in [7]. We present an example of use in Section 5. The code and documentation about the program are in open access in:

e-spacio.uned.es/fez/view/bibliuned:DptoMF-FCIE-Articulos-Acosta-5015

2. Canonical presentation

Every planar crystallographic group Δ has a canonical presentation with canonical generators: $x_1, ..., x_r$ (rotations), $c_{i1}, ..., c_{is_i+1}$, $i = 1, ..., k$ (reflections, we shall call c_{ij}, c_{ij+1} contiguous reflections), $e_1, ..., e_k$ (connecting generators) and a_j, b_j, $j = 1, ..., g$ (translations or hyperbolic isometries, in case of sign $+$ in the signature) or d_j, $j = 1, ..., g$ (glide reflections, in case of sign $-$). The relations are:

(1) $x_i^{m_i}$
(2) $c_{is_i+1}e_ic_{i1}e_i^{-1}$ (connection relation)
(3) c_{ij}^2
(4) $(c_{ij}c_{ij+1})^{n_{ij}}$
(5) $e_1^{-1}....e_k^{-1}x_1^{-1}...x_r^{-1}[a_1,b_1]....[a_g,b_g]$ in case of sign $+$ (long relation)
(6) $e_1^{-1}....e_k^{-1}x_1^{-1}...x_r^{-1}d_1^2...d_g^2$ in case of sign $-$ (long relation)

(see reference [16]).

If Δ is a planar hyperbolic crystallographic group, the hyperbolic area of the surface $\mathbb{H}^2/\Delta$ is given by:

$$S_\Delta = 2\pi\left(\eta g + k - 2 + \sum_{i=1}^{r}\left(1 - \frac{1}{m_i}\right) + \frac{1}{2}\sum_{j=1}^{k}\sum_{l=1}^{s_j}\left(1 - \frac{1}{n_{jl}}\right)\right)$$

being g the genus, $\eta = 2$ for orientable surfaces, $\eta = 1$ for non-orientable ones, and k the number of connected components of the boundary. If Δ is an Euclidean group we have:

$$0 = \eta g + k - 2 + \sum_{i=1}^{r}\left(1 - \frac{1}{m_i}\right) + \frac{1}{2}\sum_{j=1}^{k}\sum_{l=1}^{s_j}\left(1 - \frac{1}{n_{jl}}\right)$$

and the solutions to the above equation provide the 17 wall-papers groups.

A finite index subgroup Γ of Δ is also a crystallographic group and determines a $[\Delta : \Gamma]$-fold morphism $\mathbb{U}^2/\Gamma \to \mathbb{U}^2/\Delta$ ([1]), where $\mathbb{U}$ is $\mathbb{E}^2$ or $\mathbb{H}^2$. The Riemann-Hurwitz formula is satisfied: $S_\Gamma = S_\Delta[\Delta : \Gamma]$.

3. Hoare's algorithm to compute de signature of a subgroup

Let Δ be an NEC group (non-Euclidean crystallographic) group, that is, a crystallographic group in the hyperbolic setting, and let Γ be a subgroup of Δ of index n. The action of Δ on the right cosets of Γ in Δ induces a group homomorphism $\varphi : \Delta \to \Sigma\{1, ..., n\}$, where this last is the symmetric group of n symbols. In this way, $\Gamma = \varphi^{-1}(\mathrm{Stab}(\mathrm{id}))$, and we want to get the signature of Γ from a canonical presentation $\mathcal{P}$ of Δ. To that end we use the Schreier graph $\mathcal{H}(\Delta, \Gamma, \mathcal{P})$ whose vertices are the right cosets of Γ in Δ and whose edges are labelled by the elements of $\mathcal{P}$. Two vertices i, j are joined by an edge with label a, which is a generator of $\mathcal{P}$, if and only if $\varphi(a)(i) = j$. Let's explain $\mathcal{H}(\Delta, \Gamma, \mathcal{P})$ in detail now.

(1) Let x_i be an elliptic canonical generator of $\mathcal{P}$ of order n. To each orbit of length m of $\varphi(x_i)$ corresponds a proper period of order n/m in the signature of Γ (of course if $n = m$ there is no such period). In terms of the graph $\mathcal{H}(\Delta, \Gamma, \mathcal{P})$, these proper periods correspond to cycles of length m in $\mathcal{H}(\Delta, \Gamma, \mathcal{P})$ which consist of edges with label x_i.
(2) Let c be a reflection of $\mathcal{P}$. If $\varphi(c)(i) = i$, we shall say that c produces a reflection $c\_{}_i$ of Γ: the $c\_{}_i$ are the loops of $\mathcal{H}(\Delta, \Gamma, \mathcal{P})$ with label c.
(3) If x, y are contiguous reflection canonical generators of $\mathcal{P}$ such that xy has order n, for each orbit of length m of the group $< \varphi(x), \varphi(y) >$, we can have two cases:
   (a) The orbit has two cosets (i, j) (may be $i = j$) that are fixed by $\varphi(x), \varphi(y)$ (one with by $\varphi(x)$ and the other by $\varphi(y)$, if m odd, or both are fixed by the same element, $\varphi(x)$ or $\varphi(y)$, if m is even). In this case we say that the corresponding reflection generators of Γ are linked and produce a period cycle of order n/m. In terms of $\mathcal{H}(\Delta, \Gamma, \mathcal{P})$ this step is directly interpreted using paths labelled alternatively with two contiguous reflections canonical generators of $\mathcal{P}$ and joining two reflection loops.
   (b) The orbit has not fixed classes by $\varphi(x), \varphi(y)$, then it yields a proper-period $2n/m$. In this case we consider bicoloured cycles (with two contiguous reflection generators of $\mathcal{P}$ labels) in $\mathcal{H}(\Delta, \Gamma, \mathcal{P})$.
(4) Now the cycles of period cycles of the signature of Γ are given by the periods obtained in (3.a) in cycles of linked reflections of Γ. This process give us an order (up to direction) for period-cycles.
(5) The sign in the signature of Γ is $+$ if and only if the vertices of $\mathcal{H}(\Delta, \Gamma, \mathcal{P})$ can be bicoloured, in such a way that edges labelled with conformal generators of $\mathcal{P}$ join vertices with the same colour and the edges labelled with anticonformal generators join vertices with two different colours (except reflection loops) (see [**13**]).
(6) Using Riemann-Hurwitz relation we can compute the genus of $\mathbb{H}^2/\Gamma$.
(7) Assume the sign of the signature of Γ is $+$. In this case we need to find the direction of the cyclic order for the periods in the period-cycles. For this step, we follow the method in [**7**]. We fix an admissible direction for the order of the period-cycles of the signature of Δ. Assume that the reflection generators of $\mathcal{P}$ corresponding to a period-cycle are:

$$c_{i1}, \ldots c_{is_i+1}$$

being $(c_{ij}c_{ij+1})^{n_{ij}} = 1$.

By point (5), there is a bipartition of the vertices in the coset graph $\mathcal{H}(\Delta, \Gamma, \mathcal{P})$, so we can assign a "colour" (black or white) to each vertex. We start on a vertex t of $\mathcal{H}(\Delta, \Gamma, \mathcal{P})$ where there is a reflection generator c_{ij} of $\mathcal{P}$ such that $\varphi(c_{ij})(t) = t$. The edge in $\mathcal{H}(\Delta, \Gamma, \mathcal{P})$ starting in t and with label c_{ij} is a loop and produces a generator c_{ij_t} of Γ.

If t is *white* then we consider the contiguous reflection generator c_{ij+1} of $\mathcal{P}$ and assume that c_{ij_t} is linked (see (3.a)) with other reflection in Γ: $c_{ij+1\_t'}$ or $c_{ij\_t'}$. This linking produces a period $\frac{n_{ij+1}}{m}$ in the period-cycle.

Now if c_{ij_t} is linked with $c_{ij+1\_t'}$ we have that t' is a white vertex and we restart the procedure to obtain the next period $\frac{n_{ij+2}}{m'}$ in the directed order that we are looking for. If c_{ij_t} is linked with $c_{ij\_t'}$ then t' is black and we use the generator c_{ij-1} of $\mathcal{P}$ to obtain the next period.

If t is *black* we start considering the contiguous generator c_{ij-1} of $\mathcal{P}$ to link with $c_{ij-1\_t''}$ or $c_{ij\_t'}$ and to obtain a period $\frac{n_{ij}}{m}$. Now if we have $c_{ij-1\_t''}$ we use c_{ij-2} of $\mathcal{P}$ to obtain the next period and if we have $c_{ij\_t''}$ we are in a white vertex and we use c_{ij+1} of $\mathcal{P}$ to obtain the next period in the cyclic order of the period-cycle.

Note that it is enough to know the order of two periods in each cycle-period to have the direction of the cyclic order.

4. Examples where direction of period-cycles is essential to know if two subgroups of a hyperbolic crystallographic group are isomorphic

In this section we shall describe examples of hyperbolic crystallographic groups Δ admitting two subgroups Γ_1 and Γ_2 that are not isomorphic, although the signatures of Γ_1 and Γ_2 only differ in the direction of the cyclic order of the periods in a period-cycle.

EXAMPLE 4.1. Let Δ be a group with signature

$$(0; +; [2, 2]; \{(6, 6, 24)\})$$

and canonical presentation:

(1) Generators:
   x_1, x_2 rotations of order 2
   c_1, c_2, c_3, c_4 reflections of the period cycle
   e connection generator.

(2) Relations:
   x_i^2, c_j^2
   $(c_1c_2)^6, (c_2c_3)^6, (c_3c_4)^{24}, c_4ec_1e^{-1}$
   $e^{-1}x_1^{-1}x_2^{-1}$.

We consider the homomorphism with transitive image: $\varphi_1 : \Delta \to \Sigma\{1, 2, 3, 4\}$ defined by:

$x_1 \to (1, 4) \quad x_2 \to (1, 4)$
$c_1 \to (1, 3) \quad c_2 \to (1, 3) \quad c_3 \to (1, 2) \quad c_4 \to (1, 3)$
$e \to id$

and $\varphi_2 : \Delta \to \Sigma\{1, 2, 3, 4\}$ defined by:

$x_1 \to (1, 4) \quad x_2 \to (1, 4)$
$c_1 \to (1, 3) \quad c_2 \to (1, 2) \quad c_3 \to (1, 2) \quad c_4 \to (1, 3)$
$e \to id$

We want to compute the signature of the subgroups $\Gamma_1 = \varphi_1^{-1}(\mathrm{Stab}(1))$ and $\Gamma_2 = \varphi_2^{-1}(\mathrm{Stab}(1))$.

In the Figures 4.1 and 4.2 we show the cosets graphs $\mathcal{H}(\Delta, \Gamma_i, \mathcal{P})$, $i = 1, 2$.

Using the method in **[11]** we obtain that the signatures of Γ_1 and Γ_2 are either

$$(0; +; [6, 2, 2, 2, 2]; \{(2, 6, 8), (6, 6, 24)\})$$
$$\text{or } (0; +; [6, 2, 2, 2, 2]; \{(2, 8, 6), (6, 6, 24)\})$$

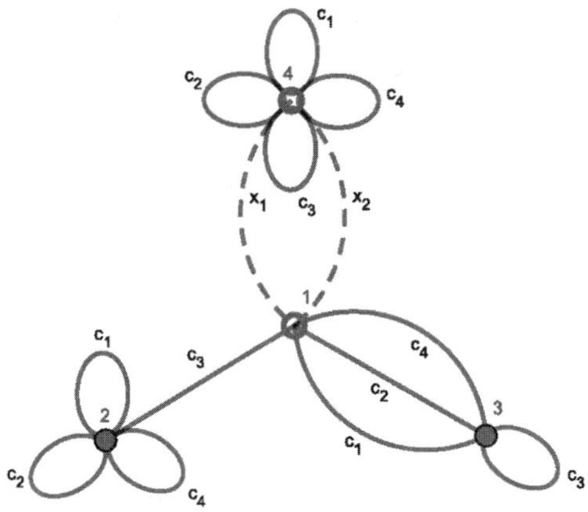

FIGURE 1. $\mathcal{H}(\Delta, \Gamma_1, \mathcal{P})$

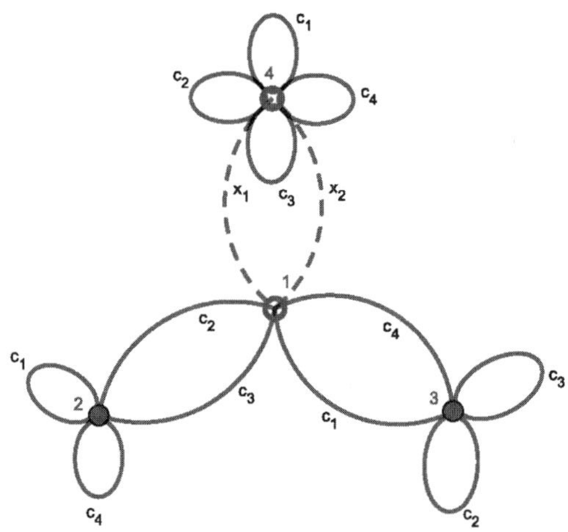

FIGURE 2. $\mathcal{H}(\Delta, \Gamma_2, \mathcal{P})$

In order to determine the cyclic order in the first period-cycles we apply the method described in the point (7) of Section 3. First we shall consider the subgroup Γ_1. The coset graph $\mathcal{H}(\Delta, \Gamma_1, \mathcal{P})$ has four vertices that we label $\{1, 2, 3, 4\}$, such that this labelling produces the monodromy φ_1 (if i and j are two adjacent vertices in $\mathcal{H}(\Delta, \Gamma_1, \mathcal{P})$ by an edge with label $g \in \mathcal{P}$, then $\varphi_1(g)(i) = j$). The set of vertices of the graph $\mathcal{H}(\Delta, \Gamma_1, \mathcal{P})$ has a bipartition $\{1, 4\}$, $\{2, 3\}$ as described in (5) of Section 3, we shall colour the vertices $1, 4$ in white and the vertices $2, 3$ in black.

To have the order of the period cycle $(2, 6, 8)$, we start with the reflection c_3 and the vertex 3 ($\varphi_1(c_3)(3) = 3$). The vertex 3 is black, then we consider now c_2, and we compute the sequence:

$$\varphi_1(c_2)(3) = 1, \ \varphi_1(c_3)(1) = 2, \ \varphi_1(c_2)(2) = 2$$

then we have a period $\frac{\text{order}(c_2 c_3)}{3} = \frac{6}{3} = 2$.

Now, since 2 is black and $\varphi_1(c_1)(2) = 2$, we have a period $\frac{\text{order}(c_1c_2)}{1} = \frac{6}{1} = 6$, then the order of the first period-cycle is $(2, 6, 8)$. Note that the second period has only one direction for the cyclic ordering.

For φ_2 we have:

- $\varphi_2(c_3)(3) = 3$, $\varphi_2(c_2)(3) = 3$, we have to period 6;
- $\varphi_2(c_2)(3) = 3$, $\varphi_2(c_1)(3) = 1$, $\varphi_2(c_2)(1) = 2$, $\varphi_2(c_1)(2) = 2$, we pass to period 2.

Hence the order is $(6, 2, 8)$.

So, the signature of Γ_1 is:

$$(0; +; [6, 2, 2, 2, 2]; \{(2, 6, 8), (6, 6, 24)\})$$

the signature Γ_2 is:

$$(0; +; [6, 2, 2, 2, 2]; \{(2, 8, 6), (6, 6, 24)\})$$

The subgroups Γ_1 and Γ_2 are not isomorphic and their signatures just differ in the direction of the cyclic order of periods of the first period-cycle.

EXAMPLE 4.2. Let Δ be a group with signature

$$(0; +; [-]; \{(6, 6, 24), (6, 6, 24)\})$$

and canonical presentation:

(1) Generators:
$c_{11}, c_{12}, c_{13}, c_{14}$ reflections of first period cycle
$c_{21}, c_{22}, c_{23}, c_{24}$ reflections of second period cycle
e_1, e_2 connection generators.

(2) Relations:
c_{ij}^2
$(c_{11}c_{12})^6, (c_{12}c_{13})^6, (c_{13}c_{14})^{24}, c_{14}e_1c_{11}e_1^{-1}$
$(c_{21}c_{22})^6, (c_{22}c_{23})^6, (c_{23}c_{24})^{24}, c_{24}e_2c_{21}e_2^{-1}$
$e_1^{-1}e_2^{-1}$.

We consider the homomorphism with transitive image: $\varphi_1 : \Delta \to \Sigma\{1, 2, 3, 4\}$ defined by:

$c_{11} \to (1, 3)$ $c_{12} \to (1, 3)$ $c_{13} \to (1, 2)$ $c_{14} \to (1, 3)$
$c_{21} \to (1, 3)$ $c_{22} \to (1, 2)$ $c_{23} \to (1, 4)$ $c_{24} \to (1, 3)$
$e_1 \to id$ $e_2 \to id$

and $\varphi_2 : \Delta \to \Sigma\{1, 2, 3, 4\}$ defined by:

$c_{11} \to (1, 3)$ $c_{12} \to (1, 2)$ $c_{13} \to (1, 2)$ $c_{14} \to (1, 3)$
$c_{21} \to (1, 3)$ $c_{22} \to (1, 2)$ $c_{23} \to (1, 4)$ $c_{24} \to (1, 3)$
$e_1 \to id$ $e_2 \to id$

Using the method in **[11]** and **[7]**, the signature of Γ_1 is:

$$(2; +; [6]; \{(2, 6, 8), (6, 6, 24), (2, 6, 8, 6, 2, 24)\})$$

and the signature of Γ_2 is:

$$(2; +; [6]; \{(2, 8, 6), (6, 6, 24), (2, 6, 8, 6, 2, 24)\})$$

The order of the two first period-cycles is obtained, exactly in the same way that in Example 4.1 and the second period-cycle has the same order for Γ_1 and Γ_2, since φ_1 and φ_2 are defined in the same way for c_{2i}, $i = 1, 2, 3, 4$ and e_2.

As in Example 4.1, the two signatures only differ in the direction of the cyclic order of periods in the first period-cycle of the signature.

EXAMPLE 4.3. Let Δ be a group with signature

$$(0;+;[-];\{(2,3,4),(-),(-)\})$$

and canonical presentation:

(1) Generators:

$c_{11}, c_{12}, c_{13}, c_{14}$ reflections of first period cycle
c_2, c_3 reflections of second and third period cycle
e_1, e_2, e_3 connection generators.

(2) Relations:

c_{1j}^2, c_2^2, c_3^2
$(c_{11}c_{12})^2, (c_{12}c_{13})^3, (c_{13}c_{14})^4, c_{14}e_1c_{11}e_1^{-1}$
$e_1^{-1}e_2^{-1}e_3^{-1}$.

We consider the homomorphism with transitive image: $\varphi_1 : \Delta \to \Sigma\{1,2,3,4\}$ defined by:

$c_{11} \to (2,4)$ $c_{12} \to (2,4)$ $c_{13} \to (2,4)$ $c_{14} \to (2,4)$
$c_2 \to (1,2)$ $c_3 \to (2,3)$
$e_1 \to id$ $e_2 \to id$ $e_3 \to id$

and the homomorphism $\varphi_2 : \Delta \to \Sigma\{1,2,3,4\}$ defined by:

$c_{11} \to (3,4)$ $c_{12} \to (3,4)$ $c_{13} \to (3,4)$ $c_{14} \to (3,4)$
$c_2 \to (1,2)$ $c_3 \to (2,3)$
$e_1 \to id$ $e_2 \to id$ $e_3 \to id$

The signature of Γ_1 is:

$$(0;+;[2,3,4];\{(2,3,4),(2,3,4),(-),(-),(-),(-)\})$$

and the signature of Γ_2 is:

$$(0;+;[2,3,4];\{(2,3,4),(2,4,3),(-),(-),(-),(-)\})$$

In this example the explanation for the cyclic orders of period-cycles is very simple. For the first subgroup the bicolouration of vertices for the coset graph is $\{1,3,4\}$, $\{2\}$ and the non-empty period-cycles correspond to $\varphi_1(c_{1i})(1) = 1$ and $\varphi_1(c_{1i})(3) = 3$, that are both in the same set of the bipartition. Then the order of the two period-cycles is $(2,3,4)$. In the second case the bipartition is $\{1,3\}$, $\{2,4\}$ and we have $\varphi_1(c_{1i})(1) = 1$ and $\varphi_1(c_{1i})(2) = 2$, note that 1 and 2 have different colours. Hence the order is $(2,3,4)$ when we are in the vertex 1 and $(2,4,3)$ when we are in the vertex with label 2.

5. Computing the signature of subgroups using a package of GAP

The NSSC package provides an implementation of the Hoare's algorithm ([**11**]), with the improvements in the calculation of the direction of the period-cycles described in [**7**].

Given a hyperbolic crystallographic group Δ and a representation φ, with transitive image, into the symmetrical group of n elements $\Sigma\{1,...,n\}$, the algorithm calculates the signature of the subgroup $\Gamma = \varphi^{-1}(Stab(1))$.

The NSSC package contains a most important function named *NSSC_CalcMain*. This function takes the information from a record with the data of the signature of Δ and the representation φ and returns a record with the signature of Γ and details on the made computations. The NSSC_CalcMain function is intended to be used in programs that generate data of crystallographic groups and need to compute data of subgroups.

For the convenience of the user that just want to analyse a particular case, the package provides a function called *NSSC_ CalcFile* that uses a text file as input and writes the result in other text file (that could be seen on the screen).

The input file must use GAP syntax since it is read using the standard GAP function *ReadAsFunction.* But it could be easily generated starting from any of the multiple examples provided in the package.

The output file is self-explanatory, and besides the signature of the subgroup contains detailed information of the calculations, although there is an option to get a terse output with just the signature.

EXAMPLE 5.1. An example of the use of the package NSSC follows. First the user must copy the package "nssc"in the directory pkg on GAP folder. The Package is called from GAP using the order:

gap > LoadPackage("nssc");

A file is necessary to provide the input data: signature of the group and monodromy epimorphism. In the documentation of the Package there are some examples to be used as templates to introduce these input data. For this example we use the data in the example descr4.g in the directory pkg/nssc/tst.

Signature of the group Δ:

$$(0; +; [6, 6]; \{(5, 8, 12)\}$$

Monodromy epimorphism $\varphi : \Delta \to \Sigma\{1, ..., 6\}$

$x_1 \to (1,4)(2,6,3)$ $\quad x_2 \to (1,3,4,5,2,6)$
$c_1 \to (1,2)(3,4)$ $\quad c_2 \to (1,3)(2,6)$ $\quad c_3 \to (1,4)(2,6)(3,5)$ $\quad c_4 \to (1,5)(2,3)$
$e \to (3,2,1)(6,5,4)$

First we tell GAP where to look for the input file:
gap> d:=NSSC_TestDir();
Now we use the main NSSC-function:
gap> NSSC_CalcFile("descr4.g":dir:=d,terse);
The result of this function is the signature of the subgroup $\varphi^{-1}(\mathrm{Stab}(1))$:
Signature:

$$(9; -; [3, 2, 6, 8]; \{(5, 2, 2)\}).$$

Acknowledgments

We wish to thank the referees for their corrections and suggestions.

References

[1] Norman L. Alling and Newcomb Greenleaf, *Foundations of the theory of Klein surfaces*, Lecture Notes in Mathematics, Vol. 219, Springer-Verlag, Berlin-New York, 1971. MR0333163

[2] Gabriel Bartolini, Antonio F. Costa, and Milagros Izquierdo, *On the connectivity of branch loci of moduli spaces*, Ann. Acad. Sci. Fenn. Math. **38** (2013), no. 1, 245–258, DOI 10.5186/aasfm.2013.3820. MR3076808

[3] Emilio Bujalance, Francisco Javier Cirre, José Manuel Gamboa, and Grzegorz Gromadzki, *Symmetries of compact Riemann surfaces*, Lecture Notes in Mathematics, vol. 2007, Springer-Verlag, Berlin, 2010. MR2683160

[4] E. Bujalance, A. F. Costa, and D. Singerman, *Application of Hoare's theorem to symmetries of Riemann surfaces*, Ann. Acad. Sci. Fenn. Ser. A I Math. **18** (1993), no. 2, 307–322. MR1234736

[5] Emilio Bujalance, José J. Etayo, José M. Gamboa, and Grzegorz Gromadzki, *Automorphism groups of compact bordered Klein surfaces*, Lecture Notes in Mathematics, vol. 1439, Springer-Verlag, Berlin, 1990. A combinatorial approach. MR1075411

[6] Cortázar I. Superficies de Riemann con morfismos p-gonales irregulares, PhD. Thesis (2015) U.N.E.D. Madrid.

[7] Ismael Cortázar and Antonio F. Costa, *Computing the signatures of subgroups of non-Euclidean crystallographic groups*, J. Algebra **477** (2017), 483–493, DOI 10.1016/j.jalgebra.2017.01.017. MR3614160

[8] Antonio F. Costa and Milagros Izquierdo, *On the connectedness of the locus of real Riemann surfaces*, Ann. Acad. Sci. Fenn. Math. **27** (2002), no. 2, 341–356. MR1922193

[9] Antonio F. Costa and Milagros Izquierdo, *Symmetries of real cyclic p-gonal Riemann surfaces*, Pacific J. Math. **213** (2004), no. 2, 231–243, DOI 10.2140/pjm.2004.213.231. MR2036917

[10] Antonio F. Costa and Milagros Izquierdo, *On real trigonal Riemann surfaces*, Math. Scand. **98** (2006), no. 1, 53–68, DOI 10.7146/math.scand.a-14983. MR2221544

[11] A. H. M. Hoare, *Subgroups of N.E.C. groups and finite permutation groups*, Quart. J. Math. Oxford Ser. (2) **41** (1990), no. 161, 45–59, DOI 10.1093/qmath/41.1.45. MR1044754

[12] A. H. M. Hoare, A. Karrass, and D. Solitar, *Subgroups of* NEC *groups*, Comm. Pure Appl. Math. **26** (1973), 731–744, DOI 10.1002/cpa.3160260515. Collection of articles dedicated to Wilhelm Magnus. MR0335657

[13] A. H. M. Hoare and D. Singerman, *The orientability of subgroups of plane groups*, Groups—St. Andrews 1981 (St. Andrews, 1981), London Math. Soc. Lecture Note Ser., vol. 71, Cambridge Univ. Press, Cambridge, 1982, pp. 221–227, DOI 10.1017/CBO9780511661884.014. MR679163

[14] Milagros Izquierdo and David Singerman, *Hypermaps on surfaces with boundary*, European J. Combin. **15** (1994), no. 2, 159–172, DOI 10.1006/eujc.1994.1018. MR1261062

[15] Bernhard Köck and David Singerman, *Real Belyi theory*, Q. J. Math. **58** (2007), no. 4, 463–478, DOI 10.1093/qmath/ham017. MR2371466

[16] A. M. Macbeath, *The classification of non-euclidean plane crystallographic groups*, Canad. J. Math. **19** (1967), 1192–1205, DOI 10.4153/CJM-1967-108-5. MR0220838

[17] A. M. Macbeath and D. Singerman, *Spaces of subgroups and Teichmüller space*, Proc. London Math. Soc. (3) **31** (1975), no. 2, 211–256, DOI 10.1112/plms/s3-31.2.211. MR0397022

[18] David Singerman, *Subgroups of Fuschian groups and finite permutation groups*, Bull. London Math. Soc. **2** (1970), 319–323, DOI 10.1112/blms/2.3.319. MR0281805

[19] Singerman, D., Watson P., Using Hoare's Theorem to find the signature of a subgroup of an NEC group, Preprint 2014, arXiv:1408.0127.

[20] David Singerman and Jürgen Wolfart, *Cayley graphs, Cori hypermaps, and dessins d'enfants*, Ars Math. Contemp. **1** (2008), no. 2, 144–153. MR2453475

[21] H. C. Wilkie, *On non-Euclidean crystallographic groups*, Math. Z. **91** (1966), 87–102, DOI 10.1007/BF01110157. MR0185013

Departamento de Matemáticas Fundamentales, Facultad de Ciencias, UNED, 28040 Madrid, Spain

Email address: `icortazar3@gmail.com`

Departamento de Matemáticas Fundamentales, Facultad de Ciencias, UNED, 28040 Madrid, Spain

Email address: `acosta@mat.uned.es`

Contemporary Mathematics
Volume **724**, 2019
https://doi.org/10.1090/conm/724/14592

On regular dessins d'enfants with $4g$ automorphisms and a curve of Wiman

Emilio Bujalance, Marston D. E. Conder, Antonio F. Costa, and Milagros Izquierdo

Abstract. In this article we show that with a few exceptions, every regular dessin d'enfant with genus g having exactly $4g$ automorphisms is embedded in Wiman's curve of type II.

1. Introduction

In 1896, Wiman [20] gave two (smooth, irreducible) complex algebraic curves for each genus $g \geq 2$: one with equation $y^2 = x^{2g+1} - 1$ admitting an automorphism of order $4g + 2$, and another with equation $y^2 = x(x^{2g} - 1)$ admitting an automorphism of order $4g$. These curves are known as Wiman's curves of type I and II respectively. In 1997 Kulkarni [17] showed that, with one exception for genus $g = 3$, Wiman's curve of type II is the only Riemann surface of given genus $g \geq 2$ admitting an automorphism of order $4g$, the exception being Picard's curve ($y^3 = x^4 - 1$); see also [16]. Wiman's curves of type II have exactly $8g$ automorphisms, except in the case $g = 2$, when the curve has 48 automorphisms (and is the curve of genus 2 having the maximum number of automorphisms). Recently Bujalance, Costa and Izquierdo [6] showed that for $g \geq 31$ the curves admitting exactly $4g$ automorphisms form an open curve $\mathcal{F}$ in moduli space. (In fact, this is the complex projective line (or Riemann sphere) with three punctures.)

The methods used to prove the results above were combinatorial. By the works of Riemann, Poincaré, Klein and others, every complex (real) algebraic curve can be uniformised by a class of Fuchsian (NEC) groups. This provides a well defined hyperbolic structure on the surface underlying the algebraic curve.

On the other hand, in 1980 Belyi [1] made an influential discovery now known as Belyi's Theorem: a complex curve X is defined on a number field if and only if X is a covering of the projective line ramified at most over three points, say 0, 1 and ∞. The covering map is called the Belyi map. In combinatorial language, a complex curve X is defined over a number field if and only if its uniformising

2010 *Mathematics Subject Classification.* Primary 30F10; Secondary 14H15, 30F60.

Key words and phrases. Riemann surface, Klein surface, Fuchsian group, non-Euclidean crystallographic group, algebraic curve, Teichmüller space, moduli space.

All authors were partially supported by the project MTM2014-55812-P, and the second author was supported by the N.Z. Marsden Fund (project UOA1626).

(surface) Fuchsian group Γ is a subgroup of a triangle group $\Delta(l,m,n)$. The Belyi map induces a cell-decomposition of X: the *dessin d'enfant* $\mathcal{H}$, also called a map (if $m=2$) or a hypermap [**11**]. The pre-images of 0 give the *hypervertices*, the pre-images of 1 the *hyperedges*, and the pre-images of ∞ the *hyperfaces* of $\mathcal{H}$. The genus of $\mathcal{H}$ is the genus of X, and the uniformising group H is called the *hypermap group*. The dessin d'enfant $\mathcal{H}$ is *regular* if the subgroup Γ is normal in $\Delta(l,m,n)$.

Given a regular dessin d'enfant $\mathcal{H}$ on a curve X, then by uniformisation one has $\mathrm{Aut}(\mathcal{H}) \leq \mathrm{Aut}(X)$. This lets us show here that dessins d'enfants have the same property as curves, namely as in the following, which is a generalisation to each genus $g \geq 2$ of an earlier result of Girondo [**12**].

THEOREM 1.1. *For all integer values of $g \geq 2$ other than* 3, 6, 12 *and* 30, *there are exactly two regular dessins d'enfant of genus g with orientation-preserving automorphism group of order $4g$. In the exceptional cases $g=3$, 6, 12 and* 30, *there are one, three, two and two additional dessins respectively. Moreover, for every $g \geq 2$ the regular map $\mathcal{W}_g$ with orientation-preserving automorphism group of order $8g$ corresponding to Wiman's curve of type II with equation $y^2 = x(x^{2g}-1)$ can be obtained as a medial subdivision of each of the two non-sporadic dessins with $4g$ orientation-preserving automorphisms.*

To prove this theorem, we follow closely the methods used in [**6**].

2. Background

2.1. Fuchsian groups and Riemann surfaces. Here we follow [**18**].

A *Fuchsian group* Γ is a discrete group of conformal isometries of the hyperbolic plane $\mathbb{D}$. We shall consider here only Fuchsian groups with compact orbit space $\mathbb{D}/\Gamma$ (which is then a closed surface). If Δ is any such group, then its algebraic structure is determined by its signature

$$(h; m_1, \ldots, m_r). \tag{2.1}$$

The number h is the topological type of $\mathbb{D}/\Gamma$, called the genus of Γ, and the integers $m_i \geq 2$ (for $1 \leq i \leq r$) are the branch indices over points of $\mathbb{D}/\Gamma$ in the natural projection $\pi : \mathbb{D} \to \mathbb{D}/\Gamma$. A Fuchsian group with signature $(g; -)$ is called a *surface Fuchsian group*.

Associated with each Fuchsian group Γ with signature $(h; m_1, \ldots, m_r)$, there exists a *canonical presentation* for Γ, with generators

$x_1, \ldots, x_r$ (elliptic elements) and
$a_1, b_1, \ldots, a_g, b_g$ (hyperbolic elements),

subject to the defining relations

$x_i^{m_i} = 1$ (for $1 \leq i \leq r$), and
$x_1 \ldots x_r\, e_1 \ldots e_k\, a_1 b_1 a_1^{-1} b_1^{-1} \ldots a_h b_h a_h^{-1} b_h^{-1} = 1.$

In the rest of this paper, we will denote by $\Delta(l,m,n)$ a Fuchsian group with signature $(0; l, m, n)$, otherwise known as the ordinary (l,m,n) triangle group. This has the somewhat simpler presentation $\langle\, x, y \mid x^l = y^m = (xy)^n = 1 \,\rangle$.

The hyperbolic area of an arbitrary fundamental region of a Fuchsian group Γ with signature (2.1) is given by

$$\mu(\Delta) = 2\pi \left(2h - 2 + \sum_{i=1}^{r} \left(1 - \frac{1}{m_i} \right) \right). \tag{2.2}$$

Furthermore, any discrete group Δ of conformal isometries of $\mathbb{D}$ containing Γ as a subgroup of finite index is also a Fuchsian group, and the hyperbolic area of a fundamental region for Λ is given by the Riemann-Hurwitz formula:

$$|\Delta : \Gamma| = \mu(\Gamma)/\mu(\Delta). \tag{2.3}$$

In particular, if Γ is a surface Fuchsian group of genus g then $\mu(\Gamma) = 2\pi(2g-2)$ and hence the Riemann-Hurwitz formula becomes

$$2g - 2 = |\Delta : \Gamma| \left(2h - 2 + \sum_{i=1}^{r} \left(1 - \frac{1}{m_i}\right)\right). \tag{2.4}$$

A Riemann surface is a surface endowed with a complex analytical structure. There is a well-known functorial equivalence between Riemann surfaces and complex algebraic smooth curves.

Let X be a compact Riemann surface of genus $g > 1$. Then there exists a surface Fuchsian group Γ such that $X = \mathbb{D}/\Gamma$, and if G is any group of automorphisms of X, then there exists a Fuchsian group Δ containing Γ and a *surface epimorphism* $\theta : \Delta \to G$ such that $\ker\theta = \Gamma$. This epimorphism θ is the monodromy of the regular (orbifold-)covering $\mathbb{D}/\Gamma \to \mathbb{D}/\Delta$. In particular, the full automorphism group $\mathrm{Aut}(X)$ is isomorphic to Δ/Γ for some Fuchsian group Δ containing Γ.

In general, given Fuchsian groups Λ and Δ with $\Lambda \leq \Delta$, Singerman's Theorem (in [**18**]) tells us that the structure of Λ (and hence also of $\mathbb{D}/\Gamma$) is determined by the structure of Δ and the monodromy $\theta : \Delta \to \Sigma_{|\Delta:\Lambda|}$, where $\Sigma_{|\Delta:\Lambda|}$ denotes the symmetric group on the cosets of Λ in Δ. In fact θ is a transitive representation, and Λ is the pre-image under θ of the stabiliser $\mathrm{Stab}(1)$ of the trivial coset.

2.2. Dessins d'enfants, maps and hypermaps. Here we follow the seminal papers on maps and hypermaps on Riemann surfaces by Jones and Singerman [**15**], and Corn and Singerman [**11**]; see also [**13**].

Belyi's Theorem (from Belyi's influential paper [**1**] in 1980) states that a plane complex curve X is defined over a number field if and only if there is a finite N-sheeted covering $\beta : X \to \widehat{\mathbb{C}}$ of the projective line ramified on at most three points $\{0, 1, \infty\}$. The meromorphic function β is called the *Belyi function.*

Translating this into the world of Fuchsian groups and hyperbolic 2-orbifolds, we have an orbifold-covering $\beta : \mathbb{D}/\Gamma \to \mathbb{D}/\Delta(l,m,n)$, where $\frac{1}{l} + \frac{1}{m} + \frac{1}{n} < 1$. The meromorphic function β induces a cell-decomposition $\mathcal{H}$ of the Riemann surface X called a *dessin d'enfant.* In general this is a *hypermap*, with the pre-images of 0 providing the *hypervertices*, the pre-images of 1 the *hyperedges*, and the pre-images of ∞ the *hyperfaces*. It can also be viewed as a bipartite graph, with 'black' vertices representing the hypervertices, and 'white' vertices representing the hyperedges, and edges between them representing the pre-images of the line segment $[0,1]$. Also if $l = 2$ then this hypermap is a *map*. From now on, we will use the terms dessin d'enfant and hypermap interchangeably. The order of the parameters l, m, n is not important for dessins d'enfants, but we will usually suppose that $l \leq m \leq n$.

The dessin $\mathcal{H}$ is said to have *type* (l,m,n), and if $\Gamma = H = \theta_\beta^{-1}(\mathrm{Stab}(1))$ then $\mathcal{H}$ has monodromy $\theta_\beta : \Delta(l,m,n) \to G$, with image G being a subgroup of $\Sigma_{|\Delta:H|}$ called the *monodromy group* of $\mathcal{H}$ and denoted by $\mathrm{Mon}(\mathcal{H})$. In particular, G has a presentation of the form $\langle\, a, s \mid a^l = s^m = (as)^n = \cdots = 1 \,\rangle$. We will be interested only when $H = \Gamma$ is a surface Fuchsian group, and in such cases $H = \Gamma$ is called the *hypermap group*, and is the uniformising group (and fundamental group) of the Riemann surface X. Note that $\mathrm{Mon}(\mathcal{H})$ is the monodromy group of the covering $\mathbb{D}/\Gamma \to \mathbb{D}/\Delta(l,m,n)$. Cycles of the permutation a are the cycles around hypervertices, while those of s are the cycles around (hyper)edges, and those of as are the cycles around hyperfaces, consistent with the orientation of X.

Two dessins d'enfants of type (l,m,n) are isomorphic if their hypermap groups are conjugate in $\Delta(l,m,n)$, in which case they define the same complex structure of $X = \mathbb{H}/H = \mathbb{H}/\Gamma$. Also note that for any dessin d'enfant $\mathcal{H}$ on a Riemann surface $X = \mathbb{H}/\Gamma$, one has $\mathrm{Aut}(\mathcal{H}) \leq \mathrm{Aut}(X)$. In particular $\mathrm{Aut}(\mathcal{H}) \leq \mathrm{Mon}(\mathcal{H}) = G$.

A dessin d'enfant $\mathcal{H}$ with hypermap group H is called *regular* if $\mathrm{Aut}(\mathcal{H})$ acts transitively on the cosets of H, so that $\mathrm{Aut}(\mathcal{H}) = \mathrm{Mon}(\mathcal{H}) = G$. In that case, $H = \Gamma$ is a normal subgroup of $\Delta(l,m,n)$, and G is isomorphic to $\Delta(l,m,n)/\Gamma$ and hence to a subgroup of $\mathrm{Aut}(X)$.

A regular dessin d'enfant $\mathcal{H}$ with monodromy group $G = \langle a, s\rangle \cong \Delta(l,m,n)/\Gamma$ is said to be *reflexible* if it is isomorphic to its mirror image, in which case the group G has an automorphism taking $a \mapsto a^{-1}$ and $s \mapsto s^{-1}$; and otherwise $\mathcal{H}$ is said to be *chiral*. Equivalently, a dessin d'enfant is reflexible if and only if it is embedded in a *symmetric* Riemann surface (which means that the surface admits an anti-conformal automorphism of order 2, called an anti-conformal involution or a *symmetry* of the surface); see [**3**]. Symmetric Riemann surfaces are also called *real* Riemann surfaces, because they correspond to real algebraic curves.

Finally, we explain how to construct a *medial* (or *medial subdivision*) $\mathrm{Med}(\mathcal{H})$ of a regular dessin or hypermap $\mathcal{H}$ of type (m,m,n), as in [**12**]. Every black or white vertex of the bipartite graph associated with $\mathcal{H}$ becomes a white vertex of $\mathrm{Med}(\mathcal{H})$, and the black vertices of $\mathrm{Med}(\mathcal{H})$ are taken as the midpoints of edges of $\mathcal{H}$. In this way, every black vertex of $\mathrm{Med}(\mathcal{H})$ is joined to just two white vertices (incident in $\mathcal{H}$ with the edge it came from), while each white vertex is joined with m black vertices, coming from its incident edges in $\mathcal{H}$. The medial $\mathrm{Med}(\mathcal{H})$ is then a regular hypermap (indeed a regular map) of type $(2,m,2n)$, and $\mathrm{Aut}(\mathcal{H})$ is isomorphic to a subgroup of index 2 in $\mathrm{Aut}(\mathrm{Med}(\mathcal{H}))$; see [**12**].

3. Regular dessins d'enfants with $4g$ automorphisms

In this paper we are interested in finding all regular hypermaps with automorphism group of order $4g$, where g is the genus. We will identify a hypermap $\mathcal{H}$ with its monodromy (or algebraic hypermap) $\theta : \Delta(l,m,n) \to G$; see [**8, 11, 14, 15**].

PROPOSITION 3.1. Every regular hypermap of genus $g \geq 2$ with automorphism group of order $4g$ is isomorphic to one of those described in the list below:

(1) $\theta : \Delta(2,4g,4g) \to C_{4g}$ for any $g \geq 2$;
(2) $\theta : \Delta(4,4,2g) \to C_{2g} \diamond C_4$ (central product), for any $g \geq 2$;
(3) $\theta : \Delta(3,4,12) \to C_{12}$, for $g = 3$;
(4) $\theta : \Delta(3,8,8) \to C_3 \rtimes C_8$, for $g = 6$;
(5) $\theta : \Delta(4,6,6) \to \mathrm{SL}(2,3)$, for $g = 6$;
(6) $\theta : \Delta(4,6,6) \to D_4 \times C_3$, for $g = 6$;
(7) $\theta : \Delta(4,6,8) \to \langle 2,3,4\rangle$ (the binary octahedral group), for $g = 12$;
(8) $\theta : \Delta(4,6,8) \to (C_3 \rtimes C_8) \rtimes C_2$, for $g = 12$;
(9) $\theta : \Delta(4,6,10) \to \mathrm{SL}(2,5)$, when $g = 30$;
(10) $\theta : \Delta(4,6,10) \to C_{15} \rtimes D_4$, for $g = 30$.

Moreover, every one of the above hypermaps is reflexible.

Note that in items (5) and (9), the group G happens to be the binary tetrahedral group and the binary icosahedral group respectively, just as G is the binary octahedral group in item (7).

PROOF. By an easy calculation using the Riemann-Hurwitz formula 2.4, all possible triples (l,m,n) and corresponding genera g with $|G| = 4g$ are given in Table 1 below.

TABLE 1. Triples (l,m,n) giving $|G| = 4g$.

(l,m,n)	genus g	(l,m,n)	genus g	(l,m,n)	genus g
$(2,4g,4g)$	any $g \geq 2$	$(3,6,2g)$	any $g \geq 2$	$(4,4,2g)$	any $g \geq 2$
$(6,6,5)$	15	$(6,6,4)$	6	$(6,6,3)$	3
$(6,11,4)$	66	$(6,10,4)$	30	$(6,9,4)$	18
$(6,8,4)$	12	$(6,4,4)$	3	$(5,19,4)$	190
$(5,18,4)$	90	$(5,16,4)$	40	$(5,15,4)$	30
$(5,12,4)$	15	$(5,10,4)$	10	$(5,5,8)$	20
$(5,5,9)$	45	$(5,5,5)$	5	$(4,7,9)$	126
$(4,7,8)$	28	$(4,7,7)$	14	$(3,4,12)$	3
$(3,11,11)$	33	$(3,11,12)$	66	$(3,11,13)$	429
$(3,10,14)$	105	$(3,10,12)$	30	$(3,10,10)$	15
$(3,9,17)$	153	$(3,9,16)$	72	$(3,9,15)$	45
$(3,9,12)$	18	$(3,9,9)$	9	$(3,8,23)$	276
$(3,8,22)$	132	$(3,8,21)$	84	$(3,8,20)$	60
$(3,8,18)$	36	$(3,8,16)$	24	$(3,8,12)$	12
$(3,8,8)$	6	$(3,7,41)$	841	$(3,7,40)$	420
$(3,7,39)$	273	$(3,7,36)$	126	$(3,7,35)$	105
$(3,7,28)$	42	$(3,7,21)$	21		

For example, if $2 = l \leq m \leq n$ then $2g-2 = 4g(0-2+3-(\frac{1}{2}+\frac{1}{m}+\frac{1}{n}))$, from which it follows that $4g(\frac{1}{m}+\frac{1}{n}) = 2$, and then since each of m and n must divide $|G| = 4g$ we find the only solution is $(m,n) = (4g,4g)$. Similarly, if $3 = l \leq m \leq n$ then 3 divides $|G| = 4g$ and $2g-2 = 4g(0-2+3-(\frac{1}{3}+\frac{1}{m}+\frac{1}{n})) = \frac{8g}{3} - 4g(\frac{1}{m}+\frac{1}{n})$, from which it follows that $4g(\frac{1}{m}+\frac{1}{n}) = 2+\frac{2g}{3}$, and hence either $\{m,n\} = \{6,2g\}$, or (m,n) is one of a number of small sporadic possibilities as given in the table.

Next, for each candidate for the parameters l, m, n and g, we need to check if there exists an epimorphism $\theta : \Delta(l,m,n) \to G$ to some group G of order $4g$.

This is easy in the first case, where $(l,m,n) = (2,4g,4g)$, because the image as of the element xy of $\Delta(2,4g,4g) = \langle\, x,y \mid x^2 = y^{4g} = (xy)^{4g} = 1\,\rangle$ has order $4g$ and so G is cyclic, and then the image a of x must be $(as)^{2g}$, and this determines the epimorphism θ uniquely. In the second case, where $(l,m,n) = (3,6,2g)$, the image as of xy has order $2g$ and so generates a subgroup of index 2 in G, but then that subgroup must contain a (since it has odd order 3) and hence also $s = a^{-1}as$, which is impossible since θ is surjective. In the third case, where $(l, m, n) = (4, 4, 2g)$, the element as generates a cyclic subgroup of index 2 containing both a^2 and s^2, and it follows that $a^2 = (as)^g = s^2$, making G a central product of C_{2g} and C_4.

These three cases were also studied in [**6**], and they give items (1) and (2) in the statement of the Proposition.

Type $(3, 4, 12)$ was dealt with in [**17, 20**] when considering Picard's curve of genus 3, and gives item (3). The other sporadic cases can be handled using the `LowIndexNormalSubgroups` facility in the MAGMA computation system [**2**] to determine whether or not the relevant triangle group $\Delta(l, m, n)$ has a smooth quotient of the expected order. This gives the remaining items (4) to (10). For the cases with genus $g \le 101$, the required computations were already done some years ago by the second author in the search for regular maps and hypermaps; see [**8–10**].

Presentations for the group G in terms of the generating pair $(a, s) = (x^\theta, y^\theta)$ in the ten items in the resulting list (for this Proposition) are as follows:

(1) $G = \langle\, a, s \mid a^2 = 1,\ a = (as)^{2g} \,\rangle \cong C_{4g}$ for every $g \ge 2$;
(2) $G = \langle\, a, s \mid a^4 = 1,\ a^2 = s^2 = (as)^g \,\rangle \cong C_{2g} \diamond C_4$ for every $g \ge 2$;
(3) $G = \langle\, a, s \mid a^3 = s^4 = [a, s] = 1 \,\rangle \cong C_{12}$;
(4) $G = \langle\, a, s \mid a^3 = s^8 = 1,\ s^{-1}as = a^{-1} \,\rangle \cong C_3 \rtimes C_8$;
(5) $G = \langle\, a, s \mid a^4 = 1,\ a^2 = s^3 = (as)^3 \,\rangle \cong \mathrm{SL}(2, 3)$;
(6) $G = \langle\, a, s \mid a^4 = s^6 = 1,\ s^2 = (as)^2 \,\rangle \cong D_4 \times C_3$;
(7) $G = \langle\, a, s \mid a^4 = 1,\ a^2 = s^3 = (as)^4 \,\rangle \cong \langle 2, 3, 4\rangle$;
(8) $G = \langle\, a, s \mid a^4 = s^6 = a^{-1}s^2as^2 = as^{-1}a^{-1}sasas^{-1} = 1 \,\rangle \cong (C_3 \rtimes C_8) \rtimes C_2$;
(9) $G = \langle\, a, s \mid a^4 = 1,\ a^2 = s^3 = (as)^5 \,\rangle \cong \mathrm{SL}(2, 5)$;
(10) $G = \langle\, a, s \mid a^4 = s^6 = (as)^{10} = [a^2, s] = a^{-1}s^2as^2 = 1 \,\rangle \cong C_{15} \rtimes D_4$.

Note that in many cases at least one of the relations $a^l = s^m = (as)^n = 1$ is missing but still holds in the group G, and is redundant. Similarly, in item (8) the relation $[a^2, s] = 1$ holds in G but is redundant.

It is now an easy exercise to verify that in each of the above cases, the group G admits an automorphism taking $a \mapsto a^{-1}$ and $s \mapsto s^{-1}$, and hence the associated hypermap $\mathcal{H}$ is reflexible, as required. (In some cases this also follows from the content of [**6, 7, 9, 10**].) □

Next, we consider in more detail the two infinite families of regular dessins given in items (1) and (2) of Proposition 3.1.

Item (1) is a family of 'cyclic' regular maps $\mathcal{M}_g$ of type $(4g, 4g)$, each with monodromy $\theta_1 : \Delta(2, 4g, 4g) \to C_{4g}$, for all $g \ge 2$. The monodromy group $\mathrm{Mon}(\mathcal{M}_g)$ can be taken as the permutation group of degree $4g$ generated by the permutations

$$a = (1, 2g+1)(2, 2g+2)(3, 2g+3)\dots(2g, 4g) \ \text{ and } \ as = (1, 2, 3, \dots, 4g).$$

Item (2) is a family of regular hypermaps $\mathcal{H}_g$ of type $(4, 4, 2g)$, each with monodromy $\theta_2 : \Delta(4, 4, 2g) \to C_{2g} \diamond C_4$, and with $\mathrm{Mon}(\mathcal{H}_g)$ generated by

$$\begin{aligned} a = {} & (1, 2g+1, g+1, 3g+1)(2, 4g, g+2, 3g)(3, 4g-1, g+3, 3g-1)\dots \\ & (g-1, 3g+3, 2g-1, 2g+3)(g, 3g+2, 2g, 2g+2). \end{aligned}$$

$$as = (1, 2, 3, \dots, 2g-1, 2g)(2g+1, 2g+2, \dots 4g-1, 4g) \ \text{ and}$$

the associated signatures $(0; 2, 4g, 4g)$ and $(0; 4, 4, 2g)$ are both in Singerman's list [**19**] of non-maximal signatures for Fuchsian groups, and each forms an index 2 'normal' pair with the signature $(0; 2, 4, 4g)$, for every $g \geq 2$. For genus $g \geq 3$ the signature $(0; 2, 4, 4g)$ is maximal, while for $g = 2$, Singerman's list of non-maximal signatures includes the pair $((0; 2, 4, 8), (2, 3, 8))$ as well.

The signature $(0; 2, 4, 4g)$ is closed related to Wiman's curve of type II with equation $y^2 = x(x^{2g} - 1)$ mentioned in Section 1. In [**17**] it was shown that this curve $\mathcal{W}_g$ is determined by a regular map of type $(4, 4g)$ with automorphism group G of order $8g$. In this case G is isomorphic to the semi-direct product $C_{4g} \rtimes_{2g-1} C_2$, with presentation $\langle\, a, s \mid a^2 = s^4 = (as)^{4g} = 1,\ a(as)a = (as)^{2g-1} \,\rangle$, realisable by the permutations

$$\begin{aligned} a = {} & (1, 4g+1)(3, 8g-1)(5, 8g-3)(7, 8g-5)\dots(4g-3, 4g+5)(4g-1, 4g+3) \\ & (2, 6g)(4, 6g-2)(6, 6g-4)(8, 6g-6)\dots(2g-2, 4g+4)(2g, 4g+2) \\ & (2g+2, 8g)(2g+4, 8g-2)(2g+6, 8g-4)\dots(4g-2, 6g+4)(4g, 6g+2) \end{aligned}$$

and

$$as = (1, 2, 3, \dots, 4g-2, 4g-1, 4g)(4g+1, 4g+2, 4g+3, \dots, 8g-2, 8g-1, 8g).$$

Wiman's curve of type II for genus $g = 2$ is also known as Bolza's curve, and is determined by the regular map $\mathcal{W}_2$ of type $(3, 8)$ with automorphism group $GL(2, 3)$ of order 48, having presentation $\langle\, a, s \mid a^2 = s^3 = (as)^8 = (asasas^{-1})^2 = 1 \,\rangle$.

In [**12**] Girondo showed that the dessin associated with Wiman's curve of genus 2 can be constructed as a medial of each of the dessins $\mathcal{M}_2$ and $\mathcal{H}_2$ (defined above). We can now complete the proof of Theorem 1.1, which generalises Girondo's discovery to every genus $g \geq 2$.

PROOF. By Proposition 3.1 and the comments following it, we need only show that the epimorphisms $\theta_1 : \Delta(2, 4g, 4g) \to C_{4g}$ and $\theta_2 : \Delta(4, 4, 2g) \to C_{2g} \diamond C_4$ given earlier both extend to an epimorphism $\theta : \Delta(2, 4, 4g) \to C_{2g} \rtimes_{2g-1} C_2$.

Such an extension of θ_1 was proved by Kulkarni in [**17**], and also by Bujalance and Conder in the final section of [**4**], and both extensions were proved to exist by Bujalance, Costa and Izquierdo in [**6**]. Here we give a direct verification, by showing that the epimorphism $\theta : \Delta(2, 4, 4g) \to C_{4g} \rtimes_{2g-1} C_2$ detemined by the presentation $\langle\, a, s \mid a^2 = s^4 = (as)^{4g} = 1,\ a(as)a = (as)^{2g-1} \,\rangle$ for the group $G = C_{2g} \rtimes_{2g-1} C_2$ restricts to each of the unique epimorphisms θ_1 and θ_2, using material from [**5**].

Before doing that, we note that s^2 is an involution in the index 2 subgroup generated by as, and so $s^2 = (as)^{2g} = (as)^{-2g}$.

Now let x, y and z be the standard generators for $\Delta = \Delta(2, 4, 4g)$, satisfying $x^2 = y^4 = z^{4g} = xyz = 1$. Then we can proceed as follows:

Case (1). By case N8 in [**5**, Section 3], there is a unique Fuchsian subgroup of index 2 in Δ with signature $(0; 2, 4g, 4g)$, namely the subgroup Λ_1 generated by y^2 and z^{-1}. The images of these elements in $G = C_{2g} \rtimes_{2g-1} C_2$ are s^2 and as, which generate a cyclic group of order $4g$, and hence we have a restriction to the given epimorphism $\theta_1 : \Delta(2, 4g, 4g) \to C_{4g}$. In particular, also the map $\mathcal{W}_g$ of type $(2, 4, 4g)$ corresponding to Wiman's curve of type II and genus g is obtained from the map $\mathcal{M}_g$ of type $(2, 2g, 2g)$ by a $(1, \infty)$-subdivision.

Case (2). By a different application of case N8 in [**5**, Section 3], there is a unique Fuchsian subgroup of index 2 in Δ with signature $(0; 4, 4, 2g)$, namely the subgroup Λ_2 generated by y and z^2. The images of these elements in $G = C_{2g} \rtimes_{2g-1} C_2$

are s and $(as)^{-2}$, which generate a central product of C_4 and C_{2g} of order $4g$ (with the involution $s^2 = ((as)^{-2})^g$ generating the centre). Hence we also have a restriction to the given epimorphism $\theta_2 : \Delta(4,4,2g) \to C_{2g} \diamond C_4$. In particular, also the map $\mathcal{W}_g$ of type $(2,4,4g)$ corresponding to Wiman's curve is obtained from the hypermap $\mathcal{H}_g$ of type $(4,4,2g)$ by a $(0,1)$-subdivision.

This completes the proof of Theorem 1.1. □

References

[1] G. V. Belyĭ, *Galois extensions of a maximal cyclotomic field* (Russian), Izv. Akad. Nauk SSSR Ser. Mat. **43** (1979), no. 2, 267–276, 479. MR534593

[2] W. Bosma, J. Cannon, and C. Playoust, *The Magma algebra system. I. The user language*, J. Symbolic Comput. **24** (1997), no. 3-4, 235–265, DOI 10.1006/jsco.1996.0125. Computational algebra and number theory (London, 1993). MR1484478

[3] A. Breda D'Azevedo, G. Jones, R. Nedela, and M. Škoviera, *Chirality groups of maps and hypermaps*, J. Algebraic Combin. **29** (2009), no. 3, 337–355, DOI 10.1007/s10801-008-0138-z. MR2496311

[4] E. Bujalance and M. Conder, *On cyclic groups of automorphisms of Riemann surfaces*, J. London Math. Soc. (2) **59** (1999), no. 2, 573–584, DOI 10.1112/S0024610799007115. MR1709666

[5] E. Bujalance, F. J. Cirre, and M. Conder, *On extendability of group actions on compact Riemann surfaces*, Trans. Amer. Math. Soc. **355** (2003), no. 4, 1537–1557, DOI 10.1090/S0002-9947-02-03184-7. MR1946404

[6] E. Bujalance, A. F. Costa, and M. Izquierdo, *On Riemann surfaces of genus g with $4g$ automorphisms*, Topology Appl. **218** (2017), 1–18, DOI 10.1016/j.topol.2016.12.013. MR3598603

[7] E. Bujalance, J. J. Etayo, J. M. Gamboa, and G. Gromadzki, *Automorphism groups of compact bordered Klein surfaces*, Lecture Notes in Mathematics, vol. 1439, Springer-Verlag, Berlin, 1990. A combinatorial approach. MR1075411

[8] M. D. E. Conder, *Regular maps and hypermaps of Euler characteristic −1 to −200*, J. Combin. Theory Ser. B **99** (2009), no. 2, 455–459, DOI 10.1016/j.jctb.2008.09.003. MR2482963

[9] Conder, M.: `www.math.auckland.ac.nz/~conder/OrientableProperHypermaps101.txt`.

[10] Conder, M.: `www.math.auckland.ac.nz/~conder/ChiralProperHypermaps101.txt`.

[11] D. Corn and D. Singerman, *Regular hypermaps*, European J. Combin. **9** (1988), no. 4, 337–351, DOI 10.1016/S0195-6698(88)80064-7. MR950053

[12] E. Girondo, *Multiply quasiplatonic Riemann surfaces*, Experiment. Math. **12** (2003), no. 4, 463–475. MR2043996

[13] E. Girondo and G. González-Diez, *Introduction to compact Riemann surfaces and dessins d'enfants*, London Mathematical Society Student Texts, vol. 79, Cambridge University Press, Cambridge, 2012. MR2895884

[14] M. Izquierdo and D. Singerman, *Hypermaps on surfaces with boundary*, European J. Combin. **15** (1994), no. 2, 159–172, DOI 10.1006/eujc.1994.1018. MR1261062

[15] G. A. Jones and D. Singerman, *Theory of maps on orientable surfaces*, Proc. London Math. Soc. (3) **37** (1978), no. 2, 273–307, DOI 10.1112/plms/s3-37.2.273. MR0505721

[16] R. S. Kulkarni, *A note on Wiman and Accola-Maclachlan surfaces*, Ann. Acad. Sci. Fenn. Ser. A I Math. **16** (1991), no. 1, 83–94, DOI 10.5186/aasfm.1991.1615. MR1127698

[17] R. S. Kulkarni, *Riemann surfaces admitting large automorphism groups*, Extremal Riemann surfaces (San Francisco, CA, 1995), Contemp. Math., vol. 201, Amer. Math. Soc., Providence, RI, 1997, pp. 63–79, DOI 10.1090/conm/201/02610. MR1429195

[18] D. Singerman, *Subgroups of Fuschian groups and finite permutation groups*, Bull. London Math. Soc. **2** (1970), 319–323, DOI 10.1112/blms/2.3.319. MR0281805

[19] D. Singerman, *Finitely maximal Fuchsian groups*, J. London Math. Soc. (2) **6** (1972), 29–38, DOI 10.1112/jlms/s2-6.1.29. MR0322165

[20] Wiman, A. Über die hyperelliptischen Curven und diejenigen von Geschlechte $p = 3$, welche eindeutige Tiansformationen in sich zulassen. *Bihang till Kongl. Svenska vetenskaps-akademiens handlingar, Stockholm* **21** (1895), 1–23.

Departamento de Matemáticas Fundamentales, Facultad de Ciencias, UNED, Senda del rey, 9, 28040 Madrid, Spain
Email address: ebujalance@mat.uned.es

Department of Mathematics, University of Auckland, Private Bag 92019, Auckland 1142, New Zealand
Email address: m.conder@auckland.ac.nz

Departamento de Matemáticas Fundamentales, Facultad de Ciencias, UNED, Senda del rey, 9, 28040 Madrid, Spain
Email address: acosta@mat.uned.es

Matematiska institutionen, Linköpings Universitet, 581 83 Linköping, Sweden
Email address: milagros.izquierdo@liu.se

Contemporary Mathematics
Volume **724**, 2019
https://doi.org/10.1090/conm/724/14593

An explicit descent of real algebraic varieties

Rubén A. Hidalgo

ABSTRACT. Let X be an smooth complex affine algebraic variety admitting a symmetry L, that is, an antiholomorphic automorphism of order two. If both, X and L are defined over $\overline{\mathbb{Q}}$, then Koeck, Lau and Singerman showed the existence of a complex smooth algebraic variety Z admitting a symmetry T, both defined over $\mathbb{R} \cap \overline{\mathbb{Q}}$, and of an isomorphism $R : X \to Z$ so that $R \circ L \circ R^{-1} = T$. The provided proof is existential and, if explicit equations for X and L are given over $\overline{\mathbb{Q}}$, then it is not described how to get the explicit equations for Z and T over $\mathbb{R} \cap \overline{\mathbb{Q}}$. In this paper we provide an explicit rational map R defined over $\overline{\mathbb{Q}}$ so that $Z = R(X)$ is defined over $\mathbb{R} \cap \overline{\mathbb{Q}}$, $R : X \to Z$ is an isomorphisms and $T = R \circ L \circ R^{-1}$ being the usual conjugation map.

1. Introduction

Let $X \subset \mathbb{C}^n$ be an smooth complex affine algebraic variety, say defined by the polynomials $P_1, \ldots, P_m \in \mathbb{C}[x_1, \ldots, x_n]$. If $J_r(x_1, \ldots, x_r) = (\overline{x_1}, \ldots, \overline{x_r})$, the usual conjugation map on $\mathbb{C}^r$, then $\overline{X} = J_n(X)$ is defined by the polynomials $\overline{P_1}, \ldots, \overline{P_m} \in \mathbb{C}[x_1, \ldots, x_n]$, where $\overline{P_j} = J_1 \circ P_j \circ J_n$. By a *symmetry* of X we mean an antiholomorphic automorphism $L : X \to X$ of order two with $J_n \circ L : X \to \overline{X}$ being a biregular isomorphism; in this case, we say that X is *symmetric* and that the pair (X, L) is a *real algebraic variety.* Note that $J_n \circ L$ is a rational map defined over every point of X and its inverse is a rational map defined over every point of $\overline{X}$.

A symmetric algebraic variety may have different symmetries, even different conjugacy classes of them (inside its group of biholomorphic automorphisms). In complex dimension one (i.e., Riemann surfaces) the number of (conjugacy classes of) symmetries is well known (see, for instance, [**2**, **3**]).

Two real algebraic varieties (X, L) and (Y, T) are *isomorphic* if there is a biholomorphism $R : X \to Y$, being also a biregular isomorphism between algebraic varieties, with $T = R \circ L \circ R^{-1}$.

Let $\mathcal{Q}$ be a subfield of $\mathbb{C}$ and (X, L) a real algebraic variety. We say that (X, L) is (i) *defined* over $\mathcal{Q}$ if both, X and L, are defined over $\mathcal{Q}$, and (ii) *definable* over $\mathcal{Q}$ if there is a real algebraic variety (Y, T) defined over $\mathcal{Q}$ and isomorphic to (X, L).

2010 *Mathematics Subject Classification.* Primary 14E05, 14A10, 14P05.
Key words and phrases. Real algebraic varieties, fields of definition.
The author was partially supported by Project Fondecyt 1150003 and Anillo ACT 1415 PIA-CONICYT.

As a consequence of Weil's descent theorem [**19**] (see also [**17**]), every real algebraic variety is definable over $\mathbb{R}$.

In [**10, 11**] Koeck, Lau and Singerman showed that a real algebraic variety (X, L) which is definable over the field $\overline{\mathbb{Q}}$ of algebraic numbers is also definable over $\mathbb{R} \cap \overline{\mathbb{Q}}$. More generally, as a consequence of Weil's descent theorem [**19**], if (X, L) is defined over a subfield $\mathcal{Q}$ of $\mathbb{C}$ so that $J_1(\mathcal{Q}) = \mathcal{Q}$ (i.e., $\mathcal{Q}$ is *conjugate-invariant*), then it is also definable over $\mathbb{R} \cap \mathcal{Q}$ (Theorem 1). The proof of such fact is existential in the sense that if we are given the explicit equations for X and L over $\mathcal{Q}$, then there is no explained how to construct explicitly R. In this paper, we describe an explicit rational isomorphism map R, defined over $\mathcal{Q}$, so that $R : X \to Z = R(X)$ is an isomorphism, Z is defined over $\mathbb{R} \cap \mathcal{Q}$ and with $T = R \circ L \circ R^{-1}$ being the usual conjugation map (see Theorem2). We hope that this may be of use in the construction of explicit examples, in special, at the level of real dessins d'enfants; the original motivation of this paper.

2. Main results

In this section, we fix a conjugate-invariant subfield $\mathcal{Q}$ of $\mathbb{C}$ and a real algebraic variety (X, L) defined over $\mathcal{Q}$. We first proceed to observe that (X, L) is definable over $\mathcal{Q} \cap \mathbb{R}$, a mild generalization of the results in [**10, 11**], and then we proceed to describe an explicit rational map R.

THEOREM 1. *The real algebraic variety* (X, L) *is definable over* $\mathcal{Q} \cap \mathbb{R}$.

A short proof of this fact is provided at the beginning of Section 3.

Remark 1. Assume that $\mathcal{K}$, $\mathcal{N}$ and $\mathcal{L}$ are subfields of $\mathbb{C}$ so that $\mathcal{L}$ contains $\mathcal{K}$ as an algebraically closed subfield (i.e., the only $\mathcal{K}$-algebraic numbers of $\mathcal{L}$ belongs to $\mathcal{K}$) and $\mathcal{N}$ is a finite Galois extension of $\mathcal{K}$. Let X be a complex smooth algebraic variety, which is definable over $\mathcal{L}$ and also over $\mathcal{N}$ (maybe by different models). In [**10**], as a consequence of Weil's descent theorem, it is shown the existence of an isomorphism $R : X \to Z$, where Z is defined over $\mathcal{K}$ (the given proof does not provide a method to obtain explicitly an isomorphism $R : X \to Z$). We should observe that this result does not imply Theorem 1; for instance take $\mathcal{Q} = \mathbb{Q}(i)$.

If the conjugate-invariant subfield $\mathcal{Q}$ is also algebraically closed (for instance, $\mathcal{Q} = \overline{\mathbb{Q}}$, then Theorem 1 can be written as follows.

COROLLARY 1. *Let* $\mathcal{Q}$ *be an algebraically closed conjugate-invariant subfield of* $\mathbb{C}$ *and let* X *be a symmetric variety definable over* $\mathcal{Q}$. *If the group of birational automorphisms of* X *is finite, then* X *is definable over* $\mathcal{Q} \cap \mathbb{R}$.

PROOF. Let $L : X \to X$ be a symmetry of X. We claim that L is defined over $\mathcal{Q}$. In fact, if $\eta \in \mathrm{Gal}(\mathbb{C}/\mathcal{Q})$, then we have the symmetry $L^\eta : X \to X$. So, there is a birational automorphism t of X so that $L^\eta = L \circ t$. Set $K = \{\eta \in \mathrm{Gal}(\mathbb{C}/\mathcal{Q}) : L = L^\eta\}$ and let $\mathcal{U}$ be the fixed field of K. Assume $L = (L_1, \ldots, L_n)$, where L_j is a rational map of the form r_j/s_j, where r_j and s_j are relatively prime polynomials. We may assume the leading coefficient of s_j to be equal to 1. The equality $r_j^\eta/s_j^\eta = r_j/s_j$, for $\eta \in K$, implies that the set of zeroes of r_j (and the set of zeroes of s_j) is invariant under K. In particular, $r_j^\eta = a_\eta r_j$ and $s_j^\eta = b_\eta s_j$, for suitable $a_\eta, b_\eta \neq 0$. Since $r_j^\eta/s_j^\eta = r_j/s_j$, we also have that $b_\eta = a_\eta$. As the leading coefficient of s_j is equal to 1, we must have $b_\eta = 1$; so r_j and s_j are polynomials defined over $\mathcal{U}$. We conclude that L is defined over $\mathcal{U}$. Now, as the group of birational automorphisms of X is finite, it follows that K is a subgroup of finite index of $\mathrm{Gal}(\mathbb{C}/\mathcal{Q})$; so $\mathcal{U}$ is a finite extension of $\mathcal{Q}$. As $\mathcal{Q}$ is assumed to be algebraically closed, it follows that $\mathcal{U} = \mathcal{Q}$ and we obtain the desired result for L. Now, we may apply Theorem 1. □

2.1. An explicit construction. Let $\mathcal{Q}$ be a conjugate-invariant subfield of $\mathbb{C}$ and let (X, L) be a real algebraic variety defined over $\mathcal{Q}$. Theorem 1 asserts the existence of a real algebraic variety (Z, T), defined over $\mathcal{Q} \cap \mathbb{R}$, and of an isomorphism $R : X \to Z$ so that $T = R \circ L \circ R^{-1}$. Of course, if $\mathcal{Q}$ is a subfield of $\mathbb{R}$, then it suffices to choose R as the identity. So, from now on, we assume that $\mathcal{Q}$ is not a subfield of $\mathbb{R}$ and that (X, L) is not already defined over $\mathcal{Q} \cap \mathbb{R}$. Under these assumptions, we proceed to the construction of a rational map R defined over $\mathcal{Q}$ so that: $R : X \to Z = R(X)$ is an isomorphism, Z is defined over $\mathcal{Q} \cap \mathbb{R}$ and $R \circ T \circ R^{-1}$ is the complex conjugation.

Assume $X \subset \mathbb{C}^n$ is defined by the polynomials $P_1(x_1, \ldots, x_n), \ldots, P_s(x_1, \ldots, x_n) \in \mathcal{Q}[x_1, \ldots, x_n]$.

2.1.1. **Step 1: Can assume $X \cap \overline{X} = \emptyset$.** In fact, if $X \cap \overline{X} \neq \emptyset$, then we may change (X, L) by an isomorphic real algebraic variety (X_0, L_0) so that $X_0 \cap \overline{X_0} = \emptyset$ as follows. Choose a point $\alpha \in \mathcal{Q} - \mathbb{R}$ and consider the algebraic variety $X_0 \subset \mathbb{C}^{n+1}$ (by adding the extra coordinate x_{n+1}) defined by the polynomials defining X and the extra polynomial $P_{s+1}(x_1, \ldots, x_{n+1}) = x_{n+1} - \alpha$ (that is, $X_0 = X \times \{\alpha\}$). The map $H(x_1, \ldots, x_n) = (x_1, \ldots, x_n, \alpha)$ induces a biregular isomorphism between X and X_0 (the projection $(x_1, \ldots, x_{n+1}) \mapsto (x_1, \ldots, x_n)$ induces its inverse). It is now clear that $X_0 \cap \overline{X_0} = \emptyset$. In this new model X_0 the symmetry is given as

$$L_0 = H \circ L \circ H^{-1} : X_0 \to X_0 : (x_1, \ldots, x_n, \alpha) \mapsto (L(x_1, \ldots, x_n), \alpha).$$

2.1.2. **Step 2: Construction of R.** By Step 1, we may assume $X \cap \overline{X} = \emptyset$. Consider the following polynomials in $\mathbb{Z}[x_1, \ldots, x_n, z_1, \ldots, z_n]$:

(1)
$$\left\{ \begin{array}{lcll} t_{1,j} & = & x_j + z_j; & j = 1, \ldots, n, \\ t_{2,j} & = & x_j z_j; & j = 1, \ldots, n, \\ t_{k,j} & = & x_{k-2}x_{k-2+j} + z_{k-2}z_{k-2+j}; & k = 3, \ldots, n, \ j = 1, \ldots, n+2-k, \\ t_{n+1,1} & = & x_{n-1}x_n + z_{n-1}z_n. & \end{array} \right\}$$

The reason of using these polynomials $t_{i,j}$ is clarified in the forthcoming Lemma 4.

For $x = (x_1, \ldots, x_n) \in X$, set $z = (z_1, \ldots, z_n) = J_n(L(x)) \in \overline{X}$ and consider the rational map

$$R : X \to \mathbb{C}^{(n^2+3n)/2} : x \mapsto t = (t_{1,1}, \ldots, t_{n+1,1}).$$

Observe that R is defined over $\mathcal{Q}$. Next result states that R as constructed above is the one we are searching.

THEOREM 2. *Under the above assumptions, the following hold.*

(1) *$Z = R(X)$ is defined over $\mathcal{Q} \cap \mathbb{R}$.*
(2) *$R : X \to Z$ is a birational isomorphism (biregular if $J \circ L$ is a polynomial), defined over $\mathcal{Q}$.*
(3) $R \circ L \circ R^{-1} = J_{(n^2+3n)/2}$.

3. Proof of Theorems 1 and 2

We only need to take care of the case when the conjugate-invariant subfield $\mathcal{Q}$ is not already a subfield of $\mathbb{R}$. So $\mathcal{Q}$ is a Galois extension of degree two of $\mathcal{K} = \mathcal{Q} \cap \mathbb{R}$. Let $\Gamma := \mathrm{Gal}(\mathcal{Q}/\mathcal{K}) = \langle \sigma(z) = \overline{z} \rangle \cong \mathbb{Z}_2$.

Let $P_1, \ldots, P_s \in \mathcal{Q}[x_1, \ldots, x_n]$ be a set of generators of the ideal of $X \subset \mathbb{C}^n$ and $L = (L_1, \ldots, L_n)$, where $J_1 \circ L_j \in \mathcal{Q}(x_1, \ldots, x_n)$. Observe that $\overline{X} = J_n(X) = X^\sigma$. If $\sigma_2 = \sigma$ and $\sigma_1 = e$ is the identity, then we set $f_2 = f_{\sigma_2} = J_n \circ L : X \to \overline{X}$, which is an isomorphism defined over $\mathcal{Q}$ (since L is defined over $\mathcal{Q}$), and $f_1 = f_{\sigma_1} = I$ is the identity.

3.1. Proof of Theorem 1. As the collection $\{f_1, f_2\}$ satisfies Weil's co-cycle conditions $f_{\sigma_i \sigma_j} = f_{\sigma_j}^{\sigma_i} \circ f_{\sigma_i}$, for $i, j \in \{1, 2\}$, it follows from Weil's descent theorem [**19**] the existence of a complex algebraic variety Z, defined over $\mathcal{K}$, and an isomorphism $R : X \to Z$, defined over $\mathcal{Q}$, so that $R = R^{\sigma_j} \circ f_{\sigma_j}$, for $j = 1, 2$. Now, if $T = R \circ L \circ R^{-1}$, then

$$\begin{aligned} T^\sigma &= R^\sigma \circ L^\sigma \circ (R^{-1})^\sigma = R^\sigma \circ L^\sigma \circ (R^\sigma)^{-1} = R^\sigma \circ L^\sigma \circ (f_2 \circ R^{-1}) \\ &= R^\sigma \circ L^\sigma \circ (J_n \circ L \circ R^{-1}) = R^\sigma \circ (J_n \circ L \circ J_n) \circ J_n \circ L \circ R^{-1} \\ &= R^\sigma \circ (J_n \circ L) \circ L \circ R^{-1} = (R^\sigma \circ f_2) \circ L \circ R^{-1} = R \circ L \circ R^{-1} \\ &= T, \end{aligned}$$

that is, T is defined over $\mathcal{K}$.

3.2. Proof of Theorem 2. As above, we fix the notation $\mathcal{K} := \mathbb{Q} \cap \mathbb{R}$, since it is constantly employed.

Now we are assuming that $\overline{X} \cap X = \emptyset$. Let us consider the explicit rational map, defined over $\mathcal{Q}$,

$$\Phi : X \subset \mathbb{C}^n \to \mathbb{C}^n \times \mathbb{C}^n = \mathbb{C}^{2n} : \quad x \mapsto (f_1(x), f_2(x)) = (x, z).$$

As each f_j is an isomorphism, $\Phi : X \to \Phi(X)$ is a birational isomorphism whose inverse is given by the restriction to $\Phi(X)$ of the projection map $\pi : \mathbb{C}^n \times \mathbb{C}^n \to \mathbb{C}^n : (x, z) \mapsto x$.

Remark 2. If f_2 is polynomial (i.e. when L is polynomial), the map $\Phi : X \to \Phi(X)$ is a biregular isomorphism and $\Phi(X)$ is described by the equations

$$\left\{ \begin{array}{c} z = f_2(x), \\ P_1(x) = 0, \ldots, P_s(x) = 0. \end{array} \right\}.$$

If

$$f_2(x) = \left(\frac{r_1(x)}{s_1(x)}, \ldots, \frac{r_n(x)}{s_n(x)} \right),$$

where $r_j, s_j \in \mathcal{Q}[x_1, \ldots, x_n]$ are relatively prime, then $\Phi(X)$ is defined by the equations

$$\left\{ \begin{array}{c} z_1 s_1(x) = r_1(x), \ldots, z_n s_n(x) = r_n(x), \\ P_1(x) = 0, \ldots, P_s(x) = 0, \\ P_1^{\sigma_2}(z) = 0, \ldots, P_s^{\sigma_2}(z) = 0, \end{array} \right\}.$$

In any of the above situations, $\Phi(X)$ still defined over $\mathcal{Q}$.

LEMMA 1. *The symmetry that induces L, by Φ, on $\Phi(X)$ is given by*

$$L_\Phi(x, z) = (\overline{z}, \overline{x}).$$

PROOF.

$$L_\Phi(x,z) = \Phi \circ L \circ \Phi^{-1}(x,z) = \Phi(L(x)) = (L(x), J_n \circ L(L(x))) = (\overline{z}, \overline{x}).$$

□

Each $\sigma_i \in \Gamma$ produces a permutation $\theta(\sigma_i) \in \mathfrak{S}_2$ by the rule $\sigma_i\sigma_j = \sigma_{\theta(\sigma_i)(j)}$. In fact, $\theta(\sigma_1) = (1)(2) = e$ is the identity and $\theta(\sigma_2) = (1,2)$. We consider the natural isomomorphism (Cayley representation)

$$\theta : \Gamma \to \mathfrak{S}_2 : \sigma \mapsto \theta(\sigma).$$

The symmetric group $\mathfrak{S}_2$ produces a natural permutation action $\eta(\mathfrak{S}_2)$ on $\mathbb{C}^n \times \mathbb{C}^n$ defined as follows. If $x, z \in \mathbb{C}^n$, then

$$\eta(e)(x,z) = (x,z), \quad \eta(1,2)(x,z) = (z,x).$$

In this way, the composition $\widehat{\theta} = \eta \circ \theta$ determines a representation of Γ as a subgroup of linear holomorphic automorphisms of $\mathbb{C}^n \times \mathbb{C}^n$ given by

$$\widehat{\theta}(\sigma_1)(x,z) = \tau_1(x,z) = (x,z), \quad \widehat{\theta}(\sigma_2)(x,z) = \tau_2(x,z) = (z,x)$$

where $x, z \in \mathbb{C}^n$. In the following, if $\eta \in \Gamma$, then $\widehat{\eta}$ is an extension of it to a field automorphism of $\mathbb{C}$.

LEMMA 2. *If $j = 1, 2$, then*
(*) $\widehat{\sigma}_j(\Phi(x)) = \tau_j \circ \Phi \circ f_{\sigma_j}^{-1}(\widehat{\sigma}_j(x))$.
(**) *If $y \in \Phi(X)$, then $\widehat{\sigma}_j(y) \in \tau_j(\Phi(X))$.*

PROOF. Let us recall that, for $\tau, \eta \in \Gamma$ we have the co-cycle relation $f_{\tau\eta} = f_\eta^\tau \circ f_\tau$. This co-cycle condition permits to see that, for $\eta \in \Gamma$, we have the following sequence of equalities

$$\begin{cases} \widehat{\eta}(f_j(x)) = f_j^\eta(\widehat{\eta}(x)) = f_{\sigma_j}^\eta(\widehat{\eta}(x)) = f_{\eta\sigma_j}\left(f_\eta^{-1}(\widehat{\eta}(x))\right) = \\ = f_{\sigma_{\theta(\eta)(j)}}\left(f_\eta^{-1}(\widehat{\eta}(x))\right) = f_{\theta(\eta)(j)}\left(f_\eta^{-1}(\widehat{\eta}(x))\right). \end{cases}$$

□

As a consequence of Lemma 2 we have the following commutative diagram (the top part is just the definition of Φ^σ and the bottom part is a consequence of the previous lemma)

$$\begin{array}{lcl} X & \overset{\Phi}{\to} & \Phi(X) \\ \downarrow\widehat{\sigma}_j & & \downarrow\widehat{\sigma}_j \\ X^{\sigma_j} & \overset{\Phi^{\sigma_j}}{\to} & \Phi^{\sigma_j}(X^{\sigma_j}) = \Phi(X)^{\sigma_j} = \widehat{\sigma}_j(\Phi(X)) = \tau_j(\Phi(X)) \\ \downarrow f_{\sigma_j}^{-1} & & \downarrow\tau_j^{-1} \\ X & \overset{\Phi}{\to} & \Phi(X) \end{array}$$

LEMMA 3. $\tau_2(\Phi(X)) \cap \Phi(X) = \emptyset$.

PROOF. Let us assume we have a point $(x,z) \in \tau_2(\Phi(X)) \cap \Phi(X)$. By the definition, $(x,z) \in \Phi(X)$ implies that $x \in X$ and $z = f_2(x) \in X^{\sigma_2} = \overline{X}$. Since $(x,z) \in \tau_2(\Phi(X))$, we also have that $(z,x) \in \Phi(X)$, that is, $z \in X$. This, in particular, ensures that $X \cap X^{\sigma_2} = X \cap \overline{X} \neq \emptyset$, a contradiction. □

Next ingredient in the computational method concerns with the algebra of invariants of a finite group of linear transformations. Let us briefly recall the general facts. Let $\mathcal{V}$ be a finite dimensional vector space over a field $\mathcal{R}$, say of dimension $n \geq 1$. Let $v_1, \ldots, v_n$ be a basis of $\mathcal{V}^*$. The symmetric algebra $S(\mathcal{V}^*)$ over $\mathcal{R}$ can be identified with the algebra $\mathcal{R}[v_1, \ldots, v_n]$. If G is a group acting linearly over $\mathcal{V}$, then it also acts linearly on $\mathcal{V}^*$. This provides a linear action on $\mathcal{R}[v_1, \ldots, v_n]$. A theorem due to Hilbert-Noether [**13, 14**] (Chap. 14 in [**15**]) states that, if G is a finite group, then the algebra of G-invariants $\mathcal{R}[\mathcal{V}]^G$ is finitely generated.

In our situation, $G = \Gamma$ is a cyclic group of order 2. Noether's bound theorem (see [**16**] for a proof) states that a set of invariant generators can be chosen on the set of polynomials of degree at most 2. More precisely, the following provides explicit generators.

LEMMA 4. *The algebra of $\widehat{\theta}(\Gamma)$-invariant polynomials is generated by the $(n^2 + 3n)/2$ ones provided in* (1) *in Section* 2.

PROOF. A polynomial $Q(x_1, \ldots, x_n, z_1, \ldots, z_n)$ is $\widehat{\theta}(\Gamma)$-invariant if and only if it is symmetric with respect to permutations of the variables x_i with z_i, for all $i = 1, \ldots, n$. By Noether's bound theorem a set of invariant generators can be chosen on the set of polynomials of degree at most 2. Now, let us assume we have a polynomial P, of degree at most 2, which is invariant under $\widehat{\theta}(\Gamma)$ and let us consider a monomial of P. It will be of the form x_i or z_j or $x_i x_j$ or $z_i z_j$ or $x_i z_j$. The invariance ensures that its symmetric monomial must also belong to P; that is, z_i or x_j or $z_i z_j$ or $x_i x_j$ or $z_i x_j$, respectively. As $x_i z_j + x_j z_i = (x_i + z_i)(x_j + z_j) - (x_i x_j + z_i z_j)$ and $x_i^2 + z_i^2 = (x_i + z_i)^2 - 2x_i z_i$, it follows that P can be generated by polynomials of the form $x_i + z_i$, $x_i z_i$ and, for $i \neq j$, $x_i x_j + z_i z_j$. $\square$

Let us consider the polynomial map

$$\Psi : \mathbb{C}^n \times \mathbb{C}^n \to \mathbb{C}^{(n^2+3n)/2}; \quad \Psi(x, z) = (t_{1,1}, \ldots, t_{n+1,1}),$$

and, as before, $\tau_1(x, z) = (x, z)$ and $\tau_2(x, z) = (z, x)$, where $x, z \in \mathbb{C}^n$.

LEMMA 5. *Keeping all the previous notations, the map Ψ satisfies the following properties:*

(1) $\Psi^{\sigma_j} = \Psi$, *for every $j = 1, 2$;*
(2) $\Psi \circ \tau_j = \Psi$, *for every $j = 1, 2$; and*
(3) $\Psi(w_1) = \Psi(w_2)$ *if and only if there is some $j \in \{1, 2\}$ so that $w_2 = \tau_j(w_1)$.*

PROOF. Properties (1) and (2) are trivial to see from the construction of Ψ. Now, if $\Psi(x, z) = \Psi(u, v)$, then, for $j = 1, \ldots, n$, it holds that

$$x_j + z_j = u_j + v_j, \quad x_j z_j = u_j v_j.$$

The first equality asserts that $v_j = x_j + z_j - u_j$ and now the second one states that

$$u_j^2 - u_j(x_j + z_j) + x_j z_j = 0.$$

As the roots of the above are x_j and z_j, we are done. $\square$

Set $V = \Psi(\mathbb{C}^n \times \mathbb{C}^n) \subset \mathbb{C}^{(n^2+3n)/2}$. Lemma 5, together classical invariant theory (see, for instance, [**7, 12**]), asserts that V is a (singular) algebraic variety of dimension $2n$ whose singular part is $\Psi(\Delta)$, where $\Delta = \{(x, z) \in \mathbb{C}^{2n} : x = z\}$.

Remark 3. If $N = (n^2 + 3n)/2$, then the algebra of regular functions on $V \subset \mathbb{C}^N$ (obtained above) is known to be isomorphic to the algebra $\mathbb{C}[x,z]^{\widehat{\theta}(\Gamma)}$ [**6**], where $\widehat{\theta}$ was defined just before Lemma 2. In fact, by taking as above the generators $t_1, \ldots, t_N$ the generators of the algebra of invariant polynomials $\mathbb{C}[x,z]^{\widehat{\theta}(\Gamma)}$, then $\xi : \mathbb{C}[V] \to \mathbb{C}[x,z]^{\widehat{\theta}(\Gamma)}$, defined by $\xi(p) = p(t_1, ..., t_N)$, defines an isomomorphism.

The map $\Psi : \mathbb{C}^n \times \mathbb{C}^n \to V$ is a degree two branched regular holomorphic covering with $\widehat{\theta}(\Gamma)$ as its deck group (see [**18**] for the general concept of a topological branched cover). The branch locus of Ψ is $\Psi(\Delta)$. In particular, $\Psi : \mathbb{C}^n \times \mathbb{C}^n - \Delta \to V - \Psi(\Delta)$ is a regular holomorphic cover map with $\widehat{\theta}(\Gamma)$ as its deck group.

LEMMA 6. *The map $\Psi : \Phi(X) \to Z = \Psi(\Phi(X))$ defines a biregular isomorphism. In particular, $R = \Psi \circ \Phi : X \to Z$ is defined over $\mathcal{Q}$ and it is a birational isomorphism, which is biregular if f_2 is polynomial.*

PROOF. Set $\Phi(X) = W$. Since $\tau_2(W) \cap W = \emptyset$, we have that the polynomial map $\Psi : W \to Z$ is bijective. Now, set $\widehat{W} = W \cup \tau_2(W)$, which is a reducible affine variety whose irreducible components are W and $\tau_2(W)$. Since these irreducible components are pairwise disjoint, we may see that the algebra of regular functions on $\widehat{W}$, say $\mathcal{C}[\widehat{W}]$, is the product of the algebras of regular functions of the components, that is,

$$\mathcal{C}[\widehat{W}] = \mathcal{C}[W] \times \mathcal{C}[\tau_2(W)].$$

The above isomorphism is just given by the restriction of each regular function of $\widehat{W}$ to each of its irreducible components. There is also natural isomorphism $\rho : \mathcal{C}[W] \to \mathcal{C}[\widehat{W}]^{\widehat{\theta}(\Gamma)}$, where $\mathcal{C}[\widehat{W}]^{\widehat{\theta}(\Gamma)}$ denotes the sub-algebra of $\widehat{\theta}(\Gamma)$-invariant regular functions on $\widehat{W}$. This isomorphism is given as follows. If $p \in \mathcal{C}[W]$, then $\rho(p) = (p, p \circ \tau_2^{-1})$ defines an injective homomorphism. It is clear that every $\widehat{\theta}(\Gamma)$-invariant regular function of $\widehat{W}$ is obtained in that way (so ρ is surjective). On the other hand, $\mathcal{C}[Z]$ is isomorphic to $\mathcal{C}[\widehat{W}]^{\widehat{\theta}(\Gamma)}$. To see this, one may consider the injective homomorphism $\chi : \mathcal{C}[Z] \to \mathcal{C}[\widehat{W}]^{\widehat{\theta}(\Gamma)}$ defined by $\chi(p) = p \circ \Psi$. Now, to see that χ is onto, we only need to note that $\rho^{-1}(\chi(\mathcal{C}[Z]))$ is a sub-algebra of $\mathcal{C}[W]$, that W is irreducible and that Z has the same dimension as W. In this way, $\chi^{-1} \circ \rho$ produces an isomorphism between $\mathcal{C}[W]$ and $\mathcal{C}[Z]$, that is, $\Psi : \Phi(X) \to Z$ is a biregular isomorphism and, in particular, $R = \Psi \circ \Phi : X \to Z$ is a birational isomorphism. As $\Psi : W \to Z$ is biregular isomorphism and $\Phi : X \to W$ is biregular if f_2 is polynomial, then $R : X \to Z$ turns out to be biregular if f_2 is polynomial. $\square$

In order to finish the proof, we only need to show that Z is defined over $\mathcal{K}$ since by Lemma 1 and the definition of Ψ we have that $R \circ L \circ R^{-1}$ is just the conjugation map.

PROPOSITION 1. *Z is defined over $\mathcal{K}$.*

PROOF. If $\eta \in \mathrm{Aut}(\mathbb{C}/\mathcal{Q})$, then, as $X^\eta = X$, $\Phi^\eta = \Phi$ (since L and X are defined over $\mathcal{Q}$) and $\Psi^\eta = \Psi$ (since Ψ is defined over the basic field $\mathbb{Q}$), one has that $R^\eta = R$; in particular, $Z^\eta = R(X)^\eta = R^\eta(X^\eta) = R(X) = Z$. So, Z is defined over $\mathcal{Q}$. We have that $\mathcal{Q}$ is a degree two extension of $\mathcal{K}$ and $\mathrm{Gal}(\mathcal{Q}/\mathcal{K})$ is generated by σ_2. It follows, from $(**)$ in Lemma 2 and (3) in Lemma 4, that $\widehat{\sigma_2} : Z \to Z$ is a bijection, that is, $Z^{\sigma_2} = Z$; so Z is defined over $\mathcal{K}$ as desired. $\square$

3.3. Some final computational remarks.

3.3.1. *Equations for V.* It is not difficult to check, for $1 \leq i < j \leq n$ (so $n(n-1)/2$ cases), the following equality

$$t_{1,i}^2 t_{2,j} - t_{1,i}t_{1,j}t_{i+2,j-i} + t_{1,j}^2 t_{2,i} - 4t_{2,i}t_{2,j} + t_{i+2,j-i}^2 = 0. \tag{2}$$

It follows that, on V, we have that $t_{i+2,j-i}$ is rationally determined by the variables $t_{1,i}$, $t_{2,i}$, $t_{1,j}$ and $t_{2,j}$. In particular, the algebraic variety V is a subvariety of the algebraic variety $\widehat{V}$ defined by the above $n(n-1)/2$ polynomial equations in (2). In worked examples we have seen that $V = \widehat{V}$. We believe that, in the general case, $V = \widehat{V}$.

3.3.2. *Finding the inverse of R.* In order to find explicitly an inverse of $R : X \to Z$, we need to search for rational expressions for the variables x_i's in the variables t_j's. In fact, we will see that we only need the variables $t_{1,i}$, $t_{2,i}$ and $t_{3,i}$ (see Section 3.3.1). We first note that

$$z_j = t_{1,j} - x_j, \quad j = 1, \ldots, n.$$

If $j = 2, \ldots, n$, then

$$t_{3,j-1} + t_{1,j}x_1 - t_{1,j}t_{1,1} = x_1x_j + z_1z_j + (x_j + z_j)x_1 - (x_j + z_j)(x_1 + z_1) =$$

$$= x_1x_j - x_jz_1 = (x_1 - z_1)x_j = (x_1 - (t_{1,1} - x_1))x_j = (2x_1 - t_{1,1})x_j,$$

that is,

$$x_j = \frac{t_{3,j-1} + t_{1,j}(x_1 - t_{1,1})}{2x_1 - t_{1,1}}, \quad j = 2, \ldots, n.$$

All the above states that each of the variables $x_2, \ldots, x_n$ can be expressed rationally in terms of the variables $t_{1,1}, \ldots, t_{3,n-1}$ and x_1. Since $t_{1,1}x_1 - t_{2,1} = (x_1 + z_1)x_1 - x_1z_1 = x_1^2$, we also have

$$x_1^2 = t_{1,1}x_1 - t_{2,1}. \tag{3}$$

Using the polynomial equations of X and/or the first coordinate polynomial equation of f_2 (which provides z_1 in terms of $x_1, \ldots, x_n$) and equality (3) (which permits to pass high powers of x_1 to a degree 1 power of it) we may obtain a linear equation on x_1 whose coefficients are rational forms of $t_{1,1}, \ldots, t_{3,n-1}$. In this way, each of the variables $x_1, \ldots, x_n$ is now rationally expressed in the variables $t_{1,1}, \ldots, t_{3,n-1}$ and the inverse map $R^{-1} : Z \to X$ is then obtained.

3.3.3. *Elimination of variables for Z.* Observe (see Section 3.3.1) that, when we are restricted to X, the coordinates $t_{2+i,j}$ (where $i < j$) are determined by the coordinates $t_{1,1}$, $t_{1,2}$, ..., $t_{1,n}$, $t_{2,n}$; so Z can be described using only these $2n$ coordinates.

4. An example

Let us consider the real algebraic curve (X, L), defined over $\mathbb{Q}(i)$, where

$$X : \left\{ \begin{array}{ccccccc} 1 & + & x_1^2 & + & x_2^2 & = & 0 \\ -1 & + & x_1^2 & + & x_3^2 & = & 0 \\ i & + & x_1^2 & + & x_4^2 & = & 0 \end{array} \right\} \subset \mathbb{C}^4,$$

and

$$L : \mathbb{C}^4 \to \mathbb{C}^4 : (x_1, x_2, x_3, x_4) \mapsto (-i\,\overline{x_1}, -i\,\overline{x_3}, -i\,\overline{x_2}, -i\,\overline{x_4}).$$

In this example,

$$f_2 = J_4 \circ L : \mathbb{C}^4 \to \mathbb{C}^4 : (x_1, x_2, x_3, x_4) \mapsto (ix_1, ix_3, ix_2, ix_4),$$

$$\tau_2 : \mathbb{C}^4 \times \mathbb{C}^4 \to \mathbb{C}^4 \times \mathbb{C}^4 : (x_1, x_2, x_3, x_4, z_1, z_2, z_3, z_4) \mapsto (z_1, z_2, z_3, z_4, x_1, x_2, x_3, x_4),$$

By Theorem 1, (X, L) is definable over $\mathbb{Q} = \mathbb{Q}(i) \cap \mathbb{R}$. Theorem 2 states that the map

$$R : X \to Z$$
$$R(x_1, x_2, x_3, x_4) = (t_1 = t_{1,1}, \dots, t_{14} = t_{5,1}), \text{where}$$

$$\begin{gathered} t_1 = (1+i)x_1,\ t_2 = x_2 + ix_3,\ t_3 = x_3 + ix_2,\ t_4 = (1+i)x_4, \\ t_5 = ix_1^2,\ t_6 = x_2 + ix_3,\ t_7 = ix_2x_3,\ t_8 = ix_4^2, \\ t_9 = x_1x_2 - x_1x_3,\ t_{10} = x_1x_3 - x_1x_2,\ t_{11} = 0,\ t_{12} = 0, \\ t_{13} = x_2x_4 - x_3x_4,\ t_{14} = x_3x_4 - x_2x_4, \end{gathered}$$

provides an isomorphism between the real algebraic curves (X, L) and (Z, J_{14}) with Z defined over $\mathbb{Q}$. It is not difficult to check that

$$R^{-1} : Z \to X : (t_1, \dots, t_{14}) \mapsto (x_1, x_2, x_3, x_4) = \left(\frac{t_1}{1+i}, \frac{t_2 - it_3}{2}, \frac{t_3 - it_2}{2}, \frac{t_4}{1+i} \right).$$

The curve Z is defined by the following equations

$$\begin{gathered} t_{14} = -t_{13} = t_4(t_3 - t_2)/2,\ t_{12} = t_{11} = 0,\ t_{10} = -t_9 = -t_1(t_2 - t_3)/2, \\ t_8 = t_4^2/2,\ t_7 = (t_2^2 + t_3^2)/4,\ t_6 = t_2,\ t_5 = t_1^2/2, \\ 4 + t_2^2 - t_3^2 = 0,\ t_1^2 + t_2t_3 = 0,\ t_1^2 + t_4^2 - 2 = 0. \end{gathered}$$

Remark 4. (1) The above also asserts that X is isomorphic to

$$Y = \left\{ \begin{array}{c} 4 + w_2^2 - w_3^2 = 0 \\ w_1^2 + w_2w_3 = 0 \\ w_1^2 + w_4^2 - 2 = 0 \end{array} \right\} \subset \mathbb{C}^4,$$

the isomorphism given by

$$\widehat{R} : X \to Y : (x_1, x_2, x_3, x_4) \mapsto (t_1, t_2, t_3, t_4) = (w_1, w_2, w_3, w_4).$$

(2) The curve X admits the group $H = \langle A_1, A_2, A_3, A_4 \rangle \cong \mathbb{Z}_2^4$ as subgroup of conformal automorphisms, where

$$\begin{gathered} A_1(x_1, x_2, x_3, x_4) = (-x_1, x_2, x_3, x_4),\ A_2(x_1, x_2, x_3, x_4) = (-x_1, -x_2, x_3, x_4), \\ A_3(x_1, x_2, x_3, x_4) = (x_1, x_2, -x_3, x_4),\ A_4(x_1, x_2, x_3, x_4) = (x_1, x_2, x_3, -x_4). \end{gathered}$$

In the model Y, these correspond to

$$\begin{gathered} A_1(w_1, w_2, w_3, w_4) = (-w_1, w_2, w_3, w_4),\ A_2(w_1, w_2, w_3, w_4) = (w_1, iw_3, -iw_2, w_4), \\ A_3(w_1, w_2, w_3, w_4) = (w_1, -iw_3, iw_2, w_4),\ A_4(w_1, w_2, w_3, w_4) = (w_1, w_2, w_3, -w_4), \end{gathered}$$

which are defined over $\mathbb{Q}(i)$. Observe that the minimal field of definition of the pair (X, H) is $\mathbb{Q}(i)$.

5. MAGMA implementation

A simple pseudo-routine in MAGMA [**4**] to make the computations as in Theorem 2, under the mild assumption that L is polynomial, is the following. Let us assume $\mathcal{Q} = \mathbb{Q}(\alpha) < \overline{\mathbb{Q}}$ and that $q(t) \in \mathbb{Q}[t]$ is the irreducible polynomial associated to α. We have each $z_j = z_j(x_1, \ldots, x_n) \in \mathcal{Q}[x_1, \ldots, x_n]$.

First we state the ambient spaces.

> Q:=Rationals( );
> $P < t >$:=PolynomialRing(Q);
> $q := q(t)$;
> $K < r >$:=SplittingField(q);
> $A < x_1, \ldots, x_n >$:=AffineSpace(K,n);
> $B < t_{1,1}, \ldots, t_{n+1,1} >$:=AffineSpace($K$, $(n^2 + 3n)/2$);

We introduce the algebraic variety $X \subset \mathbb{C}^n$

> X:=Scheme(A,[$P_1(x_1, \ldots, x_n), \ldots, P_s(x_1, \ldots, x_n)$]);

We introduce the rational map $R : X \to Z$

> R:=map< $A-> B|[x_1 + z_1, \ldots, x_{n-1} * x_n + z_{n-1} * z_n]$ >;

We now ask for equations of the image $Z = R(X)$.

> Image(R);
> $R(X)$;

Remark 5.

(1) Note that in the above one should replace the "..." by the corresponding data; for instance, if $n = 3$, then the line
"$A < x_1, \ldots, x_n >$:=AffineSpace(K,n);"
should be replaced by the line
"$A < x_1, x_2, x_3 >$:=AffineSpace(K,3);".

(2) Usually MAGMA will provide the defining polynomials over $\mathcal{Q} \cap \mathbb{R}$, but in case that some of these polynomials, say $M \in \mathcal{Q}[t_{1,1}, \ldots, t_{n+1,1}]$, is not defined over the desired field, then we may replace it by the two new trace polynomials $\mathrm{Tr}(M) = M + \overline{M}, \mathrm{Tr}(aM) = aM + \overline{a}\overline{M} \in \mathcal{Q} \cap \mathbb{R}[t_{1,1}, \ldots, t_{n+1,1}]$, where $\{1, a\}$ is a basis of $\mathcal{Q}$ as a $\mathcal{Q} \cap \mathbb{R}$-vector space and $\overline{M}$ is obtained from M after conjugating its coefficients. Since $M = \lambda_1 \mathrm{Tr}(M) + \lambda_2 \mathrm{Tr}(aM)$, where $\lambda_1 = \overline{a}/(\overline{a} - a)$ and $\lambda_2 = 1/(a - \overline{a})$, the new set of polynomials, all of them defined over $\mathcal{Q} \cap \mathbb{R}$, will generate the ideal of Z.

(3) In the example provided in Section 4, the MAGMA pseudo-routine described above with $q = t^2 + 1$ provides the following equations for Z (all of them already over $\mathbb{Q}$):

$$\left\{\begin{array}{c} t_{13} + t_{14} = 0,\ t_{12} = 0,\ t_{11} = 0,\ t_9 + t_{10} = 0, \\ t_1 + \frac{1}{4}t_3t_{10}^3 - \frac{3}{2}t_3t_{10} + \frac{1}{8}t_4t_{10}^3t_{14} + \frac{1}{8}t_4t_{10}t_{14}^3 + \frac{3}{4}t_4t_{10}t_{14} = 0, \\ t_7 - t_8 - \frac{1}{2}t_{10}^2 - \frac{1}{2}t_{14}^2 + 1 = 0, \\ t_8t_{14}^2 - t_8 + \frac{1}{4}t_{10}^2t_{14}^2 + \frac{1}{4}t_{14}^4 - t_{14}^2 = 0, \\ t_8t_{10}^2 + t_8 - \frac{1}{4}t_{10}^2t_{14}^2 - \frac{1}{4}t_{14}^4 = 0, \\ t_{10}^4 + 2t_{10}^2t_{14}^2 - 4t_{10}^2 + t_{14}^4 - 4 = 0, \\ t_6 - t_8 - \frac{1}{2}t_{10}^2 - \frac{1}{2}t_{14}^2 + 1 = 0, \\ t_5 + t_8 - 1 = 0,\ t_4^2 - 2t_8 = 0, \\ t_3t_{14} - \frac{1}{2}t_4t_{10}^2 - \frac{1}{2}t_4t_{14}^2 - t_4 = 0, \\ t_3t_8 - t_4t_8t_{14} - \frac{1}{4}t_4t_{10}^2t_{14} - \frac{1}{4}t_4t_{14}^3 + \frac{1}{2}t_4t_{14} = 0, \\ t_2 + \frac{1}{2}t_3t_{10}^2 - 2t_3 + \frac{1}{4}t_4t_{10}^2t_{14} + \frac{1}{4}t_4t_{14}^3 + \frac{3}{2}t_4t_{14} = 0, \\ t_3t_4 - 2t_8t_{14} - \frac{1}{2}t_{10}^2t_{14} - \frac{1}{2}t_{14}^3 + t_{14} = 0, \\ t_3^2 - 2t_8 - t_{10}^2 - t_{14}^2 = 0. \end{array}\right\}.$$

Acknowledgments

We are grateful to the referees for their several useful comments, corrections and their careful reading.

References

[1] A. Baker, *A concise introduction to the theory of numbers*, Cambridge University Press, Cambridge, 1984. MR781734

[2] E. Bujalance, F. J. Cirre, J. M. Gamboa, and G. Gromadzki, *Symmetries of compact Riemann surfaces*, Lecture Notes in Mathematics, vol. 2007, Springer-Verlag, Berlin, 2010. MR2683160

[3] E. Bujalance, G. Gromadzki, and M. Izquierdo, *On real forms of a complex algebraic curve*, J. Aust. Math. Soc. **70** (2001), no. 1, 134–142, DOI 10.1017/S1446788700002329. MR1808396

[4] W. Bosma, J. Cannon, and C. Playoust, *The Magma algebra system. I. The user language*, J. Symbolic Comput. **24** (1997), no. 3-4, 235–265, DOI 10.1006/jsco.1996.0125. Computational algebra and number theory (London, 1993). MR1484478

[5] A. Carocca, V. González-Aguilera, R. A. Hidalgo, and R. E. Rodríguez, *Generalized Humbert curves*, Israel J. Math. **164** (2008), 165–192, DOI 10.1007/s11856-008-0025-2. MR2391145

[6] D. Cox, J. Little, and D. O'Shea, *Ideals, varieties, and algorithms*, 3rd ed., Undergraduate Texts in Mathematics, Springer, New York, 2007. An introduction to computational algebraic geometry and commutative algebra. MR2290010

[7] D. Hilbert, *Ueber die Theorie der algebraischen Formen* (German), Math. Ann. **36** (1890), no. 4, 473–534, DOI 10.1007/BF01208503. MR1510634

[8] G. Humbert. Sur un complexe remarquable de coniques. *Jour. Ecole Polyth.* **64** (1894), 123–149.

[9] T. W. Hungerford, *Algebra*, Graduate Texts in Mathematics, vol. 73, Springer-Verlag, New York-Berlin, 1980. Reprint of the 1974 original. MR600654

[10] B. Köck and E. Lau, *A note on Belyi's theorem for Klein surfaces*, Q. J. Math. **61** (2010), no. 1, 103–107, DOI 10.1093/qmath/han034. MR2592026

[11] B. Köck and D. Singerman, *Real Belyi theory*, Q. J. Math. **58** (2007), no. 4, 463–478, DOI 10.1093/qmath/ham017. MR2371466

[12] D. Mumford, J. Fogarty, and F. Kirwan, *Geometric invariant theory*, 3rd ed., Ergebnisse der Mathematik und ihrer Grenzgebiete (2) [Results in Mathematics and Related Areas (2)], vol. 34, Springer-Verlag, Berlin, 1994. MR1304906

[13] E. Noether, *Der Endlichkeitssatz der Invarianten endlicher Gruppen* (German), Math. Ann. **77** (1915), no. 1, 89–92, DOI 10.1007/BF01456821. MR1511848

[14] E. Noether. Der Endlichkeitssatz der Invarianten endlicher linear Gruppen der Characteristik p. *Nachr. Akad. Wiss. Göttingen* (1926), 28–35.

[15] C. Procesi. *Lie Groups: An approach through invariants and representations.* (Universitex) Springer-Verlag, 2006.

[16] B. J. Schmid. Finite groups and invariant theory. Séminaire d'Algèbre (P. Dubriel et M.-P. Malliavin, 1989–1990). *Lecture Notes in Math.*, vol. **1478**, Springer-Verlag, Heidelberg, Berlin, 1991.

[17] R. Silhol. Moduli problems in real algebraic geometry. *Real Algebraic Geometry* (1972), 110–119. Ed. M. Coste et al. (Springer-Verlag, Berlin).

[18] A. W. Tucker, *Branched and folded coverings*, Bull. Amer. Math. Soc. **42** (1936), no. 12, 859–862, DOI 10.1090/S0002-9904-1936-06446-3. MR1563453

[19] A. Weil, *The field of definition of a variety*, Amer. J. Math. **78** (1956), 509–524, DOI 10.2307/2372670. MR0082726

Departamento de Matemática y Estadística, Universidad de La Frontera. Temuco, Chile

Email address: `ruben.hidalgo@ufrontera.cl`

Contemporary Mathematics
Volume **724**, 2019
https://doi.org/10.1090/conm/724/14594

Curves in isomonodromy and isospectral deformations: Painlevé VI as a case study

E. Previato

ABSTRACT. Certain integrable dynamical problems can be described by isospectral deformations; under certain restrictions, the 'spectrum' is an algebraic curve. On the other hand, certain isomonodromy deformations may have an associated algebraic curve, under certain restrictions on the monodromy group. We explore the relationship between these two occurrences of algebraic curves using Painlevé VI as a "case study".

1. Introduction

The Painlevé equations have a rich history and a vast amount of applications, we refer to [**IKSY**] for information and connections with special function and to [**SHC**, Sec. 1] for the farthest-reaching range of applications. In the original approach, the six Painlevé equations, PI to PVI, were found to be (up to holomorphic change of the independent, and linear fractional transformation of the dependent variable) the only second-order non-linear ODEs with solutions whose only movable singularities are poles ("Painlevé property", which defines equations of "P-type", for short), to exclude branch points or essential singularities. The problem was posed by E. Picard [**Pi**]: we found three slightly different versions and adopt [**IKSY**, III.1.1], where the differential equation is of the form

$$F(t, y, \frac{dy}{dt}, ..., \frac{d^n y}{dt^n}) = 0,$$

defined in a domain D of the complex plane, with coefficients meromorphic in $t \in D$ and polynomial in $(y, \frac{dy}{dt}, ..., \frac{d^n y}{dt^n})$ (in [**C2**], F is assumed to be analytic in t and rational in $(y, \frac{dy}{dt}, ..., \frac{d^n y}{dt^n})$; in [**BBT**, 8.11], F is a rational function of all arguments). Painlevé and his students found the only such ODOs of order $n = 2$ that could not be integrated in terms of previously known functions (namely, elementary or elliptic); they are now known as PI–PVI. It is also possible to produce them as (time-dependent) Hamiltonian flows, and in that context the question of complete integrability, e.g, is still to some extent open, and was settled only recently for certain classes of functions (playing the role of independent variables) [**ŻF**]. There is, however, a seemingly unrelated way to come across the same equations, namely "isomonodromic deformation". The discovery of PVI in this way is originally due to

2010 *Mathematics Subject Classification.* Primary 34M55, 37K10; Secondary 34M56, 35C06.
Key words and phrases. Painlevé equations, integrable PDEs, similarity reduction.

R. Fuchs [**Fu**]. The question he posed, following investigations of his father Lazarus Fuchs, was: find conditions on a second-order linear ODE that has four regular-singular points so that the monodromy group be independent of the coefficients (namely, one parameter t, since the other three essential singularities can be moved to $0,\ 1, \infty$ by a linear-fractional transformation). By Poincaré's theory [**Po**], such an equation must have an additional apparent singularity (where he coefficients of the equation have poles but the solutions are single-valued meromorphic functions), say λ, and Fuchs' startling discovery was that isomonodromy implies that λ as a function ot t satisfies PVI. In this way, a (genus-one) curve turns out to be related to PVI. Other curves arise, both in the problem of isomonodromy, and in the theory of isospectral deformations that gives rise to "integrable hierarchies" of non-linear PDEs. The research reported in this note was prompted by the goal of identifying relationships among these curves. The finding is puzzling: to the best of our knowledge, no theoretical understanding of any such relationship is known, despite several announcements (e.g., [**FN**, **K**], detailed below) where the question was posed and a future publication was anticipated to contain the answer (we were unable to place that publication). This note therefore is devoted to creating a context for the questions, linking references that pursued similar issues unawares of each other, exploring the difference among their methods, and highlighting connections which were not made before. We then set ourselves a seemingly less ambitious task. There is another way in which P-type equations appear within the theory of integrable hierarchies, and that is "similarity reduction". We briefly summarize the emergence of that relationship.

Integrable hierarchies are sequences of non-linear PDEs that can be viewed as integrable Hamiltonian systems in a sequence of commuting variables $t_1, ..., t_n, ...$ and as such, they admit certain symmetries. Some of these symmetries produce, by reduction, the "self-similar" solutions, which typically depend only on one variable: strikingly, in that variable the solutions were found to satisfy an ODE of P-type (possibly of order higher than two). This led Ablowitz, Ramani and Segur to propose the "Painlevé conjecture": *A nonlinear PDE is solvable by an inverse scattering transform (IST) only if every nonlinear ODE obtained by exact reduction is of P-type, perhaps after a transformation of variables* [**AS2**, 3.7.b]. The conjecture prompted a great deal of work and was confirmed in special cases, but our paper is concerned instead with finding a structure behind this phenomenon. This need arises from a lack of systematic relationships between the PDEs and the ODEs: notably, "wheras the self-similar solutions of integrable evolution equations appear to give nonautonomous ordinary differential equations, there are many members of the latter class which do not result from self-similar limits of the former" [**FN**, Subsection 2B]. To begin with, "characterizing the set of PDEs that can be solved by IST" relies largely on "clever guesswork" [**AS2**, Ch. 3 Overview]. If we already have a given IST, then further analyticity assumptions, both on the kernel of the transform and on the type of solutions does make it clear, using classical theorems, that the solutions to the ODE should only have poles as singularities [**McLO**]. In this delicate setting, then, where detecting a certain qualitative behavior depends on 'clever guesses' and changes of variables, we felt that it was worthwhile to ask the simplest possible questions, and look for all available information in one "case study", which we describe next.

In Section 2 we present all six equations via isomonodromy. In that context, coalescence (or confluence) of the singularities (and also, by a different deformation procedure, introducing into variables and parameters of the equation an additional deformation parameter and taking the limit as it approaches zero) reduces PVI to all the others, according to the following diagram [**IKSY**, Prop. 1.2.1]:

$$
\begin{array}{ccccccc}
 & & & PIV & & & \\
 & & \nearrow & & \searrow & & \\
PVI & \to & PV & & & PII \to PI. & \\
 & & \searrow & & \nearrow & & \\
 & & & PIII & & &
\end{array}
$$

On the PDE side, the Kadomtsev-Petiashvili (KP) equation:

$$(u_t + \frac{3}{2} u u_x + \frac{1}{4} u_{xxx})_x + \frac{3}{4} \sigma u_{yy} = 0, \quad \sigma = \pm 1,$$

and its hierarchy, reduce to the Korteweg-de Vries (KdV), Boussinesq equation (and their hierarchy), respectively, when the solution is independent of y, t, resp. Since KdV has self-similar solutions that satisfy PI, others PII, and Boussinesq, under the same symmetry reduction, solutions that satisfy PI, PII, or PIV (according to the choice of parameters), we originally asked a most naïve question: does KP have self-similar solutions that satisfy PVI? If so, a diagram with vertical arrows corresponding to similarity reduction, and horizontal arrows corresponding to coalescence, would have been commutative in some sense. But of course we were wrong: KP does not reduce to KdV or Boussinesq by any sort of coalescence of the poles; the relationship between the two types of isomonodromy that connect an integrable PDE and a P-type equation is much more elusive, and we survey one version of the connection for PVI in Subsection 2.4. It can also be safely stated that "[PVI yields all other PI-V by appropriate limiting procedure], yet the geometric content of this statement is still very obscure" [**FN**, Subsection 2C] In any case, heuristically, since the KP hierarchy is governed by an eigenfunction that has one essential singularity on the spectral curve, the isospectral deformation problem [**FN**, (2.12-13)] posed by Flaschka and Newell has only one essential singularity, so there is no issue of coalescence.

We refocused the goal of our search, and decided that it seemed useful, on the one hand, to bring together all the similarity reductions we could find for KP, on the other, all the ways that PVI was produced by a similarity reduction (Section 3).

The interplay between isomonodromic and isospectral deformations, envisioned in [**FN**, Section 1], is still completely mysterious. We focus on the case when both the spectral curve and the monodromy curve are algebraic. In Section 2 we present different constructions of the attendant isomonodromy curve, which to our knowledge have not been related either, focusing on obtaining PVI, which again, in one of the contexts at least, is a recent result [**B**].

In Section 4, we summarize the structural relationships between the aspects we covered and list other aspects of the P-type equations that are related to the theory of algebraic curves and integrable systems, with the ultimate goal of finding a common root for all of them. Generalizations are also outlined.

2. Connecting integrable PDEs and Painlevé-type. I: Isomonodromy aspects

As mentioned in the Introduction, Fuchs found that PVI is equivalent to an isomonodromy problem. This discovery was generalized to the Schlesinger isomonodromy problem to yield all equations PI-PVI. We briefly recall his method, setting out two steps; the second step does not appear to have been generalized to the several-essential-singularities case, posed by Garnier in his thesis [**G**], and proposed in [**ENP**].

2.1. Elliptic curve. Step one, in the case of order 2 and $g+1$ regular-singular points with $g=1$, amounted to the following statement:

THEOREM 2.1. *The isomonodromy problem for the equation*

$$\frac{d^2y}{dx^2}=\left[\frac{a}{x^2}+\frac{b}{(x-1)^2}+\frac{c}{(x-t)^2}+\frac{\alpha}{x}+\frac{\beta}{x-1}+\frac{\gamma}{x-t}+\frac{e}{(x-\lambda)^2}+\frac{\epsilon}{(x-\lambda)}\right]y$$

is solved, and admits four arbitrary constants $k_0,\ k_1,\ k_\infty,\ k_t$[1] *when λ as a function of t satisfies PVI:*

$$\frac{d^2\lambda}{dt^2}+\left[\frac{1}{t}+\frac{1}{t-1}+\frac{1}{\lambda-t}\right]\frac{d\lambda}{dt}-\frac{1}{2}\left[\frac{1}{\lambda}+\frac{1}{\lambda-1}+\frac{1}{\lambda-t}\right]\left(\frac{d\lambda}{dt}\right)^2$$

$$=\frac{1}{2}\frac{\lambda(\lambda-1)(\lambda-t)}{t^2(t-1)^2}\left[k_\infty-k_0\frac{t}{\lambda^2}+k_1\frac{t-1}{(\lambda-1)^2}-(k_t-1)\frac{t(t-1)}{(\lambda-t)^2}\right],$$

with k,a,b,c given (linearly) in terms of $k_0,\ k_1,\ k_t,\ k_\infty$:

$$-4(k-1)+4a+4b+4c=k_\infty,$$

$$4\left(a+\frac{1}{4}\right)=k_0,\ 4\left(b+\frac{1}{4}\right)=k_1,\ 4\left(c+\frac{1}{4}\right)=k_t.$$

Step two was a remarkable observation which Fuchs possibly came across while looking into the special case $k_0=k_1=k_t=k_\infty=0$; this corresponds to Gauss' hypergeometric equation, and the result is the following:

THEOREM 2.2. *The roots $y_1,\ y_2$ of Gauss' hypergeometric equation undergo a monodromy independent of t, upon circling the points $t,\ 0,\ 1,\ \infty$, if and only if the function*

$$u=\int_0^\lambda\frac{d\lambda}{\sqrt{\lambda(\lambda-1)(\lambda-t)}}$$

satisfies Legendre's equation:

$$\frac{d^2u}{dt^2}+\frac{2t-1}{t(t-1)}\frac{du}{dt}+\frac{u}{t(t-1)}=0.$$

[1] The right number of parameters, together with λ, for the fixed monodromy group to be arbitrary, cf. e.g. [**G**, p. 74].

Note that λ is an elliptic function of t. More generally, PVI is equivalent to the equation

$$\frac{d^2u}{dt^2}+\frac{2t-1}{t(t-1)}\frac{du}{dt}+\frac{u}{t(t-1)}$$
$$=\frac{\sqrt{\lambda(\lambda-1)(\lambda-t)}}{2t^2(t-1)^2}\times\left[k_\infty-k_0\frac{t}{\lambda^2}+k_1\frac{t-1}{(\lambda-1)^2}-k_t\frac{t(t-1)}{(\lambda-t)^2}\right]$$

for the elliptic integral.

Notably, Manin [**M1**] gave a beautiful interpretation of Fuchs' work in moduli, and connected it with mirror symmetry [**M2**].

2.2. Schlesinger flow. We recall this setting, not only for completeness, but also because its interpretation is still evolving and recent research [**B**] checks directly that PI-PVI are the compatibility condition of the Schlesinger flow with the initial Fuchsian system, an important way to obtain "spectral curves", in this case bi-spectral. In [**B**] PVI is produced rigorously by Hamitonian dynamics under symplectic projection, and notably, a counterexample is given, to the fact that PVI, conversely, implies isomonodromy, in this particular setting. We refer to [**B**] for more specific definitions, since our goal is just to highlight the conversion between the ODE problem and the Schlesinger monodromy problem.

The monodromy of an equation

$$\frac{d}{dz}\Psi=A(z,t)\Psi,\quad A\in \mathrm{gl}(N),\ \Psi\in \mathrm{GL}(N),$$

where z is a variable ranging over a suitable complex domain, the complex parameter t belongs to a complex disk, and N is a natural number, is defined by a conjugacy class (up to constant) of matrices in $\mathrm{GL}(N)$.

THEOREM 2.3. *The monodromy is constant in t (i.e., "isomonodromy" holds for the equation) if and only if there exists an analytic single-valued function $B(z,t)$ such that the form $Adz+Bdt=\omega$ is flat, i.e., $d\omega=\omega\wedge\omega$, also expressed by the zero-curvature condition $A_t-B_z+[A,B]=0$, or:*

$$\begin{cases}\frac{\partial}{\partial z}\Psi=A\Psi\\ \frac{\partial}{\partial t}\Psi=B\Psi.\end{cases}$$

Specifically, Babich analyzes the Fuchsian system (with rational coefficients)

$$\frac{d}{dz}\Psi=\sum_{k=1}^{M}\frac{A^{(k)}}{z-z_k}\Psi,$$

with $\Psi\in\mathrm{GL}(N)$, $A^{(k)}\in\mathrm{gl}(N)$, $\sum_k A^{(k)}=0$ and the poles z_k viewed as deformation parameters t_k. He proves that the Schlesinger equations

$$dA^{(k)}=-\left[A^{(k)},\sum_{i\neq k}A^{(i)}d\log(z_k-z_i)\right]$$

are a sufficient condition for isomonodromy (but gives a counterexample to the converse). He also gives an equivalent Hamiltonian formulation over a space of orbits of matrices with natural symplectic structure; under Hamiltonian reduction (with respect to diagonal conjugation by constant matrices), he obtains the (multi-time) Schlesinger system for the Hamiltonian:

$$\omega - \sum_{i<j} d\mathrm{tr}(A^{(i)}A^{(j)}) \wedge d\log(z_i - z_j),$$

and this allows him to derive PVI, as the Lagrange equation corresponding to the above-mentioned Hamiltonian on $\prod_{k=1}^4 \mathcal{O}_{J_k}//\mathrm{PSL}(2)$, with the four $A^{(k)} \in \mathrm{End}(\mathbb{C}^2)$, J_k an orbit representative, for example the Jordan form, and the deformation parameter t taken to be the cross-ratio of the four singularities z_k; the four arbitrary constants in PVI are essentially the eigenvalues of the four orbits.

However, we are unable to construct Fuchs' elliptic curve from this Hamiltonian.

We also believe it is very interesting that an elliptic curve arises from a different monodromy aspect, which we describe in the next subsection.

2.3. Algebraic monodromy curve: Hitchin's example. Hitchin's approach [**Hi**] is to interpret the Schlesinger equation as the holonomy of a flat connection for a parabolically-stable rank-two vector bundle over the Riemann sphere, with a weighted flag structure at four marked points. Under the assumption that the holonomy group $\Gamma \subset \mathrm{SL}(2,\mathbb{Z})$ of the connection is finite, the corresponding solution of PVI is algebraic. Hitchin assumes the group to be the dihedral group D_k, and in that case $y(x)$ satisfies PVI with parameters $(\alpha,\beta,\gamma,\delta) = (1/8,-1/8,1/8,3/8)$, which are respectively $(k_\infty, -k_0, k_1, -(k_t - 1))$ in the previous notation of Subsection 2.1. Hitchin's proof consists of a staggering set of identifications that range from twistor spaces to k-sided plane polygons inscribed in a fixed conic C and circumscribed about a conic B (a case of Poncelet porism [**BM**]), so that the elliptic curve is the incidence correspondence in $C \times B^*$, with choice of origin. Hitchin gives the explicit equation of the conics in terms of the poles of the connection, thus the elliptic curve. We do not see any way to generalize this construction to Fuchs' curve (PVI with arbitrary parameters) because in that case the curve defived by the holonomy of the connection is not algebraic. But one could ask this question for other algebraic solutions of PVI (which have been classified), particularly in view of significant relationships between reflection groups and algebraic solutions of PVI [**DM**].

2.4. Sato's tau function. On the other hand, equations of statistical mechanics were brought to isomonodromic form: the earliest instances we could find are [**BMcCW**] and [**SMJ**]). Thus, the question [**AS2**, Section 3.7]: "Is there some connection between the two-dimensional Ising model and [Inverse Scattering Transform]?" This connection is indeed our main goal, namely: an explicit relation between isospectral and isomonodromy curves.

Flaschka and Newell [**FN**] state precisely this goal, take it so far as to work out two examples and announce "results will be presented in a sequence of future papers" that never appeared. Again, the elusiveness of this goals motivated the research we present in this note.

We review the framework in [**FN**] in Subsection 3.3 below, because the way they obtain an ODE of P-type is to impose a similarity reduction; in fact, they do not obtain PVI; here, we follow the expository [**BBT**, Ch. 8], after [**SMJ**] (cf. also [**KaKa**] for the setting that links isomonodromy and isospectrality), noting that PVI is derived in a way which resembles Fuchs' yet differs in the *ad hoc* introduction of parameters to serve as new variables.

What we find most interesting in [**BBT**, Ch. 8] is the explicit relationship between the isomonodromy and the isospectral curve provided by Sato's tau function; however, what becomes more mysterious is the appearance of PI-PVI [**BBT**, Section 8.11] "which can be interpreted as isomonodromic deformations but not as isospectral deformations". This seems to negate our attempt at linking isomonodromy and isospectrality, but we don't think the authors intend this negation, since they do exhibity a wave function for an isospectral hierarchy. We found a clue to what we believe they mean in [**SHC**, Section 1], "It is worth making some comments on the use of the word "hierarchy." In the theory of soliton equations, (...) all of the equations can be derived from a sequence of Lax pairs $\mathcal{L}\phi = \lambda\phi$, $\partial\phi/\partial t_j = \mathcal{M}_j\phi$, where the Lax operator $\mathcal{L}$ is the same for each j," giving rise to a sequence of commuting flows $\partial/\partial t_j$, $j = 1, 2, ...$ whereas the corresponding flows of P-type equations that arise as similarity reductions of the PDE hierarchy may not commute even when each has a Lax pair. "To add to the confusion," (*ibid.*) there is an additional sequence of P-type equations, also called "hierarchy", obtained through a sequence of Bäcklund transformations, "[t]herefore, for each N the corresponding [P-type] equation is a scaling similarity reduction of a flow in a different hierarchy of (...) PDEs." In summary, the wave function undergoes a hierarchy of commuting isospectral flows, but as it also keeps track of isomonodromy equations, those do not organize in a sequence of ODEs whose independent variables commute. This is one of the reasons why we pose the final question of Section 4.

For the case study of PVI, we specify the general construction to 2×2 matrices in $\mathfrak{sl}(2)$ whose entries are rational functions of λ:

$$M_\lambda(\lambda) = \sum_{k=1}^{K}\left(\frac{A_1^{(k)}}{\lambda-\lambda_k} + ... + \frac{A_{n_k+1}^{(k)}}{(\lambda-\lambda_k)^{n_k+1}}\right) - A_0^{(\infty)} - ... - A_{n_\infty-1}^{(\infty)}\lambda^{n_\infty-1}$$

(set $M_\lambda(\lambda) = \sum_k M_\lambda^{(k)}(\lambda)$ where $M_\lambda^{(k)}(\lambda)$ is the polar part of M at λ_k, including ∞), with one singularity at $\lambda = \infty$ and three regular singularities at $\lambda_0 = 0, \lambda_1 = 1, \lambda_t = t$.

Goal: Study the isomonodromy-deformation problem with respect to the parameter t.

The explicit link with isospectral deformations is the following: view the solution of $\partial_\lambda\Psi = M_\lambda\Psi$ as a "wave function":

$$\Psi(\lambda) = g(\lambda)e^{\sum_i \xi_i(\lambda)t_i},\ i = (k, n, \alpha),$$

$g(\lambda)$ a regular matrix, $\xi_i(\lambda) = \sum_\alpha e_{\alpha\alpha}/(\lambda-\lambda_k)^n$, with

$$\partial_{t_i}\Psi(\lambda) = M_i(\lambda)\Psi(\lambda),\ M_i(\lambda) = \left(g(\lambda)\xi_i(\lambda)g^{-1}(\lambda)\right)_-$$

(the notation $()_-$ means taking the polar part at λ_k).

This gives a hierarchy of commuting flows, satisfying the zero-curvature equations:

$$\partial_{t_i}M_j - \partial_{t_j}M_i - [M_i, M_j] = 0.$$

Now the $\Psi(\lambda)$ have essential singularities at the $\lambda = \lambda_k$ and at $\lambda = \infty$ and in general have non-trivial monodromy around the singularities (thus, are only defined on a Riemann surface),

$$\Psi(\lambda) \sim g^{(k)}(\lambda)e^{\xi^{(k)}(\lambda)},$$

where

$$g^{(k)}(\lambda) = g_0^{(k)} + g_1^{(k)}(\lambda - \lambda_k) + \dots$$

is regular at λ_k, and

$$\xi^{(k)} = B_0^{(k)} \log(\lambda - \lambda_k) + \sum_{\alpha;n=1}^{n_k} \frac{t_{(k,n,\alpha)}}{(\lambda - \lambda_k)^n} e_{\alpha\alpha}$$

$$\xi^{(\infty)} = B_0^{(\infty)} \log\left(\frac{1}{\lambda}\right) + \sum_{\alpha;n=1}^{n_\infty} t_{(\infty,n,\alpha)} \lambda^n e_{\alpha\alpha}$$

Goal: write evolution equations that describe the deformations of $M_\lambda(\lambda)$ under which the monodromy of $\Psi(\lambda)$ around a singularity of $M(\lambda)$ is fixed. The isomonodromic deformation parameters will include the $t_{(k,n,\alpha)}$, and we have an enriched family of commuting flows:

$$\partial_{\lambda_k}\Psi(\lambda) = M_{\lambda_k}\Psi(\lambda),\ M_{\lambda_k} = \left((g^{(k)}\partial_{\lambda_k}\xi_i^{(k)} g^{(k)-1}(\lambda)\right)_- .$$

In our illustration we choose the singularity data:

$$\xi^{(k)}(\lambda) = \begin{bmatrix} \theta_k & 0 \\ 0 & 0 \end{bmatrix} \log(\lambda - \lambda_k),\ k = 0, 1, t,$$

we diagonalize M at ∞ by global gauge transformation by a matrix constant in λ:

$$M(\lambda) = \frac{1}{\lambda}\begin{bmatrix} \kappa_1 & 0 \\ 0 & \kappa_2 \end{bmatrix} + O(\lambda^{-2}),\ \lambda \to \infty.$$

The equation $\partial_\lambda \Psi = M\Psi$ has a regular singularity at ∞ and the Fuchs condition reads: $\kappa_1 + \kappa_2 = \theta_0 + \theta_1 + \theta_t$. Set:

$$M(\lambda) = \begin{bmatrix} m_{11}(\lambda) & m_{12}(\lambda) \\ m_{21}(\lambda) & m_{22}(\lambda) \end{bmatrix} = \frac{1}{\lambda(\lambda-1)(\lambda-t)} A(\lambda),$$

with the entries of the matrix A polynomial in λ, in particular, setting: $A_{12}(\lambda) = \gamma(\lambda - y)$ introduces the parameter y which will turn out to satisfy PVI. We also parametrize $A_{11}(\lambda)$ by the values it takes at $\lambda = t$ and $\lambda = y$, in other words we interpolate as follows:

$$A_{11}(\lambda) = (\lambda - t)(\lambda - y)\kappa_1 + \frac{(\lambda - y)A_{11}(t) - (\lambda - t)A_{11}(y)}{t - y}.$$

Since $A_{21}(\lambda)$ is a linear function of λ,

$$A_{21}(t) = (1-t)A_{21}(0) + tA_{21}(1)$$

and from the condition that the determinant of A is zero at the three singularities $\lambda = 0, 1, t$ we derive three constraints:

$$A_{21}(\lambda) = \frac{A_{11}(\lambda)A_{22}(\lambda)}{\gamma(\lambda - y)},$$

which by substitution of the interpolation formula give a quadratic condition in $A_{11}(y)$ and "unexpectedly" (*loc. cit.*) linear in $A_{11}(y)$. The dynamical variables are now $y, \gamma, A_{11}(y)$ We use the equation of motion: $\partial_t M_\lambda = \partial_\lambda M_t + [M_t, M_\lambda]$ and afterwards set $\lambda = y$:

$$\begin{bmatrix} \dot m_{11}(\lambda) & \dot m_{12}(\lambda) \\ \dot m_{21}(\lambda) & \dot m_{22}(\lambda) \end{bmatrix}_{\lambda=y} = \frac{1}{t(t-1)(y-t)^2} \cdot \left\{ A(t) - \frac{1}{y(y-1)}[A(t), A(y)] \right\}.$$

Taking the (1,2) entry of this equation and evaluating it at $\lambda = y$ gives:

$$\dot{y} = \frac{y(y-1)(y-t)}{t(t-1)}\left[2m_{11}(y) - \frac{\theta_0}{y} - \frac{\theta_1}{y-1} - \frac{\theta_t - 1}{y-t}\right].$$

Taking the (1,1) entry of this equation and evaluating it at $\lambda = y$ gives:

$$\dot{m}_{11}(\lambda)|_{\lambda=y} = \frac{A_{11}(t)}{t(t-1)(y-t)^2} + \frac{\gamma}{t(t-1)}m_{21}(y).$$

Introduce the dynamical variable $z = m_{11}(y)$ and compute:

$$\dot{z} = \dot{m}_{11}(\lambda)|_{\lambda=y} + (\partial_\lambda m_{11})|_{\lambda=y} \cdot \dot{y}.$$

By eliminating z between the equations for $\dot{y}$ and $\dot{z}$ we obtain PVI for y, with appropriate values for the parameters: $\alpha = 1/2(\kappa_1 - \kappa_2 + 1)^2$, $\beta = -(1/2)\theta_0^2$, $\gamma = 1/2\theta_1^2$, $\delta = (1/2)(1-\theta_t)^2$.

In this context, the link between isomonodromy and integrable hierarchies of isospectral type arises as follows:

The τ function. The deformation equations imply that the following 1-form is closed:

$$Y = -\sum_k \mathrm{Res}_{\lambda=\lambda_k} \mathrm{Tr}(g^{(k)-1}\partial_\lambda g^{(k)} d\xi^{(k)})d\lambda,$$

summed over all the singularities including ∞.

Example. In the case of the Schlesinger equations,

$$Y = \frac{1}{2}\sum_{k\neq l} \mathrm{Tr}(A_k A_l)\frac{d\lambda_k - d\lambda_l}{\lambda_k - \lambda_l}.$$

DEFINITION 2.4. To define Sato's tau-function, set: $Y = d\log\tau$.

THEOREM 2.5. *The tau-function satisfies the Hirota equations, in the following sense.*

Define the bilinear operator

$$D_i^n f\cdot g = \left(\frac{\partial}{\partial y_i}\right)^n f(x+y)g(x-y)|_{y=0},$$

then for the polynomial operator

$$(D_1^4 + 3D_2^2 - 4D_1D_3)\tau\cdot\tau = 0,$$

implies the KP equation:

$$3u_{t_2t_2} + (-4u_{t_3} - 6uu_{t_1} + u_{t_1t_1t_1})_{t_1} = 0, \quad u = \frac{\partial^2}{\partial t_1^2}\log\tau.$$

The proof consists in showing that the tau-function obeys the Schlesinger transformations.

3. Connecting integrable PDEs and Painlevé-type. II: Symmetry reduction

One feature of integrability for a dynamical system is that it is invariant under a continuous group. There are many possible definitions of invariance, even more (computational) techniques to search for such a group, and a remaining gaping question whether all possible symmetries were found [**C1**, Example]. Our concern is limited to the following issues: Bringing together results by different authors who are apparently unaware of each other; Focusing on "soliton equations", and even more narrowly, on hierarchies that can be defined as isospectral deformations by a Lax pair, in particular have a spectral curve under certain finiteness conditions (which we will specify); And attempting to find a unifying theory for these symmetries, ideally by solving an 'inverse spectral problem', namely starting with geometric data to produce the deformations, their solutions, and their groups of symmetries at the same time. Much of the geometric ground at the basis of dynamical symmetries is covered in [**O**]; [**AC**] is a rich reference for examples and interconnections, although we were unable to satisfy ourselves that certain references specifically relevant to PVI (given, as most other results, without proofs or summaries) achieved the wanted results (we will highlight this in Subsection 3.4 below).

The somewhat surprising link between integrable equations and the Painlevé condition for ODEs was pointed out in [**AS1**], where the focus was on inverse scattering as a kind of "non-linear Fourier transform". Pursuant the linearized dispersion relation for KdV

$$u_t + 6uu_x + u_{xxx},$$

solutions to KdV were rescaled, following the observation that the scattering data have a homogeneity, one power for x versus cube powers for t and "self-similar" solutions were constructed cf. [AS2, §3.4]

$$u = \frac{1}{(3t)^{2/3}}w(z), \quad z = \frac{x}{(3t)^{1/3}},$$

and $w(z)$ satisfies $w''' + 6ww' - 2w - zw = 0$; a 'Miura transformation'[2] $w(z) = V'(z) - V(z)^2$ reduces this ODE:

$$(-2V + \partial_z)(V''' - 6V^2V' - V - zV') = 0.$$

Integration of the latter factor yields PII: $V'' - 2V^3 - zV - \alpha = 0$, where α is a constant.

On the other hand, $x' = x + 3t\lambda,\ t' = t + \lambda/\alpha,\ u' = u + \lambda$ reduces KdV to PI; the technique is to make $t' = 0$, or $\lambda = -\alpha t$, $z := x' = x - 3\alpha t^2$, $U(z) := u' = u - \alpha t$, and put $u_t,\ u_x$ in KdV to obtain PI: $U''' + 6UU' + \alpha = 0$.

These discoveries led to many further results linking specific PDEs to specific Painlevé equations, and most notably to conjectures for integrability which will not be reviewed here (cf. [**AS2, AC**] for surveys), except for mentioning that the Non-Linear Schrödinger equation was reduced to PIV [**BoPe**] with the interesting consequence that Bäcklund transformations for PIV correspond to the addition rule for the Weierstrass $\wp$-function.

Our question is, how to give a systematic understanding of the correspondence. For example, as recalled in the Introduction, given that PVI is the 'most general' of the six Painlevé equations in the sense that it can be reduced to the previous five in suitable limits, and the fact that the KP hierarchy reduces to both KdV and Boussinesq, one might naïvely ask whether KP is connected to PVI. However, in the treatments of KP that we found, this was not the case. In [**KL**], and in [**R**] (cf. [**AS2**, Exercise 12, Section 3.7] and [**TNK**], by similarity KP is reduced to PI or PII. None of these works cites any of the others, and the method is slightly different. In

[2]Recall that under the transform $u = v_x - v^2$, KdV becomes $(-2v + \partial_x)(v_t - 6v^2 v_x + v_{xxx}) = 0$, giving the mKdV equation. Solutions to mKdV therefore always transform back to solutions of KdV, but as is well known, the converse does not always hold.

both [**KL**] and [**TNK**], the general similarity method is implemented by postulating an infinitesimal deformation, and Lie-invariance is determined (see Subsection3.1 below), one reduction is found, but the invariance equations being non-linear, some simplifying assumptions lead to seemingly different results, namely only Boussinesq in [**KL**], Boussinesq or KdV in [**TNK**]. However, since both Boussinesq or KdV can be reduced once more to either PI or PII, from the point of vew of our question, the outcome is essentially the same. The method in [**R**] achieves both reductions at once, indeed Redekopp uses the linearized dispersion relation, as had Ablowitz and Segur:

$$u_{xt} + \frac{1}{4}u_{xxxx} + \frac{3}{4}u_{yy} = 0$$

for KP:

$$(u_t + \frac{3}{2}uu_x + \frac{1}{4}u_{xxx})_x + \frac{3}{4}\sigma u_{yy} = 0, \quad \sigma = \pm 1,$$

hence the phase

$$kx + my + \frac{t}{4}\left(k^3 - 3\frac{m^2}{k}\right) = (lx - \imath l^2 y + l^3 t) + nx + \imath n^2 y + n^3 t)$$

in the Fourier transform, with separation: $k = l + n$, $m = -\imath(l^2 - n^2)$, sets

$$p = (3t)^{1/3}l + \imath\frac{y}{(3t)^{2/3}}, \quad s = (3t)^{1/3}n - \imath\frac{y}{(3t)^{2/3}}$$

and achieves similarity by:

$$z = (3t)^{-1/3}x + \sigma(3t)^{-4/3}y^2,$$

integrating once the resulting ODE for $u = F(z)$ and multiplying by F, F^2 is seen to satisfy PII. We are unable to find a proof, however, that one or the other method yields potentially more possibilities, namely reductions to different Painlevé equations. We believe that the problem is revealed by a phenomenon observed, apparently for the first time, in [**OR**]: there may be transformation groups that act on the (independent and dependent) variables of an equation but are only "conditional symmetries", in the sense that they do not preserve the equation but still give rise to a condition of ODE type. Thus, both methods of similarity reduction and "nonclassical-reduction" were implemented by several authors, starting from the theory in [**BC**] and up to currently active computational advances; our point here is just that there is no way to find 'all' possible P-type equations that can be obtained from a given integrable PDE; to the best of our understanding, even when reaching a point in the reduction which where no special assumptions were made, the problem, being non-linear or overdetermined, may not have a "general solution" computable by available algorithms. Almost as a footnote, another work on self-similar reductions of KP (later than [**R**, **TNK**] and referring to neither) adopts an idea of P.A. Clarkson and M.D. Kruskal which creates self-similar reductions 'directly', and depending on arbitrary functions of time [**L**]. Ultimately, we must endorse [**FN**] as a statement of our understanding: "not even the correct general definition of [the class of "multiphase similarity solution"] has been found."

3.1. Aspect I: Lie symmetry reduction. Close to the previous similarity method but somehow working 'from the top down' is the theory of (Lie) symmetries of the PDE: these give rise to self-similar solutions, though as we saw above other similarity reductions may also be relevant; the interest of this method is that it gives a complete and systematic way to study the symmetry group of a PDE, in analogy with the differential Galois group of an ODE. For our purposes, the criterion is given in [**McLO**],

THEOREM 3.1. *A vector field*

$$v = \sum_{i=1}^{m} \xi^i(x,u)\frac{\partial}{\partial x_i} + \sum_{j=1}^{n} \phi_j(x,u)\frac{\partial}{\partial u_j}$$

is the infinitesimal generator of a symmetry group G of a system of differential equations $\Delta(x,u) = 0$, in the sense that the action of G on the space $\mathbb{R}^m \times \mathbb{R}^n \ni (x,u)$ sends solutions of the system to other solutions, if and only if the prolongation of v, acting on the spaces of partial derivatives of u with respect to x induced by the action of G on functions $u = f(x)$, is 0 whenever $\Delta = 0$.

Under technical conditions, it is then possible to obtain the G-invariant solutions by integrating a reduced system.

3.2. Aspect II: Inverse scattering. Originally, Ablowitz and Segur propose a method for solving the KdV equation, cf. [**AS2**]. The more general point of view of symmetry groups of PDEs was brought to bear in [**McLO**], where the focus is the Gel'fand-Levitan representation for an integrable equation; indeed, in [**FA**] the authors show that if φ satisfies the linear integral equation

$$\varphi(k) + \imath \exp\left[\imath(kx + k^3t)\right] \int_C \frac{\varphi(l)}{l+k}\, d\lambda(l) = \exp\left[\imath(kx + k^3t)\right]$$

where $d\lambda(l)$ and C are a measure and a contour, respectively, then $u = \partial_x \int_C \varphi(k)\, dk$ is a solution of the KdV equation. Furthermore, it is shown that the usual Gel'fand-Levitan equation is contained in the above as a special case. More general initial value problems are not solved. This result, however, generates a three-parameter family of solutions for Painlevé II, which is the nonlinear ordinary differential equation $d^2V/dx^2 - xV - 2V^3 - \alpha = 0$, where α is a constant. The generated solutions are only valid for $-\frac{1}{2} < \alpha < \frac{1}{2}$, as noted by D. Kaup [**FA**].

Following this program, namely a linearization of Painlevé-type problems, which can be used not only to give criteria for (non-linear PDE) integrability but also to detect special solutions of Painlevé equations and transformations between them, [**ARS**] went back from the (linear) integral equation to PIII and PIV; in this way one could aim to find PVI first, then connect it to an integrable hierarchy, and lastly 'geometrize' the connection.

The cleanest way of phrasing the appearance of the similarity in the Gel'fand-Levitan transform of inverse scattering that we found is given in [**AvdL**, Section 1], "a self-similarity reduction is closely related to the scaling behavior of [the] solutions". Let us illustrate this statement more precisely: in the example we gave at the beginning of the section, where substituting such function $u(x,t)$ into the mKdV equation yields PII, notice that the single similarity variable z ensures the scaling property;

$$u(\lambda x, \lambda^3 t) = \lambda^{-1} u(x,t)$$

"which characterizes the function $u(x,t)$ in the self-similarity condition." Now notice that this is also equivalent to the linear condition

$$(x\partial_x + 3t\partial_t)u(x,t) = -u(x,t),$$

a special case of the L_{-1} Virasoro constraint. The authors generalize to other integrable hierarchies this "equivalence of introducing a self-similarity limit by either imposing a certain scaling behavior or a certain Virasoro condition." This again begs the question whether all the self-similar reductions can be so characterized, particularly those that are "non-classical" and therefore do not come from a Lie algebra.

3.3. Aspect III: Isomonodromy. We now review the Flaschka-Newell procedure for deriving self-similar solutions of a zero-curvature deformation problem, in which they exhibit a link with P-type equations by specifically showing that the zero-curvature problem is equivalent to isomonodromy. However, they apply self-similarity first.

In [**FN**], in one of the two specific examples given, putting the scattering problem for mKdV $q_t + 6q^2q + q_{xxx} = 0$,

$$\begin{aligned} v_{1x} &= -\imath\zeta v_1 + qv_2 \\ v_{2x} &= \imath\zeta v_2 + rv_1, \end{aligned}$$

$$\begin{aligned} v_{1t} &= Av_1 + Bv_2 \\ v_{2t} &= Cv_1 + Dv_2 \end{aligned}$$

(A, B, C, D functions independent of the vector (v_1, v_2)) in self-similar form: $z = x/(3t)^{1/3}, \chi = \zeta(3t)^{1/3}, q(x,t) = (3t)^{-1/3}f(x(3t)^{-1/3})$, using the invariance of the (Baker) eigenvector $\underline{v} = (v_1, v_2)$, namely $\underline{v}(\beta x, \beta^3 t, \beta^{-1}\zeta)$ giving a solution to the same compatibility condition (note that this is the property for which Krichever [**K**] solves explicitly), then $\underline{w}(x(3t)^{-1/3}, \zeta(3t)^{1/3}) = \underline{v}(x,t,\zeta)$

$$\begin{aligned} w_{1\zeta} &= -\imath(4\zeta^2 + x + 2f^2)w_1 + (4\zeta f + 2\imath f')w_2 \\ w_{1\zeta} &= (4\zeta f - 2\imath f')w_1 + \imath(4\zeta^2 + x + 2f^2)w_2 \end{aligned}$$

(using x for $x(3t)^{-1/3}$ and ζ for $\zeta(3t)^{1/3}$) gives a system of ODE's whose Stokes multipliers near an irregular singular point are independent of z only if $w(z)$ satisfies a form of PII. This is a specific example of irregular singular point; the general theory runs along the same lines. A deformation of the system is said to be monodromy-preserving if the Stokes multipliers associated with formal solutions at $\zeta = \infty$, the monodromy matrix at $\zeta = 0$ and the matrix connecting fundamental solutions at $\zeta = 0$ and ∞ are unchanged when x is varied and q, r change as functions of x. For this, it is necessary and sufficient that $r = q_x$ and that q satisfy the second Painlevé equation $q_{xx} = 2q^3 + xq - \nu$. In order to prove this result, the authors give properties of the transform data. The functions r and q are derived via a linear singular integral equation for the columns of the fundamental solution matrix of the system. In the particular case $\nu = 0$ and $f(x) \sim \rho\mathrm{Ai}(x)$, the authors reproduce the solution found in [**AS1**]. When $\nu = n$, the integral equation yields the known rational solutions. This discovery agrees with the theme set by R. Fuchs, when he found an equivalence between PVI and an isomonodromy condition for a linear 2nd-order ODE of Fuchsian (L. Fuchs) type. In particular, the authors give comprehensive references to early examples relating Painlevé equations and isomonodromy (R. Fuchs, G.D. Birkhoff). In the spirit of the three proposals for generalization of PVI in Garnier's doctoral dissertation [**G**], they ask what is the condition on the x-deformation so that the two equations

$$y_x = Ay + y_\zeta, \quad y_{\zeta\zeta} = \left[\sum_0^m a_j \zeta^j + \sum_0^n \left(\frac{3/4}{(\zeta-\lambda_k)^2} + \frac{\rho_k}{\zeta-\lambda_k}\right)\right] y$$

form an integrable system (in the sense of Frobenius, namely compatibility), with A, B assumed rational in ζ; for $n = 0$ and $\lambda_0 = \lambda$ the answer turns out to be PII: $\lambda'' = 2\lambda^3 + x\lambda - \nu$.

What is more intriguing and presented as a speculation in [**FN**] is the idea that both problems should reduce to an algebraic condition. Through examples, [**FN**] advocate associating to an integrable PDE not only a similarity solution in one variable (and its attendant Painlevé problem) but also a multivariable P-type problem, which expresses an isomonodromic deformation.

3.4. The case study. As we were looking for similarity reduction of PDEs that yield PVI in the literature, we also found an early reference for the discovery of the link between integrability and the Painlevé property [**FY**], where the focus is not so much on 'general' similarity reduction but on the applications (Bäcklund transformations, cf. [**FA**] for a survey of applications of inverse scattering to the relevant ODEs). We also note that in [**AC**, Example 6.5.15] there are references, e.g. [**FLMS**], aiming to similarity reduction to PVI, but they appear instead to be relating the similarity ODE to PVI, short of showing equivalence; our understanding was vindicated by [**CGM**], "at the present time, various reductions of this system have been integrated with most of the six Painlevé functions [including the reference given in [**AC**]], but no explicit link with the generic sixth Painlevé equation has been found." In [**CGM**, Section 8], these authors then express their gratitude to the referee for providing reference [**KK**], which we had found earlier: we choose to report that calculation below, although PVI is likewise obtained in[**CGM**]. Lastly, we note that in [**AvdL**] the same integrable equation is reduced to PVI also, but the four PVI parameters are not arbitrary; indeed, 2α, -2β, 2γ and $1-2\delta$ are squares of integers.

The program that we briefly sketch [**KK**] extends (in principle) to the whole Drinfel'd-Sokolov integrable hierarchy the symmetry-group reduction leading to ODEs of Painlevé type. However, we see no way to predict which specific Painlevé equation one obtains.

The authors express the three-wave resonant system:

$$\partial_\tau u_i + c_i \partial_\chi u_i = \sqrt{-1}\gamma_i u_j^* u_k^*, \quad \{i,j,k\} = \{1,2,3\}$$

as the $(1,1,1)$-reduction of the 3-component KP hierarchy, equivalently the Drinfel'd-Sokolov hierarchy associated with the (1,1,1) homogeneous Heisenberg subalgebra of the affine Lie-algebra $\widehat{\mathfrak{gl}}_3$. The link with isomonodromy (also explored by other authors listed in the references of [**KK**]) is provided by a scaling symmetry of the $\widehat{\mathfrak{gl}}_3$ homogenoeous hierarchy, which sends solutions to solutions viewed as elements $g(z;\underline{t}) = \Psi_0(z;\underline{t})g(z;0)$, $g(z;0)$ being an element of the exponentiated algebra (its Kac-Moody group) and

$$\Psi_0(z;\underline{t}) := \exp\left(\sum_{a=1}^n \left(\sum_{j>0} z^j t_j^a\right) E_{aa}\right)$$

(E_{aa} are the diagonal matrices with just one entry equal to 1). The rescaling, by $\lambda \in \mathbb{C}^*$, acting on the Gauss decomposition $g(z;\underline{t}) = g_{<0}(z;\underline{t})^{-1}g_{\geq 0}(z;\underline{t})$ (with respect to the homogeneous gradation,

$$\widehat{\mathfrak{g}} = \oplus_j \in \mathbb{Z}\widehat{\mathfrak{g}}_j,\ \widehat{\mathfrak{g}}_j = \{x \in \widehat{\mathfrak{g}}|[d,x] = jx\},$$

is:

$$\underline{t}_\lambda = (\lambda^j t_j^1, ..., \lambda^j t_j^3)_{j>0}$$
$$\bar{g}_{<0}(z,\underline{t}) = \lambda^{D(\alpha)} g_{<0}(\lambda^{-1}z;\underline{t}_\lambda)\lambda^{-D(\beta)}\bar{g}_{\geq 0}(z,\underline{t}) = \lambda^{D(\alpha)} g_{\geq 0}(\lambda^{-1}z;\underline{t}_\lambda)\lambda^{-D(\beta)}$$

where $D(\alpha) = \mathrm{diag}(\alpha_1, \alpha_2, \alpha_3)$, $D(\beta) = \mathrm{diag}(\beta_1, \beta_2, \beta_3)$, and a choice $\beta_3 = -1$ will be made later to obtain the Schlesinger equations, while the parameters

$$\alpha = \frac{(\beta_1 - \beta_2 - 1)^2}{2}, \beta = \frac{(\alpha_1 + 1)^2}{2}, \gamma = \frac{(\alpha_2 + 1)^2}{2}, \delta\frac{1 - (\alpha_3 + 1)^2}{2}$$

will yield the general PVI equation. This permits reducing the equations so that the Baker functions with parameters,

$$\Psi = \Psi^{(0)}(z;\underline{t},\beta) = g_{\geq 0}(0;\underline{t})z^{D(\beta)},\ \text{ or } \Psi = \Psi^{(\infty)}(z;\underline{t},\alpha) = g_{<0}(0;\underline{t})\Psi_0(z;\underline{t})z^{D(\alpha)}$$

both satisfy a (nonautonomous) ODE, compatible with the time evolutions:

$$z\frac{\partial \Psi}{\partial z} = \left(D(\alpha) + \textstyle\sum_{a=1}^3 \sum_{j>0} j t_j^a B_j^a\right)\Psi$$
$$\frac{\partial \Psi}{\partial t_j^a} = B_j^a \Psi,\ 1 \leq a \leq 3,\ j > 0.$$

The key to transform this into a Schlesinger system (setting $t_i := t_1^i, t_j^i = 0$ for $j > 1$, is a Laplace transform[3]!

$$\Psi(z) \mapsto \Phi(x) = L[\Psi(z)](x) = \int_\gamma e^{-zx}\Psi(z)dz,\ \gamma \text{ such that } \int_\gamma \frac{\partial f}{\partial z}dz = 0.$$

The transformed (Schlesinger) equations have 4 regular singular points (t_1, t_2, t_3 and ∞); the constraint (rather, a WLOG choice) $\beta = -1$ makes the third column of the matrices equal zero, PVI is as customary the isomonodromy condition for the reduction to a 2×2-system on the space $(y_1, y_2, 0)$.

4. Comments and further aspects

We conclude with a discussion of what was and was not achieved, and related directions.

4.1. Our goals. We note that to some extent, in [**BePr**] we answered our question of finding a systematic way to link an integrable hierarchy, through its most general similarity reduction, and a P-type equation, or–in the opposite direction–guess from a P-type equation the attendant integrable hierarchies: we combined two unrelated constructions, sketched in short announcements in the 1980s by H. Flaschka and I.M. Krichever [**Fl, K**]: both authors stated that they would then provide the full technique in later papers, which never appeared. Roughly, the concept is this: use the PDE as in Subsection 3.3 to produce monodromy data [**Fl**]; use the monodromy data to build a Baker function and therefore the ODE of isomonodromy, as well as the PDE related to it, as in Subsection 3.3, by Riemann-Roch data on the curve of isospectrality [**K**]. In all this, one gets stuck because

[3]The authors refer to a 1994 paper by J. Harnad for this tool [**Ha**], which here they apply to a general monodromy problem for a 3×3 system with a simple pole at 0 and a double pole at ∞).

solving non-linear equations is not always possible, and they are produced by "clever guesswork" (cf. the Introduction), not by systematic association. The advantage is that the P-type equation is produced by differential algebra techniques, and we developed a "Sato-operator" type of isomonodromy. In fact, in [**BePr**] we were only able to go from Boussinesq to a new third-order equation of P-type: we were able to go from a P-type ODE to a integrable hierarchy which has symmetries yielding the ODE one begins with.

In conclusion, this article collects the connections we found between integrable hierarchies and P-type ODOs, hoping to facilitate further analyses of the exchange of geometric information.

4.2. Computational aspects. There are powerful computational aids that can detect the "classical symmetries" of ODEs and PDEs, namely actions of groups on the independent and dependent variables that leave the equation invariant, referred to as "Lie symmetry algebra" of the (system of) differential equations. References in [**C1**] discuss the available symbolic software, largely based on the method of differential Gröbner bases. However, computational methods for Gröbner bases over non-commutative coefficients only seem to have been proposed in [**Z**] (and published versions of various chapters of this dissertation). This would be a very important technique to study the ring of differential operators, which is non-commutative, and whose maximal-commutative subrings correspond to spectral curves/vector bundles as we illustrated above. One question, for example, is to check for the commuting Ordinary Differential Operators (ODOs) L_4 and L_{4g+2} (of order 4g+2) of [**Mi**], where $L_4 = \left(\partial_x^2 + \sum_{i=0}^3 \alpha_i x^i\right)^2 + g(g+1)\alpha_3 x$, with generic $\alpha_i \in \mathbb{C}$, $i = 0, ..., 3$, whether the $\mathbb{C}$-algebra generated by L_4 and L_{4g+2} is maximal-commutative in the Weyl algebra.

In addition, Clarkson and Mansfield have focused on the "nonclassical method of symmetry reduction" [**CM**], namely additional vector fields (obtained by solving in fact fewer conditions) that leave the equations invariant, but do not in general form a Lie algebra. There exists also an algorithmic "direct method" for finding symmetry reductions, but the situation is complicated enough to say that "the precise relationships between these methods has yet to be ascertained" [**C1**].

In our vein of posing overarching questions for this theory to come together, we would rather pose the following:

After symmetry reduction of a PDE whose Bäcklund transformations are known, once an ODE of P-type is found, as well as its Bäcklund transformations, or more generally a Weyl group acting on it, is the reduction diagram equivariant?

Acknowledgment

This report is based on research conducted in collaboration with G.N. Benes of Image Insight, Inc. GammaPix East Hartford, Connecticut.

References

[AC] M. J. Ablowitz and P. A. Clarkson, *Solitons, nonlinear evolution equations and inverse scattering*, London Mathematical Society Lecture Note Series, vol. 149, Cambridge University Press, Cambridge, 1991. MR1149378

[ARS] (series by the authors) M.J. Ablowitz, A. Ramani and H. Segur, Nonlinear evolution equations and ordinary differential equations of Painlevé type, *Lett. Nuovo Cimento (2)* **23** (1978), no. 9, 333–338; A connection between nonlinear evolution equations and ordinary differential equations of P-type. I. *J. Math. Phys.* **21** (1980), no. 4, 715–721; A connection between nonlinear evolution equations and ordinary differential equations of P-type. II. *J. Math. Phys.* **21** (1980), no. 5, 1006–1015.

[AS1] M. J. Ablowitz and H. Segur, *Exact linearization of a Painlevé transcendent*, Phys. Rev. Lett. **38** (1977), no. 20, 1103–1106, DOI 10.1103/PhysRevLett.38.1103. MR0442981

[AS2] M. J. Ablowitz and H. Segur, *Solitons and the inverse scattering transform*, SIAM Studies in Applied Mathematics, vol. 4, Society for Industrial and Applied Mathematics (SIAM), Philadelphia, Pa., 1981. MR642018

[AvdL] H. Aratyn and J. van de Leur, *Solutions of the Painlevé VI equation from reduction of integrable hierarchy in a Grassmannian approach*, Int. Math. Res. Not. IMRN, posted on 2008, Art. ID rnn 080, 41 pp., DOI 10.1093/imrn/rnn080. MR2439558

[BBT] O. Babelon, D. Bernard, and M. Talon, *Introduction to classical integrable systems*, Cambridge Monographs on Mathematical Physics, Cambridge University Press, Cambridge, 2003. MR1995460

[B] M. V. Babich, *Isomonodromic deformations and Painlevé equations*, Constr. Approx. **41** (2015), no. 3, 335–356, DOI 10.1007/s00365-015-9286-2. MR3346713

[BMcCW] E. Barouch, B.M. McCoy and T.T. Wu, Zero-field susceptibility of 2-dimensional Ising-model near TC, *Phys. Rev. Lett.* **31** (1973), no. 23, 1409-1411.

[BM] W. Barth and J. Michel, *Modular curves and Poncelet polygons*, Math. Ann. **295** (1993), no. 1, 25–49, DOI 10.1007/BF01444875. MR1198840

[BePr] G. N. Benes and E. Previato, *Differential algebra of the Painlevé property*, J. Phys. A **43** (2010), no. 43, 434006, 14 pp., DOI 10.1088/1751-8113/43/43/434006. MR2727780

[BoPe] M. Boiti and F. Pempinelli, *Nonlinear Schrödinger equation, Bäcklund transformations and Painlevé transcendents* (English, with Italian and Russian summaries), Nuovo Cimento B (11) **59** (1980), no. 1, 40–58, DOI 10.1007/BF02739045. MR598413

[BC] G. W. Bluman and J. D. Cole, *Similarity methods for differential equations*, Springer-Verlag, New York-Heidelberg, 1974. Applied Mathematical Sciences, Vol. 13. MR0460846

[C1] P. A. Clarkson, *Nonclassical symmetry reductions of nonlinear partial differential equations*, Math. Comput. Modelling **18** (1993), no. 10, 45–68, DOI 10.1016/0895-7177(93)90214-J. Similarity, symmetry and solutions of nonlinear boundary value problems (Wollongong, 1992). MR1253590

[C2] P. A. Clarkson, *Painlevé equations—nonlinear special functions*, Proceedings of the Sixth International Symposium on Orthogonal Polynomials, Special Functions and their Applications (Rome, 2001), J. Comput. Appl. Math. **153** (2003), no. 1-2, 127–140, DOI 10.1016/S0377-0427(02)00589-7. MR1985685

[CM] P. A. Clarkson and E. L. Mansfield, *Algorithms for the nonclassical method of symmetry reductions*, SIAM J. Appl. Math. **54** (1994), no. 6, 1693–1719, DOI 10.1137/S0036139993251846. MR1301278

[CGM] R. Conte, A. M. Grundland, and M. Musette, *A reduction of the resonant three-wave interaction to the generic sixth Painlevé equation*, J. Phys. A **39** (2006), no. 39, 12115–12127, DOI 10.1088/0305-4470/39/39/S07. MR2266215

[DM] B. Dubrovin and M. Mazzocco, *Monodromy of certain Painlevé-VI transcendents and reflection groups*, Invent. Math. **141** (2000), no. 1, 55–147, DOI 10.1007/PL00005790. MR1767271

[ENP] V.Z. Enolskii, F. Nijhoff and E. Previato, Note on isomonodromic Garnier systems: Lagrangian structure and higher-genus analogues of the Painlevé VI equation Preprint Institut Mittag-Leffler (ISSN 1103-467X ISRN IML-R-28-05/06–SE+fall).

[Fl] H. Flaschka, *A commutator representation of Painlevé equations*, J. Math. Phys. **21** (1980), no. 5, 1016–1018, DOI 10.1063/1.524549. MR574873

[FN] H. Flaschka and A. C. Newell, *Monodromy- and spectrum-preserving deformations. I*, Comm. Math. Phys. **76** (1980), no. 1, 65–116. MR588248

[FA] (series by the authors) A.S. Fokas and M.J. Ablowitz, Linearization of the Korteweg-de Vries and Painlevé II equations, *Phys. Rev. Lett.* **47** (1981), no. 16, 1096–1100. A.S.

Fokas and M.J. Ablowitz, On a unified approach to transformations and elementary solutions of Painlevé equations, *J. Math. Phys.* **23** (1982), no. 11, 2033–2042.

[FLMS] A. S. Fokas, R. A. Leo, L. Martina, and G. Soliani, *The scaling reduction of the three-wave resonant system and the Painlevé* VI *equation*, Phys. Lett. A **115** (1986), no. 7, 329–332, DOI 10.1016/0375-9601(86)90624-9. MR838679

[FY] A. S. Fokas and Y. C. Yortsos, *The transformation properties of the sixth Painlevé equation and one-parameter families of solutions*, Lett. Nuovo Cimento (2) **30** (1981), no. 17, 539–544. MR621591

[Fu] R. Fuchs, Sur quelques équations differentielles lineaires du second ordre, *C. R. Acad. Sci. (Paris)* 141:555–558 (1906). *Math. Ann.* 63:301–321 (1907).

[G] R. Garnier, *Sur des équations différentielles du troisième ordre dont l'intégrale générale est uniforme et sur une classe d'équations nouvelles d'ordre supérieur dont l'intégrale générale a ses points critiques fixes* (French), Ann. Sci. École Norm. Sup. (3) **29** (1912), 1–126. MR1509146

[Ha] J. Harnad, *Dual isomonodromic deformations and moment maps to loop algebras*, Comm. Math. Phys. **166** (1994), no. 2, 337–365. MR1309553

[Hi] N. J. Hitchin, *Poncelet polygons and the Painlevé equations*, Geometry and analysis (Bombay, 1992), Tata Inst. Fund. Res., Bombay, 1995, pp. 151–185. MR1351506

[IKSY] K. Iwasaki, H. Kimura, S. Shimomura, and M. Yoshida, *From Gauss to Painlevé*, Aspects of Mathematics, E16, Friedr. Vieweg & Sohn, Braunschweig, 1991. A modern theory of special functions. MR1118604

[KK] S. Kakei and T. Kikuchi, *The sixth Painlevé equation as similarity reduction of* $\widehat{\mathfrak{gl}}_3$ *generalized Drinfel'd-Sokolov hierarchy*, Lett. Math. Phys. **79** (2007), no. 3, 221–234, DOI 10.1007/s11005-007-0144-4. MR2309777

[KL] P. Kaliappan and M. Lakshmanan, *Kadomtsev-Petviashvile and two-dimensional sine-Gordon equations: reduction to Painlevé transcendents*, J. Phys. A **12** (1979), no. 10, L249–L252. MR545388

[KaKa] M. Kashiwara and T. Kawai, *Monodromy structure of solutions of holonomic systems of linear differential equations is invariant under the deformation of the system*, Publ. Res. Inst. Math. Sci. **15** (1979), no. 3, 741–747, DOI 10.2977/prims/1195187873. MR566078

[K] I. M. Krichever, *Self-similar solutions of equations of Korteweg - de Vries type* (Russian), Funktsional. Anal. i Prilozhen. **14** (1980), no. 3, 83–84. MR583811

[L] S. Y. Lou, *Similarity solutions of the Kadomtsev-Petviashvili equation*, J. Phys. A **23** (1990), no. 13, L649–L654. MR1062976

[McLO] J. B. McLeod and P. J. Olver, *The connection between partial differential equations soluble by inverse scattering and ordinary differential equations of Painlevé type*, SIAM J. Math. Anal. **14** (1983), no. 3, 488–506, DOI 10.1137/0514042. MR697525

[M1] Yu. I. Manin, *Rational curves, elliptic curves, and the Painlevé equation*, Surveys in modern mathematics, London Math. Soc. Lecture Note Ser., vol. 321, Cambridge Univ. Press, Cambridge, 2005, pp. 24–33, DOI 10.1017/CBO9780511614156.004. MR2166923

[M2] Yu. I. Manin, *Sixth Painlevé equation, universal elliptic curve, and mirror of* $\mathbf{P}^2$, Geometry of differential equations, Amer. Math. Soc. Transl. Ser. 2, vol. 186, Amer. Math. Soc., Providence, RI, 1998, pp. 131–151, DOI 10.1090/trans2/186/04. MR1732409

[Mi] A. E. Mironov, *Self-adjoint commuting ordinary differential operators*, Invent. Math. **197** (2014), no. 2, 417–431, DOI 10.1007/s00222-013-0486-8. MR3232010

[O] P. J. Olver, *Applications of Lie groups to differential equations*, 2nd ed., Graduate Texts in Mathematics, vol. 107, Springer-Verlag, New York, 1993. MR1240056

[OR] (series by the authors:) P. J. Olver and P. Rosenau, The construction of special solutions to partial differential equations, *Phys. Lett. A* **114** (1986), no. 3, 107–112; Group-invariant solutions of differential equations, *SIAM J. Appl. Math.* **47** (1987), no. 2, 263–278.

[Pi] E. Picard, *Remarques sur les équations différentielles* (French), Acta Math. **17** (1893), no. 1, 297–300, DOI 10.1007/BF02391996. Extrait d'une lettre adressée à M. Mittag-Leffler. MR1554841

[Po] H. Poincaré, *Sur les groupes des équations linéaires* (French), Acta Math. **4** (1884), no. 1, 201–312, DOI 10.1007/BF02418420. MR1554639

[R] L. G. Redekopp, *Similarity solutions of some two-space-dimensional nonlinear wave evolution equations*, Stud. Appl. Math. **63** (1980), no. 3, 185–207, DOI 10.1002/sapm1980633185. MR595454

[SMJ] (series by the authors:) M. Satō, T. Miwa and M. Jimbō, Studies in holonomic quantum fields, Research on microlocal analysis, in *Proc. Sympos., Res. Inst. Math. Sci., Kyoto Univ., Kyoto, 1977)*, Sūrikaisekikenkyūsho Kókyūroku No. 295 (1977), 77–87; Studies on holonomic quantum fields. I. *Proc. Japan Acad. Ser. A Math. Sci.* **53** (1977), no. 1, 6–10; Studies on holonomic quantum fields. II. *Proc. Japan Acad. Ser. A Math. Sci.* **53** (1977), no. 5, 147–152; Studies on holonomic quantum fields. III. *Proc. Japan Acad. Ser. A Math. Sci.***53** (1977), no. 5, 153–158; Studies on holonomic quantum fields. IV. *Proc. Japan Acad. Ser. A Math. Sci.* **53** (1977), no. 6, 183–185; Studies on holonomic quantum fields. V. *Proc. Japan Acad. Ser. A Math. Sci.* **53** (1977), no. 7, 219–224; Studies on holonomic quantum fields. VI. *Proc. Japan Acad. Ser. A Math. Sci.* **54** (1978), no. 1, 1–5; Studies on holonomic quantum fields. VII. *Proc. Japan Acad. Ser. A Math. Sci.* **54** (1978), no. 2, 36–41.

[SHC] A. Sen, A. N. W. Hone, and P. A. Clarkson, *On the Lax pairs of the symmetric Painlevé equations*, Stud. Appl. Math. **117** (2006), no. 4, 299–319, DOI 10.1111/j.1467-9590.2006.00356.x. MR2263633

[TNK] M. Tajiri, T. Nishitani, and S. Kawamoto, *Similarity solutions of the Kadomtsev-Petviashvili equation*, J. Phys. Soc. Japan **51** (1982), no. 7, 2350–2356, DOI 10.1143/JPSJ.51.2350. MR673893

[Z] Y. Zhang, *Algorithms for noncommutative differential operators*, ProQuest LLC, Ann Arbor, MI, 2004. Thesis (Ph.D.)–The University of Western Ontario (Canada). MR2706667

[ŻF] H. Żołądek and G. Filipuk, *Painlevé equations, elliptic integrals and elementary functions*, J. Differential Equations **258** (2015), no. 4, 1303–1355, DOI 10.1016/j.jde.2014.10.018. MR3294349

Department of Mathematics and Statistics, Boston University, Boston, Massachusetts 02215

Email address: `ep@bu.edu`

Contemporary Mathematics
Volume **724**, 2019
https://doi.org/10.1090/conm/724/14595

Quasi-quadratic residue codes and hyperelliptic curves

Nigel Boston and Jing Hao

ABSTRACT. We introduce a family of codes called *quasi-quadratic residue (QQR) codes* in algebraic coding theory. We are interested in these codes because their weight distributions play an important role in a famous conjecture in coding theory known as *Goppa's conjecture*. We summarize some existing results on their weight distributions. Furthermore, the weight of their codewords have a close relation with the number of points on corresponding hyperelliptic curves. In the second part, we implement a heuristic model to estimate the limiting behaviour of number of points on related hyperelliptic curves over $\mathbb{F}_p$ as p goes to infinity. For primes $p < 50$, explicit calculations are given. These results provide evidence that Goppa's conjecture is dubious.

1. Quasi-quadratic residue codes

1.1. Introduction to algebraic coding theory. In this section, we will give a brief introduction to coding theory in the setting of *linear codes*. For a more general introduction to coding theory, check eg. [**7**].

DEFINITION 1.1 (Linear code). Let $\mathbb{F}_q$ be a finite field. A linear code C is a subspace of $\mathbb{F}_q^n$. The *length* of C is n. Elements of C are called *codewords*.

DEFINITION 1.2 (Hamming distance). Let $u=(u_1, \cdots, u_n)$ and $v=(v_1, \cdots, v_n)$. Then

$$d(u, v) = |\{i|1 \leq i \leq n, u_i \neq v_i\}|$$

is called the *Hamming distance* between u and v.

We can also define the weight of a codeword.

DEFINITION 1.3 (Hamming weight). The *weight* of a codeword c is

$$wt(c) = d(c, 0)$$

For a linear code, it's easy to see that the minimum distance (between distinct codewords) is the same as the minimum weight of a nonzero codeword. A code C is said to be an $[n, k, d]-$code if its length is n, its dimension is k as a vector space, and its minimum distance is d.

2010 *Mathematics Subject Classification.* Primary 94B65.

1.2. Goppa's conjecture. For a code C with parameters $[n, k, d]$, we define its *information rate* R to be k/n. This measures the efficiency of the code. Let $\delta = d/n$. A code C with minimum distance d can correct up to $\frac{d-1}{2}$ errors, and therefore δ measures the error correcting ability of the code.

Ideally, we want R and δ both to be large, and we are interested in the asymptotic behavior of (δ, R) as the length of the code tends to infinity. More precisely, we want to understand pairs (δ, R) such that there is a family of codes $\{C_i | i \in I\}$ of increasing length with corresponding (δ_i, R_i) approaching (δ, R). There is a fundamental tradeoff captured by the following theorem.

THEOREM 1.4 (Manin). [**8**, Section 3] *Fix q, and let U_q be the set of limit points of pairs (δ, R) for all linear codes over $\mathbb{F}_q$. There exists a continuous function $\alpha_q(x)$, $x \in [0, 1]$, such that*

$$U_q = \{(\delta, R) | 0 \le R \le \alpha_q(\delta)\}$$

Moreover,

$$\alpha_q(0) = 1;$$
$$\alpha_q(\delta) \le \max\{1 - \frac{q}{q-1}\delta, 0\}$$

Manin's theorem established the existence of a curve that separates the dense and isolated pairs (δ, R). From $\alpha_q(\delta) \le \max\{1 - \frac{q}{q-1}\delta, 0\}$, one can deduce that $\alpha_q(\delta) = 0$ for $\delta \ge \frac{q-1}{q}$. This also follows from the *Plotkin bound* (see e.g. [**7**, Chapter 2, Theorem 1]). For δ strictly between 0 and $\frac{q-1}{q}$, not a single value of $\alpha_q(\delta)$ is known.

There are, however, known upper and lower bounds for $\alpha_q(\delta)$. Gilbert ([**3**]) and Varshamov ([**11**]) showed that $\alpha_q(\delta) \ge 1 - \delta \log_q(q-1) + \delta \log_q(\delta) + (1-\delta)\log_q(1-\delta)$. For $q = 2$, it is believed that this is an equality.

DEFINITION 1.5 (Binary entropy function). The binary entropy function $H_2(x)$ is

$$H_2(x) = -x \log_2(x) - (1 - x) \log_2(1 - x)$$

CONJECTURE 1.6 (Goppa's conjecture [**4**]).

$$\alpha_2(x) = 1 - H_2(x)$$

This conjecture is related to hyperelliptic curves via a family of codes called *quasi-quadratic residue codes*. We give an introduction of these codes in the following section.

1.3. Quasi-quadratic residue codes. We first introduce *cyclic codes*.

DEFINITION 1.7 (Cyclic code). A linear code is called *cyclic* if any cyclic shift of a codeword is also a codeword.

For a cyclic code of length n over $\mathbb{F}_q$, a codeword $c = (c_0, c_1, \cdots, c_{n-1})$ can be identified with the polynomial $c_0 + c_1 x + \cdots + c_{n-1}x^{n-1}$ in the quotient ring $R_n := \mathbb{F}_q[x]/(x^n - 1)$. Via this identification, one can check that a cyclic code of length n is an ideal of $\mathbb{F}_q[x]/(x^n - 1)$ (see e.g. [**7**, Chapter 7, Section 3]).

Since R_n is a PID, every ideal can be generated by a single element of R_n. This polynomial is called a *generator polynomial* of the code.

Let p be an odd prime, and S a subset of $\mathbb{F}_p$. We can assign to S a polynomial in $\mathbb{F}_2[x]/(x^p - 1)$ by

$$r_S := \sum_{a \in S} x^a$$

Denote the quadratic residues and the quadratic non-residues of $\mathbb{F}_p$ by Q and N respectively. The following codes are well studied.

DEFINITION 1.8 (Quadratic residue codes). Let $p \equiv \pm 1 \pmod 8$.

$$\{(r_Q r_S) | S \subseteq \mathbb{F}_p\}$$
$$\{(r_N r_S) | S \subseteq \mathbb{F}_p\}$$

are the *quadratic residue codes* associated with p.

Quasi-quadratic residue codes form a family of double-circulant codes.

DEFINITION 1.9 (Quasi-quadratic residue (QQR) codes). For an odd prime p,

$$C := \{(r_Q r_S, r_N r_S) | S \subset \mathbb{F}_p\}$$

is called the *quasi-quadratic residue code* associated with p. $(r_Q r_S, r_N r_S)$ is identified with an element in $\mathbb{F}_2^{2p}$ in the usual way.

The QQR code associated with p has length $2p$. Its dimension is p if $p \equiv 3 \pmod 4$ and $p - 1$ if $p \equiv 1 \pmod 4$ (see [**5**, page 128]).

The following proposition indicates the relation between QQR codes and hyperelliptic curves.

PROPOSITION 1.10 (Joyner [**2**]). *Let C be the QQR code associated with p, and $c = (r_Q r_S, r_N r_S) \in C$, where $S \subseteq \mathbb{F}_p$. Define $f_S(x) := \prod_{a \in S}(x - a)$. Let $X_S(\mathbb{F}_p)$ be the set of points on the hyperelliptic curve $y^2 = f_S(x)$ over $\mathbb{F}_p$.*

(1) *If $|S|$ is even,*
$$wt(c) = 2p - |X_S(\mathbb{F}_p)|$$

(2) *If $|S|$ is odd and $p \equiv 1 \pmod 4$,*
$$wt(c) = 2p - |X_{S^c}(\mathbb{F}_p)|$$

(3) *If $|S|$ is odd and $p \equiv 3 \pmod 4$,*
$$wt(c) = |X_{S^c}(\mathbb{F}_p)|$$

REMARK 1.11. Throughout this paper, by the points on hyperelliptic curves we always mean affine points. The original result by Joyner included points at infinity.

This result built a bridge between coding theory and hyperelliptic curves. By counting points on hyperelliptic curves over $\mathbb{F}_p$, we can obtain results on the weight distribution of QQR codes. This is illustrated by the following relation between Goppa's conjecture and hyperelliptic curves.

THEOREM 1.12 (Joyner [**2**]). *Let $B(c, p)$ be the statement: $|X_S(\mathbb{F}_p)| \leq c \cdot p$ for all $S \subset \mathbb{F}_p$. If $B(1.77, p)$ is true for infinitely many primes p with $p \equiv 1 \pmod 4$, then Goppa's conjecture is false.*

PROOF. QQR codes have rate approaching $\frac{1}{2}$ as $p \to \infty$, and we check that $(0.115, \frac{1}{2})$ is above the Gilbert-Varshamov curve

$$R = 1 - H_2(\delta)$$

If $B(1.77, p)$ is true for infinitely many primes p with $p \equiv 1 \pmod 4$, then according to Proposition 1.10, for any codeword c in the code associated with p, we must have

$$wt(c) \geq 2p - 1.77p \geq 0.23p$$

Then

$$\delta = \frac{d}{2p} \geq 0.115$$

This contradicts Goppa's conjecture. □

1.77 in the original theorem is a rounded number. We can deduce the best possible value by examining the proof.

COROLLARY 1.13. *Let γ_0 be the solution to $\frac{1}{2} = H_2(\frac{2-x}{2})$ in interval $[1, 2]$, where H_2 is the binary entropy function. If for $\gamma < \gamma_0$, $B(\gamma, p)$ is true for infinitely many primes p with $p \equiv 1 \pmod 4$, then Goppa's Conjecture is false.*

PROOF. If γ_0 satisfies $\frac{1}{2} = H_2(\frac{2-\gamma_0}{2})$, then $(\frac{2-\gamma_0}{2}, \frac{1}{2})$ is on the Gilbert-Varshamov curve. The rest is essentially the same as the proof of Theorem 1.12. □

REMARK 1.14. Since the binary function $H_2(x)$ is symmetric about $x = 1/2$, there are two solutions to equation $\frac{1}{2} = H_2(\frac{2-x}{2})$ and we take the solution in $[1, 2]$. For reference, the rounded value for γ_0 to 10 decimal places is $\gamma_0 \approx 1.7799442711$.

From Theorem 1.12 and Corollary 1.13 we see that the following function is of great interest.

$$M(p) := \max_{S \subseteq \mathbb{F}_p} \frac{|X_S(\mathbb{F}_p)|}{p}$$

We denote the set of p satisfying the condition in Corollary 1.13 as $\mathfrak{S}$.

$$\mathfrak{S} = \{p | p > 2 \text{ is prime}, M(p) < \gamma\}$$

for $\gamma < \gamma_0$. We will give evidence that $\mathfrak{S}$ is huge, casting doubt on Goppa's conjecture.

1.4. Weight distribution of QQR codes. In this section, we summarize some previous results on QQR codes.

From the definition of quadratic residue code and QQR code, it is obvious that when $p \equiv 7 \pmod 8$, by only taking the first p bits of a QQR code, we can obtain the corresponding quadratic residue code with the same p. Moreover, we can show a stronger connection between quadratic residue codes and QQR codes in this case.

PROPOSITION 1.15 (Boston-Hao [**1**]). *When $p \equiv 7 \pmod 8$, let C be the QQR code associated with p, and let Q and N denote the corresponding quadratic residue codes. Then C is the even subcode of $Q \oplus N$. Moreover, their minimum distances satisfy*

$$d(C) = d(Q) + 1$$

The *even subcode* of a linear code is the subset of codewords with even weight, which forms a linear code.

For a linear code C, the *weight enumerator* counts the number of codewords of each weight. We give the formal definition below.

DEFINITION 1.16 (Weight enumerator). The *weight enumerator* of a code of length n (or a subset of such a code) is

$$A_C := \sum_{i=0}^{n} A_i x^{n-i} y^i$$

where

$$A_i = |\{c \in C | wt(c) = i\}|$$

Similarly to minimum distance, the weight enumerator measures the error-correcting ability of a code. Whilst one can use a computational algebra system such as MAGMA to compute the weight enumerator by exhaustive search, this is only practical for small n. If, however, we know some properties of the weight enumerator of a code, the amount of work will be significantly reduced.

We have shown that the weight enumerators of quadratic residue codes and QQR codes have the following properties.

THEOREM 1.17 (Boston-Hao [**1**]). *The weight enumerator of a quadratic residue code is divisible by $(x+y)^d$, where d is its minimum distance.*

THEOREM 1.18 (Boston-Hao [**1**]). *The weight enumerator of a QQR code is divisible by $(x^2+y^2)^{d-1}$, where d is its minimum distance.*

Using Theorem 1.17, we were able to correct the weight enumerators for $p = 113, 127$ of quadratic residue codes posted on OEIS [**10**].

As an alternative to weight enumerator, we can also define a distribution function for the weights of a code.

DEFINITION 1.19 (Cumulative distribution function). The *cumulative distribution function* $A(z)$ of a code C of dimension k is given by

$$A(z) = \sum_{i \geq \mu - \sigma z} a_i$$

where

$$a_i = A_i / 2^k$$

$$\mu = \sum_{i=0}^{n} i a_i$$

We have the following result on the weight distributions of QQR codes.

THEOREM 1.20 (Boston-Hao [**1**]). *The weight distributions of QQR codes are asymptotically normal as p tends to infinity.*

More precisely, let $\Phi(z)$ be the cumulative density function of the normal distribution defined by

$$\Phi(z) = \frac{1}{\sqrt{2\pi}} \int_{-\infty}^{z} e^{-t^2/2} dt$$

Then $A(z)$ converges to $\Phi(z)$ uniformly as p tends to infinity.

Using Theorem 1.20 and Proposition 1.10, we can show that for a family of hyperelliptic curves, if we symmetrize the distribution of number of points, the result will converge to a normal distribution as p tends to infinity.

Interested readers can find more details in [**1**].

2. Hyperelliptic curves

2.1. Heuristics. As discussed in the previous section, the validity of Goppa's conjecture is closely linked with number of points on hyperelliptic curves over finite fields. A good estimate of $M(p)$ when p is sufficiently large, will suggest whether Goppa's conjecture is true or not.

This raises some interesting questions. For small primes p, we can compute $M(p)$ explicitly. As p grows, the computation load is significantly larger. Instead of increasing our computational capacity, we give an estimate of this value using a heuristic model.

Recall that for $S \subseteq \mathbb{F}_p$,

$$f_S(x) = \prod_{a \in S} (x - a)$$

and the hyperelliptic curve X_S is defined by

$$y^2 = f_S(x)$$

We are interested in $|X_S(\mathbb{F}_p)|$, i.e. the number of points on X_S over $\mathbb{F}_p$. As x varies, there are three possibilities for $f_S(x)$:

(1) If $x \in S$, then $f_S(x) = 0$. $(x, 0) \in X_S(\mathbb{F}_p)$.
(2) If $x \notin S$ and $f_S(x) \in Q$, then $(x, \pm\sqrt{f_S(x)}) \in X_S(\mathbb{F}_p)$.
(3) If $x \notin S$ and $f_S(x) \in N$, there are no points in $X_S(\mathbb{F}_p)$ with the first coordinate equal to x.

Therefore

$$|X_S(\mathbb{F}_p)| = |S| + 2|\{x | f_S(x) \in Q\}| \tag{2.1}$$

Q and N contain the same number of elements. The number of points on hyperelliptic curves over $\mathbb{F}_p$ follow a generalized Sato-Tate distribution. However, this distribution is quite complex. Recent work by Larsen[**6**] showed that the limit of the generalized Sato-Tate distribution tends to normal when the genus tends to infinity. Therefore, we use a simplified model instead.

We hypothesize that, for a given $x \notin S$, the probability that $f_S(x)$ is in Q is half. This leads to the following considerations.

Therefore, we use a binomial model $Bin(p - |S|, \frac{1}{2})$ to estimate $M(p)$. For each set S, let

$$n := |S|$$
$$m := |\{f_S(x) \in Q, x \notin S\}|$$

Then

$$|X_S(\mathbb{F}_p)| = 2m + n$$

The probability for any combination of m and n is $\frac{\binom{p-n}{m}}{2^{p-n}}$. There are $\binom{p}{n}$ sets with size n, and hence the total probability is

$$Prob(p, n, m) := \frac{\binom{p}{n}\binom{p-n}{m}}{2^{p-n}} \tag{2.2}$$

We assume that when $Prob(p, n, m)$ is greater than 1, there are at least m values of x such that $f_S(x) \in Q$. We use MAGMA to compare the result of this assumption and the actual calculation for small primes. For each prime p, we take

$$\max\{2m + n | n + m \leq p, Prob(p, n, m) \geq 1\} \tag{2.3}$$

and compare to the actual number. See Table 1.

TABLE 1. Heuristics

p	$\max(2m+n)$	$\max_{S\subseteq\mathbb{F}_p}\lvert X_S(\mathbb{F}_p)\rvert$	$\max(2m+n)/p$
3	4	3	1.33
5	8	7	1.6
7	11	11	1.57
11	18	15	1.64
13	22	19	1.69
17	29	29	1.70
19	32	31	1.68
23	39	39	1.69

For $p \leq 23$, we can see that the actual number of points and the experimental data match well. Moreover, the heuristic results seem to be slightly larger than the actual number. This difference is in our favor as we will show later that the heuristics suggest that $M(p) \approx \gamma_0$ for most large p, which indicates that the actual $M(p)$ for larger p might be smaller. Besides, this model will approximate the number of points better as p gets larger. Next, we will give an estimate of $M(p)$ when p is sufficiently large based on this model.

We are interested in when the following holds.

$$\frac{\binom{p}{n}\binom{p-n}{m}}{2^{p-n}} \geq 1 \tag{2.4}$$

Since binomial coefficients are symmetric, for each p and given n, there are two solutions to the equality

$$\frac{\binom{p}{n}\binom{p-n}{m}}{2^{p-n}} = 1 \tag{2.5}$$

These two solutions sum to $p-n$, and the larger solution is the largest m can be in order that (2.4) holds. Taking logarithms to base 2, (2.5) can be rewritten as

$$\log_2\binom{p}{p-n} + \log_2\binom{p-n}{m} = p-n$$

Using Stirling's approximation on $\log_2\binom{p}{r}$ and $\log_2\binom{r}{m}$,

$$\log_2\binom{p}{p-n} \approx pH_2(\frac{p-n}{p})$$

$$\log_2\binom{p-n}{m} \approx (p-n)H_2(\frac{m}{p-n})$$

where $H_2(t) = -t\log_2(t) - (1-t)\log_2(1-t)$ is the binary entropy function. Letting

$$x := \frac{p-n}{p}$$

$$y := \frac{m}{p-n}$$

We obtain

$$H_2(x) + xH_2(y) = x$$

and it's easy to check $(2m+n)/p = 2xy - x + 1$. Our goal is to maximize $2xy - x + 1$ under the condition that $H_2(x) + xH_2(y) = x$. Applying the method of Lagrange multipliers and letting

$$f(x,y) = 2xy - x + 1$$
$$g(x,y) = H_2(x) + xH_2(y) - x$$

Then f and g satisfy

$$\nabla f = \lambda \cdot \nabla g$$
$$g = 0$$

We have the following two equations

$$\log_2(y) + \log_2(1-y) = 2(-\log_2(x) + \log_2(1-x) - 1)$$
$$H_2(x) + xH_2(y) = x$$

This yields the numerical solutions

$$x = 0.804156533029013$$
$$y = 0.9849455541854943$$

This gives $2xy - x + 1 = 1.77994427112$. By comparing the number in decimals, this looks exactly like γ_0! Note for certain special primes p, using advanced algebraic number theory, $M(p)$ can be shown to exceed γ_0. In fact, for every positive integer ℓ, we can find a sequence of primes p satisfying $M(p) > (2 - \frac{1}{\ell})p$ [**5**, page 142]. We, however, are concerned with what happens for generic primes p and this is where the heuristics should apply.

2.2. Calculations for small primes. In this section, we take $\gamma = 1.775 < \gamma_0$ and provide explicit calculations to show that for $p < 50$, only 41 has $M(p) > \gamma$. In other words, $\mathfrak{S}$ contains every prime less than 50 other than 41, providing some initial evidence that $\mathfrak{S}$ is large. For $p \leq 23$, we have given the exact values of $M(p)$ in Table 1. For $p > 23$, we will explain why $M(p) < \gamma$.

The extreme case where $\forall x \in \mathbb{F}_p$, either $f_S(x) = 0$ (when $x \in S$) or $f_S(x) \in Q$ (whenever $x \notin S$) is of interest. We define this family of curves as the following.

DEFINITION 2.1 ($k-$extremal curves). For $S \subseteq \mathbb{F}_p$, X_S is called a *k-extremal* curve over $\mathbb{F}_p$ if $|S| = k$ and the values of $f_S(x)$ is always a square, i.e. a quadratic residue in $\mathbb{F}_p$ or 0. p is called *k-extremal* if there exists a set $S \subseteq \mathbb{F}_p$ such that X_S is a $k-$extremal curve over $\mathbb{F}_p$.

We can show $k-$extremal primes satisfy the following two propositions.

PROPOSITION 2.2. *If $k < p$ is even, then p is k-extremal if and only if p is $(k-1)$-extremal.*

PROPOSITION 2.3. *If $k \equiv 1 \pmod 4$ and $p \equiv 3 \pmod 4$, then p is not k-extremal.*

PROOF OF PROPOSITION 2.2. Let S be a set such that X_S is a k-extremal curve over $\mathbb{F}_p$. Then there are $2p - k$ affine points on X_S over $\mathbb{F}_p$. We can make the following assumptions about S.

(1) $0 \in S$: If we shift every element by a constant, $|X_S(\mathbb{F}_p)|$ does not change. Therefore, we can assume $0 \in S$.

(2) $\prod_{a\in S, a\neq 0}(-a) \in Q$: Letting g be the genus of this curve, $k = 2g+2$. Expanding $f_S(x)$, we can write the curve as

$$y^2 = x^{2g+2} + c_{2g+1}x^{2g+1} + \cdots + c_1 x$$

Then $c_1 = \prod_{a\in S, a\neq 0}(-a)$. If $c_1 \notin Q$, take any $n \in N$, and denote nS as the set $\{ns | s \in S\}$. For every $x \in \mathbb{F}_p$,

$$\prod_{a\in nS}(x-a) = \prod_{x\in S}(x-na) = n^{2g+2}\prod_{x\in S}(x/n-a) \in Q$$

Since $n^{2g+2} \in Q$, $n^{2g+2}\prod_{x\in S}(x/n-a) \in Q$ if and only if $\prod_{x\in S}(x/n-a) \in Q$. The latter is true according to our assumption. Therefore X_{nS} has the same number of points as X_S. Moreover, $\prod_{a\in nS, a\neq 0}(-a) = n^{2g+1}\prod_{a\in S, a\neq 0}(-a) \in Q$ since the product of two non-quadratic residues is a quadratic residue. Therefore, we can assume that $c_1 \in Q$.

We can now construct a $(k-1)$-extremal curve based on X_S. Dividing $y^2 = x^{2g+2} + \cdots + c_1 x$ by $c_1 x^{2g+2}$,

$$\frac{y^2}{c_1 x^{2g+2}} = 1/c_1 + \cdots + \frac{1}{x^{2g+1}}$$

Letting

(2.6) $$u = 1/x$$

(2.7) $$v = y/\sqrt{c_1}x^{g+1}$$

we obtain the following curve

$$C_2 : v^2 = 1/c_1 + \cdots + u^{2g+1}$$

We will show that C_2 has exactly one more point than X_S using the relationships defined in (2.6) and (2.7).

(1) For every (x, y) on C_1, where $x \neq 0$, there is a one-to-one mapping to a (u, v) on C_2 with $u \neq 0$.
(2) If $x = 0$, then we have $(0, 0)$.
(3) If $u = 0$, then $v = \pm\sqrt{\frac{1}{c_1}}$.

Therefore there are $2p - k + 1$ points on C_2, so C_2 is $k-1$-extremal. On the other hand, let

$$C_3 : v^2 = d_0 + \cdots + u^{2g+1}$$

be a $(k-1)$-extremal curve, we will show there exists a k-extremal curve. Similarly to before, we can make the following assumptions about C_3.

(1) $d_0 \neq 0$: Firstly, note there exists an element $e \in \mathbb{F}_p$, such that if we shift every element in S by e, then all the elements become non-zero. Otherwise $S = \mathbb{F}_p$, contradicting $k < p$. Thus we can assume $d_0 \neq 0$.
(2) $d_0 \in Q$: Dividing the expression of C_3 by u^{2g+2},

$$\frac{v^2}{u^{2g+2}} = \frac{d_0}{u^{2g+2}} + \cdots + \frac{1}{u}$$

Let $x = 1/u$ and $y = u/v^{g+1}$, then

$$y^2 = x + \cdots + d_0 x^{2g+2}$$

Note that when $u = 0$, $v = d_0$. Therefore we have $d_0 \in Q$.

Letting $z = y/\sqrt{d_0}$, we have a new curve

$$C_4 : z^2 = x^{2g+2} + \cdots + \frac{x}{d_0} \tag{2.8}$$

With a similar point-counting argument as before, we can show this new curve is a k-extremal curve. □

REMARK 2.4. The construction for C_2 is inspired by Poonen's lecture on rational points on curves [**9**, Page 36, Remark 2.8.1].

PROOF OF PROPOSITION 2.3. We assume $0 \in S$ as before. Let $R = S\backslash\{0\}$. We have the following equality

$$\sum_{a\in\mathbb{F}_p} \chi(f_R(a)) + \sum_{a\in\mathbb{F}_p} \chi(f_S(a)) + p - 2 \equiv 0 \pmod 4$$

When $S = 0$, we explicitly compute that

$$\sum_{a\in\mathbb{F}_p} \chi(f_S(a)) = \sum_{a\in\mathbb{F}_p} \chi(a) = 0$$

Therefore when $p \equiv 1 \pmod 4$,

$$\sum_{a\in\mathbb{F}_p} \chi(f_S(a)) \equiv \begin{cases} 0 \pmod 4 & |S| \text{ is odd} \\ 1 \pmod 4 & |S| \text{ is even} \end{cases}$$

If $p \equiv 1 \pmod 4$, $k \equiv 3 \pmod 4$, and p is k-extremal, then there exists $S \subseteq \mathbb{F}_p$, such that

$$\sum_{a\in\mathbb{F}_p} \chi(f_S(a)) = p - k \equiv 2 \pmod 4$$

Contradiction.

□

The first part of the proof for Proposition 2.2 also applies to non-extremal curves, and we have the following corollary.

COROLLARY 2.5. *If $|S| = k$ is even, then there exists a set $R \subseteq \mathbb{F}_p$ with $|R| = k - 1$, s.t. $|X_R(\mathbb{F}_p)| = |X_S(\mathbb{F}_p)| + 1$.*

The Hasse-Weil bound is well-known for hyperelliptic curves.

THEOREM 2.6 (Hasse-Weil). *Let $|X(\mathbb{F}_p)|$ be the number of points of curve X over $\mathbb{F}_p$, then*

$$|X(\mathbb{F}_p) - p| \leq 2g\sqrt{p}$$

where g is the genus of the curve.

REMARK 2.7. We are using the version of Hasse-Weil bound that only counts affine points so as to be consistent with our own calculations.

PROPOSITION 2.8. *Let $k < p$ be odd. If p is $k-$extremal, then $k \geq \sqrt{p}$.*

PROOF. If p is $k-$extremal, let $S \subseteq \mathbb{F}_p$ be the corresponding set. Then

$$|X_S(\mathbb{F}_p)| = k + 2(p - k) = 2p - k$$

The genus g of X_S satisfies $2g + 1 = k$, so

$$g = \frac{k-1}{2}$$

Applying the Hasse-Weil bound,

$$p-(k-1)\sqrt{p} \leq 2p-k \leq p+(k-1)\sqrt{p}$$

Solving above equation gives

$$k \geq \sqrt{p}$$

□

Combining Proposition 2.8 and Proposition 2.2, we have the following corollary.

COROLLARY 2.9. *Let l be the smallest odd number greater than $\sqrt{p}$. If p is $k-$extremal, then $k \geq l$.*

In the following part, we will compute $M(p)$ for primes $p > 23$ as these are related to Goppa's conjecture.

EXAMPLE 2.10 ($p = 29$). When $p = 29$, for $M(p) > \gamma$ we would need at least 52 points on X_S for some $S \subseteq \mathbb{F}_p$. If $|S| \leq 5$, then $g \leq 2$. Using the Hasse-Weil bound, we have

$$|X_S(\mathbb{F}_p)| \leq 29 + 2 \cdot 2 \cdot \sqrt{29} \leq 51$$

If $|S| \geq 6$, then

$$|X_S(\mathbb{F}_p)| = 2p - |S| \leq 2 \cdot 29 - 6 = 52$$

with equality taking place when $|S| = 6$ and X_S is $|S|-$extremal. Therefore $M(p) > \gamma$ if and only if 29 is 6-extremal. Since $\sqrt{29} \approx 5.38$, this is not possible according to Corollary 2.9.

We can do likewise for $p = 31, 37, 43, 47$. We briefly outline the calculations here for completeness.

EXAMPLE 2.11 ($p = 31$). For $p = 31$ to satisfy $M(p) > \gamma_0$ we need at least 56 points on a curve X_S. If $|S| \leq 5$, $|X_S(\mathbb{F}_p)| < 56$ according to the Hasse-Weil bound. If $|S| \geq 6$, then the only way for $M(31) > \gamma_0$ is for 31 to be 6-extremal. According to Corollary 2.9, this is impossible.

EXAMPLE 2.12 ($p = 37$). For $p = 37$ to satisfy $M(p) > \gamma_0$, we need at least 66 points on a curve X_S. If $|S| \leq 5$, $|X_S(\mathbb{F}_p)| < 66$ according to the Hasse-Weil bound. If $|S| \geq 9$, then $|X_S(\mathbb{F}_p)| \leq 2 \cdot 37 - 9 = 65$. Therefore we just need to focus on $|S| = 6, 7, 8$. Exhaustive search using MAGMA shows no curve attains 66 points.

EXAMPLE 2.13 ($p = 43$). For $p = 43$ to satisfy $M(p) > \gamma_0$, we need at least 77 points on a curve X_S. If $|S| \leq 5$, $|X_S(\mathbb{F}_p)| \leq 70$ according to the Hasse-Weil bound. When $|S| \geq 10$, then $|X_S(\mathbb{F}_p)| \leq 2 \cdot 43 - 10 = 76$. Therefore we just need to consider $|S| = 7, 9$. According to Proposition 2.3, 43 cannot be 9-extremal. Exhaustive computation shows no set with size 7 attains 77 points.

EXAMPLE 2.14 ($p = 47$). For $p = 47$, we need at least 84 points. For $|S| \leq 5$, $|X_S(\mathbb{F}_p)| < 84$ according to the Hasse-Weil bound. For $|S| \geq 11$, $|X_S(\mathbb{F}_p)| \leq 2 \cdot 47 - 11 = 83$. Therefore we only need to consider $|S| = 7, 9$. 47 cannot be 9-extremal according to Proposition 2.3. If $|S| = 9$, we must have $|X_S(\mathbb{F}_p)| \leq 9 + 2(47 - 9 - 1) = 83$. Exhaustive computation shows sets with size 7 do not attain 84 points either.

EXAMPLE 2.15 ($p = 41$). As the only prime that has $M(p) > \gamma_0$, 41 is 9-extremal. For 9-extremal curves, $|X_S(\mathbb{F}_p)| = 2 \cdot 41 - 9 = 73$. Therefore $M(p) = 73/41 \approx 1.78$.

In general, if $k < (\gamma_0 - 1)\sqrt{p} + 1$, according to the Hasse-Weil bound, we have $|X_S(\mathbb{F}_p)| < \gamma_0 p$. When $k > (2-\gamma)p$, we also have $|X_S(\mathbb{F}_p)| < 2p - (2-\gamma)p < \gamma p$. Also for $k = (2-\gamma)p$, p must be $k-$extremal to achieve γp points. We can use Proposition 2.2, 2.3 and 2.8 to tackle those case by case, and we only need to consider k odd due to Corollary 2.9.

There is obviously a gap between between $(\gamma-1)\sqrt{p}+1$ and $(2-\gamma)p$. For small p, this gap is small, as seen in previous examples. When p gets bigger however, this gap also increases. Exhaustive search is not practical, and we are really interested in the limiting behavior of $M(p)$ as p gets large.

The fact that for primes $p < 50$, only $p = 41$ has $M(p) > \gamma$ suggests that primes achieving this inequality are rare. Since Theorem 1.12 only requires an infinite sequence of $p \equiv 1 \pmod 4$ satisfying $M(p) < 1.77$, in conjunction with the other heuristics we have presented evidence that this is achievable and so Goppa's conjecture can be considered dubious.

References

[1] N. Boston and J. Hao, *The weight distribution of quasi-quadratic residue codes*, Adv. Math. Commun. **12** (2018), no. 2, 363–385, DOI 10.3934/amc.2018023. MR3808234

[2] D. Joyner, *On quadratic residue codes and hyperelliptic curves*, Discrete Math. Theor. Comput. Sci. **10** (2008), no. 1, 129–146. MR2393232

[3] E. N. Gilbert, *A comparison of signalling alphabets*, Bell System Technical Journal **31** (1952), no. 3, 504–522.

[4] V.D. Goppa, *Bounds for codes*, Dokl. Acad. Nauk SSSR **333** (1993).

[5] D. Joyner and J.-L. Kim, *Selected unsolved problems in coding theory*, Applied and Numerical Harmonic Analysis, Birkhäuser/Springer, New York, 2011. MR2838861

[6] Michael Larsen, *The Normal Distribution as a Limit of Generalized Sato-Tate Measures*, 2008, arXiv:0810.2012.

[7] F. J. MacWilliams and N. J. A. Sloane, *The theory of error-correcting codes. I*, North-Holland Publishing Co., Amsterdam-New York-Oxford, 1977. North-Holland Mathematical Library, Vol. 16. MR0465509

[8] Yu. I. Manin, *What is the maximum number of points on a curve over* $\mathbf{F}_2$*?*, J. Fac. Sci. Univ. Tokyo Sect. IA Math. **28** (1981), no. 3, 715–720 (1982). MR656047

[9] B. Poonen, *Lectures on rational points on curves*, 2006.

[10] N. J. A. Sloane, *The on-line encyclopedia of integer sequences.*

[11] R. R. Varshamov, *Estimate of the number of signals in error correcting codes*, Doklady Akademii Nauk SSSR **117** (1957), 739–741.

DEPARTMENT OF MATHEMATICS, DEPARTMENT OF ELECTRICAL AND COMPUTER ENGINEERING, UNIVERSITY OF WISCONSIN-MADISON, MADISON, WISCONSIN 53706

Email address: boston@math.wisc.edu

DEPARTMENT OF MATHEMATICS, UNIVERSITY OF WISCONSIN-MADISON, MADISON, WISCONSIN 53706

Email address: jing.hao@wisc.edu

Contemporary Mathematics
Volume **724**, 2019
https://doi.org/10.1090/conm/724/14596

Curves, Jacobians, and cryptography

Gerhard Frey and Tony Shaska

ABSTRACT. The main purpose of this paper is to give an overview over the theory of abelian varieties, with main focus on Jacobian varieties of curves reaching from well-known results till to latest developments and their usage in cryptography. In the first part we provide the necessary mathematical background on abelian varieties, their torsion points, Honda-Tate theory, Galois representations, with emphasis on Jacobian varieties and hyperelliptic Jacobians. In the second part we focus on applications of abelian varieties on cryptography and treating separately, elliptic curve cryptography, genus 2 and 3 cryptography, including Diffie-Hellman Key Exchange, index calculus in Picard groups, isogenies of Jacobians via correspondences and applications to discrete logarithms. Several open problems and new directions are suggested.

Preface

There has been a continued interest on Abelian varieties in mathematics during the last century. Such interest is renewed in the last few years, mostly due to applications of abelian varieties in cryptography. In these notes we give a brief introduction to the mathematical background on abelian varieties and their applications on cryptography with the twofold aim of introducing abelian varieties to the experts in cryptography and introducing methods of cryptography to the mathematicians working in algebraic geometry and related areas.

A word about cryptography. Information security will continue to be one of the greatest challenges of the modern world with implications in technology, politics, economy, and every aspect of everyday life. Developments and drawbacks of the last decade in the area will continue to put emphasis on searching for safer and more efficient crypto-systems. The idea and lure of the quantum computer makes things more exciting, but at the same time frightening.

There are two main methods to achieve secure transmission of information: *secret-key cryptography* (*symmetric-key*) and *public-key cryptography* (*asymmetric-key*). The main disadvantage of symmetric-key cryptography is that a shared key must be exchanged beforehand in a secure way. In addition, managing keys in a large public network becomes a very complex matter. Public-key cryptography is used as a complement to secret-key cryptography for signatures , authentication and key-exchange. There are two main methods used in public-key cryptography, namely RSA and the discrete logarithm problem (DLP) in cyclic groups of prime

2010 *Mathematics Subject Classification.* Primary 14H10, 14H45.

order which are embedded in rational points of Abelian varieties, in particular of Jacobian varieties of curves. The last method is usually referred to as *curve-based cryptography.*

In addition, there is always the concern about the post-quantum world. What will be the crypto-systems which can resist the quantum algorithms? Should we develop such systems now? There is enthusiasm in the last decade that some aspects of curve-based cryptography can be adapted successfully to the post-quantum world. Supersingular Isogeny Diffie-Hellman (SIDH), for example, is based on isogenies of supersingular elliptic curves and is one of the promising schemes for post-quantum cryptography. Isogenies of hyperelliptic Jacobians of dimension 2 or 3 have also been studied extensively in the last decade and a lot of progress has been made. In this paper we give an overview of recent developments in these topics.

Audience. Computer security and cryptography courses for mathematics and computer science majors are being introduced in all major universities. Curve-based cryptography has become a big part of such courses and a popular area even among professional mathematicians who want to get involved in cryptography. The main difficulty that these newcomers is the advanced mathematical background needed to be introduced to curve-based cryptography.

Our target audience is advanced graduate students and researchers from mathematics or computer science departments who work with curve-based cryptography. Many researchers from other areas of mathematics who want to learn about abelian varieties and their use in cryptography will find these notes useful.

Notations and bibliography. The symbols $\mathbb{N}$ and $\mathbb{Z}$ will denote the natural numbers and the ring of integers while $\mathbb{Q}$, $\mathbb{R}$, and $\mathbb{C}$ the fields of rationals, reals, and complex numbers. $\mathbb{F}_p = \mathbb{Z}/p\mathbb{Z}$ will denote the field of p elements, for a prime number p and $\mathbb{F}_q$ a field of q elements, where $q = p^n$ is a power of p.

In general a field will be denoted by k. We shall always assume that k is a perfect field, i.e., every algebraic extension of k is separable. Denote its characteristic by char k, and its algebraic closure by $\bar{k}$. The Galois group $\mathrm{Gal}(\bar{k}/k)$ will be denoted by G_k. A number field will be denoted by K and its ring of integers by $\mathcal{O}_K$.

By a "curve" we mean an irreducible, smooth, projective algebraic curve. A genus $g \geq 2$ curve defined over k will be denoted by $\mathcal{C}/k$ or sometimes by $\mathcal{C}_g$ and its Jacobian by $\mathrm{Jac}_\mathcal{C}$. The automorphism group of $\mathcal{C}$ GF is denoted by $\mathrm{Aut}\,\mathcal{C}$ and it means the full group of automorphisms of $\mathcal{C}$ over the algebraic closure $\bar{k}$.

We will use $\mathcal{A}$, $\mathcal{B}$ to denoted Abelian varieties defined over a field k and $k(\mathcal{A})$, $k(\mathcal{B})$ their function fields. The set of n-torsion points of $\mathcal{A}$ will be denoted by $\mathcal{A}[n]$. For a subscheme G of $\mathcal{A}$, the quotient variety is denoted by $\mathcal{A}/G$.

Background and preliminaries. We assume the reader is familiar with the basic tools from algebra and algebraic geometry. Familiarity with algebraic curves is expected and the ability to read some of the classical works on the subject [**56**].

Organization of these notes. In Part 1 we give the mathematical background on Abelian varieties, their torsion points, endomorphisms and isogenies. We focus mostly on Abelian varieties defined over fields of positive characteristic. The main references here are [**56**], and [**21**].

We give a brief introduction of abelian varieties as complex tori with period matrices in Section 1. Of course, of special interest for us are Jacobian varieties, hence we define in detail algebraic curves, constant field extensions, group schemes, and principally polarized varieties.

In Section 2 we focus on endomorphism rings of abelian varieties and isogenies, the characteristic polynomial of the Frobenius, l-adic Tate module, and Tate's result on determining necessary and sufficient conditions for two Abelian varieties to be isogenous.

Algebraic curves and Jacobian varieties are treated in detail in Section 3, including Picard groups on curves, the group of divisors, canonical divisors, Riemann-Roch theorem, and the definition of Jacobians of curves.

In Section 4 we focus on applications of the Riemann-Roch theorem, including the Hurwitz genus formula, gonality of curves and Hurwitz spaces, Cantor's algorithm on Jacobians of hyperelliptic curves, and automorphism groups of curves and their Jacobians. As illustration and for later applications curves of small genus are discussed in more detail.

In Section 5 we give a brief description of the theory of modular curves over $\mathbb{C}$, modular polynomials, and the arithmetic theory of modular curves.

In Part 2 we focus on applications of abelian varieties on cryptography. Our main reference is [**7**] and the material provided in Part 1.

In Section 6 we go over the preliminaries of the Diffie-Hellman Key Exchange and the mathematical challenges including Q-bit security, Key Exchange with G-sets, and the abstract setting of Key Exchange.

In Section 7 we describe the methods of index calculus in Picard groups and their use in cryptography. Such methods have been quite successful due to work of Diem, Gaudry, et al. As consequence one sees that only elliptic and hyperelliptic curves of genus ≤ 3 provide candidates for secure crypto systems based on discrete logarithms. Hence we shall discuss these curves in detail. In Section 8 we focus on isogenies of Jacobians via correspondences. We discuss the Weil descent, modular correspondences, and correspondences via monodromy groups. It is an open and difficult problem to find interesting correspondences of low degree between Jacobian varieties induced by correspondences between curves.

In Section 9 we study hyperelliptic Jacobians of dimension 3. We give a short introduction of non-hyperelliptic and hyperelliptic genus 3 curves and their plane equations. Then we define Picard groups of genus 3 curves and their use in cryptography and results of Diem and Hess. In the following part we describe the index-calculus attacks applied to genus 3 and results of Diem, Gaudry, Thomé, Thériault. We also discuss isogenies via S_4-covers and work of Frey and Kani [**22**], [**23**] and Smith [**66**].

In Section 4.2.7 we focus on dimension 2 Jacobians and their use in cryptography. Methods based on [**47**] of how to compute the endomorphism ring of a dimension 2 Jacobian are described and in particular isogenies of Abelian surfaces via Donagi-Livné approach and some recent results of Smith [**66**]. Further we give details of point counting algorithms on genus 2 Jacobians and explicit formulas for $[n]D$, when D is a reduced divisor. Work of Gaudry, Harley, Schost and others is briefly described. In Section 11 we focus on the elliptic curves and elliptic curve cryptography. We give an explicit description of the methods used in supersingular isogeny-based cryptography. We describe the necessary background including Velu's formula, ordinary and supersingular elliptic curves and the more recent results [**11**], [**10**], [**12**] among others.

Part 1. Abelian varieties

In the first part of these notes we give the basic theory of abelian varieties, their endomorphisms, torsion points, characteristic polynomial of the Frobenius, Tate models, and then focus on Jacobian varieties and hyperelliptic Jacobians. While there are many good references on the topic, we mostly use [**56**], [**69**].

1. Definitions and basic properties

We shall use projective respectively affine *schemes* defined over k. Let $n \in \mathbb{N}$ and I_h (respectively I) be a homogeneous ideal in $k[Y_0, \cdots, Y_n]$ different from $\langle Y_0, \dots Y_n \rangle$ (respectively an arbitrary ideal in $k[X_1, \dots, X_n]$).

Let $R_h := k[Y_0, \dots, Y_n]/I_h$ (respectively $R := k[X_1, \dots, X_n]/I$) be the quotients. By assumption, R_h is a graded ring, and so localizations $R_{h,\mathfrak{A}}$ with respect to homogeneous ideals $\mathfrak{A}$ are graded, too. Let $R_{h,\mathfrak{A}\ 0}$ be the ring of elements of grade 0.

The projective scheme $\mathcal{S}_h$ (respectively the affine scheme $\mathcal{S}$) defined by I_h (I) consists of

(1) the topological space $V_h := \mathrm{Proj}(R_h)$ ($V := \mathrm{Spec}(\mathrm{R})$) consisting of homogeneous prime ideals in R_h with pre-image in $k[Y_0, \dots, Y_n]$ different from $\langle Y_0, \dots Y_n \rangle$ (prime ideals in R) endowed with the Zariski topology and
(2) the sheaf of rings of holomorphic functions given on Zariski-open sets $U \subset V_h$ ($U \subset V$) as elements of grade 0 in localization of $R_{h,0}$ (R) with respect to the elements that become invertible when restricted to U.

Examples:

(1) The projective space $\mathbb{P}^n$ over k of dimension n is given by the ideal $\langle 0 \rangle \subset k[Y_0, \dots, Y_n]$. The ring of holomorphic functions on $\mathbb{P}^n$ (take $U = \mathbb{P}^n$) is k.

    Next take $U = \emptyset$ to get the ring of *meromorphic* functions on $\mathbb{P}^n$: It consists of the quotients

$$f/g \text{ with } f, g \text{ homogeneous of degree d with } g \neq 0.$$

(2) The affine space $\mathbb{A}^n$ of dimension n over k is the topological space

$$\mathrm{Spec}(k[X_1, \dots X_n]).$$

    The ring of holomorphic functions on A^n is $k[X_1, \dots, X_n]$, where polynomials are interpreted as polynomial functions. The ring of meromorphic functions on $\mathbb{A}^n$ (take $U = \emptyset$) is the field of rational functions $k(X_1, \dots, X_n)$.
(3) The easiest but important example for an affine scheme: Take $n = 1$, $I = \langle X_1 \rangle$, $V = \mathrm{Spec}(k) = \{(0)\}$ and $O_{(0)} = k^*$.

Morphisms of affine or projective schemes are continuous maps between the underlying topological spaces induced (locally) by (in the projective case, quotients of the same degree) of polynomial maps of the sheaves.

Rational maps f between affine or projective schemes $\mathcal{S}$ and $\mathcal{T}$ are equivalence classes of morphisms defined on open subschemes U_i of $\mathcal{S}$ with image in $\mathcal{T}$ and compatible with restrictions to $U_i \cap U_j$. If f is invertible (as rational maps from $\mathcal{T}$ to $\mathcal{S}$), then f is *birational*, and $\mathcal{S}$ and $\mathcal{T}$ are birationally equivalent.

The k-rational points $\mathcal{S}(k)$ of a scheme $\mathcal{S}$ is the set of morphisms from $\mathrm{Spec}(k)$ to $\mathcal{S}$. The reader should verify that for projective schemes defined by the ideal I_h the set $\mathcal{S}(k)$ is, in a natural way, identified with points $(y_0 : y_1 \cdots : y_n)$ with k-rational homogeneous coordinates in the projective space of dimension n which are common zeros of the polynomials in I_h, and an analogous statement holds for affine schemes.

Constant field extensions: Let $k \overset{\iota}{\hookrightarrow} L$ be an embedding of k into a field L (or $k \subset L$ if the embedding is clear) of k. Let $\mathcal{S}$ be a projective (affine) scheme defined over k with ring R. ι induces a morphism $\mathfrak{f}_\iota$ from R in $R \otimes_k L =: R_\iota$ given by the interpretation via ι of polynomials with coefficients in k as polynomials with coefficients in L. The ideal $I_{\mathcal{S}}$ extends to a ideal in R_ι and hence we get in a natural way a projective scheme $\mathcal{S}_\iota$ with a morphism

$$\mathcal{S}_\iota \to \mathcal{S}$$

as Spec(k) schemes. $\mathcal{S}_\iota$ is again a projective (affine) scheme now defined over L, which is denoted as *scalar extension* by ι. If there is no confusion possible (for instance if $k \subset L \subset \bar{k}$ and ι is the inclusion) we denote $\mathcal{S}_\iota$ by $\mathcal{S}_L$.

A scheme $\mathcal{S}$ is irreducible if the ideal I_h (respectively I) is a prime ideal. $\mathcal{S}$ is absolutely irreducible if $\mathcal{S}_{\bar{k}}$ is irreducible. This is the case if and only if k is algebraically closed in R. Classically, irreducible schemes are called *irreducible varieties.*

Affine covers There are many possibilities to embed $\mathbb{A}^n$ into $\mathbb{P}^n$, and there is no "canonical" way to do this. But after having chosen coordinates there is a standard way to construct a covering of $\mathbb{P}^n$ by $n+1$ copies of $\mathbb{A}^n$. Every homogeneous polynomial $P(Y_0, ..., Y_n)$ can be transformed into $n+1$ polynomials $p_j(X)$ $(j = 0, ..., n)$ in n variables by the transformation

$$t_j : Y_i \mapsto X_i := Y_i/Y_j.$$

Define U_j as open subscheme of $\mathbb{P}^n$ which is the complement of the projective scheme attached to the ideal $\langle Y_j \rangle$. Then $t_j\,_{|U_j}$ is holomorphic and bijective and its image is isomorphic to $\mathbb{A}^n$.

By the inverse transform ι_j we embed $\mathbb{A}^n$ into $\mathbb{P}^n$ and so U_j is isomorphic to $\mathbb{A}^n$ as affine variety. Taking the collection $(\iota_0, \ldots \iota_n)$ we get a finite open covering of $\mathbb{P}^n$ by $n+1$ affine subspaces.

Having an affine cover U_j of $\mathbb{P}^n$ one can intersect it with projective varieties V and get

$$V = \bigcup_j V_{j,a} \text{ with } V_{j,a} := V \cap U_j$$

as union of affine varieties.

Converse process: Given a polynomial $p(X_1, ..., X_n)$ of degree d we get a homogeneous polynomial $p^h(Y_0, ..., Y_n)$ of degree d by the transformation

$$X_i \mapsto Y_i/Y_0 \text{ for } i = 1, \ldots, n$$

and then clearing denominators. Assume that V_a is an affine variety with ideal $I_a \subset k[X_1, \ldots, X_n]$. By applying the homogenization explained above to all polynomials in I_a we get a homogeneous ideal $I_a^h \subset k[Y_0, ..., Y_n]$ and a projective variety V with ideal I_a^h containing V_a in a natural way. V is called a projective closure of V_a. By abuse of language one calls $V \cap U_0 = V \setminus V_a$ "infinite points" of V_a.

Function fields: Let $\mathcal{S} \subset \mathbb{A}^n$ be an affine irreducible variety with ring R. In particular, R is an integral domain. The function field $k(\mathcal{S})$ is the quotient field of R. It consists of the meromorphic functions of $\mathbb{A}^n$ restricted to $\mathcal{S}$. $\mathcal{T}$ is birational equivalent to $\mathcal{S}$ if and only if $k(\mathcal{S}) = k(\mathcal{T})$.

If $U \neq \emptyset$ is affine and open in a projective variety $\mathcal{S}$ then the field of meromorphic functions $k(\mathcal{S}) = k(U)$. In particular, it is independent of U.

DEFINITION 1. Let $\mathcal{S}$ be an irreducible variety. The dimension of $\mathcal{S}$ is the transcendental degree of $k(\mathcal{S})$ over k.

Group schemes: A projective (affine) group scheme G defined over k is a projective (affine) scheme over k endowed with

i) addition, i.e., a morphism

$$m: G \times G \to G$$

ii) inverse, i.e., a morphism

$$i: G \to G$$

iii) the identity, i. e., a k-rational point $0 \in G(k)$,

such that it satisfies group laws. The group law is uniquely determined by the choice of the identity element. A morphism of group schemes that is compatible with the addition law is a homomorphism.

Let L be a field extension of k. $G(L)$ denotes the set of L-rational points of G and it is also a group. A homomorphism between groups schemes induces a homomorphism between the group of rational points. If G is an absolutely irreducible projective variety, then the group law m is commutative.

DEFINITION 2. An Abelian variety defined over k is an absolutely irreducible projective variety defined over k which is a group scheme.

We will denote an Abelian variety defined over a field k by $\mathcal{A}_k$ or simply $\mathcal{A}$ when there is no confusion. From now on the addition $m(P, Q)$ in an abelian variety will be denoted by $P \oplus Q$ or simply $P + Q$ and the inversion $i(P)$ by $\ominus P$ or simply by $-P$.

Fact: A morphism from the Abelian varieties $\mathcal{A}_1$ to the Abelian variety $\mathcal{A}_2$ is a homomorphism if and only if it maps the identity element of $\mathcal{A}_1$ to the identity element of $\mathcal{A}_2$.

An abelian variety over a field k is called **simple** if it has no proper nonzero Abelian subvariety over k, it is called **absolutely simple** (or **geometrically simple**) if it is simple over the algebraic closure of k.

1.1. Complex tori and abelian varieties.

Though we are interested in Abelian varieties over arbitrary fields k or in particular, over finite fields, it is helpful to look at the origin of the whole theory, namely the theory of Abelian varieties over the complex numbers. Abelian varieties are connected, projective algebraic group schemes. Their analytic counterparts are the connected compact Lie groups.

Let d be a positive integer and $\mathbb{C}^d$ the complex Lie group (i.e., with vector addition as group composition). The group $\mathbb{C}^d$ is not compact, but we can find quotients which are compact. Choose a lattice $\Lambda \subset \mathbb{C}^d$ which is a $\mathbb{Z}$-submodule of rank $2d$. The quotient $\mathbb{C}^d/\Lambda$ is a complex, connected Lie group which is called a *complex d-dimensional torus.* Every connected, compact Lie group of dimension d is isomorphic to a torus $\mathbb{C}^d/\Lambda$.

A Hermitian form H on $\mathbb{C}^d \times \mathbb{C}^d$ is a form that can be decomposed as

$$H(x,y) = E(ix,y) + i\,E(x,y),$$

where E is a skew symmetric real form on $\mathbb{C}^d$ satisfying $E(ix,iy) = E(x,y)$. E is called the imaginary part $\mathrm{Img}(H)$ of H. The torus $\mathbb{C}^d/\Lambda$ can be embedded into a projective space if and only if there exists a positive Hermitian form H on $\mathbb{C}^d$ with $E = \mathrm{Img}(H)$ such that restricted to $\Lambda \times \Lambda$ has values in $\mathbb{Z}$. Let $\mathbb{H}_g$ be the Siegel upper half plane

$$\mathbb{H}_d = \{\tau \in \mathrm{Mat}_d(\mathbb{C}) \mid \tau^T = \tau,\, \mathrm{Img}(\tau) > 0\}.$$

Then, we have the following.

LEMMA 1. *Let $\mathbb{C}^d/\Lambda$ be a complex torus attached to an abelian variety $\mathcal{A}$. Then Λ is isomorphic to $\mathbb{Z}^d \oplus \Omega \cdot \mathbb{Z}^d$, where $\Omega \in \mathbb{H}_d$.*

The matrix Ω is called the **period matrix** of $\mathcal{A}$. The lattice $\hat{\Lambda}$ given by

$$\hat{\Lambda} := \{x \in \mathbb{C}^d \,|\, E(x,y) \in \mathbb{Z}, \quad \text{for all} \quad y \in \Lambda\}$$

is called the **dual lattice** of Λ. If $\hat{\Lambda} = \Lambda$ then E is called a *principal polarization* on $\mathcal{A}$ and the pair $(\mathcal{A}, E)$ is called a **principally polarized** abelian variety; we may also say that $\mathcal{A}$ admits a principal polarization.

For a principally polarized abelian variety $(\mathcal{A}, E)$ there exists a basis $\{\mu_1, \dots, \mu_{2d}\}$ of Λ such that

$$J := [E(\mu_i, \mu_j)]_{1 \le i,j \le 2d} = \begin{bmatrix} 0 & I_d \\ -I_d & 0 \end{bmatrix}.$$

The symplectic group

$$Sp(2d,\mathbb{Z}) = \{M \in GL(2d,\mathbb{Z}) \mid MJM^T = J\}$$

acts on $\mathbb{H}_d$, via

$$\begin{aligned} &Sp(2d,\mathbb{Z}) \times \mathbb{H}_d \to \mathbb{H}_d \\ &\begin{bmatrix} a & b \\ c & d \end{bmatrix} \times \tau \to (a\tau + b)(c\tau + d)^{-1} \end{aligned}$$

where a, b, c, d, τ are $d \times d$ matrices. The moduli space of d-dimensional abelian varieties is

$$\mathbf{A}_g := \mathbb{H}_d / Sp(2d,\mathbb{Z}).$$

Jacobian varieties of projective irreducible nonsingular curves admit canonical principal polarizations. These Abelian varieties are in the center of our interest and will be discussed in detail in Section 3.

1.1.1. *Elliptic curves over* $\mathbb{C}$. Take $d = 1$ and a lattice $\Lambda_\tau := \mathbb{Z} + \mathbb{Z}\tau$ with τ in the upper half plane $\mathbb{H}_1 = \mathbb{H}$. The torus $\mathbb{C}/\Lambda_\tau$ is a compact Riemann surface and so an algebraic projective curve $\mathcal{E}_\tau$ over $\mathbb{C}$.

The function field $\mathbb{C}(\mathcal{E}_\tau)$ is generated by the Weierstrass function $\wp_\tau$ and its derivative $\wp_\tau'$, which are meromorphic functions on $\mathbb{C}$ with periods $1, \tau$ and poles of order 2 respectively 3 in Λ_τ. $\wp_\tau$ satisfies a differential equation

$$W_\tau : \wp_\tau'^2 = \wp_\tau^3 - g_2(\tau)\wp_\tau^2 - g_3(\tau).$$

This is an affine equation for $\mathcal{E}_\tau$, by introducing homogeneous coordinates $(X : Y : Z)$ by $\wp_\tau = X/Z, \wp'_\tau = Y/Z$, we get the projective plane curve $\mathcal{E}_\tau$ with equation

$$Y^2Z = X^3 - g_2(\tau)XZ^2 - g_3(\tau)Z^3$$

with coefficients $g_2(\tau), g_3(\tau)$ depending on Λ_τ in a very specific way: g_2 and g_3 are Eisenstein series in τ. It follows that $\Delta_\tau = 4g_2(\tau)^3 - 27g_3(\tau)^2 \neq 0$ and so $\mathcal{E}$ is without singularities. We get a parametrization

$$\phi : \mathbb{C} \to \mathcal{E}_\tau(\mathbb{C})$$

by

$$z \mapsto (\wp_\tau(z) : \wp'_\tau(z) : 1) \text{ if and only if } z \notin \Lambda_\tau$$

and $\phi(\Lambda_\tau) = (0 : 1 : 0)$, the point at infinity. This parametrization yields on $\mathcal{E}_\tau$ an addition and makes $\mathcal{E}_\tau$ to an Abelian variety of dimension 1 over $\mathbb{C}$.

DEFINITION 3. An Abelian variety of dimension 1 is called an **elliptic curve**.

We have seen that we can attach to every elliptic curve $\mathcal{E}$ an element $\tau \in \mathbb{H}$ such that $\mathcal{E}$ is isomorphic to $\mathcal{E}_\tau$. Let $\mathcal{E}_{\tau'}$ be another elliptic curve. Then $\mathcal{E}_\tau$ is isomorphic to $\mathcal{E}_{\tau'}$ if and only if τ is equivalent to τ' under the action of $Sp(2d, \mathbb{Z}) = Sl(2, \mathbb{Z})$ on $\mathbb{H}$.

Since $\mathbb{H}/Sl(2, \mathbb{Z})$ is as Riemann surface isomorphic to $\mathbb{A}^1$ we get a one-to-one correspondence between isomorphism classes of elliptic curves over $\mathbb{C}$ and points on the affine line. This correspondence is given by a modular function (i.e. a holomorphic function on $\mathbb{H}$ invariant under $Sl(2, \mathbb{Z})$): the j-function.

DEFINITION 4. The absolute invariant of $\mathcal{E}_\tau$ is given by $j(\tau) := 12^3 \frac{4g_2^3(\tau)}{\Delta_\tau}$.

THEOREM 5. *$\mathcal{E}_\tau$ is isomorphic to $\mathcal{E}_{\tau'}$ if and only it $j(\tau) = j(\tau')$. Hence the j-function is an analytic map from $\mathbb{A}^1$ to $\mathbb{A}^1$.*

We remark that we shall define elliptic curves $\mathcal{E}$ in a purely algebraic setting over arbitrary fields k (cf. Section 4.2.3) and that we shall define an absolute invariant j for such curves, which coincides with $j(\tau)$ if $k = \mathbb{C}$, and which also has the property: If $\mathcal{E}$ is isomorphic to $\mathcal{E}'$ then $j_\mathcal{E} = j'_\mathcal{E}$, and the converse holds if k is algebraically closed.

2. Endomorphisms and isogenies

Let $\mathcal{A}$, $\mathcal{B}$ be abelian varieties over a field k. We denote the $\mathbb{Z}$-module of homomorphisms $\mathcal{A} \mapsto \mathcal{B}$ by $\mathrm{Hom}(\mathcal{A}, \mathcal{B})$ and the ring of endomorphisms $\mathcal{A} \mapsto \mathcal{A}$ by $\mathrm{End}\,\mathcal{A}$. In the context of Linear Algebra it can be more convenient to work with the $\mathbb{Q}$-vector spaces $\mathrm{Hom}^0(\mathcal{A}, \mathcal{B}) := \mathrm{Hom}(\mathcal{A}, \mathcal{B}) \otimes_\mathbb{Z} \mathbb{Q}$, and $\mathrm{End}^0\,\mathcal{A} := \mathrm{End}\,\mathcal{A} \otimes_\mathbb{Z} \mathbb{Q}$. Determining $\mathrm{End}\,\mathcal{A}$ or $\mathrm{End}^0\,\mathcal{A}$ is an interesting problem on its own; see [**58**].

For any abelian variety $\mathcal{A}$ defined over a number field K, computing $\mathrm{End}_K(\mathcal{A})$ is a harder problem than computation of $\mathrm{End}_{\bar{K}}(\mathcal{A})$; see [**47**, lemma 5.1] for details.

2.1. Isogenies. A homomorphism $f : \mathcal{A} \to \mathcal{B}$ is called an **isogeny** if $\mathrm{Img}\, f = \mathcal{B}$ and $\ker f$ is a finite group scheme. If an isogeny $\mathcal{A} \to \mathcal{B}$ exists we say that $\mathcal{A}$ and $\mathcal{B}$ are isogenous. We remark that this relation is symmetric, see Lem. 5.

The degree of an isogeny $f : \mathcal{A} \to \mathcal{B}$ is the degree of the function field extension

$$\deg f := [k(\mathcal{A}) : f^\star k(\mathcal{B})].$$

It is equal to the order of the group scheme $\ker(f)$, which is, by definition, the scheme theoretical inverse image $f^{-1}(\{0_{\mathcal{A}}\})$.

The group of $\bar{k}$-rational points has order $\#(\ker f)(\bar{k}) = [k(A) : f^{\star}k(B)]^{sep}$, where $[k(A) : f^{\star}k(B)]^{sep}$ is the degree of the maximally separable extension in $k(\mathcal{A})/f^{\star}k(\mathcal{B})$. f is a **separable isogeny** if and only if

$$\# \ker f(\bar{k}) = \deg f.$$

Equivalently: The group scheme $\ker f$ is étale. The following result should be compared with the well known result of quotient groups of abelian groups.

LEMMA 2. *For any Abelian variety $\mathcal{A}/k$ there is a one to one correspondence between the finite subgroup schemes $\mathcal{K} \leq \mathcal{A}$ and isogenies $f : \mathcal{A} \to \mathcal{B}$, where $\mathcal{B}$ is determined up to isomorphism. Moreover, $\mathcal{K} = \ker f$ and $\mathcal{B} = \mathcal{A}/\mathcal{K}$.*

f is separable if and only if $\mathcal{K}$ is étale, and then $\deg f = \#\mathcal{K}(\bar{k})$.

Isogenous Abelian varieties have commensurable endomorphism rings.

LEMMA 3. *If $\mathcal{A}$ and $\mathcal{B}$ are isogenous then* $\operatorname{End}^0(\mathcal{A}) \cong \operatorname{End}^0(\mathcal{B})$.

LEMMA 4. *If $\mathcal{A}$ is a absolutely simple Abelian variety then every endomorphism not equal 0 is an isogeny.*

We can assume that $k = \bar{k}$. Let f be a nonzero isogeny of $\mathcal{A}$. Its kernel $\ker f$ is a subgroup scheme of $\mathcal{A}$ (since it is closed in the Zariski topology because of continuity and under $\oplus$ because of homomorphy). It contains $0_{\mathcal{A}}$ and so its connected component, which is, by definition, an Abelian variety.

Since $\mathcal{A}$ is simple and $f \neq 0$ this component is equal to $\{0_{\mathcal{A}}\}$. But it has finite index in $\ker f$ (Noether property) and so $\ker f$ is a finite group scheme.

2.1.1. *Computing isogenies between Abelian varieties.* Fix a field k and let $\mathcal{A}$ be an Abelian variety over k. Let H denote a finite subgroup scheme of $\mathcal{A}$. From the computational point of view we have the following problems:

- Given $\mathcal{A}$ and H, determine $\mathcal{B} := \mathcal{A}/H$ and the isogeny $\mathcal{A} \to \mathcal{B}$.
- Given two Abelian varieties $\mathcal{A}$ and $\mathcal{B}$, determine if they are isogenous and compute a rational expression for an isogeny $\mathcal{A} \to \mathcal{B}$.

There is a flurry of research activity in the last decade to solve these problems explicitly for low dimensional Abelian varieties; see [**48**], [**49**] among many others. For a survey and some famous conjectures on isogenies see [**21**].

REMARK 1. *For elliptic curves (Abelian varieties of dimension* 1*) and for Jacobians of curves of genus* 2 *we shall come back to these questions in more detail.*

2.2. Torsion points and Tate modules. The most classical example of an isogeny is the scalar multiplication by n map $[n] : \mathcal{A} \to \mathcal{A}$. The kernel of $[n]$ is a group scheme of order $n^{2\dim \mathcal{A}}$ (see [**56**]). We denote by $\mathcal{A}[n]$ the group $\ker[n](\bar{k})$. The elements in $\mathcal{A}[n]$ are called n-**torsion points** of $\mathcal{A}$.

LEMMA 5. *Let $f : \mathcal{A} \to \mathcal{B}$ be a degree n isogeny. Then there exists an isogeny $\hat{f} : \mathcal{B} \to \mathcal{A}$ such that*

$$f \circ \hat{f} = \hat{f} \circ f = [n].$$

COROLLARY 1. *Let $\mathcal{A}$ be an absolutely simple Abelian variety. Then* $\operatorname{End}(\mathcal{A})^0$ *is a skew field.*

PROOF. Every endomorphism $\neq 0$ of $\mathcal{A}$ is an isogeny, hence invertible in $\mathrm{End}(\mathcal{A})^0$. □

THEOREM 6. *Let $\mathcal{A}/k$ be an Abelian variety, $p = char\ k$, and* $\dim \mathcal{A} = g$.

i) *If $p \nmid n$, then $[n]$ is separable, $\#\mathcal{A}[n] = n^{2g}$ and $\mathcal{A}[n] \cong (\mathbb{Z}/n\mathbb{Z})^{2g}$.*

ii) *If $p \mid n$, then $[n]$ is inseparable. Moreover, there is an integer $0 \leq i \leq g$ such that*

$$\mathcal{A}[p^m] \cong (\mathbb{Z}/p^m\mathbb{Z})^i, \text{ for all } m \geq 1.$$

If $i = g$ then $\mathcal{A}$ is called **ordinary**. If $\mathcal{A}[p^s](\bar{k}) = \mathbb{Z}/p^{ts}\mathbb{Z}$ then the abelian variety has p-**rank** t. If $\dim \mathcal{A} = 1$ (elliptic curve) then it is called **supersingular** if it has p-rank 0.[1] An abelian variety $\mathcal{A}$ is called **supersingular** if it is isogenous to a product of supersingular elliptic curves.

REMARK 2. *If* $\dim \mathcal{A} \leq 2$ *and $\mathcal{A}$ has p-rank 0 then $\mathcal{A}$ is supersingular. This is not true for* $\dim \mathcal{A} \geq 3$.

Let l be a prime that is (here and in the following) different from $p = \mathrm{char}\ k$ and $k \in \mathbb{N}$. Then,

$$[l]\mathcal{A}\left[l^{k+1}\right] = \mathcal{A}[l^k].$$

Hence, the collection of groups

$$\ldots, \mathcal{A}[l^{k+1}], \ldots, \mathcal{A}[l^k], \ldots$$

forms a projective system. The l-adic Tate module of $\mathcal{A}$ is

$$T_l(\mathcal{A}) := \varprojlim \mathcal{A}[l^k].$$

LEMMA 6. *The Tate module $T_l(\mathcal{A})$ is a $\mathbb{Z}_l$-module isomorphic to $\mathbb{Z}_l^{2\dim \mathcal{A}}$.*

2.3. l-adic representations and characteristic polynomials.

2.3.1. *Galois representations.* Torsion points on abelian varieties are used to construct very important representations of the Galois group of k. Let n be relatively prime to p and $g = \dim \mathcal{A}$. Then G_k acts on $\mathcal{A}[n]$ which gives rise to a representation

$$\rho_{\mathcal{A},n} : G_k \to \mathrm{Aut}\,(\mathcal{A}[n])$$

and after a choice of basis in $\mathcal{A}[n]$ yields a representation

$$\rho_{\mathcal{A},n} : G_k \to GL_{2g}(\mathbb{Z}/n\mathbb{Z})$$

This action extends in a natural way to $T_l(\mathcal{A}) \otimes \mathbb{Q}_\ell$ and therefore to a ℓ-adic representation $\tilde{\rho}_{\mathcal{A},l}$ which is called the **l-adic Galois representation attached to $\mathcal{A}$**.

2.3.2. *Representations of endomorphisms.* Let ϕ be an endomorphism of the g-dimensional Abelian variety $\mathcal{A}$. By restriction ϕ induces a $\mathbb{Z}$-linear map ϕ_n on $\mathcal{A}[n]$. Since the collection (ϕ_{ℓ^k}) is compatible with the system defining $T_\ell(\mathcal{A})$ it yields a $\mathbb{Z}_\ell$-linear map $\tilde{\phi}_\ell$ on $T_\ell(\mathcal{A})$.

Applying this construction to all elements in $\mathrm{End}(\mathcal{A})$ we get an injection (since $\mathcal{A}[\lambda^\infty] := \cup_{k\in\mathbb{N}}\mathcal{A}[\ell^k]$) is Zariski-dense in $\mathcal{A}$) from $\mathrm{End}(\mathcal{A})$ into $Gl(2g, \mathbb{Z}_\ell)$. By tensorizing with $\mathbb{Q}_\ell$ we get the ℓ-adic representation

$$\tilde{\eta}_\ell : \mathrm{End}(\mathcal{A}) \otimes \mathbb{Q}_\ell \to Gl_{2g}(\mathbb{Q}_\ell).$$

THEOREM 7. *$\tilde{\eta}_\ell$ is injective.*

[1]For an alternative definition see Thm. 43.

For a proof see [**56**, Theorem 3, p.176]. This result has important consequences for the structure of $\mathrm{End}^0(\mathcal{A})$, more precisely $\mathrm{End}^0(\mathcal{A})$ is a $\mathbb{Q}$-algebra of dimension $\leq 4\dim(\mathcal{A})^2$.

Adding more information (see Corollary 2 in [**56**]) one gets that $\mathrm{End}^0(\mathcal{A})$ is a semi-simple algebra, and by duality (key word Rosati-involution) one can apply a complete classification due to Albert of *possible* algebra structures on $\mathrm{End}^0(\mathcal{A})$, which can be found on [**56**, pg. 202].

The question is: Which algebras occur as endomorphism algebras? The situation is well understood if k has characteristic 0 (due to Albert) but wide open in characteristic $p > 0$. For $g = 1$ (elliptic curves) everything is explicitly known due to M. Deuring. We describe the results in Thm. 43. For curves of genus 2 we give an overview on results in Section 4.2.7

Characteristic Polynomial: For $\phi \in \mathrm{End}^0(\mathcal{A})$ let $\tilde{\phi}_\ell$ be its ℓ-adic representation. Denote its characteristic polynomial by $\chi_{\ell,\phi}(T) \in \mathbb{Z}_\ell[T]$.

THEOREM 8 (Weil). *$\chi_{\ell,\phi}(T)$ is a monic polynomial $\chi_\phi(T) \in \mathbb{Z}[T]$ which is independent of ℓ. We have*

$$\chi_\phi(\phi) \equiv 0 \text{ on } \mathcal{A},$$

*and so it is justified to call $\chi_\phi(T)$ the **characteristic polynomial** of ϕ.*

The degree of $\chi_\phi(T)$ is $2\dim(\mathcal{A})$, the second-highest coefficient is the negative of the trace of ϕ, and the constant coefficient is equal to the determinant of ϕ.

2.3.3. *Frobenius representations.* Let $\mathcal{A}$ be a g-dimensional Abelian variety defined over $\mathbb{F}_q$, where $q = p^d$ for a prime p and $\bar{\mathbb{F}}_q$ the algebraic closure of $\mathbb{F}_q$. Let $\pi \in \mathrm{Gal}(\bar{\mathbb{F}}_q/\mathbb{F}_q)$ be the Frobenius automorphism of $\mathbb{F}_q$, given by

$$\pi : x \to x^q.$$

Since $\mathrm{Gal}(\bar{\mathbb{F}}_q/\mathbb{F}_q)$ is topologically generated by π and because of continuity $\rho_{\mathcal{A},n}$ is determined by $\rho_{\mathcal{A},n}(\pi)$. We define

$$\chi_{\mathcal{A},q}(T) := \chi(T)\,(\tilde{\rho}_{\mathcal{A},l}(\pi)) \in \mathbb{Z}_\ell[T] \tag{1}$$

as the characteristic polynomial of the image of π under $\tilde{\rho}_{\mathcal{A},l}$.

LEMMA 7 (Weil). *$\chi_{\mathcal{A},q}(T)$ is a monic polynomial of degree $2g$ in $\mathbb{Z}[T]$, independent of ℓ, and for all $n \in \mathbb{N}$ we get*

$$\chi_{\mathcal{A},q}(T) \equiv \chi(\rho_{\mathcal{A},n}(\pi)) \mod n.$$

LEMMA 8 (Tate). *Let $k = \mathbb{F}_q$. The ℓ-adic representation $\tilde{\rho}_{\mathcal{A},l}$ is semi-simple and so is determined by $\chi(T)\,(\tilde{\rho}_{\mathcal{A},l}(\pi))$.*[2]

Geometric interpretation: We continue to assume that $\mathcal{A}$ is an Abelian variety defined over $\mathbb{F}_q$. Hence π acts on the algebraic points of $\mathcal{A}$ by exponentiation on coordinates with q. This action induces an action on the function field $\mathbb{F}_q(\mathcal{A})$ given again by exponentiation by q.

This action is polynomial, and so it induces a morphism on $\mathcal{A}$. Without loss of generality we can assume that this morphism fixes $0_\mathcal{A}$ and so is an endomorphism ϕ_q called the **Frobenius endomorphism**.

[2]An analogous result for $k = K$ a number field is the main result of Faltings on his way to prove Mordell's conjecture.

By definition it follows that the characteristic polynomial of the ℓ-adic representation of ϕ_q is equal to the characteristic polynomial $\chi_{\mathcal{A},q}(T)$ of the $\ell - adic$ Galois representation of π.

So for given $\mathcal{A}$, the Frobenius automorphism plays a double role as Galois element and as endomorphism, and this is of great importance for the arithmetic of Abelian varieties over finite fields.

The explicit knowledge of ϕ_q yields immediately that it is purely inseparable and

$$\deg \phi_q = [k(\mathcal{A}) : \pi^\star k(\mathcal{A})] = q^g.$$

DEFINITION 9. $\chi_{\mathcal{A},q}(T)$ is the characteristic polynomial of the Frobenius endomorphism ϕ_q of $\mathcal{A}$.

Its importance for the arithmetic of Abelian varieties over finite fields becomes evident by the following theorem.

THEOREM 10 (Tate). *Let $\mathcal{A}$ and $\mathcal{B}$ be Abelian varieties over a finite field $\mathbb{F}_q$ and $\chi_\mathcal{A}$ and $\chi_\mathcal{B}$ the characteristic polynomials of their Frobenius endomorphism and $l \neq p$ a prime. The following are equivalent.*

i) *$\mathcal{A}$ and $\mathcal{B}$ are isogenous.*
ii) $\chi_{\mathcal{A},q}(T) \equiv \chi_{\mathcal{B},q}(T)$
iii) *The zeta-functions for $\mathcal{A}$ and $\mathcal{B}$ are the same. Moreover, $\#\mathcal{A}(\mathbb{F}_{q^n}) = \#\mathcal{B}(\mathbb{F}_{q^n})$ for any positive integer n.*
iv) $T_l(\mathcal{A}) \otimes \mathbb{Q} \cong T_l(\mathcal{B}) \otimes \mathbb{Q}$

$\chi_{\mathcal{A},q}(T)$ is the most important tool for **counting points** on $\mathcal{A}(\mathbb{F}_q)$: Since ϕ_q is purely inseparable the endomorphism $\phi_q - id_\mathcal{A}$ is separable, and hence $\deg \ker(\phi_q - id_\mathcal{A})$ is equal to the number of elements in its kernel. Since π fixes exactly the elements of $\mathbb{F}_q$ the endomorphism ϕ_q fixes exactly $\mathcal{A}(\mathbb{F}_q)$ and so $\ker(\phi_q - id_\mathcal{A})(\bar{\mathbb{F}}_q) = \mathcal{A}(\mathbb{F}_q)$. By linear algebra it follows that:

THEOREM 11. *The number of points over $\mathbb{F}_q$ is given by*

$$\#(\mathcal{A}(\mathbb{F}_q)) = \chi_{\mathcal{A},q}(1).$$

The importance of this observation for *algorithms* for the computation of $\#(\mathcal{A}(\mathbb{F}_q)$ is due to one of the deepest results (**Hasse** for $g = 1$ and **Weil** for general g) in the arithmetic of Abelian varieties over finite fields, which is the analogue of the Riemann Hypothesis in number theory.

THEOREM 12. *Let $\mathcal{A}$ be an Abelian variety of dimension g over $\mathbb{F}_q$. The zeroes $\lambda_1, \ldots, \lambda_{2g}$ of the characteristic polynomial of the Frobenius endomorphism $\chi_{\mathcal{A},q}(T)$ have the following properties:*

- *Each λ_i is an algebraic integer.*
- *After a suitable numeration one gets for $1 \leq i \leq g$*

$$\lambda_i \cdot \lambda_{i+g} = q.$$

- *The complex absolute value $|\lambda_i|$ is equal to $\sqrt{q}$.*

For the proof we refer to [**56**].

It is evident that this theorem yields bounds for the size of the coefficients of $\chi_{\mathcal{A},q}(T)$ deepening only on g and q and so estimates for the size of $\#(\mathcal{A}(\mathbb{F}_q))$. We state the following Corollary.

COROLLARY 2.
$$|\#(\mathcal{A}(\mathbb{F}_q)) - q^g| = \mathcal{O}(q^{g-1/2}).$$

REMARK 3. *If $\mathcal{A}$ is the Jacobian of a curve $\mathcal{C}$ of genus g one can use this result to prove the Riemann Hypothesis for curves over finite fields:*
$$||\mathcal{C}(\mathbb{F}_q)| - q - 1| \leq 2g\sqrt{q}.$$

For a proof and refinements see [**67**].

In the next few sections we will focus on some special cases of Abelian varieties, namely Jacobian varieties and more specifically Jacobians of hyperelliptic curves.

3. Projective curves and Jacobian varieties

3.1. Curves. First let us establish some notation and basic facts about algebraic curves. In this paper the notion *curve* is an absolutely irreducible projective variety of dimension 1 without singularities.

At some rare points of the following discussion it is convenient to have that $\mathcal{C}(k) \neq \emptyset$, and without loss of generality we then can assume that there is a point P_∞ "at infinity", i.e. in $\mathcal{C}(k) \setminus U_0$. If we have to study curves with different properties (like being affine or having singularities) we shall state this explicitly.

Let $\mathcal{C}$ be a curve defined over k. Hence there is $n \in \mathbb{N}$ and a homogeneous *prime* ideal $I_\mathcal{C} \subset k[X_0, \ldots, X_n]$ such that, with $R = k[X_0, \ldots, X_n]/I_\mathbb{C}$, we have

(1) $\mathcal{C}$ is the scheme consisting of the topological space $\mathrm{Proj}(R)$ and the sheaf of holomorphic functions given on open subsets U of $\mathrm{Proj}(R)$ by the localization with respect to the functions in R not vanishing on U.
(2) The dimension of $\mathcal{C}$ is one, i.e. for every non-empty affine open subset $U \subset \mathrm{Proj}(R)$ the ring of holomorphic functions R_U on U is a ring with Krull dimension 1.
(3) $\mathcal{C}$ is regular, i.e. the localization of R with respect to every maximal ideal M in R is a discrete valuation ring R_M of rank 1. The equivalence class of the valuations attached to R_M is the *place* $\mathfrak{p}$ of $\mathcal{C}$, in this class the valuation with value group $\mathbb{Z}$ is denoted by w_M. Alternatively we use the notation $R_\mathfrak{p}$ and $w_\mathfrak{p}$. A place $\mathfrak{p}$ of $\mathcal{C}$ is also called *prime divisor* of $\mathcal{C}$.
(4) (Absolute irreducibility) $I_\mathcal{C} \cdot \bar{k}[X_0, \ldots, X_n]$ is a prime ideal in $\bar{k}[X_0, \ldots, X_n]$. This is equivalent with: k is algebraically closed in $\mathrm{Quot}(R)$.

As important consequence we note that for all open $\emptyset \neq U \neq \mathcal{C}$ the ring R_U is a *Dedekind domain.*

3.1.1. *Prime divisors and points.* The set of all places $\mathfrak{p}$ of the curve $\mathcal{C}$ is denoted by $\Sigma_\mathcal{C}(k)$. The *completeness* of projective varieties yields:

PROPOSITION 1. *There is a one-to-one correspondence between $\Sigma_\mathcal{C}(k)$ and the equivalence classes of valuations of $k(\mathcal{C})$, which are trivial on k.*

Let $\mathfrak{p} \in \Sigma_\mathcal{C}(k)$ be a prime divisor with corresponding maximal ideal $M_\mathfrak{p}$ and valuation ring $R_\mathfrak{p}$. We have a homomorphism
$$r_\mathfrak{p} : R_\mathfrak{p} \to R_p/M_\mathfrak{p} =: L$$
where L is a finite algebraic extension of k.

DEFINITION 13. The **degree** of the prime divisor $\mathfrak{p}$ is $\deg(\mathfrak{p}) := [L : k]$.

If $\deg(\mathfrak{p}) = 1$ then $L = k$ and $r_{\mathfrak{p}}$ induces a morphism from $\text{Spec}(k)$ into $\mathcal{C}$ and so corresponds to a point $P \in \mathcal{C}(k)$, uniquely determined by $\mathfrak{p}$. More explicitly, the point P has the homogeneous coordinates $(y_0 : y_1 : \ldots, : y_n)$ with $y_i = r_{\mathfrak{p}}(Y_i)$.

LEMMA 9. *The set $\Sigma^1_{\mathcal{C}}(k)$ of prime divisors of $\mathcal{C}$ of degree 1 is in bijective correspondence with the set of k-rational points $\mathcal{C}(k)$ of the curve $\mathcal{C}$.*

Now look at $\mathcal{C}_{\bar{k}}$, the curve obtained from $\mathcal{C}$ by constant field extension to the algebraic closure of k. Obviously, every prime divisor of $\mathcal{C}_{\bar{k}}$ has degree 1.

COROLLARY 3. *The set of prime divisors of $\mathcal{C}_{\bar{k}}$ corresponds one-to-one to the points in $\mathcal{C}_{\bar{k}}(\bar{k})$.*

Let's go back to k. Since $\bar{k}/k$ is separable, every equivalence class $\mathfrak{p}$ of valuations of $k(\mathcal{C})$ trivial on k has $\deg(\mathfrak{p}) = d$ extensions to $\bar{k}$ and these extensions are conjugate under the operation of G_k (Hilbert theory of valuations). Denote these extension by $(\tilde{\mathfrak{p}}_1, \ldots, \tilde{\mathfrak{p}}_d)$ and the corresponding points in $\mathcal{C}_{\bar{k}}(\bar{k})$ by $(P_1, \ldots, P_d)$. Then $\{P_1, \ldots, P_d\}$ is an orbit under the action of G_k and we have:

COROLLARY 4. *$\Sigma_{\mathcal{C}}(k)$ corresponds one-to-one to the G_k-orbits of $\mathcal{C}_{\bar{k}}(\bar{k})$.*

3.2. Divisors and Picard groups. Given a curve $\mathcal{C}/k$, the group of k-rational divisors $\text{Div}_{\mathcal{C}}(k)$ is defined as follows.

DEFINITION 14. $\text{Div}_{\mathcal{C}}(k) = \bigoplus_{\mathfrak{p} \in \Sigma_{\mathcal{C}}(k)} \mathbb{Z} \cdot \mathfrak{p}$, i.e. $\text{Div}_{\mathcal{C}}(k)$ is the free abelian group with base $\Sigma_{\mathcal{C}}(k)$.

Hence a **divisor** D of $\mathcal{C}$ is a formal sum

$$D = \sum_{\mathfrak{p} \in \Sigma_{\mathcal{C}}(k)} z_{\mathfrak{p}} P$$

where $z_{\mathfrak{p}} \in \mathbb{Z}$ and $z_{\mathfrak{p}} = 0$ for all but finitely many prime divisors $\mathfrak{p}$. The degree of a divisor is defined as

$$\deg(D) := \sum_{\mathfrak{p} \in \Sigma_{\mathcal{C}}(k)} z_{\mathfrak{p}}.$$

As we have seen in Cor. 4 we can interpret divisors as formal sums of G_k-orbits in $\mathcal{C}_{\bar{k}}(\bar{k})$. But we remark that taking points in $\mathcal{C}(k)$ is in general not enough to get all k-rational divisors of $\mathcal{C}$. The map

$$D \mapsto \deg(D)$$

is a homomorphism from $\text{Div}_{\mathcal{C}}(k)$ to $\mathbb{Z}$. Its kernel is the subgroup $\text{Div}_{\mathcal{C}}(k)^0$ of divisors of degree 0.

EXAMPLE 1. *Let $f \in k(\mathcal{C})^*$ be a meromorphic function on $\mathcal{C}$. For $\mathfrak{p} \in \Sigma_{\mathcal{C}}(k)$ we have defined the normalized valuation $w_{\mathfrak{p}}$. The* divisor of f *is defined as*

$$(f) = \sum_{\Sigma_{\mathcal{C}}(k)} w_{\mathfrak{p}} \cdot \mathfrak{p}.$$

It is not difficult to verify that (f) is a divisor, and that its degree is 0, see [**67**]. *Moreover $(f \cdot g) = (f) + (g)$ for functions f, g, and $(f^{-1}) = -(f)$. The completeness of $\mathcal{C}$ implies that $(f) = 0$ if and only if $f \in k^*$, and so (f) determines f up to scalars $\neq 0$.*

Thus, the set of principal divisors $\mathrm{PDiv}_{\mathcal{C}}(k)$ consisting of all divisors (f) with $f \in k(\mathcal{C})^*$ is a subgroup of $\mathrm{Div}_{\mathcal{C}}^0(k)$.

DEFINITION 15. The group of divisor classes of $\mathcal{C}$ is defined by

$$\mathrm{Pic}_{\mathcal{C}}(k) := \mathrm{Div}_{\mathcal{C}}(k)/\,\mathrm{PDiv}_{\mathcal{C}}(k)$$

and is called the **divisor class group** of $\mathcal{C}$. The group of divisor classes of degree 0 of $\mathcal{C}$ is defined by

$$\mathrm{Pic}_{\mathcal{C}}^0(k) := \mathrm{Div}_{\mathcal{C}}^0(k)/\,\mathrm{PDiv}_{\mathcal{C}}(k)$$

and is called the **Picard group** (of degree 0) of $\mathcal{C}$.

The Picard functor: Let L be a finite algebraic extension of k and $\mathcal{C}_L$ the curve obtained from $\mathcal{C}$ by constant field extension. Then places of $k(\mathcal{C})$ can be extended to places of $L(\mathcal{C}_L)$. By the conorm map we get an injection of $\mathrm{Div}_{\mathcal{C}}(k)$ to $\mathrm{Div}_{\mathcal{C}_L}(L)$. The well known formulas for the extensions of places yield that

$$\mathrm{conorm}_{L/k}(\mathrm{Div}_{\mathcal{C}}^0(k)) \subset \mathrm{Div}_{\mathcal{C}_L}^0(L)$$

and that principal divisors are mapped to principal divisors. Hence we get a homomorphism

$$\mathrm{conorm}_{L/k} : \mathrm{Pic}_{\mathcal{C}}^0(k) \to \mathrm{Pic}_{\mathcal{C}_L}^0(L)$$

and therefore a functor

$$\mathrm{Pic}^0 : L \mapsto \mathrm{Pic}_{\mathcal{C}_L}^0(L)$$

from the category of algebraic extension fields of k to the category of abelian groups. Coming "from above" we have a Galois theoretical description of this functor. Clearly,

$$\mathrm{Div}_{\mathcal{C}_L}(L) = \mathrm{Div}_{\mathcal{C}_{\bar{k}}}(\bar{k})^{G_L}$$

and the same is true for functions. With a little bit of more work one sees that an analogue result is true for $\mathrm{PDiv}_{\mathcal{C}_L}(L)$ and for $\mathrm{Pic}_{\mathcal{C}_L}^0(L)$.

THEOREM 16. *For any curve $\mathcal{C}_k$ and any extension L/k with $k \subset L \subset \bar{k}$ the functor*

$$L \mapsto \mathrm{Pic}_{\mathcal{C}_L}^0(L)$$

is the same as the functor

$$L \mapsto \mathrm{Pic}_{\mathcal{C}_{\bar{k}}}^0(\bar{k})^{G_L}.$$

In particular, we have

$$\mathrm{Pic}_{\mathcal{C}_{\bar{k}}}^0(\bar{k}) = \bigcup_{k \subset L \subset \bar{k}} \mathrm{Pic}_{\mathcal{C}_L}^0(L),$$

where inclusions are obtained via conorm maps.

REMARK 4. *For a finite extension L/k we also have the norm map of places of $\mathcal{C}_L$ to places of $\mathcal{C}_k$ induces a homomorphism from $\mathrm{Pic}_{\mathcal{C}_L}^0(L)$ to $\mathrm{Pic}_{\mathcal{C}}^0(k)$. In general, this map will be neither injective nor surjective.*

It is one of the most important facts for the theory of curves that the functor Pic^0 can be represented: There is a variety $\mathcal{J}_{\mathcal{C}}$ defined over k such that for all extension fields L of k we have a functorial equality

$$\mathcal{J}_{\mathcal{C}}(L) = \mathrm{Pic}_{\mathcal{C}_L}^0(L).$$

$J_{\mathcal{C}}$ is the **Jacobian variety** of $\mathcal{C}$. This variety will be discussed soon.

3.3. Riemann-Roch theorem. Here we take as guideline the book [**67**] of H. Stichtenoth.

3.3.1. *Riemann-Roch spaces.* We define a partial ordering of elements in $\mathrm{Div}_{\mathcal{C}}(k)$ as follows;

$$D = \sum_{\mathfrak{p} \in \Sigma_{\mathcal{C}}(k)} z_{\mathfrak{p}} P$$

is *effective* $(D \geq 0)$ if $z_{\mathfrak{p}} \geq 0$ for every $\mathfrak{p}$, and $D_1 \geq D_2$ if $D_1 - D_2 \geq 0$.

DEFINITION 17. The **Riemann-Roch space** associated to D is

$$\mathcal{L}(D) = \{f \in k(\mathcal{C})^* \text{ with } (f) \geq -D\} \cup \{0\}.$$

So the elements $x \in \mathcal{L}(D)$ are defined by the property that $w_{\mathfrak{p}}(x) \geq -z_{\mathfrak{p}}$ for all $\mathfrak{p} \in \Sigma_{\mathcal{C}}(k)$. Basic properties of valuations imply immediately that $\mathcal{L}(D)$ is a vector space over k. This vector space has positive dimension if and only if there is a function $f \in k(\mathcal{C})^*$ with $D + (f) \geq 0$, or equivalently, $D \sim D_1$ with $D_1 \geq 0$.

Here are some immediately obtained facts: $\mathcal{L}(0) = k$ and if $\deg(D) < 0$ then $\mathcal{L}(D) = \{0\}$. If $\deg(D) = 0$ then either D is a principal divisor or $\mathcal{L}(D) = \{0\}$. The following result is easy to prove but fundamental.

PROPOSITION 2. *Let $D = D_1 - D_2$ with $D_i \geq 0$. Then*

$$\dim(\mathcal{L}(D)) \leq \deg(D_1) + 1.$$

We remark that for $D \sim D'$ we have $\mathcal{L}(D) \sim \mathcal{L}(D')$. In particular $\mathcal{L}(D)$ is a finite-dimensional k-vector space.

DEFINITION 18. $\ell(D) := \dim_k(\mathcal{L}(D))$.

To compute $\ell(D)$ is a fundamental problem in the theory of curves. It is solved by the Theorem of Riemann-Roch. For all divisors D we have the inequality

$$\ell(D) \leq \deg(D) + 1.$$

For a proof one can assume that $\ell(D) > 0$ and so $D \sim D' > 0$. The important fact is that one can estimate the interval given by the inequality.

THEOREM 19 (**Riemann**). *For given curve $\mathcal{C}$ there is a minimal number $g_{\mathcal{C}} \in \mathbb{N} \cup \{0\}$ such that for all $D \in \mathrm{Div}_{\mathcal{C}}$ we have*

$$\ell(D) \geq \deg(D) + 1 - g_{\mathcal{C}}.$$

For a proof see [**67**, Proposition 1.4.14]. Therefore,

$$g_{\mathcal{C}} = \max\{\deg D - \ell(D) + 1;\ D \in \mathrm{Div}_{\mathcal{C}}(k)\}$$

exists and is a non-negative integer independent of D.

DEFINITION 20. The integer $g_{\mathcal{C}}$ is called the **genus** of $\mathcal{C}$.

We remark that the genus does not change under constant field extensions because we have assumed that k is perfect. This can be wrong in general if the constant field of $\mathcal{C}$ has inseparable algebraic extensions.

COROLLARY 5. *There is a number $n_{\mathcal{C}}$ such that for* $\deg(D) > n_{\mathcal{C}}$ *we get equality*

$$\ell(D) = \deg(D) + 1 - g_{\mathcal{C}}.$$

Thm. 19 together with its corollary is the "Riemann part" of the Theorem of Riemann-Roch for curves. To determine $n_{\mathcal{C}}$ and to get more information about the inequality for small degrees one needs canonical divisors.

3.3.2. *Canonical divisors.* Let $k(\mathcal{C})$ be the function field of a curve $\mathcal{C}$ defined over k. To every $f \in k(\mathcal{C})$ we attach a symbol df, the *differential* of f lying in a $k(\mathcal{C})$-vector space $\Omega(k(\mathcal{C}))$ generated by the symbols df modulo the following relations: For $f, g \in k(\mathcal{C})$ and $\lambda \in k$ we have:

i) $d(\lambda f + g) = \lambda df + dg$
ii) $d(f \cdot g) = fdg + gdf$.

The relation between derivations and differentials is given by the

DEFINITION 21 (Chain rule). Let x be as above and $f \in k(\mathcal{C})$. Then $df = (\partial f/\partial x)dx$.

As in calculus one shows that the $k(\mathcal{C})$-vector space of differentials $\Omega(k(\mathcal{C}))$ has dimension 1 and it is generated by dx for any $x \in k(\mathcal{C})$ for which $k(\mathcal{C})/k(x)$ is finite and separable. We use a well known fact from the theory of function fields F in one variable.

Let $\mathfrak{p}$ be a place of F, i.e. an equivalence class of discrete rank one valuations of F trivial on k). Then there exist a function $t_\mathfrak{p} \in F$ with $w_\mathfrak{p}(t_\mathfrak{p}) = 1$and $F/k(t_\mathfrak{p})$ separable. We apply this to $F = k(\mathcal{C})$. For all $\mathfrak{p} \in \Sigma_\mathcal{C}(k)$ we choose a function $t_\mathfrak{p}$ as above. For a differential $0 \neq \omega \in \Omega(k(\mathcal{C}))$ we get $\omega = f_\mathfrak{p} \cdot dt_\mathfrak{p}$. The divisor (ω) is given by

$$(\omega) := \sum_{\mathfrak{p} \in \Sigma_\mathfrak{p}} w_\mathfrak{p}(f_\mathfrak{p}) \cdot \mathfrak{p}$$

and is a called a **canonical divisor** of $\mathcal{C}$.

The chain rule implies that this definition is independent of the choices, and the relation to differentials yields that (ω) is a divisor. Since $\Omega(k(\mathcal{C}))$ is one-dimensional over $k(\mathcal{C})$ it follows that the set of canonical divisors of $\mathcal{C}$ form a divisor class $k_\mathcal{C} \in \mathrm{Pic}_\mathcal{C}(k)$ called the **canonical class** of $\mathcal{C}$. We are now ready to formulate the Riemann-Roch Theorem.

THEOREM 22 (**Riemann-Roch theorem**). *Let W be a canonical divisor of $\mathcal{C}$. For all $D \in \mathrm{Div}_\mathcal{C}(k)$ we have*

$$\ell(D) = \deg(D) + 1 - g_\mathcal{C} + \ell(W - D).$$

For a proof see [**67**, Section 1.5].

A differential ω is *holomorphic* if (ω) is an effective divisor. The set of holomorphic differentials is a k-vector space denoted by $\Omega^0_\mathcal{C}$ which is equal to $\mathcal{L}(W)$. If we take $D = 0$ respectively $D = W$ in the theorem of Riemann-Roch we get the following:

COROLLARY 6. *$\Omega^0_\mathcal{C}$ is a $g_\mathcal{C}$-dimensional k-vector space and* $\deg(W) = 2g_\mathcal{C} - 2$.

For our applications there are two further important consequences of the Riemann-Roch theorem.

COROLLARY 7. *The following are true:*

(1) *If* $\deg(D) > 2g_\mathcal{C} - 2$ *then* $\ell(D) = \deg(D) + 1 - g_\mathcal{C}$.
(2) *In every divisor class of degree g there is a positive divisor.*

PROOF. Take D with $\deg(D) \geq 2g_\mathcal{C} - 1$. Then $\deg(W - D) \leq -1$ and therefore $\ell(W - D) = 0$. Take D with $\deg(D) = g_\mathcal{C}$. Then $\ell(D) = 1 + \ell(W - D) \geq 1$ and so there is a positive divisor in the class of D. □

4. Applications of the Riemann-Roch theorem

4.1. The Hurwitz genus formula. In the theory of curves the notion of a cover is important.

DEFINITION 23. Let $\mathcal{C}$, $\mathcal{D}$ be curves defined over k, with $\mathcal{D}$ not necessarily absolutely irreducible. A finite surjective morphism

$$\eta : \mathcal{D} \to \mathcal{C}$$

from $\mathcal{D}$ to $\mathcal{C}$ is a *cover morphism*, and if such a morphism exists we call $\mathcal{D}$ a **cover** of $\mathcal{C}$.

As usual, we denote by

$$\eta^* : k(\mathcal{C}) \hookrightarrow k(\mathcal{D})$$

the induced monomorphism of the function fields and identify $k(\mathcal{C})$ with its image. η is separable if and only if $k(\mathcal{D})$ is a separable extension of $k(\mathcal{C})$, and η is Galois with Galois group G if $k(\mathcal{D})/k(\mathcal{C})$ is Galois with group G. The cover η is geometric if k is algebraically closed in $k(\mathcal{D})$.

Assume in the following that η is separable. We shall use the well known relations between prime divisors of $k(\mathcal{C})$ and those of $k(\mathcal{D})$ such as extensions, ramifications and sum formulas for the degrees. In particular we get:

Let $\mathfrak{p} \in \Sigma_{\mathcal{C}}(k)$ be a prime divisor and $\mathfrak{P}_1, \dots, \mathfrak{P}_r$ the primes divisors in $\Sigma_{\mathcal{D}}(k)$ which extend $\mathfrak{p}$, written as $\mathfrak{P}/\mathfrak{p}$. Let $t_{\mathfrak{p}}$ be an element in $k(\mathcal{C})$ with $w_{\mathfrak{p}}(t_{\mathfrak{p}}) = 1$. The ramification index $e(\mathfrak{P}_i/\mathfrak{p}) =: e_i$ is defined as $e_i = w_{\mathfrak{P}_i}(t_{\mathfrak{p}})$, hence there is a function $t_{\mathfrak{P}_i}$ on $\mathcal{D}$ such that $t_{\mathfrak{P}_i}^{e_i} = t_{\mathfrak{p}} \cdot u$ with $w_{\mathfrak{P}_i}(u) = 0$. The *conorm* of $\mathfrak{p}$ is the divisor

$$\operatorname{conorm}(\mathfrak{p}) = \sum_i \mathfrak{P}_i^{e_i}$$

and its degree is $[k(\mathcal{D}) : k(\mathcal{C})]$, the *norm* of $\mathfrak{P}_i$ is $\mathfrak{p}$. The cover η is *tamely ramified* if for all $\mathfrak{p} \in \Sigma_{\mathcal{C}}(k)$ the ramification numbers of all extensions are prime to char(k).

We want to relate the genus of $\mathcal{D}$ to the genus of $\mathcal{C}$. Let $x \in k(\mathcal{C})$ be such that $k(\mathcal{C})/k(x)$ is finite separable, and let $dx_{\mathcal{C}}$ respectively $dx_{\mathcal{D}}$ be corresponding differentials with divisors $(dx)_{\mathcal{C}}$ and $dx_{\mathcal{D}}$. We know that

$$2g_{\mathbb{C}} - 2 = \deg(dx)_{\mathcal{C}} \text{ and } 2g_{\mathcal{D}} - 2 = \deg(dx)_{\mathcal{D}}.$$

We compute the value $z_{\mathfrak{p}}$ respectively $z_{\mathfrak{P}_i}$ of these divisors at $\mathfrak{p} \in \Sigma_{\mathcal{C}}(k)$ and in the extensions $\mathfrak{P}_1, \dots \mathfrak{P}_r$ with ramification numbers e_i. To ease notation we take $\mathfrak{P} := \mathfrak{P}_i$, $e_i = e_{\mathfrak{P}}$ and $t_{\mathfrak{P}} \in k(\mathcal{D})$ with $w_{\mathfrak{P}} = 1$. Then we can choose

$$t_{\mathfrak{p}} = u \cdot t_{\mathfrak{P}}^{e_{\mathfrak{P}}} \in k(\mathcal{C}),$$

with $w_{\mathfrak{p}}(u) = 0$. By the rules for differentials we get $dt_p = (e_{\mathfrak{P}} \cdot u \cdot t_{\mathfrak{P}}^{e_{\mathfrak{P}}-1} + u' \cdot t_{\mathfrak{P}}^{e_{\mathfrak{P}}})dt_{\mathfrak{P}}$ and so

$$w_{\mathfrak{P}}(dx) = e_{\mathfrak{P}} \cdot w_{\mathfrak{p}}(dx) + e_{\mathfrak{P}} - 1.$$

Summing up over $\mathfrak{P}_1, \dots, \mathfrak{P}_r$ we get that

$$\deg\left(\sum_{\mathfrak{P}|\mathfrak{p}} z_{\mathfrak{P}}\right) = \deg\left(\sum_{i=1}^r z_{\mathfrak{p}} \mathfrak{P}_i^e\right) + \sum_{i=1}^r (e_i - 1).$$

Summing up over all $\mathfrak{p} \in \Sigma_{\mathcal{C}}(k)$ we get the Hurwitz theorem.

THEOREM 24 (Hurwitz). *Any separable, tamely ramified degree n cover $\eta : \mathcal{D} \to \mathcal{C}$ with $e_{\mathfrak{P}}$ the ramification index of $\mathfrak{P} \in \Sigma_{\mathcal{D}}(k)$ satisfies*

$$2g_{\mathcal{D}} - 2 = n \cdot (2g_{\mathcal{C}} - 2) + \sum_{\mathfrak{P} \in \Sigma_{\mathcal{D}}} (e_{\mathfrak{P}} - 1).$$

Let us illustrate the theorem with a classical example.

EXAMPLE 2. *Assume that $\mathcal{C} = \mathbb{P}^1$, the genus $g_{\mathbb{P}^1} = 0$ curve. Let $\mathcal{D}$ be tamely ramified cover of degree n of $\mathbb{P}^1$. Then*

$$g_{\mathcal{D}} = 1 - n + \frac{1}{2} \sum_{\mathfrak{P} \in \Sigma_{\mathcal{D}}(k)} (e_{\mathfrak{P}} - 1).$$

In particular $\mathbb{P}^1$ has no unramified extensions.

The special case $n = 2$ will be important for us. Assume that char(k) $\neq 2$. Then we can apply the Hurwitz formula and get

$$g_{\mathcal{D}} = \frac{1}{2} r - 1,$$

where r is the number of prime divisors of $\mathbb{P}^1$ (or of $\mathcal{D}$) which are ramified (i.e. ramification order is larger than 1) under η.

4.2. Gonality of curves and Hurwitz spaces. Let $\mathcal{C}$ be a curve defined over k and $\eta : \mathcal{C} \to \mathbb{P}^1$ a degree n cover. We assume that $\mathcal{C}$ has a k-rational point P_∞ and hence a prime divisor $\mathfrak{p}_\infty$ of degree 1.

DEFINITION 25. The gonality $\gamma_{\mathcal{C}}$ of $\mathcal{C}$ is defined by

$$\gamma_{\mathcal{C}} = \min\left\{\deg(\eta) : \mathcal{C} \to \mathbb{P}^1\right\} = \min\left\{[k(\mathcal{C}) : k(x)] \mid x \in k(\mathcal{C})\right\}.$$

For $x \in k(\mathcal{C})^*$, define the pole divisor $(x)_\infty$ by

$$(x)_\infty = \sum_{\mathfrak{p} \in \Sigma_{\mathcal{C}}(k)} \max(0, -w_{\mathfrak{p}}(x)) \cdot \mathfrak{p}.$$

By the property of conorms of divisors we get $\deg((x)_\infty) = [k(\mathcal{C}) : k(x)]$ if $x \notin k$ and so

$$\gamma_{\mathcal{C}} = \min\left\{\deg{(x)_\infty}, \mid x \in k(\mathcal{C}) \setminus k.\right\}$$

PROPOSITION 3. *For $\gamma_{\mathcal{C}} \geq 2$ we have $\gamma_{\mathcal{C}} \leq g$.*

PROOF. By Riemann-Roch theorem

$$\ell(g_{\mathcal{C}} \cdot P_\infty) = 1 + \ell(W - g_{\mathcal{C}} \cdot P_\infty)$$

and since $g_{\mathcal{C}} \geq 2$ the divisor $(W - g_{\mathcal{C}} \cdot P_\infty)$ has degree ≥ 0 and so $\ell(W - g_{\mathcal{C}} \cdot P_\infty) \geq 1$. But then $\ell(g_{\mathcal{C}} \cdot P_\infty) \geq 2$ and there is a non-constant function x whose pole divisor is a multiple of $\mathfrak{p}_\infty$ of order $\leq g_{\mathcal{C}}$.

□

This proves more than the proposition.

COROLLARY 1. *For curves $\mathcal{C}$ of genus ≥ 2 with prime divisor $\mathfrak{p}_\infty$ of degree 1 there exists a cover*

$$\eta : \mathcal{C} \to \mathbb{P}^1$$

with $\deg(\eta) = n \leq g_{\mathcal{C}}$ such that $\mathfrak{p}_\infty$ is ramified of order n and so the point $P_\infty \in \mathcal{C}(k)$ attached to $\mathfrak{p}_\infty$ is the only point on $\mathcal{C}$ lying over the infinite point $(0 : 1)$ of $\mathbb{P}^1$.

In general, the inequality in the proposition is not sharp but of size $g/2$ as we shall see below. Curves with smaller gonality are special and so per se interesting.

4.2.1. *Gonality of the generic curve.* Let us assume that k is algebraically closed. We are interested in the classification of isomorphism classes of projective irreducible regular curves of genus $g \geq 2$.

The moduli scheme $\mathcal{M}_g$ is a scheme defined over k with the property that it parametrizes these classes (i.e., to every point P there is a unique class of a curve $\mathcal{C}$ of genus g). The coordinates of $\mathcal{C}$ (chosen in an appropriate affine open neighborhood) are the invariants of $\mathcal{C}$. It is a classical task to determine such systems of invariants and then to find the curve $\mathcal{C}$ with these invariants. We shall come back to this in the case of curves of small genus.

REMARK 5. *The scheme $\mathcal{M}_g$ is defined over non-algebraically closed fields k but then it is only a coarse moduli scheme.*

The construction of $\mathcal{M}_g$ is done over $\mathbb{C}$ either by Teichmüller theory or, more classically, by Hurwitz spaces (see below), and so over algebraically closed fields of characteristic 0 by the so-called Lefschetz principle. Its existence in the abstract setting of algebraic geometry uses deep methods of geometric invariant theory as developed and studied by Deligne and Mumford in [**13**].

One knows that $\mathcal{M}_g$ is irreducible and so there exists a generic curve of genus g. Moreover the dimension of $\mathcal{M}_g$ is equal to $3g-3$. Curves with special properties (i.e. non-trivial automorphisms or small gonality) define interesting subschemes of $\mathcal{M}_g$. Here is one example.

DEFINITION 26. A curve $\mathcal{C}$ with genus ≥ 2 is hyperelliptic if and only if it has gonality 2.

The subspace of hyperelliptic curves in $\mathcal{M}_g$ is the *hyperelliptic locus* $\mathcal{M}_{g,h}$. We shall see below that this locus has dimension $2g-1$.

Hurwitz spaces: We continue to assume that k is algebraically closed and consider separable covers $\eta : \mathcal{C} \to \mathbb{P}^1$ of degree n. Then η^* allows to identify $k(\mathbb{P}^1) =: k(x)$ with a subfield of $k(\mathcal{C})$. First, we introduce the equivalence: $\eta \sim \eta'$ if there are isomorphisms $\alpha : \mathcal{C} \to \mathcal{C}'$ and $\beta \in \mathrm{Aut}(\mathbb{P}^1)$ with

$$\beta \circ \eta = \eta' \circ \alpha.$$

The *monodromy group* of η is the Galois group of the Galois closure L of $k(\mathcal{C})/k(x)$. We embed G into S_n, the symmetric group with n letters, and fix the ramification type of the covers η. We assume that exactly $r \geq 3$ points in $\mathbb{P}^1(k)$ are ramified (i.e. the corresponding prime divisors have at least one extension to $k(\mathcal{C})$ with ramification order > 1) and that all ramification orders are prime to char(k). It follows that the ramification groups are cyclic.

By the classical theory of covers of Riemann surfaces, which can be transferred to the algebraic setting by the results of Grothendieck (here one needs tameness of ramification) it follows that there is a tuple $(\sigma_1, \dots, \sigma_r)$ in S_n such that $\sigma_1 \cdots \sigma_r = 1$, $\mathrm{ord}(\sigma_i) = e_i$ is the ramification order of the i-th ramification point P_i in L and $G := \langle \sigma_1, \dots, \sigma_r \rangle$ is a transitive group in S_n. We call such a tuple the **signature** σ of the covering η and remark that such tuples are determined up to conjugation in S_n, and that the genus of $\mathcal{C}$ is determined by the signature because of the Hurwitz genus formula.

Let $\mathcal{H}_\sigma$ be the set of pairs $([\eta], (p_1, \ldots, p_r)$, where $[\eta]$ is an equivalence class of covers of type σ, and $p_1, \ldots, p_r$ is an ordering of the branch points of ϕ modulo automorphisms of $\mathbb{P}^1$. This set carries the structure of a algebraic scheme, in fact it is a quasi-projective variety, the *Hurwitz space* $\mathcal{H}_\sigma$. We have the forgetful morphism

$$\Phi_\sigma : \mathcal{H}_\sigma \to \mathcal{M}_g$$

mapping $([\eta], (p_1, \ldots, p_r)$ to the isomorphic class $[\mathcal{C}]$ in the moduli space $\mathcal{M}_g$. Each component of $\mathcal{H}_\sigma$ has the same image in $\mathcal{M}_g$.

Define the **moduli dimension of** σ (denoted by $\dim(\sigma)$) as the dimension of $\Phi_\sigma(\mathcal{H}_\sigma)$; i.e., the dimension of the locus of genus g curves admitting a cover to $\mathbb{P}^1$ of type σ. We say σ has **full moduli dimension** if $\dim(\sigma) = \dim \mathcal{M}_g$; see [**50**] for details.

EXAMPLE 3. *Take* $n = 2$, *so* $G = S_2$, $r \geq 6$ *and* $\mathrm{char(k)} \neq 2$ *and the notations from above. A signature* σ *is completely determined by the* r *ramification points* $P_1, \cdots, P_r$. *Hence* $\mathcal{H}_\sigma$ *consists of classes of hyperelliptic curves of genus* $g_r = r/2 - 1$ *(so* r *is even). Since we can apply automorphisms of* $\mathbb{P}^1$ *we can assume that* $P_1 = (1:0), P_2 = (1,1), P_3 = (0:1)$ *and so we have* $r - 3$ *free parameters modulo a finite permutation group.*

So the moduli dimension is $r - 3 = 2g + 2$, *and the hyperelliptic locus* $\mathcal{M}_{g,h}$ *has dimension* $2g - 1$ *and codimension* $g - 2$. *Hence all curves of genus* 2 *are hyperelliptic, and for* $g \geq 3$ *the locus of the hyperelliptic curves has positive codimension.*

For a fixed $g \geq 3$, we want to find σ of full moduli dimension and of minimal degree. This would give a generic covering of minimal degree for a generic curves of genus g and so its gonality.

A first condition is that $r = 3g$. Because of the Hurwitz genus formula this yields conditions for the ramification cycles, which have to have minimal order. This is worked out in [**61**].

LEMMA 10. *For any* $g \geq 3$ *there is a minimal degree* $d = \lfloor \frac{g+3}{2} \rfloor$ *generic cover*

$$\psi_g : \mathcal{C}_g \to \mathbb{P}^1$$

of full moduli dimension from a genus g *curve* $\mathcal{C}_g$ *such that it has* $r = 3g$ *branch points and signature:*

i) If g *is odd, then* $\sigma = (\sigma_1, \ldots, \sigma_r)$ *such that* $\sigma_1, \ldots, \sigma_{r-1} \in S_d$ *are transpositions and* $\sigma_r \in S_d$ *is a 3-cycle.*

ii) If g *is even, then* $\sigma = (\sigma_1, \ldots, \sigma_r)$ *such that* $\sigma_1, \ldots, \sigma_r \in S_d$ *are transpositions.*

4.2.2. *Equations for curves.* There is a one-to-one correspondence between function fields F of transcendence degree 1 over the field of constants k with k algebraically closed in F and isomorphism classes of projective regular absolutely irreducible curves $\mathcal{C}$ with $k(\mathcal{C}) = F$. The natural question is: Given F, how can one find $\mathcal{C}$ as embedded projective curve in an appropriate $\mathbb{P}^n$?

The main tool to solve this question are Riemann-Roch systems. Let D with $\ell(D) = d + 1 > 0$ and $(f_0, f_1, \ldots, f_d)$ a base of $\mathcal{L}(D)$. Then

$$\begin{aligned} \Phi_D : \mathcal{C}(\bar{k}) &\to \mathbb{P}^d(\bar{k}) \\ P &\mapsto (f_0(P) : f_1(p) : \cdots : f_d(P)) \end{aligned}$$

is a rational map defined in all points for which $f_0, \ldots, f_d$ do not vanish simultaneously. $\mathcal{L}(D)$ is without base points if this set is empty, and then Φ_D is a morphism from $\mathcal{C}$ in $\mathbb{P}^d$.

LEMMA 11. *For $g \geq 3$ and $D = W_\mathcal{C}$ the space $\mathcal{L}(W) = \Omega^0_\mathcal{C}$ is without base points, and so Φ_W is a morphism from $\mathcal{C}$ to $\mathbb{P}^{g_\mathcal{C}-1}$.*

Φ_W may not be an embedding but the only exception is that $\mathcal{C}$ is hyperelliptic, and than the image of Φ_W is the projective line.

THEOREM 27. *Let $\mathcal{C}$ be a curve of genus $g_\mathcal{C} > 2$ and assume that $\mathcal{C}$ is not hyperelliptic. Then Φ_W is an embedding of $\mathcal{C}$ into $\mathbb{P}^{g_\mathcal{C}-1}$ and the image is a projective regular curve of degree $2g_\mathcal{C} - 2$ (i.e. the intersection with a generic hyperplane has $2g_\mathcal{C} - 2$ points).*

So having determined a base of the canonical class of $\mathcal{C}$ one gets a parameter representation of $\mathcal{C}$ and then one can determine the prime ideal in $k[Y_0, \ldots, Y_{g_\mathcal{C}}]$vanishing on $\Phi_W(\mathcal{C})$. Φ_W is the **canonical embedding** of $\mathcal{C}$.

EXAMPLE 4. *Take $g_\mathcal{C} = 3$ and assume that $\mathcal{C}$ is not hyperelliptic. Then the canonical embedding maps $\mathcal{C}$ to a regular projective plane curve of degree 4. In other words: All non-hyperelliptic curves of genus 3 are isomorphic to non-singular quartics in $\mathbb{P}^2$.*

Plane curves: Only very special values of the genus of $\mathcal{C}$ allow to find plane regular projective curves isomorphic to $\mathcal{C}$. We have just seen that $g = 3$ is such a value. The reason behind is the Plücker formula, which relates degree, genus and singularities of plane curves. But of course, there are many projective plane curves which are birationally equivalent to $\mathcal{C}$.

Take $x \in k(\mathcal{C}) \setminus k$ with $k(\mathcal{C})/k(x)$ separable. Then there is an element $y \in k(\mathcal{C})$ with $k(x, y) = k(\mathcal{C})$, and by clearing denominators we find a polynomial $G(x, y) \in k[X, Y]$ with $G(x, y) = 0$. Then the curve $\mathcal{C}'$ given by the homogenized polynomial

$$G_h(X, Y, Z) = 0$$

is a plane projective curve birationally equivalent to $\mathcal{C}$ but, in general, with singularities. Using the gonality results we can chose $G(X, Y)$ such that the degree in Y is $\lfloor \frac{g+3}{2} \rfloor$. Using the canonical embedding for non hyperelliptic curves and general projections we can chose $G_h(X, Y, Z)$ as homogeneous polynomial of degree $2g_\mathcal{C} - 2$.

In the next subsection we shall describe a systematic way to find plane equations for hyperelliptic curves.

4.2.3. *Plane equations for elliptic and hyperelliptic curves, Weierstrass normal forms.* We first focus on elliptic curves.

Elliptic curves: We assume that $\mathcal{E}$ is a curve of genus 1 with a k-rational point P_∞ and corresponding prime divisor $\mathfrak{p}_\infty$. By definition, $\mathcal{E}$ is an *elliptic curve defined over k*. We look at the Riemann-Roch spaces $\mathcal{L}_i := \mathcal{L}(i \cdot \mathfrak{p}_\infty)$ and denote their dimension by ℓ_i. Since $2g_\mathcal{E} - 2 = 0$ we can use the theorem of Riemann-Roch to get that $\ell_i = i$. Hence $\mathcal{L}_1 = \langle 1 \rangle$, $\mathcal{L}_2 = \langle 1, x \rangle$ with a function $x \in k(\mathcal{E})$ with $(x)_\infty = 2\mathfrak{p}_\infty$, $\mathcal{L}_3 = \langle 1, x, y \rangle$ with $(y)_\infty = 3\mathfrak{p}_\infty$ and $\mathcal{L}_5 = \langle 1, x, x^2, y, xy \rangle$ with 5 linearly independent functions.

Now look at $\mathcal{L}_6$. This is a vector space of dimension 6 over k. It contains the seven elements $\{1, x, x^2, x^3, y, xy, y^2\}$ and hence there is a non-trivial linear relation

$$\sum_{0 \leq i \leq 3;\, 0 \leq j \leq 2} a_{i,j} x^i y^2.$$

Because of the linear independence of $(1, x, x^2, y, xy)$ we get that either $a_{3,0}$ or $a_{0,2}$ are not equal 0, and since x^3 and y^2 have a pole of order 6 in $\mathfrak{p}_\infty$ it follows that $a_{0,2} \cdot a_{3,0} \neq 0$. By normalizing we get x and y satisfy the equation

$$Y^2 + a_1 X \cdot Y + a_3 Y = a_0 X^3 + a_2 X^2 + a_4 X + a_6.$$

By multiplying with a_0^2 and substituting (X, Y) by $(a_0 X, a_0 Y)$ we get an **affine Weierstrass equation** for $\mathcal{E}$:

$$W_{\mathcal{E}_{aff}} : Y^2 + a_1 X \cdot Y + a_3 Y = X^3 + a_2 X^2 + a_4 X + a_6.$$

The homogenization give the cubic equation

$$W_{\mathcal{E}} : Y^2 \cdot Z + a_1 X \cdot Y \cdot Z + a_3 Y \cdot Z^2 = a_0 X^3 + a_2 X^2 \cdot Z + a_4 X \cdot Z^2 + a_6 \cdot Z^3,$$

which defines a plane projective curve.

The infinite points of this curve have $Z = 0$, and so the only infinite point is $P_\infty = (0, 1, 0)$ corresponding to the chosen $\mathfrak{p}_\infty$. Looking at the partial derivatives one verifies that $\mathcal{E}$ has no singularities if and only if the discriminant of the affine equation $W_{\mathcal{E}_{aff}}$ as polynomial in X is different from 0, and that this is equivalent with the condition that $k(\mathcal{E})$ is not a rational function field.

THEOREM 28. *Elliptic curves defined over k correspond one-to-one the isomorphism classes of plane projective curves without singularities given by Weierstrass equations*

$$W_{\mathcal{E}} : Y^2 \cdot Z + a_1 X \cdot Y \cdot Z + a_3 Y \cdot Z^2 = X^3 + a_2 X^2 \cdot Z + a_4 X \cdot Z^2 + a_6 \cdot Z^3$$

with non-vanishing X-discriminant.

Since we are dealing with isomorphism classes of such curves we can further normalize the equations and finally find invariants for the isomorphism class of a given $\mathcal{E}$. This is a bit tedious if char(k) | 6. In this case we refer to [**65**].

Assume that char(k) $\neq 2, 3$. Then we can use Tschirnhausen transformations to get an equation

$$W_{\mathcal{E}} : Y^2 \cdot Z = X^3 - g_2 X \cdot Z^2 - g_3 \cdot Z^3$$

and the reader should compare this equation with the differential equation satisfied by the Weierstrass $\wp$-function.

We use this analogy and define $\Delta(\mathcal{E}) = 4g_2^3 - 27g_3^2$ and this is, because of the regularity of $\mathcal{E}$, an element $\neq 0$, and

$$j_E = 12^3 \frac{4g_2^3}{\Delta_{\mathcal{E}}}.$$

If k is algebraically closed then $j_{\mathcal{E}}$ determines the isomorphic class of $\mathcal{E}$. For an arbitrary k the curve E is determined up to a *twist*, which is quadratic if char(k) is prime to 6 (see [**65**]).

Weierstrass equations for hyperelliptic curves: Let $\mathcal{C}$ be a curve over k of genus $g \geq 2$ with a degree 2 cover

$$\eta : \mathcal{C} \to \mathbb{P}^1.$$

We assume that there is a point $P_\infty \in \mathcal{C}(k)$ corresponding to a prime divisor $\mathfrak{p}_\infty$ of $\mathcal{C}$ of degree 1. Take $Q_\infty = \eta(P_\infty) \in \mathbb{P}^1(k)$ and $x \in k(\mathbb{P}^1)$ with $(x)_\infty = \mathfrak{p}_{0,\infty}$ with $\mathfrak{p}_{0,\infty}$ a prime divisor of degree 1 of $\mathbb{P}^1$. Thus, $\operatorname{conorm}(\mathfrak{p}_{0,\infty}) = 2 \cdot \mathfrak{p}_\infty$ and so η is ramified in Q_0, or $\operatorname{conorm}(\mathfrak{p}_{0,\infty}) = \mathfrak{p}_\infty \cdot \mathfrak{p}'_\infty$. In any case $\operatorname{conorm}(\mathfrak{p}_{0,\infty}) =: D$ is a positive divisor of degree 2. We define the Riemann-Roch spaces $\mathcal{L}_i = \mathcal{L}(i \cdot D)$ and $\ell_i = \dim_k(\mathcal{L}_i)$.

By assumption $\mathcal{L}_1$ has as base $(1, x)$ and so $\ell_1 = 2$. Since $\deg(g+1)\cdot D > 2g-2$ the theorem of Riemann- Roch implies that $\ell_{g+1} = 2(g+1) - g + 1 = g+3$. Hence there is a function $y \in \mathcal{L}_{g+1}$ linearly independent from powers of x. So $y \notin k[x]$. The space $\mathcal{L}_{2(g+1)}$ has dimension $3g+3$ and contains the $3g+4$ functions

$$\{1, x, x^{g+1}, y, x^{g+2}, xy, \ldots, x^{2(g+1)}, x^{g+1}y, y^2\}.$$

So there is a nontrivial k-linear relation between these functions, in which y^2 has to have a non-trivial coefficient. We can normalize and get the equation

$$y^2 + h(x)y = f(x) \quad \textit{with} \quad h(x), f(x) \in k[x]$$

and $\deg(h(x)) \leq g+1$, $\deg(f) \leq 2g+2$. So

$$W_{\mathcal{C}_{aff}} : Y^2 + h(X)Y = f(X)$$

is the equation for an affine part $\mathcal{C}_{aff}$ of a curve birationally equivalent to $\mathcal{C}$. It is called an *affine Weierstrass equation* for $\mathcal{C}$, and its homogenization is the equation of a projective plane curve $\mathcal{C}'$ birationally equivalent to $\mathcal{C}$.

The prime divisors of $\mathcal{C}$ are extensions of prime divisors of $k(x)$ and hence correspond (over $\bar{k}$) to points (x, y) in $\mathbb{A}^2$ or the points lying over $\mathfrak{p}_{0,\infty}$. To get more information we use the Hurwitz genus formula and assume for simplicity that char(k) $\neq 2$ and so η is tamely ramified and separable; for the general case see [**7**, Section 14.5.1].

Then we can apply the Tschirnhausen transformation and can assume that $h(x) = 0$. We know that η has to have $2g+2$ ramification points. Ramification points of η are fixed points under the hyperelliptic involution ω which generates $\operatorname{Gal}(k(\mathcal{C})/k(x))$. Since ω acts on points (x, y) by sending it to $(x, -y)$ the affine ramification points correspond to the zeros of $f(X)$. If $\mathfrak{p}_{0,\infty}$ is unramified then it follows that $f(X)$ has to have $2g+2$ zeros, and so $\deg(f(X)) = 2g+2$ and all zeros are simple.

Assume that $\mathfrak{p}_{0,\infty}$ is ramified. Then there have to be $2g-1$ places with norm $\neq \mathfrak{p}_{0,\infty}$ and so $\deg(f(X)) = 2g+1$ and again all zeros are different. Hence in both cases we have that $\mathcal{C}_{aff}$ is without singularities. This is not true for the point $(0, 1, 0)$, the only point at infinity of $\mathcal{C}'$. It is a singular point, and it corresponds to two points (over $\bar{k}$) on $\mathcal{C}$ if $\mathfrak{p}_{0,\infty}$ is unramified, and to one point on $\mathcal{C}(k)$ if $\mathfrak{p}_{0,\infty}$ is ramified. For computational purposes the latter case is more accessible. The arithmetic in $k(\mathcal{C})$ is analogue to the arithmetic in imaginary quadratic fields.

The automorphism group of $\mathcal{C}$: We assume that k is algebraically closed. We identify the places of $k(x)$ with the points of $\mathbb{P}^1 = k \cup \{\infty\}$ by their X-coordinate. As seen $k(\mathcal{C})$ is a quadratic extension field of $k(x)$ ramified exactly at $2g+2$ places

$\alpha_1, \ldots, \alpha_{2g+2}$ of $k(x)$. The corresponding places of $k(\mathcal{C})$ are called the *Weierstrass points* $Q_1, \ldots, Q_{2g+2}$ of $k(\mathcal{C})$, the set formed by these points is denoted by $\mathcal{P}$.

Weierstrass points play a very important role in the arithmetic of curves. For a detailed discussion see [**67**]. In particular, Weierstrass points of $\mathcal{C}$ are uniquely determined up to permutations. So, every automorphism of $\mathcal{C}$ and equivalently, of $k(\mathcal{C})/k$, fixes $\mathcal{P}$ and so fixes $k(x)$, and therefore $k(x)$ is the unique subfield of index 2 in $k(\mathcal{C})$.

It follows that $\langle \omega \rangle$ is central in $\mathrm{Aut}(k(\mathcal{C})/k)$, and $\bar{G} := G/\langle \tau \rangle$ is naturally isomorphic to the subgroup of $\mathrm{Aut}(k(x)/k)$ induced by G. We have a natural isomorphism $\Gamma := PGL_2(k) \stackrel{\cong}{\to} \mathrm{Aut}(k(x)/k)$. The action of Γ on the places of $k(x)$ corresponds under the above identification to the usual action on $\mathbb{P}^1$ by fractional linear transformations $t \mapsto \frac{at+b}{ct+d}$. Since G permutes the Weierstrass points and $2g+2 \geq 6$ its action determines G and so we get an embedding $\bar{G} \to S_n$.

Since $k(\mathcal{C})$ is the unique degree 2 extension of $k(x)$ ramified exactly at $a_1, \ldots, a_{2g+2}$, each automorphism of $k(x)$ permuting these $2g+2$ places extends to an automorphism of $k(\mathcal{C})$. Hence under the isomorphism $\Gamma \mapsto \mathrm{Aut}(k(x)/k)$, $\bar{G}$ corresponds to the stabilizer $\Gamma_{\mathcal{P}}$ in Γ of the $2g+2$-set $\mathcal{P}$. By a theorem of Klein, $\bar{G}$ is isomorphic to a cyclic group, a dihedral group, or A_4, S_4 or A_5. Hence, we can determine $\mathrm{Aut}(\mathcal{C})$ as a degree 2 central extension of $\bar{G}$ for any fixed genus $g \geq 2$.

Minimal degrees: We have seen above that non-hyperelliptic curves of genus ≥ 3 are birational equivalent to plane projective curves of degree $\leq 2g+2$. But in general, this is not the minimal degree one can achieve. On the other side one has an estimate from below for the degree of plane curves birational equivalent to a hyperelliptic curve of genus $g \geq 3$; see [**8**] for details.

PROPOSITION 4. *Let $\mathcal{C}$ be a hyperelliptic curve of genus g and let $\mathcal{C}'$ be a plane projective curve birationally equivalent to $\mathcal{C}$. Then the degree of the equation of $\mathcal{C}'$ is $\geq g+2$.*

4.2.4. *Addition in Picard groups over $\mathbb{F}_q$.* We take $k = \mathbb{F}_q$ and $\mathcal{C}$ a curve of genus g defined over $\mathbb{F}_q$. By a result of F. K. Schmidt (proved by using Zeta-functions) curves over finite fields have a rational divisor D_0 of degree 1 (Caution: Only for curves of genus ≤ 1 this implies that they have a rational point.) It is not difficult to show that this divisor can be computed effectively. We use this to present divisor classes c of degree 0 of $\mathcal{C}$.

Let $\mathcal{D}_{\mathcal{C}}(\mathbb{F}_q)^g_{>0}$ denote the positive divisors of degree g of $\mathcal{C}$. A consequence of the theorem of Riemann-Roch is that the map

$$\begin{aligned} \varphi : \mathcal{D}_{\mathcal{C}}(\mathbb{F}_q)^g_{>0} &\to \mathrm{Pic}^0_{\mathcal{C}}(\mathbb{F}_q) \\ D &\mapsto \varphi(D) = D - g \cdot D_1 \end{aligned}$$

is surjective. A first consequence is that $\mathrm{Pic}^0_{\mathcal{C}}(\mathbb{F}_q)$ is a finite abelian group since there are only finitely many positive divisors of degree D rational over $\mathbb{F}_q$. Our aim is to find an algorithm, which computes the addition in $\mathrm{Pic}^0_{\mathcal{C}}(\mathbb{F}_q)$ fast. The main task is the following *reduction*:

Given $D, D' \in \mathcal{D}_{\mathcal{C}}(\mathbb{F}_q)^g_{>0}$ find a divisor $S \in \mathcal{D}_{\mathcal{C}}(\mathbb{F}_q)^g_{>0}$ with

$$D + D' - 2gD_1 \sim S - gD_1.$$

Then $S - \mathcal{D}_1$ lies in the divisor class that is the sum of the divisor class of $D - D_1$ with the class of $D' - D_1$. An analogue reduction is well-known from computational number theory and ideal classes of orders. There one uses Minkowski's theorem instead of the Riemann-Roch theorem.

The idea of F. Heß in [**34**] and worked out with many additional details in [**16**] is to use the fact the holomorphic functions in affine open parts of $\mathcal{C}$ are Dedekind domains and that divisors with support on these parts can be identified with ideals of these rings. As first step compute (e.g. from the function field $k(\mathcal{C})$) a plane curve $\mathcal{C}'$ birationally equivalent to $\mathcal{C}$ of a degree d of size $\mathcal{O}(g)$ (see our arguments above).

The next step is to go to an affine part of $\mathcal{C}'$ which is without singularities and for which divisors can be identified with ideals in its coordinate ring (approximation properties of functions in function fields can be used since we are only interested in divisor classes). Now the algorithms known from number theory are applicable. The result is given by the following theorem.

THEOREM 29. *[Heß, Diem] Let $\mathcal{C}$ be a curve of genus g over $\mathbb{F}_q$. The addition in the degree 0 class group of C can then be performed in an expected time which is polynomially bounded in g and* $\log(q)$.

This result is a highlight in algorithmic arithmetic geometry and it opens the access to the Picard groups as abelian groups for arbitrary curves. Of course, it will be a challenge to implement it. In our context, namely to construct crypto systems, its importance is the *existence* of the algorithm which make certain attacks thinkable!

In the next sections we shall see how we can find explicit algorithms and even formulas to perform group operations in Picard groups of hyperelliptic curve very rapidly.

4.2.5. *The Jacobian variety of a curve.* In Section 3.2 we defined the *Picard functor* $\mathrm{Pic}^0_{\mathcal{C}}$ from the category of extension fields L/k into the category of abelian groups given by

$$L \mapsto \mathrm{Pic}^0_{\mathcal{C}_L}(L).$$

In addition we stated that $\mathrm{Pic}^0_{\mathcal{C}}$ is a Galois functor, i.e. that if $k \subset L \subset \bar{k}$ then $\mathrm{Pic}^0_{\mathcal{C}L}(L) = \mathrm{Pic}^0_{\mathcal{C}\bar{k}}(\bar{k})^{G_L}$. We also announced that this functor is *representable* in terms of algebraic geometry.

More precisely: Let $\mathcal{C}$ be a curve of positive genus and assume that there exists a k-rational point $P_0 \in \mathcal{C}(k)$ with attached prime divisor $\mathfrak{p}_0$. There exists an abelian variety $\mathcal{J}_{\mathcal{C}}$ defined over k and a uniquely determined embedding

$$\phi_{P_0} : \mathcal{C} \to \mathcal{J}_{\mathcal{C}} \text{ with } \phi_{P_0}(P_0) = 0_{\mathcal{J}\mathcal{C}}$$

such that

(1) for all extension fields L of k we get $J_{\mathcal{C}}(L) = \mathrm{Pic}^0_{\mathcal{C}_L}(L)$ where this equality is given in a functorial way and
(2) if $\mathcal{A}$ is an Abelian variety and $\eta : \mathcal{C} \to \mathcal{A}$ is a morphism sending P_0 to $0_{\mathcal{A}}$ then there exists a uniquely determined homomorphism $\psi : \mathcal{J}_{\mathcal{C}} \to \mathcal{A}$ with $\psi \circ \phi_{P_0} = \eta$.

$\mathcal{J}_\mathcal{C}$ is uniquely determined by these conditions and is called the **Jacobian variety** of $\mathcal{C}$. The map ϕ_{P_0} is given by sending a prime divisor $\mathfrak{p}$ of degree 1 of $\mathcal{C}_\mathcal{L}$ to the class of $\mathfrak{p} - \mathfrak{p}_0$ in $\mathrm{Pic}^0_{\mathcal{C}_L}(L)$.

Properties of Jacobian varieties: From functoriality and universality of the Jacobian it follows that we can introduce coordinates for divisor classes of degree 0 such that the group law in $\mathrm{Pic}^0_{\mathcal{C}_L}(L)$ is given by rational functions defined over k and depending only on $\mathcal{C}$ (and not on L). Moreover, we can interpret the norm and conorm maps on divisor classes geometrically.

Let L/k be a finite algebraic extension. Then the Jacobian variety $\mathcal{J}_{\mathcal{C}_L}$ of $\mathcal{C}_L$ is the scalar extension of $\mathcal{J}_\mathcal{C}$ with L, hence a fiber product with projection p to $\mathcal{J}_\mathcal{C}$. The norm map is p_*, and the conorm map is p^*.

Proposition 5. *If $f : \mathcal{C} \to \mathcal{D}$ is a surjective morphism of curves sending P_0 to Q_0, then there is a uniquely determined surjective homomorphism*

$$f_* : \mathcal{J}_\mathcal{C} \to J_\mathcal{D}$$

such that $f_* \circ \phi_{P_0} = \phi_{Q_0}$.

Proof. Apply the universal property to the morphism $\phi_{Q_0} \circ f$ to get f_*. The surjectivity follows from the fact that for $k = \bar{k}$ the sums of divisor classes of the form $\mathfrak{p} - \mathfrak{p}_0$ with $\mathfrak{p} \in \Sigma_\mathcal{C}(k)$ generate $\mathrm{Pic}^0_\mathcal{C}(\bar{k})$. □

A useful observation is

Corollary 2. *Assume that $\mathcal{C}$ is a curve of genus ≥ 2 such that FG $\mathcal{J}_\mathcal{C}$ is a simple abelian variety, and that $\eta : \mathcal{C} \to \mathcal{D}$ is a separable cover of degree > 1. Then $\mathcal{D}$ is the projective line.*

For the proof use the Hurwitz genus formula and the universal properties of Jacobians.

What about the **existence** of Jacobian varieties? Over the complex numbers the classical theory of curves (key words: Riemann surfaces and the Theorem of Abel-Jacobi) is used to prove the existence of Jacobian varieties already in the 19-th century. In fact, this notion is historically earlier than the notion "Abelian variety" introduced by A. Weil as most important tool for his proof of the geometric Riemann hypothesis. By the Lefschetz principle the existence of Jacobian varieties follows for algebraically closed fields of characteristic 0.

For a proof in the framework of Algebraic Geometry (and so over arbitrary ground fields k) see Lang [**45**]. The important fact is that we "know" a birational affine model of $J_\mathcal{C}$.

By the Theorem of Riemann-Roch we have a surjective map from $\Sigma^g_\mathcal{C}(L)$ to $\mathrm{Pic}^0_\mathcal{C}(L)$ by sending any positive divisor D of degree g to $D - g \cdot \mathfrak{p}_0$. We can interpret such positive divisors geometrically. Take the g-fold cartesian product $\mathcal{C}^g$ of the curve $\mathcal{C}$ of genus g and embed it (via Segre's map) into a projective space. On this variety we can permute the factors and so have an action of S_g, the symmetric group with g letters. Define the g-fold symmetric product $\mathcal{C}^{(g)}$ by $\mathcal{C}^g/S_g$. Then we can identify $\mathcal{C}^{(g)}(L)$ with $\Sigma^g_\mathcal{C}(L)$ and so define a birational map from $\mathcal{C}^{(g)}$ to $\mathcal{J}_\mathcal{C}$. Taking an affine part of $\mathcal{C}$ (e.g. found as a regular part of a plane model of $\mathcal{C}$) we get an affine variety which is birational equivalent to $\mathcal{J}_\mathcal{C}$.

The Jacobian varieties connect the arithmetic in divisor classes of curves (which is very accessible to algorithms) with the very rich geometric structure of abelian varieties (e.g. isogenies, endomorphisms and ℓ-adic representations).

4.2.6. *Construction of curves by period matrices.* It is convenient to assume in the following that k is algebraically closed. We look at the following task: Assume that a point P in the moduli scheme $\mathcal{M}_g(k)$ is given by coordinates in a certain coordinate system. How can we find an equation for a curve $\mathcal{C}$ corresponding to P?

It is useful to look at the case that $k = \mathbb{C}$ and at the parametrization of isomorphism classes principally polarized abelian varieties by period matrices. We reformulate the question and ask whether we can find a curve such that the Jacobian has a given period matrix. Of course, the first problem is that not every principally polarized abelian variety is the Jacobian of a curve, and the decision for this is the well-known *Schottky Problem* which is unsolved till now.

There are two cases where the situation is better: If the dimension of the Abelian variety is ≤ 3 then such a curve exists, and if we are looking for hyperelliptic curves we can solve the Schottky Problem and determine a Weierstrass equation if we know the period matrix.

This latter result is based on *invariant theory.* Details are worked out in the thesis of H.J. Weber [**72**] (explicitly for curves up to genus 5). Important cases for our applications are curves of genus 1 (use the j-invariant), genus 2 and genus 3. We remark that this method works very well over number fields and by reduction, over finite fields, too.

We shall give more details in the interesting case that the genus of $\mathcal{C}$ is equal to 2.

4.2.7. *Example: Curves of genus 2.* Let $\mathcal{C}$ be a genus 2 curve defined over a field k. By Prop. 3 we have that its gonality is $\gamma_{\mathcal{C}} = 2$. Hence, genus 2 curves are hyperelliptic and we denote the hyperelliptic projection by $\pi : \mathcal{C} \to \mathbb{P}^1$. By the Hurwitz's formula this covering has $r = 6$ branch points which are images of the Weierstrass points of $\mathcal{C}$. The moduli space has dimension $r - 3 = 3$; see Example 3.

The arithmetic of the moduli space of genus two curves was studied by Igusa in his seminal paper [**35**] expanding on the work of Clebsch, Bolza, and others. Arithmetic invariants by $J_2, J_4, J_6, J_8, J_{10}$ determine uniquely the isomorphism class of a genus two curve. Two genus two curves $\mathcal{C}$ and $\mathcal{C}'$ are isomorphic over $\bar{k}$ if and only if there exists $\lambda \in \bar{k}^\star$ such that $J_{2i}(\mathcal{C}) = \lambda^{2i} J_{2i}(\mathcal{C}')$, for $i = 1, \dots, 5$. If char $k \neq 2$ then the invariant J_8 is not needed.

From now on we assume char $k \neq 2$. Then $\mathcal{C}$ has an affine Weierstrass equation

$$y^2 = f(x) = a_6 x^6 + \cdots + a_1 x + a_0, \tag{2}$$

over $\bar{k}$, with discriminant $\Delta_f = J_{10} \neq 0$. The moduli space $\mathcal{M}_2$ of genus 2 curves, via the Torelli morphism, can be identified with the moduli space of the principally polarized abelian surfaces $\mathbb{A}_2$ which are not products of elliptic curves. Its compactification $\mathbb{A}_2^\star$ is the weighted projective space $\mathbb{WP}^3_{(2,4,6,10)}(k)$ via the Igusa invariants J_2, J_4, J_6, J_{10}. Hence,

$$A_2 \cong \mathbb{WP}^3_{(2,4,6,10)}(k) \setminus \{J_{10} = 0\}.$$

A point $\mathfrak{p} \in \mathbb{WP}^3_{(2,4,6,10)}$ for $J_2 \neq 0$ can be written as

$$\left[1 : \frac{1}{2^4 3^2} \mathbf{x}_1 : \frac{1}{2^6 3^4} \mathbf{x}_2 + \frac{1}{2^4 3^3} \mathbf{x}_1 : \frac{1}{2 \cdot 3^5} \mathbf{x}_3 \right]$$

where $\mathbf{x}_1, \mathbf{x}_2, \mathbf{x}_3$ are given as ratios of Siegel modular forms and are called *absolute invariants* and denoted by i_1, i_2, i_3 by other authors; see [**36**]. Two genus 2 curves are isomorphic over $\bar{k}$ if and only if they have the same absolute invariants. Notice that the absolute invariants are not defined for $J_2 = 0$. There are different sets of absolute invariants used by many authors, but all of them are not defined over $J_2 = 0$.

Recovering the curve from invariants. Given a moduli point $\mathfrak{p} \in \mathcal{M}_2$, with automorphism group of order 2, we can recover the equation of the corresponding curve over a minimal field of definition following Mestre's approach [**55**], where the point is given in terms of the absolute invariants. The case of automorphism group of order > 2 was done in [**5**] and [**62**]. In all these papers the case when the absolute invariants are not defined had to be treated differently, introducing a new set of invariants. In [**51**] it is given an equation of the curve in terms of J_2, J_4, J_6, J_{10} without using any absolute invariants.

In [**52**], for any number field K, a *height* on the moduli space $\mathbb{WP}^3_{(2,4,6,10)}(K)$ is introduced. This makes it possible to store the *smallest* tuple of invariants in a unique way. This is used in [**2**] to create a database of all genus 2 curves with small height and defined over K including all the twists of minimal moduli height.

4.2.8. *Example: Elliptic curves.* Let $\mathcal{E}$ be an elliptic curve over k, i.e. a curve of genus 1 with a k-rational point. Its isomorphy class over $\bar{k}$ is uniquely determined by the j-invariant. As seen above, $\mathcal{E}$ is isomorphic to a plane curve $\mathcal{E}'$ given by a Weierstrass equation.

We choose one k-rational point P_∞ with prime divisor $\mathfrak{p}_\infty$ and projective coordinates such that $P_\infty = (0 : 1 : 0)$ is the infinite point of the curve $\mathcal{E}'$ with equation

$$Y^2Z + a_1XYZ + a_3YZ^2 = X^3 + a_2X^2Z + a_4XZ^2 + a_6Z^3$$

and identify $\mathcal{E}$ with $\mathcal{E}'$.

Let $\mathcal{J}_\mathcal{E}$ be the Jacobian variety of $\mathcal{E}$. We look at $\phi_{P_\infty} : \mathcal{E} \to \mathcal{J}_\mathcal{E}$ given by

$$P \mapsto [\mathfrak{p} - \mathfrak{p}_\infty]$$

where [.] means divisor class. Since $2g_\mathcal{E} - 2 = 0$ the Riemann-Roch theorem implies that for all extension fields L of k in each L-rational divisor class of degree 1 there is exactly one prime divisor $\mathfrak{p}$ of degree 1 corresponding to a point $P \in \mathcal{E}(L)$, and to each divisor class c of degree 0 there is exactly one prime divisor $\mathfrak{p}$ of degree 1 with $c = [\mathfrak{p} - \mathfrak{p}_\infty]$. So ϕ_{P_∞} is injective and surjective and hence an isomorphism of projective varieties. By transport of structure we endow $\mathcal{E}$ with a group structure:

For extension fields L of k and $P_1, P_2 \in \mathcal{E}(L)$ define $P_1 \oplus P_2$ as the point belonging to the prime divisor in the class $\mathfrak{p}_1 + \mathfrak{p}_2 - 2\mathfrak{p}_\infty$.

It is obvious that this makes $\mathcal{E}(L)$ to an abelian group with neutral element P_∞. We conclude: Three points P_1, P_2, P_3 sum up to 0 if $\mathfrak{p}_1 + \mathfrak{p}_2 + \mathfrak{p}_3 - 3\mathfrak{p}_\infty = (f)$ with $f \in L(\mathcal{E})$.

Now recall that $\mathcal{E}$ has degree 3 and so lines intersect with $\mathcal{E}$ in 3 points (counted with multiplicities) and so f defines a line in $\mathbb{P}^2$. Hence $P_1 + P_2 + P_3 = 0$ if and only if the three points are collinear, and then $P_1 \oplus P_2 = \ominus P_3$. Using coordinates we get an algebraic recipe for addition:

For given $P \neq P\infty$ take the line through P and P_∞ to get: $\ominus(P)$ is the third intersection point of the line with $\mathcal{E}$ (if this point is equal to P the line is a tangent *and $P = \ominus P$ is an element of order 2). Given two points $P_1 \neq P_2$ compute the line through these two points, take its third intersection point P_3 with $\mathcal{E}$ to get $P_1 \oplus P_2 = \ominus P_3$.*

By elementary algebra one can perform this recipe by writing down formulas in rational functions in (X, Y, Z) and so we get

THEOREM 30. *After the choice of a base point P_∞ the elliptic curve $\mathcal{E}$ is an Abelian variety of dimension 1 with neutral element P_∞ which is equal to $\mathcal{J}_\mathcal{E}$.*

Division polynomials for elliptic curves: To simplify we assume that char $k \neq 2, 3$ and so we can take the affine Weierstrass equation of E as

$$E: \quad Y^2 = X^3 + aX + b,$$

for $a, b \in k$. Recall that for $m \in \mathbb{N}$ the endomorphism $[m]$ of $\mathcal{E}$ is induced by the scalar multiplication by m. We want to give formulas for this endomorphism.

LEMMA 12. *For any integer m and point $P(x, y) \neq \mathcal{O}$ in E, the point $[m]P$ has coordinates*

$$[m]P = \left(\frac{\phi_m(x,y)}{\psi_m(x,y)^2}, \frac{\omega_m(x,y)}{\psi_m(x,y)^3} \right)$$

where the polynomials ϕ_m, ψ_m, ω_m are given by the recurrences

$$\begin{aligned}
&\psi_1 = 1,\\
&\psi_2 = 2Y^2,\\
&\psi_3 = 3X^4 + 6aX^2 + 12bX - a^2,\\
&\psi_4 = (2X^6 + 10aX^4 + 40bX^3 - 10a^2X^2 - 8abX - 2a^3 - 16b^2)2Y^2,\\
&\dots\\
&\psi_{2m+1} = \psi_{m+2}\psi_m^3 - \psi_{m-1}\psi_{m+1}^3 \quad \text{for } m \geq 2,\\
&\psi_2\psi_{2m} = (\psi_{m+2}\psi_{m-1}^2 - \psi_{m-2}\psi_{m+1}^2)\psi_m \quad \text{for } m \geq 3.
\end{aligned} \tag{3}$$

and

$$\begin{aligned}
\phi_m &= x\psi_m^2 - \psi_{m+1}\psi_{m-1},\\
\omega_m &= \psi_{m-1}^2\psi_{m+2} + \psi_{m-2}\psi_{m+1}^2.
\end{aligned} \tag{4}$$

The proof follows from classical identities of the Weierstrass function $\wp$ if $k = \mathbb{C}$ and is then transferred to arbitrary perfect fields (see [**65**]). The polynomial ψ_m is called the **m-th division polynomial** and it vanishes in $E[m]$

COROLLARY 8. *All m-torsion points $P(x, y)$ of E have coordinates satisfying $\psi_m(x, y) = 0$*

This provides a computational approach on how to determine the m-torsion points for any given $m \geq 2$.

4.3. Cantor's algorithm. Inspired by the group law on elliptic curves and its geometric interpretation we give an *explicit* algorithm for the group operations on Jacobian varieties of hyperelliptic curves.

Take a genus $g \geq 2$ hyperelliptic curve $\mathcal{C}$ with at least one rational Weierstrass point given by the affine Weierstrass equation

$$W_\mathcal{C}: \ y^2 + h(x)\,y = x^{2g+1} + a_{2g}x^{2g} + \cdots + a_1x + a_0 \tag{5}$$

over k. We denote the prime divisor corresponding to $P_\infty = (0:1:0)$ by $\mathfrak{p}_\infty$. The affine coordinate ring of $W_{\mathcal{C}}$ is

$$\mathcal{O} = k[X,Y]/(Y^2 + h(X)\,\langle Y - (X^{2g+1} + a_{2g}X^{2g} + \cdots + a_1 X + a_0)\rangle)$$

and so prime divisors $\mathfrak{p}$ of degree d of $\mathcal{C}$ correspond to prime ideals $P \neq 0$ with $[\mathcal{O}/P : k] = d$. Let ω be the hyperelliptic involution of $\mathcal{C}$. It operates on $\mathcal{O}$ and on Spec($\mathcal{O}$) and fixes exactly the prime ideals which "belong" to Weierstrass points, i.e. split up in such points over $\bar{k}$.

Following Mumford [**56**] we introduce polynomial coordinates for points in $J_{\mathcal{C}}(k)$. The first step is to normalize representations of divisor classes. In each divisor class $c \in \mathrm{Pic}^0(k)$ we find a unique *reduced* divisor

$$D = n_1\mathfrak{p}_1 + \cdots + n_r\mathfrak{p}_r - d\,\mathfrak{p}_\infty$$

with $\sum_{i=1}^r n_i \deg(\mathfrak{p}_i) = d \leq g$, $\mathfrak{p}_i \neq \omega(\mathfrak{p}_j)$ for $i \neq j$ and $\mathfrak{p}_i \neq \mathfrak{p}_\infty$ (we use Riemann-Roch and the fact that ω induces $-id_{J_{\mathcal{C}}}$).

Using the relation between divisors and ideals in coordinate rings we get that $n_1\mathfrak{p}_1 + \cdots + n_r\mathfrak{p}_r$ corresponds to an ideal $I \subset \mathcal{O}$ of degree d and the property that if the prime ideal P_i is such that both P and $\omega(P)$ divide I then it belongs to a Weierstrass point. The ideal I is a free $\mathcal{O}$-module of rank 2 and so

$$I = k[X]u(X) + k[x](v(X) - Y).$$

Fact: $u(X), v(X) \in k[X]$, u monic of degree d, $\deg(v) < d$ and u divides $v^2 + h(X)v - f(X)$; see [**7**, Theorem 4.143].

Moreover, c is uniquely determined by I, I is uniquely determined by (u,v) and so we can *take* (u,v) *as coordinates for* c.

THEOREM 31 (Mumford representation). *Let $\mathcal{C}$ be a hyperelliptic curve of genus $g \geq 2$ with affine equation*

$$y^2 + h(x)\,y \;=\; f(x),$$

where $h, f \in k[x]$, $\deg f = 2g+1$, $\deg h \leq g$.

Every non-trivial group element $c \in \mathrm{Pic}^0_{\mathcal{C}}(k)$ can be represented in a unique way by a pair of polynomials $u, v \in k[x]$, such that

i) u is a monic

ii) $\deg v < \deg u \leq g$

iii) $u \mid v^2 + vh - f$

How to find the polynomials u, v? We can assume without loss of generality that $k = \bar{k}$ and identify prime divisors $\mathfrak{p}_i$ with points $P_i = (x_i, y_i) \in k \times k$. Take the reduced divisor $D = n_1\mathfrak{p}_1 + \cdots + n_r\mathfrak{p}_r - d\mathfrak{p}_\infty$ now with $r = d \leq g$. Then

$$u(X) = \prod_{i=1}^{r}(X - x_i)^{n_i}.$$

Since $(X - x_i)$ occurs with multiplicity n_i in $u(X)$ we must have for $v(X)$ that

$$\left(\frac{d}{dx}\right)^j \left[v(x)^2 + v(x)\,h(x) - f(x)\right]_{x=x_i} = 0,$$

and one determines $v(X)$ by solving this system of equations.

Addition: Take the divisor classes represented by $[(u_1, v_1)]$ and $[(u_2, v_2)]$ and in "general position". Then the product is represented by the ideal $I \in \mathcal{O}$ given by

$$\langle u_1u_2, u_1(y-v_2), u_2(y-v_1), (y-v_1)(y-v_2)\rangle.$$

We have to determine a base, and this is done by Hermite reduction. The resulting ideal is of the form $\langle u_3'(X), v_3'(X) + w_3'(X)Y\rangle$ but not necessarily reduced. To reduce it one uses recursively the fact that $u \mid (v^2 - hv - f)$.

The formalization of this procedure and the treatment of special cases is called **Cantor's algorithm**. For readers acquainted with algorithmic number theory it may be enlightening to compare this algorithm with the well known method to add in class groups of imaginary quadratic number fields, going back to Gauss and based on the theory of definite quadratic forms with fixed discriminant. The very explicit and efficient "generic" algorithm can be found in [**7**, Algorithm 14.7]. For curves of genus 2 a detailed analysis including all special cases is done in [**7**, Section 14.3.2], including a determination of complexity (see Table 14.2 and Table 14.13). For curves of genus 3 we refer to Section 14.6 in[**7**].

Addition by interpolation Another approach to describe addition in the Jacobians of hyperelliptic curves is to use approximation by rational functions; see [**46**]. This is analogous to the geometric method used for elliptic curves.

For simplicity we assume that $k = \bar{k}$. Let D_1 and D_2 be reduced divisors on $\mathrm{Jac}_k\,\mathcal{C}$ given by

$$\begin{aligned} D_1 &= \mathfrak{p}_1 + \mathfrak{p}_2 + \cdots + \mathfrak{p}_{h_1} - h_1\mathfrak{p}_\infty,\\ D_2 &= \mathfrak{q}_1 + \mathfrak{q}_2 + \cdots + \mathfrak{q}_{h_2} - h_2\mathfrak{p}_\infty, \end{aligned} \tag{6}$$

where $\mathfrak{p}_i$ and $\mathfrak{q}_j$ can occur with multiplicities, and $0 \le h_i \le g$, $i = 1, 2$. As usual we denote by P_i respectively Q_j the points on $\mathcal{C}$ corresponding to $\mathfrak{p}_i$ and $\mathfrak{q}_j$.

Let $g(X) = \frac{b(X)}{c(X)}$ be the unique rational function going through the points P_i, Q_j. In other words we are determining $b(X)$ and $c(X)$ such that $h_1 + h_2 - 2r$ points P_i, Q_j lie on the curve

$$Y\,c(X) - b(X) = 0.$$

This rational function is uniquely determined and has the form

$$Y = \frac{b(X)}{c(X)} = \frac{b_0X^p + \dots b_{p-1}X + b_p}{c_0X^q + c_1X^{q-1} + \cdots + c_q} \tag{7}$$

where

$$p = \frac{h_1 + h_2 + g - 2r - \epsilon}{2}, \quad q = \frac{h_1 + h_2 - g - 2r - 2 + \epsilon}{2},$$

ϵ is the parity of $h_1 + h_2 + g$. By replacing Y from Eq. (7) in Eq. (5) we get a polynomial of degree $\max\{2p,\, 2q(2g-1)\}$, which gives $h_3 \le g$ new roots apart from the X-coordinates of P_i, Q_j. Denote the corresponding points on $\mathcal{C}$ by $R_1, \dots, R_{h_3}$ and $\bar{R}_1, \dots, \bar{R}_{h_3}$ are the corresponding symmetric points with respect to the $y = 0$ line. Then, we define

$$D_1 + D_2 = \bar{R}_1 + \dots \bar{R}_{h_3} - h_3\mathcal{O}.$$

For details we refer the reader to [**46**].

REMARK 6. *For $g = 1, 2$ we can take $g(X)$ to be a cubic polynomial.*

EXAMPLE 5 (Curves of genus 2.). *Let $\mathcal{C}$ be a genus 2 curve defined over a field k with a rational Weierstrass point. If char $k \neq 2, 3$ the $\mathcal{C}$ is birationally isomorphic to an affine plane curve with equation*

$$Y^2 = a_5X^5 + a_4X^4 + a_3X^3 + a_2X^2 + a_1X + a_0. \tag{8}$$

Let $\mathfrak{p}_\infty$ be the prime divisor corresponding to the point at infinity. Reduced divisors in generic position

are given by

$$D = \mathfrak{p}_1 + \mathfrak{p}_2 - 2\mathfrak{p}_\infty$$

where $P_1(x_1, y_1)$, $P_2(x_2, y_2)$ are points in $\mathcal{C}(k)$ (since k is algebraically closed) and $x_1 \neq x_2$. For any two divisors $D_1 = \mathfrak{p}_1 + \mathfrak{p}_2 - 2\mathfrak{p}_\infty$ and $D_2 = \mathfrak{q}_1 + \mathfrak{q}_2 - 2\mathfrak{p}_\infty$ in reduced form, we determine the cubic polynomial

$$Y = g(X) = b_0X^3 + b_1X^2 + b_2X + b_3, \tag{9}$$

going through the points $P_1(x_1, y_1)$, $P_2(x_2, y_2)$, $Q_1(x_3, y_3)$, and $Q_2(x_4, y_4)$. This cubic will intersect the curve $\mathcal{C}$ at exactly two other points R_1 and R_2 with coordinates

$$R_1 = (x_5, g(x_5)) \quad \textit{and} \quad R_2 = (x_6, g(x_6)), \tag{10}$$

where x_5, x_6 are roots of the quadratic equation

$$x^2 + \left(\sum_{i=1}^4 x_i\right) x + \frac{b_3^2 - a_5}{b_0^2 \prod_{i=1}^4 x_i} = 0. \tag{11}$$

Let us denote by $\overline{R}_1 = (x_5, -g(x_5))$ and $\overline{R}_2 = (x_6, -g(x_6))$. Then,

$$[D_1] \oplus [D_2] = [\overline{R}_1 + \overline{R}_2 - 2\mathfrak{p}_\infty]. \tag{12}$$

After having defined explicitly the addition in $\mathrm{Jac}_\mathcal{C}$ it is a natural problem that given a reduced divisor $D \in \mathrm{Jac}_\mathcal{C}$, determine explicitly the formulas for $[n]D$, at least in generic cases similarly as in the case of elliptic curves. Hence, one wants to determine explicitly division polynomials (i.e polynomials that have torsion points of order n as zeroes) or more generally, ideals which define zero-dimensional schemes containing $J_\mathcal{C}[n]$). There has been a lot of activity on this area lately; see [**57**], [**38**], [**39**].

4.4. Automorphisms of curves and their Jacobians. Let $\mathcal{C}$ be an algebraic curve defined over k and $\mathrm{Jac}_k(\mathcal{C})$ its Jacobian. What are the automorphism groups of $\mathcal{C}$ and $\mathrm{Jac}_k(\mathcal{C})$?

4.4.1. *Automorphisms of curves.* Let $\mathrm{Aut}(\mathcal{C})$ be the automorphism of the curve $\mathcal{C}$. If $\mathcal{C} = \mathbb{P}^1$ the automorphism group over an infinite field k is infinite, as seen in Section 4.2.3. The same is true for elliptic curves $\mathcal{E}$, if $E(k)$ is infinite, i.e. $k = \bar{k}$, for then there are infinitely many translations. But caution: If we look only at automorphisms with a fixed point (which are automatically homomorphisms with respect to the group structure with the fixed point as origin) then the group is finite, well understood, and for generic elliptic curves $\mathcal{E}$ equal to $\{id_\mathcal{E}, -id_\mathcal{E}\}$.

For curves of genus ≥ 2 the picture changes completely. The automorphism group $\mathrm{Aut}(\mathcal{C})$ is a finite group. The reason is the existence of $2g + 2$ Weierstrass points and the faithful action of $\mathrm{Aut}(\mathcal{C})$ on these points, which gives an injection of $\mathrm{Aut}(\mathcal{C})$ into S_{2g+2}. In fact, one can describe all occurring groups. If char(k) = 0 one has the *Hurwitz bound*

$$\#\,\mathrm{Aut}(\mathcal{C}) \leq 84(g-1).$$

One gets a stratification of $\mathcal{M}_g$ by strata of curves with the same automorphism group, and the generic curve of genus $G > 2$ has trivial automorphism group.

It is very interesting to study curves with large automorphism group; for instance a curve of genus 3 with group $PSl(2,7)$ (168 automorphisms) is the famous Klein Quartic, which also occurs as modular curve (see below). For more details and a full account of automorphisms of curves see [**50**].

4.4.2. *Automorphisms of Jacobian varieties.* By functoriality it follows that automorphism of $\mathcal{C}$ induce automorphisms of $\mathcal{J}_\mathcal{C}$, or, to be more precise, of $(\mathrm{Jac}_\mathcal{C}, \iota)$ where ι is the principal polarization of $\mathcal{J}_\mathcal{C}$ attached to $\mathcal{C}$.

THEOREM 32. *Let $\mathcal{C}$ be an algebraic curve and $\mathcal{A} := \mathrm{Jac}(\mathcal{C})$ with canonical principal polarization ι. Then,*

$$\mathrm{Aut}\,\mathcal{C} \cong \begin{cases} \mathrm{Aut}(\mathcal{A},\iota), & \textit{if } \mathcal{C} \textit{is hyperelliptic} \\ \mathrm{Aut}(\mathcal{A},\iota)/\{\pm 1\}, & \textit{if } \mathcal{C} \textit{is non-hyperelliptic} \end{cases}$$

The above result can be used to find Jacobians of genus 3 hyperelliptic curves; see [**73**] and Section 9.

4.4.3. *Endomorphism of Abelian varieties.* The ring of endomorphisms of generic Abelian varieties is "as small as possible". For instance, if char(k) = 0 $\mathrm{End}(\mathcal{A}) = \mathbb{Z}$ in general. If k is a finite field, the Frobenius endomorphism will generate a larger ring, but again, this will be all in the generic case. A concrete result is the following [**75**]:

THEOREM 33 (Zarhin). *Let $\mathcal{C}$ be a hyperelliptic curves with affine equation $y^2 = f(x)$, $n = \deg f$, and $f \in \mathbb{Q}[x]$. If $\mathrm{Gal}(f)$ is isomorphic to A_n or S_n then $\mathrm{End}_{\overline{\mathbb{Q}}}(\mathrm{Jac}\,\mathcal{C}) \cong \mathbb{Z}$.*

The theorem is actually true over any number field K. See [**76**] for detailed results on endomorphisms of Jacobians of hyperelliptic and superelliptic curves.

From this point of view it will be interesting to find Abelian varieties with larger endomorphism rings. This leads to the theory of real and complex multiplication. For instance, the endomorphism ring of the Jacobian of the Klein quartic contains an order in a totally real field of degree 3 over $\mathbb{Q}$. We shall see in Section 8 that the Jacobians of modular curves have real multiplication.

4.5. Endomorphism ring of an abelian surface. For char $k \neq 2$, a point $\mathfrak{p}$ in the moduli space $\mathcal{M}_2$ is determined by the tuple (J_2, J_4, J_6, J_{10}), for discriminant $D := J_{10} \neq 0$. In the case of char $k = 2$ another invariant J_8 is needed; see [**35**].

Humbert surfaces: For every $D := J_{10} > 0$ there is a Humbert hypersurface H_D in $\mathcal{M}_2$ which parametrizes curves $\mathcal{C}$ whose Jacobians admit an optimal action on $\mathcal{O}_D$; see [**33**]. Points on H_{n^2} parametrize curves whose Jacobian admits an (n,n)-isogeny to a product of two elliptic curves.

Shimura curves: For every quaternion ring R there are irreducible curves $S_{R,1}, \ldots, S_{R,s}$ in $\mathcal{M}_2$ that parametrize curves whose Jacobians admit an optimal action of R. Those $S_{R,1}, \ldots, S_{R,s}$ are called *Shimura curves.*

Curves with complex multiplication: Curves whose Jacobians admit complex multiplication correspond to isolated points in $\mathcal{M}_2$. We have the following:

PROPOSITION 6. $\mathrm{Jac}(\mathcal{C})$ *is a geometrically simple Abelian variety if and only if it is not* (n,n)*-decomposable for some* $n>1$.

A more detailed discussion is given in [**47**, Section 2.5]. The endomorphism rings of Abelian surfaces can be determined by the Albert's classification and results in [**58**]. We summarize in the following:

PROPOSITION 7. *The endomorphism ring* $\mathrm{End}^0_{\overline{\mathbb{Q}}}(\mathrm{Jac}\,\mathcal{C})$ *of an abelian surface is either* $\mathbb{Q}$, *a real quadratic field, a CM field of degree 4, a non-split quaternion algebra over* $\mathbb{Q}$, $F_1 \oplus F_2$, *where each* F_i *is either* $\mathbb{Q}$ *or an imaginary quadratic field, the Mumford-Tate group* F, *where* F *is either* $\mathbb{Q}$ *or an imaginary quadratic field.*

REMARK 7. *Genus 2 curves with extra involutions have endomorphism ring larger than* $\mathbb{Z}$. *Let* $\mathcal{C}$ *be a genus 2 curve defined over* $\mathbb{Q}$. *If* $\mathrm{Aut}(\mathcal{C})$ *is isomorphic to the Klein 4-group* V_4,*then* $\mathcal{C}$ *is isomorphic to a curve* $\mathcal{C}'$ *with equation*

$$y^2 = f(x) = x^6 - ax^4 + bx^2 - 1.$$

We denote $u = a^3+b^3$ *and* $v = ab$. *The discriminant* $\Delta_f = -2^6\cdot\left(27-18v+4u-u^2\right)^2$, *is not a complete square in* $\mathbb{Q}$ *for any values of* $a, b \in \mathbb{Q}$. *In this case* $\mathrm{Gal}_{\mathbb{Q}}(f)$ *has order 24. There is a twist of this curve, namely* $y^2 = f(x) = x^6 + a'x^4 + b'x^2 + 1$, *in which case* Δ_f *is a complete square in* $\mathbb{Q}$ *and* $\mathrm{Gal}_{\mathbb{Q}}(f)$ *has order 48. In both cases, from Thm. 33 we have that* $\mathrm{End}_{\overline{\mathbb{Q}}}(\mathrm{Jac}\,\mathcal{C}') \neq \mathbb{Z}$.

Next we turn our attention to determining the endomorphism ring of abelian surfaces. Let us first recall a few facts on characteristic polynomials of Frobenius for abelian surfaces. The Weil q-polynomial arising in genus 2 have the form

$$f(T) = T^4 - aT^3 + (b+2q)T^2 - aqT + q^2, \tag{13}$$

for $a, b \in \mathbb{Z}$ satisfying the inequalities

$$2|a|\sqrt{q} - 4q \leq b \leq \frac{1}{4}a^2 \leq 4q.$$

We follow the terminology from [**3**]. Let $\mathcal{C}$ be a curve of genus 2 over $\mathbb{F}_q$ and $\mathcal{J} = \mathrm{Jac}_{\mathcal{C}}$. Let f be the Weil polynomial of J as in Eq. (13). We have that $\#\mathcal{C}(\mathbb{F}_q) = q+1-a$, $\#J(\mathbb{F}_q) = f(1)$ and it lies in the genus-2 Hasse interval

$$\mathcal{H}_q^{(2)} = \left[(\sqrt{q}-1)^4, (\sqrt{q}+1)^4\right]$$

In [**3**] are constructed decomposable $(3,3)$-jacobians with a given number of rational points by glueing two elliptic curves together.

Next we describe some of the results obtained in [**47**] for $\mathrm{End}_K(\mathcal{A})$ in terms of the characteristic polynomial of the Frobenius. We let K be a number field and M_K the set of norms of K. Let $\mathcal{A}$ be an abelian surface defined over K and f_v the characteristic Frobenius for every norm $v \in M_K$.

LEMMA 13. *Let* v *be a place of characteristic* p *such that* $\mathcal{A}$ *has good reduction. Then* $\mathcal{A}_v$ *is ordinary if and only if the characteristic polynomial of the Frobenius*

$$f_v(x) = x^4 + ax^3 + bx^2 + apx + p^2,$$

satisfies $b \not\equiv 0 \mod p$.

Then from [**47**, Lemma 4.3] we have the following.

LEMMA 14. *Let $\mathcal{A}$ be an absolutely simple abelian surface. The endomorphism algebra* $\mathrm{End}^0_{\bar{K}}(\mathcal{A})$ *is non-commutative (thus a division quaternion algebra) if and only if for every $v \in M_K$, the polynomial $f_v(x^{12})$ is a square in $\mathbb{Z}[x]$.*

The following gives a condition for geometrically reducible abelian surfaces.

PROPOSITION 8 ([**47**]). *i) If $\mathcal{A}/K$ is geometrically reducible then for all $v \in M_k$ for which $\mathcal{A}$ has good reduction the polynomial $f_v(x^{12})$ is reducible in $\mathbb{Z}[x]$.*

ii) If $\mathcal{C}$ is a smooth, irreducible genus 2 curve with affine equation $y^2 = f(x)$ such that $f(x) \in K[x]$ is an irreducible polynomial of degree 5 then $\mathrm{Jac}\,\mathcal{C}$ *is absolutely irreducible.*

In [**47**] is given a detailed account of all the cases and an algorithm how to compute $\mathrm{End}_K \mathcal{A}$.

5. Modular curves

As stated in Section 2 we are interested in isogenies between Abelian varieties. Of special interest is the case of elliptic curves, and it turns out that their isogenies are fairly well accessible both from the theoretical and algorithmic point of view. The reason for this is is the very rich and well understood structure of *modular curves*, which parametrize isogenies of elliptic curves.

5.1. Modular curves over $\mathbb{C}$. As we have seen in Section 1 the isomorphism classes of elliptic curves $\mathcal{E}$ over $\mathbb{C}$ correspond one-to-one to isomorphism classes of lattices Λ_τ with τ in $\mathbb{H} = \{z = x + iy \in \mathbb{C} \mid y > 0\} \subset \mathbb{C}$. Moreover, $\Lambda_\tau \cong \Lambda_{\tau'}$ if and only if $\tau' = \frac{a\tau+b}{c\tau+d}$ with $\begin{pmatrix} a & b \\ c & d \end{pmatrix}$ in $SL_2(\mathbb{Z})$.

So the moduli space of isomorphism classes of elliptic curves over $\mathbb{C}$ is in a natural way equal to $\mathbb{H}/\,\mathrm{SL}_2(\mathbb{Z})$, which is, via the j-function, identified with $\mathbb{A}^1$. It is more convenient to work with compact Riemann surfaces and so, with projective curves. Hence we compactify by adjoining *cusps*.

Define $\mathbb{H}^* = \mathbb{H} \cup \mathbb{Q} \cup \{i \cdot \infty\}$ and extend the action of $\mathrm{SL}_2(\mathbb{Z})$ in an obvious way (e.g.: $1/z$ interchanges the cusp 0 with $i \cdot \infty$). Then $\mathbb{H}^* /\,\mathrm{SL}_2(\mathbb{Z}) = \mathbb{P}^1$, the j-function extends to a meromorphic function on $\mathbb{H}^*$ with a simple pole at the cusps (which are all equivalent) and all points $P \in \mathbb{P}^1 \setminus \{j(i \cdot \infty)\}$ have a *modular interpretation*: to $P = j(\tau)$ there corresponds the isomorphism class of the elliptic curve $\mathcal{E}_\tau$ with lattice $\mathbb{Z} + \tau\mathbb{Z}$. In this interpretation we call $\mathbb{P}^1 = X(1)$ and $\mathbb{A}^1 = Y(1)$ modular curves of level 1.

We introduce now a special family of congruence subgroups of $\mathrm{SL}_2(\mathbb{Z})$, which are linked to isogenies. For $N \in \mathbb{N}$ define

$$\Gamma_0(N) := \left\{ \begin{pmatrix} a & b \\ c & d \end{pmatrix} \in SL(2, \mathbb{Z}) \mid c \equiv 0 \mod N \right\}$$

This is a subgroup of $SL_2(\mathbb{Z})$ of finite index, and so $\mathbb{H}^*/\Gamma_0(N)$ is a compact Riemann surface $X_0(N)$ with a natural cover morphism

$$\eta_N : X_0(N) \to X(1)$$

of degree $\varphi(N)$, the value of the Euler function of N. As affine part in $X_0(N)$ we find $Y_0(N) = X_0(N) \setminus \eta^{-1}(j(i \cdot \infty))$ and points P in $Y_0(N)$ have the following modular interpretation.

Let $P \in Y(1)$ corresponding to an isomorphism class of the lattice $\Lambda_\tau = \mathbb{Z}+\tau\mathbb{Z}$. The inverse image $\eta^{-1}(P)$ consists of the equivalence classes of lattices isomorphic to Λ_τ *with the additional information* consisting of a lattice $\Lambda_\tau(N) \supset \Lambda_\tau$ with $\Lambda_\tau(N)/\Lambda_\tau \cong \mathbb{Z}/N$. Hence a point $P_N \in Y_0(N)$ is the $\mathbb{C}$-isomorphism class of a *pair* $(\mathcal{E}_\tau, C_N)$ where C_N is a cyclic group of order N in $\mathcal{E}[N]$.

In other words, $Y_0(N)$ parametrizes *isomorphism classes of elliptic curves together with cyclic isogenies η_N of order N* and so they are moduli spaces for the pairs $(\mathcal{E}, \eta_N)$ over $\mathbb{C}$. Using Hurwitz genus formula and the well known fixed points of $\Gamma_0(N)$ one can compute the genus g_N of $X_0(N)$. We have that $g_N \sim N$. For $N = p$ a prime we get

$$g_N = \begin{cases} 0 \quad if \quad p = 2,3 \\ \dfrac{(p-13)}{12} \quad if \quad p \equiv 1 \mod 12 \\ \dfrac{(p-5)}{12} \quad if \quad p \equiv 5 \mod 12 \\ \dfrac{(p-7)}{12} \quad if \quad p \equiv 7 \mod 12 \\ \dfrac{(p-11)}{12} \quad if \quad p \equiv 11 \mod 12 \end{cases}$$

5.1.1. *Modular functions and forms.* Because of the construction as quotient of $\mathbb{H}^*$ we can identify algebraic-geometric objects on X_0 with analytic objects attached to $\mathbb{H}^*$ with specific symmetry properties. For instance, rational functions on $X_0(N)$ come from those meromorphic functions on $\mathbb{H}^*$ that are invariant under the action of $\Gamma_0(N)$, ie.

$$f\left(\frac{az+b}{cz+d}\right) = f(z) \text{ for all } \begin{pmatrix} a & b \\ c & d \end{pmatrix} \in \Gamma_0(N)$$

and are called *modular functions* of level N. A basic example is $j(z)$, which is modular of level 1 and hence of all levels. Since for all τ the elliptic curve the lattice $\mathbb{Z} + \frac{\tau}{N}\mathbb{Z}$ gives rise to an isogeny of $\mathcal{E}_\tau$ with cyclic kernel of order N the function $j\left(\frac{z}{N}\right) =: j_n(z)$ is a modular function of level N, and one checks that $\mathbb{C}(X_0(N))$ is isomorphic to $\mathbb{C}(j(z), j(\frac{z}{N}))$.

Differentials ω on $X_0(N)$ are of the form $f(z)dz$ with $f(z)$ meromorphic on $\mathbb{H}^*$ and, because of their invariance under $\Gamma_0(N)$, they satisfy a functional equation

$$f\left(\frac{az+b}{cz+d}\right) = (cz+d)^2 f(z) \text{ for all } \begin{pmatrix} a & b \\ c & d \end{pmatrix} \in \Gamma_0(N).$$

Such functions f are called *modular forms of weight* 2. If ω is holomorphic then $f(z)$ is holomorphic on $\mathbb{H}$ and vanishes on the cusps: $f(z)$ is a cusp form of weight 2 and level N. We know from the Riemann-Roch theorem that the cusp forms $\mathcal{S}_0(N)(\mathbb{C})$ are a $\mathbb{C}$-vector space of dimension g_N. This space plays a very important role in the arithmetic of modular forms.

5.1.2. *q-expansion.* Modular forms and functions are invariant under the transformation $z \mapsto z+1$. Define as new variable $q = e^{2\pi i z}$ (hence transform neighborhoods of $i \cdot \infty$ to neighborhoods of 0). Then we can form the Fourier series of modular forms and functions $f(q)$ and get the **q-expansion**

$$f(q) = \sum_{j=k}^{\infty} a_j q^j$$

with $k \in \mathbb{Z}$. If f is a cusp form then $k \geq 1$. Modular forms and functions are uniquely determined by their Fourier expansion.

5.2. Modular polynomials. The important fact is that in many cases one can compute the coefficients of the q-expansion effectively. This is so for $j(q)$ (and hence for j_N in the variable $q^{1/N}$). In particular, $j(q)$ has a pole of order 1 at 0 and has coefficients in $\mathbb{Z}$, and for the exact expression see [**7**, Proposition 5.41].

We know that j_N satisfies a polynomial over $\mathbb{C}(j)$ of degree $\varphi(N)$. To find this polynomial (and so an equation for $X_0(N)$) one compares sufficiently many coefficients of the Fourier expansion and finds the *modular polynomials* $\phi(j, j_N)$ which are monic and symmetric in (j, j_N) and have degree $\varphi(N)$ in j and j_N.

So two elliptic curves with j-invariants j_1 and j_2 are isogenous under a cyclic isogeny of degree N if and only if $\phi_N(j_1, j_2) = 0$. The equation $\phi_N(X, Y) = 0$ is the canonical equation of the affine modular curve $Y_0(N)$, and $X_0(N)$ is given by the closure in $\mathbb{P}^2$. We display $\phi_N(x, y)$ for $N = 2, 3$.

$$
\begin{aligned}
\phi_2 &= x^3 - x^2y^2 + y^3 + 1488xy(x+y) + 40773375xy - 162000(x^2+y^2) \\
&\quad + 8748000000(x+y) - 157464000000000 \\
\phi_3 &= -x^3y^3 + 2232x^3y^2 + 2232y^3x^2 + x^4 - 1069956x^3y + 2587918086x^2y^2 \\
&\quad - 1069956y^3x + y^4 + 36864000x^3 + 8900222976000x^2y + 8900222976000y^2x \\
&\quad + 36864000y^3 + 452984832000000x^2 - 770845966336000000xy + 452984832000000y^2 \\
&\quad + 1855425871872000000000x + 1855425871872000000000y
\end{aligned}
$$

There are tables of modular polynomials for large N.

REMARK 8. *There have been some attempts in the last decade to generalize modular polynomials for higher dimensional varieties. The interested reader should consult* [**4**] *for abelian surfaces.*

5.3. The arithmetical theory of modular curves. A general reference for the following discussions is [**53**]. One observes that the modular polynomials have coefficients in $\mathbb{Z}$. This is no accident.

Looking at the modular interpretation one sees that the curves $X_0(N)$ represent over $\mathbb{C}$ a moduli problem: "Parametrize the isomorphism classes of pairs $(\mathcal{E}, \eta_N)$ of elliptic curves with cyclic isogeny of degree N".

This problem makes sense over any ring R with invertible element N and so over $\mathbb{Z}[1/N]$. Because of the existence of twists χ and the fact that the pair $(\chi(\mathcal{E}), \chi(\eta_N))$ is isomorphic to $(\mathcal{E}, \eta_N)$ over $\bar{k}$ but in general not over k we cannot expect to find a fine moduli scheme for our problem. But a coarse moduli scheme exists and so $\mathbb{H}^*/\Gamma_0(N)$ is the constant field extension of a curve $X_0(N)$ (i.e a scheme of relative dimension 1) over $\mathbb{Z}[1/N]$ with affine part given by the equation $\phi_N(X, Y)$.

A deep analysis of reductions of modulo prime numbers dividing N due to Deligne-Rapoport ([**14**]) and Katz-Mazur ([**40**]) shows that one can define $X_0(N)$ over $\mathbb{Z}$. Hence all geometric objects like holomorphic differentials (and so cusp forms) have a $\mathbb{Z}$-structure and so it makes sense to speak of $\mathcal{S}_0(N)(R)$, the R-module of cusp forms over the commutative ring R (to avoid complications assume that N is invertible in R).

How to find these cusp forms? Following a ground breaking idea of **J. Tate** one replaces the theory of elliptic curves over $\mathbb{C}$ by the theory of elliptic curves in the realm of rigid p-adic spaces and finds the elliptic Tate curve in $\mathbb{Z}((q))$, the ring of formal Laurent series in one variable in q. Using this curve and exploiting functions and differentials on this curve one finds the q-expansion over all rings R and gets for instance:

THEOREM 34 (q-**expansion principle**). *Let R be a commutative ring with $N \in R^*$. Then the q-expansion determines uniquely holomorphic differentials on $X_0(N)$, and in particular we have:*

(1) *$f \in \mathcal{S}_0(N)(R)$ if and only if the Fourier expansion of f has coefficients in R, and*
(2) *there exists a $\mathbb{Z}$-base $f_1, \dots, f_{g_N}$ of $\mathcal{S}_0(N)(\mathbb{Z})$ such that*

$$\mathcal{S}_0(N)(R) = \langle f_1, \dots, f_{g_N} \rangle \otimes R.$$

We can apply these results to the Jacobian variety and its minimal model over $\mathbb{Z}$, denoted by $\mathcal{J}_0(N)_{\mathbb{Z}}$. Endomorphism of this variety are uniquely determined by their action on $\mathcal{S}_0(N)(\mathbb{C})$ and so accessible to computation after we have computed a base of the cusp forms over $\mathbb{Z}$. This computation is done in an effective way by using *modular symbols* (see [**54**]). It turns out that $\mathrm{End}_{\mathbb{Z}}(\mathcal{J}_0(N)_{\mathbb{Z}})$ is large, it contains the Hecke algebra $\mathcal{T}_N$ generated by Hecke operators constructed in Section 8. In particular, it follows that simple factors of $\mathcal{J}_0(N)$ have real multiplication.

Galois representations of $G_{\mathbb{Q}}$ attached to the Tate modules of these factors split up in a sum of 2-dimensional representations. One of the greatest results in Arithmetic Geometry is the theorem of Khare-Kisin-Wintenberger (see [**41**], [**42**]) which confirms the conjecture of Serre and states that all odd two-dimensional Galois representations are obtained (up to twists by characters) by the operations on appropriate factors of $\mathcal{J}_0(N)$ with an explicit recipe how to find N and the twist character. From this, the Fermat conjecture FLT is an easy consequence (a "five line proof").

Part 2. Cryptography

For communication in the modern world it is crucial that one can use open communication channels to

- exchange keys,
- sign messages
- authenticate entities, and
- encrypt and decrypt (not too large) messages

with simple protocols, clear and easy to follow implementation rules based on *cryptographic primitives*, which rely on (hopefully) hard mathematical tasks. The part of cryptology, which is devoted to solve these challenges is *public key cryptography* and relies on the ground breaking ideas of Diffie and Hellman [**17**]. In this paper we shall concentrate to the first item in its simplest form (and not using a secure protocol). The reader is encouraged to look at the other aspects, too, e.g. in [**7**].

6. Diffie-Hellman key exchange

6.1. The classical case. The task to solve is: find a protocol such that two partners P_1 and P_2 can agree on a common secret by using public channels and algorithms.

A groundbreaking solution was found by W. Diffie and H. E. Hellman in [**17**] with the idea to use (computational) one-way functions. They suggest to use the multiplicative group of finite fields $\mathbb{F}_q$ and, for a prime number $\ell | q-1$ choose a primitive root ζ_ℓ, private keys $k_i \in \mathbb{Z}$ and public keys $p_i = \zeta_\ell^{k_i}$. The common secret is

$$s_{1,2} = \zeta_\ell^{k_1 \cdot k_2}.$$

All computations are very fast (polynomial in $\log(q)$). The security is measured by the hardness of the **Diffie-Hellman** computational problem (CDH):

For random elements $a, b \in \{0, \ldots, \ell-1\}$ and given ζ_ℓ^a, ζ_ℓ^b compute $\zeta_\ell^{a \cdot b}$.

Let ζ be a primitive root of unity in $\mathbb{F}_q^*$. Define the (classical) discrete logarithm (DL) of an element $x \in \mathbb{F}_q^*$ with respect to the base ζ by

$$\log_\zeta(x) = \min\{n \in \mathbb{N} \text{ such that } \zeta^n = x.\}$$

It is obvious that an algorithm that computes discrete logarithms (e.g. in ζ_ℓ) solves (CDH). This problem is rather old (going back at least to the 19-th century). C.F. Gauss introduced the term "index" in the Disquisitiones Arithmeticae (1801) for the discrete logarithm modulo p, and there are tables for primes up to 1000 by C. G. Jacobi (1839).

A systematic algorithm is given in the book of Kraichik (1922) [**43**]; in fact this is the index-calculus algorithm reinvented and refined in cryptography from 1980 till today [**37**]. As result one gets algorithms of subexponential complexity (with relatively small constants, see [**37**]), which are even dramatically faster if q is not a prime. The reason for these fast algorithms is the fact that it is easy to lift elements in $\mathbb{F}_q$ to elements in rings of integers of number fields.

6.2. A first abstraction. Obviously we can use the Diffie-Hellman key exchange scheme if we have

- a finite cyclic group $(C, \circ)$ with a generator g_0,
- a numeration, i.e. an injective map

$$f : C \to \mathbb{N},$$

- an addition law $\oplus$ on $f(C)$ with

$$f(f^{-1}(a) \circ f^{-1}(b)) = a \oplus b \text{ for all } a, b \in f(C).$$

$f(C)$ becomes a $\mathbb{Z}$-module with the usual scalar multiplication: $0 \cdot a = f(0_C)$, $n \cdot a = (n-1)$-fold addition of a to itself, $(-n) \cdot a = n \cdot (\ominus a)$ for $a \in f(C)$, $n \in \mathbb{N}$.

The private keys are again $k_i \in \mathbb{Z}$, the public keys are $k_i \cdot f(g_0)$, and the common secret is $k_1 \cdot (k_2 \cdot f(g_0))$. The CDH problem is: for random a_1, $a_2 \in f(C)$ with publicly unknown k_1, k_2 such that $k_i \cdot f(g_0) = a_i$ compute $c = (k_1 \cdot k_2) \cdot f(g_0)$.

Define the discrete logarithm (DL) by

$$\log_{g_0}(a) := \min\{n \in \mathbb{N} \mid \text{ such that } n \cdot f(g_0) = a.\}$$

Again, the computation of the (DL) solves CDH. By elementary number theory (CRT and p-adic expansion) one sees immediately that the computation of (DL) is reduced to the computation of the discrete logarithms in all $f(C_\ell)$ with C_ℓ the subgroup of C of elements of order dividing ℓ and ℓ dividing $|C|$. Hence we can and will assume from now on that C is cyclic of prime order ℓ. For shortness, we denote the task to compute the discrete logarithm by (DLP).

6.2.1. *Black box groups.* A "generic" object of the situation above is given by a black box group C of prime order ℓ.

(1) There are algorithms that compute (DL) (probabilistically) with $\mathcal{O}(\sqrt{\ell})$ group operations in C (e.g. Shank's bay-step giant step algorithm, Pollard's ρ algorithm et.al.), these algorithms are applicable for all finite cyclic groups, and one cannot do better.
(2) Up to algorithms with subexponential complexity, the computation of (DL) in C is equivalent with (CDH)(Maurer-Wolf).

6.3. Mathematical task. In order that we can use (a family of) groups C for crypto systems based on discrete logarithms they have to satisfy four crucial conditions:

(1) C has a known large prime order ℓ and a numeration $f : C \to \mathbb{N}$.
(2) Condition for the numeration: The elements in C can be stored in a computer in a compact way (e.g. $\mathcal{O}(\log \ell)$ bits needed).
(3) The group composition $\oplus$ induced by f is given by an algorithm that is easily and efficiently implemented and very fast.
(4) The computation of the DL in $f(C)$ (for random elements) is very hard and so infeasible in practice (ideally the bit-complexity should be exponential in $\log \ell$).

It is surprisingly hard to construct such groups. All known examples today are related with subgroups of Picard groups of hyperelliptic curves of genus ≤ 3 over prime fields $\mathbb{F}_p$. It will be one of the main aims of the paper to explain this statement.

6.4. Q-bit security. As said, we shall describe below DL-systems for which we have good reasons to believe that the bit-complexity is exponential and so the task in Section 6.3 is solved. But the possibility that *quantum computing* may be realizable in foreseeable time yields new aspects for the discussion of security of crypto systems. By Shor's algorithm it follows that the q-bit complexity of discrete logarithms in **all** finite groups is polynomial!.

So it is challenging to find key exchange systems that are not based on discrete logarithms in groups but still are near to the original idea of Diffie and Hellman. In the quantum world new relations between crypto primitives arise, and it seems that hidden subgroup problem and connected to it, the hidden shift problem related to groups G are central ([**60**] and [**44**]). Here the state of the art is that for abelian groups G the problems can be solved in subexponential time and space, for dihedral groups there is "hope".

6.5. Key exchange with G-sets. The DL-system in Section 6.3 can be seen in the following way: By scalar multiplication $f(C)$ becomes a $\mathbb{Z}$-set, and so elements of $\mathbb{Z}$ induce commuting endomorphisms on $f(C)$.

Denote by $\mathbb{Z}'$ the semigroup of elements in $\mathbb{Z}$ prime to $|f(C)|$. Then the set A of generators of $f(C)$ becomes a $\mathbb{Z}'$-set, and elements in $\mathbb{Z}'$ induce commuting endomorphisms of A. A next step to generalize the Diffie-Hellman key exchange is to replace $\mathbb{Z}'$ by a (semi-) group G and the set of generators of $f(C)$ by a G-set $A \subset \mathbb{N}$ on which G operates transitively. For $g \in G$, define $t_g \in \mathrm{End}_{set}(A)$ by

$$a \mapsto t_g(a) := g \cdot a.$$

Let G_1 be a semi-subgroup of G and $G_2 = Z(G_1)$ the centralizer of G_1 in G (if G is abelian then $G = G_1 = G_2$). Because of

$$g_1 \cdot (g_2 \cdot a_0) = (t_{g_1} \circ t_{g_2}) \cdot a_0 = (t_{g_2} \circ t_{g_1}) \cdot a_0$$

for $g_1 \in G_1$ and $g_2 \in G_2$ we can use (A, a_0, G_1, G_2) for key exchange by defining an obvious analogue of the scheme in Section 6.3.

The security of this exchange depends on the difficulty to find the translations t_{g_i}. We remark that though the security of such systems is, in general, not related to discrete logarithms, it may happen that the generic algorithms from Section 6.3 can still be applied.

What about quantum security? One breaks the system if one can determine t_{g_1}. This is a typical problem for the hidden shift. Take the maps

$$f_0 : G_1 \to A, \text{ such that } f_0(g) = t_g \cdot a_0$$
$$f_1 : G_1 \to A, \text{ such that } f_1(g) = t_g \cdot (t_{g_1} \cdot a_0)$$

and find the shift. For G_1 abelian and finite there is an algorithm of Kuperberg [**44**], which solves this task in subexponential time. In particular we see that every Diffie-Hellman key exchange based on $\mathbb{Z}$-sets has at best subexponential security.

6.6. Abstract setting of key exchange. On our way to generalization we get rid of the algebraic structures. Assume $A \subset \mathbb{N}$ and let $B_1, B_2 \subset \mathrm{End}_{set}(A)$. Choose $a_0 \in A$. We need the **centralizing condition**. The elements of B_1 commute with the elements of B_2 on $B_i\{a_0\}$. Then

$$b_1(b_2(a_0)) = b_2(b_1(a_0))$$

and this is all we need for key exchange.

The effectiveness of this exchange is given if for $b_i \in B_i, b_j \in B_j$ the value $b_i(b_j(a_0))$ can be quickly evaluated (i.e., calculated and represented). The analogue of the Computational Diffie-Hellman problem is

CDH: For randomly given $a_1, a_2 \in A$, compute (if exists) a_3 with
$$a_3 = b_{a_1} \cdot (b_{a_2} \cdot a_0),$$

where $b_{a_i} \in B_i$ such that $b_{a_i} \cdot a_0 = a_i$. It is clear that CDH can be solved if one can calculate for random $a \in B_i \cdot \{a_0\}$ an endomorphism $b_a \in B_i$ with $b_a(a_0) = a$. We remark that b_a may not be uniquely determined by a.

Problem:

(1) Find a "genuine" usable instance for the abstract setting!
(2) What can one say about quantum computing security?

6.7. Key exchange in categories. We make a final step of abstraction. As always we assume that we have two partners P_1 and P_2 who want to have a common secret key.

Let $\mathcal{C}_i$, $i = 1, 2$ be two categories whose objects are the same sets A_j and with morphisms $B^i_{j,k} = \mathrm{Mor}^i(A_j, A_k)$. We fix a "base" object A_0 and assume that $\mathcal{C}_1, \mathcal{C}_2, A_0$ satisfies the following conditions:

(1) For every $\varphi \in B^1(A_0, A_j)$ and every $\psi \in B^2(A_0, A_k)$ the pushout exists, i.e. there is a uniquely (up to isomorphisms) determined triple

$$(A_l,\ \gamma_1 \in B^1(A_k, A_l),\ \gamma_2 \in B^2(A_j, A_l))$$

with

$$\gamma_2 \circ \varphi = \gamma_1 \circ \psi$$

such that this triple is minimal (universality condition).

(2) P_1 can determine A_l if he knows φ, A_k and an additional (publicly known) information $P(\psi)$ (which is often a subset of A_k), and an analogue fact holds for P_2.

Key exchange. Given such categories $\mathcal{C}_1, \mathcal{C}_2$ the partners can chose φ, ψ, send A_j, A_k and $P(\psi)$ respectively $P(\varphi)$ and compute the ***common secret*** A_l.

Effectiveness. We assume that all the objects concerning $\mathcal{C}_i$ can be handled by computers in a fast and compact way, in particular, for chosen φ, ψ the objects A_j, A_k as well as the additional information can be computed rapidly. Moreover, using the given information, P^i can compute of A_l quickly.

Security. The scheme is broken if (CDH) is weak: For randomly given A_j, A_k determine A_l, which is the pushout of

$$A_0 \stackrel{\varphi}{\to} A_j$$

and

$$A_0 \stackrel{\psi}{\to} A_k.$$

For this, it is allowed to use the additional information. We shall see an example for this categorial key exchange in Section 11.3.3, and till now all algorithms for breaking this system have exponential complexity.

7. Index calculus in Picard groups

We want to use systems based on discrete logarithms in groups G and so find groups which satisfy the conditions formulated in Section 6.3. Motivated by ideas of V. Miller and N. Koblitz we want to use subgroups of Picard groups of curves over finite fields. Thm. 29 of Hess-Diem implies that at least in principle for such groups the conditions 2 and 3 are satisfied. For finding subgroups of large prime order one has to be able to determine the order of Picard groups rapidly. Here the key word is point counting, and again there is, in principle, a solution by a polynomial time algorithm due to Pila generalizing the Schoof algorithm for elliptic curves. But these algorithms are much too slow, and an acceleration is only known for elliptic curves (AES-algorithm), for curves of genus 2 (Gaudry, Schost) and for curves with special endomorphism rings (complex multiplication or real multiplication).

But before investing a lot of work in point counting it is useful to look at the security aspect. We want to compare the hardness of the computation of the DL in the specific groups with the generic hardness, i.e. $\sim |G|^{1/2}$. We recall that a main reason against the classical DL was the index-calculus algorithm, which is based on the (easy) lifting of finite fields to integers in number fields or function fields over finite fields. This kind of attack is not possible in Picard groups of curves of positive genus as pointed out by Miller and Koblitz: The "golden shield" of the Néron-Tate quadratic form prevents a (easy) lifting of elements in Abelian varieties over finite fields to global fields. But unfortunately there are very effective variants of the index- calculus attack to Picard groups.

7.1. Introduction to index calculus. Let $(G, \oplus)$ be a cyclic group of order N with generator g_0.

First step: Find a "*factor base*" consisting of relatively few elements and compute G as $\mathbb{Z}-$module given by the free abelian group generated by the base elements modulo relations.

So choose a subset $\mathcal{B} = \{g_1, \ldots, g_r\}$ of G generating G and look for relations

$$R_j : \oplus_{i=1}^{r} [n_i] g_i = 0_G. \tag{14}$$

Obviously R_j yields the relation

$$\sum_{i=1}^{r} n_i \log_{g_0}(g_i) \equiv 0 \mod N \tag{15}$$

for discrete logarithm.

We assume that we can find sufficiently many independent relations as in Eq. (14) for solving the system in Eq. (15) via linear algebra for $\log_g g_i$, $i = 1, \ldots, r$. Then we have an explicit presentation of G as $\mathbb{Z}$-module by

$$G \cong \mathbb{Z}^r / \langle \ldots, R_j, \ldots \rangle.$$

Second step: Take $g \in G$ randomly and chose a "random walk" with steps $g^0 = g, \ldots, g^j = [k_j] g^{j-1}$ and assume that after a few steps j we find a tuple $e_1, \ldots, e_r$ with e_i small and

$$g^j = [e_1] g_1 + \cdots [e_r] g_r.$$

"To find" means: There is a fast algorithm to decide whether such e_i exist, and then the computation of these e_i is also fast. This boils down to a smoothness condition. (Recall: A number $n \in \mathbb{N}$ is B-smooth if all prime divisors of n are $\leq B$, and results from analytic number theory by Canfield, Erdös, Pomerance determine the probability for n being smooth.) The second step is usually done by an appropriate sieving method.

The important task in this method is to balance the number of elements in the factor base to make the linear algebra over $\mathbb{Z}$ manageable and to guarantee "smoothness" of arbitrary elements with respect to this base. Usually one finds a kind of *size* in G (size of lifted elements in $\mathbb{Z}$ or degree in polynomial rings, degree of reduced divisors , ...) to define factor bases. Typically , successful index-calculus approaches give rise to algorithms for the computation of the DL in G which have *subexponential* complexity and so, for large enough order of G, the DL-system has a poor security.

For an axiomatic approach of index-calculus algorithms we refer to [**19**]. This principle is refined in concrete situations with enormous effect as we shall see below.

7.2. Index calculus for hyperelliptic Jacobians. Index calculus can be applied to a DL in Jacobians of hyperelliptic curves. Let $\mathcal{C}$ be a hyperelliptic curve of genus $g \geq 2$ over a finite field $\mathbb{F}_q$ of characteristic p and G a cyclic subgroup in $\mathrm{Pic}^0_{\mathcal{C}}$.

We can represent every element in G in a unique way by the Mumford representation $[u(x), v(x)]$, where $u(x)$ is a polynomial in $\mathbb{F}_q[X]$ of degree $\leq g$.

As factor base we choose points in $\mathrm{Pic}^0_{\mathcal{C}}$ with $u(X)$ irreducible of degree at most B, a chosen smoothness bound. A divisor is said to be *B-smooth* if all the prime divisors in its decomposition have degree at most B. This leads to the historically first algorithm to compute discrete logarithms in Picard groups of hyperelliptic curves. It is due to Adleman, Demarrais, and Huang. For an explicit description of the algorithm see [**7**, pg. 525]. For $N \in \mathbb{Z}^{>0}$, $s, c \in \mathbb{R}$, with $0 \leq s \leq 1$ denote

$$L_N(s,c) = \exp\left((c+o(1))\left(\log N\right)^s \left(\log\log N\right)^{1-s}\right),$$

as $N \to \infty$. Then we have:

THEOREM 35. *For* $\log q \leq (2g+1)^{1-\epsilon}$, *there exists a constant* $c \leq 2.18$ *such that the discrete logarithms in* $\mathrm{Jac}_C(\mathbb{F}_q)$ *can be computed in expected time* $L_{q^{2g+1}}(1/2, c)$.

This remarkable result gives an subexponential algorithm for "large" genus. But much more important for practical applications are *exponential* algorithms, which weaken the DLP for small but realistic genus. The first groundbreaking result is

THEOREM 36 (Gaudry). *Let* $\mathcal{C}$ *be a a hyperelliptic curve of genus* $g \geq 2$ *defined over a finite field* $\mathbb{F}_q$. *If* $q > g!$ *then discrete logarithms in* $\mathrm{Jac}_{\mathbb{F}_q}(\mathcal{C})$ *can be computed in expected time* $O(g^3 q^{2+\epsilon})$.

Since the expected size of $\mathcal{J}_\mathcal{C}(\mathbb{F}_q)$ is q^g (see Weil's result, Thm. 12) we are, for $g > 4$, far away from the generic security bound, and so we have to exclude hyperelliptic curves of genus ≥ 5 if we want a DL-system in Picard groups. But Gaudry's result can be sharpened. N. Thériault suggested to use "large primes" as well as the original elements of the factor base consisting of points on the curve of small degree. With many more refinements [**28**] one gets

THEOREM 37. *There exists a (probabilistic) algorithm which computes the DL, up to* log*-factors, in the divisor class group of hyperelliptic curves of genus* g *in expected time of* $\mathcal{O}(q^{(2-2/g)})$.

This rules out $g = 4$ for hyperelliptic curves since the ratio of the expected group order to time complexity, $\mathcal{O}(q^g)/\mathcal{O}(q^{2-2/g})$, gets too big.

7.3. Index-calculus in Picard groups in curves with plane models of small degree. The following is mainly work of C. Diem. He gives an algorithm for computing discrete logarithms in $J_C(\mathbb{F}_q)$ assuming that one has a plane curve C' of degree d. We recall that for non-hyperelliptic curves $d = 2g_C - 2$ is possible but that for hyperelliptic curves $d \geq g_C + 1$.

So the minimal degree of plane models of hyperelliptic curves of genus ≥ 3 is larger than the degree of such models for non-hyperelliptic curves. Using factor bases constructed with the help of Semaev polynomials and using a large amount of ingredients from abstract algebraic geometry (e.g. membership tests for zero-dimensional schemes) Diem succeeds to prove:

THEOREM 38. *Fix* $d \geq 4$. *Then the DLP in* Pic^0_C *of curves birationally equivalent to plane curves of degree* d *can be solved, up to* log*-factors, in expected time* $\mathcal{O}(q^{2-\frac{2}{d-2}})$.

For genus 4 and non-hyperelliptic curve $\mathcal{C}$ we get $d = 6$ and so the hardness of D is bounded, up to log-factors, by $\mathcal{O}(q^{3/2})$. Since the expected group size is q^4 this is too far away from the generic complexity, and it is not advisable to use (hyperelliptic or not hyperelliptic) curves of genus 4 for DL-systems.

For non-hyperelliptic curves of genus 3 we get $d = 4$ and so the complexity of the DL is $\mathcal{O}(g)$ and again such curves can not be used for DL-systems. Hence our discussion for fulfilling conditions 3 and 4 in Section 6.3 can be restricted to hyperelliptic and elliptic curves of genus $1, 2, 3$. Before doing this in detail we have one more general section, interesting both from theoretical and practical point of view.

8. Isogenies of Jacobians via correspondences and applications to discrete logarithms

We describe a general construction of isogenies between abelian varieties closely attached to Jacobians of curves. The crypto-graphical relevance of these constructions is that every computable isogeny yields a transfer of the (DLP), and it may be easier to solve the problem after the application of the isogeny.

As always, k is assumed to be a perfect field. Let L be a finite algebraic extension field of k. Let $\mathcal{D}_1$ be a regular projective curve over L and $\mathcal{D}_2$ a regular projective curve defined over k. We recall some properties of cover morphisms of curves and attached norm and conorm homomorphisms of Jacobians. Let $\mathcal{H}$ be a curve over L and

$$\varphi_1 : \mathcal{H} \to \mathcal{D}_1,$$

respectively

$$\varphi_2 : \mathcal{H} \to \mathcal{D}_2 \times_{\mathrm{Spec}(k)} \mathrm{Spec}(L) =: \mathcal{D}_{2,L},$$

be L-rational morphisms. The morphism φ_1 induces the L-rational **conorm morphism**

$$\varphi_1^* : \mathcal{J}_{\mathcal{D}_1} \to \mathcal{J}_{\mathcal{H}}$$

and the morphism φ_1 induces the **norm morphism**

$$\varphi_{2,*} : \mathcal{J}_{\mathcal{H}} \to \mathcal{J}_{\mathcal{D}_{2,L}}.$$

By composition we get a homomorphism

$$\eta_L : \mathcal{J}_{\mathcal{D}_1} \to \mathcal{J}_{\mathcal{D}_{2,L}}$$

defined over L.

Let $\mathcal{W}_{L/k}$ be the Weil restriction of the Jacobian of $\mathcal{D}_1$ to k. This is an abelian variety defined over k with $\mathcal{W}_{L/k}(k) = \mathrm{Pic}^0_{\mathcal{D}_1}$. Applying the norm map from L to k and using the functorial properties of the Weil restriction we get a homomorphism

$$\eta : \mathcal{W}_{L/k} \to \mathcal{J}_{\mathcal{D}_2}.$$

In general, neither the kernel nor the cokernel of η will be finite. But under, usually mild, conditions one can assure that that η has a finite kernel, and so it induces an isogeny of $\mathcal{W}_{L/k}$ to an abelian subvariety of $\mathcal{J}_{\mathcal{D}_2}$.

As application we get a transfer of the discrete logarithm problem from $\mathrm{Pic}^0_{\mathcal{D}_1}$ (defined over L) to the DL-problem in a subvariety of $\mathcal{J}_{\mathcal{D}_2}$ (defined over k). Of course, the efficiency of this transfer depends on the complexity of the algorithms computing the norm and conorm maps (hence φ_i and $[L : k]$ must have reasonably small degrees), and an attack makes sense only if the DL-problem after the transfer is easier than before.

8.1. Weil descent. Take $k = \mathbb{F}_q$ and $L = \mathbb{F}_{q^d}$ with $d > 1$ and $\mathcal{H} = \mathcal{D}_{2,L}$, i.e. a given curve $\mathcal{C}$ defined over $\mathbb{F}_{q^d}$ is covered by a curve $\mathcal{D}_{\mathbb{F}_{q^d}}$, which is the scalar extension of a curve $\mathcal{D}$ defined over k.

This yields a k-rational homomorphism from the Weil restriction $\mathcal{W}_{L/k}$ of $\mathcal{J}_\mathcal{C}$ to $\mathcal{J}_\mathcal{D}$. Then $\mathcal{D}$ will (in all non-trivial cases) be a curve of a genus larger than the genus of $\mathcal{C}$ but since it is defined over the smaller field $\mathbb{F}_q$ one can hope that one can apply fast algorithms to compute the discrete logarithm in $\mathcal{J}_\mathcal{D}(\mathbb{F}_q)$, e.g. by methods of index-calculus in Section 7. Indeed, if $\mathcal{C}$ is not defined over a proper subfield of $\mathbb{F}_{q^d}$ this is the principle of the so-called GHS-attack in (see [**25**] and [**7**, Section 22.3.2]), which is successful in remarkably many cases.

If $\mathcal{C}$ is already defined over $\mathbb{F}_q$ one is lead to the so-called trace-zero varieties in $\mathrm{Jac}_\mathcal{C}(\mathbb{F}_{q^d})$ (see [**7**, Section 7.4.2]) and again correspondences induced by covers of curves can be used for attacks on crypto systems based on discrete logarithms on these varieties by work of Diem [**7**, 22.3.4]. These results already indicate that the use of Picard groups of curves (e.g. elliptic curves) over non-prime fields $\mathbb{F}_{q^d}$ with $d \geq 4$ is not advisable for cryptographic use.

By more recent work of C. Diem this "feeling" is reinforced for instance for families of elliptic curves in towers of finite fields. The methods used in these papers use the Weil restriction method explained above only as a "guideline" and sometimes as tools for proof. The real heart of the methods of Diem is the use of Semaev's summation polynomials. In this context and in particularly because of suggestions of pairing based cryptography using (supersingular) elliptic curves it is important to mention the enormous progress made in the computation of discrete logarithm in the multiplicative group of finite non-prime fields [**37**].

8.2. Modular correspondences. We recall from Section 5 that for N prime to char (k) the modular curve $X_0(N)$ is a regular projective curve, defined over $\mathbb{Z}[1/N]$ and so in particular over $\mathbb{Q}$ and over $\mathbb{F}_p$ with p prime to N. As explained in Section 5 there is an affine part $Y_0(N)$, which is a (coarse) moduli scheme for the isomorphism classes of pairs (E, η_N) of elliptic curves with cyclic isogeny of degree N. This means that for every point $P = (j_E, j_\eta)$ in $Y_0(N)(k)$ there is an elliptic curve E defined over k and an isogeny $\eta_N : E \to E'$ with $\ker(\eta_N)$ invariant under the action of G_k and as abelian group isomorphic to $\mathbb{Z}/N$ such that the invariants of E and E' are (j_E, j_η).

The points in $X_0(N) \setminus Y_0(N)$ are the cusps, and it is important that these points have a modular interpretation, too. For example, if N is squarefree, then there is one cusp point at ∞ (in the upper half plane) which corresponds to the pair (Néron polygon with N vertices, $\langle \zeta_N \rangle$) where ζ_N is a primitive N-th root of unity.

Let ℓ be a prime not dividing char(k)$\cdot$N. By the splitting $\mathbb{Z}/\ell \cdot N \cong \mathbb{Z}/\ell \times \mathbb{Z}/N$ and an analogous splitting of the kernel of a cyclic isogeny of degree $\ell \cdot N$ in $C_\ell \times C_N$ we get a natural k-morphism

$$\varphi_\ell : X_0(\ell \cdot N) \to X_0(N).$$

Let ω_ℓ be the involution of $X_0(\ell \cdot N)$ induced by the map that sends the pair (E, η) with $\ker(\eta) = C_\ell \times C_N$ to the pair (E, η') where the kernel of η' is $E[\ell]/C_\ell \times C_N$. Define

$$\psi_\ell := \varphi_\ell \circ \omega_\ell : X_0(\ell \cdot N) \to X_0(N).$$

We are in the situation described above (with $k = L$) and can define the Hecke correspondence

$$T_\ell : \mathcal{J}_0(N) \to \mathcal{J}_0(N),$$

by $T_\ell := \varphi_{\ell *} \circ \psi_\ell^*$.

The Hecke ring of $X_0(N)$ is $\mathcal{T}_N = \langle T_\ell$ with ℓ prime to $N\rangle$, the ring generated by the endomorphisms T_ℓ. It is a commutative ring, which is very near to $\mathrm{End}(\mathcal{J}_0(N))$ (see [**53**]). It acts on the vector space of holomorphic differentials of $X_0(N)$ which can be identified with the k-vector space of k-rational cusp forms $\mathcal{S}_0(k)$ of level N (and trivial nebentype). By classical theory one knows that $\mathcal{T}_N$ is endowed with an Hermitian structure due to the Peterson scalar product, and so the eigenvalues of the operators T_ℓ are totally real numbers.

REMARK 9. *Assume that $\mathcal{A}$ is a simple factor of $\mathcal{J}_0(N)$. Then* $\mathrm{End}^0(\mathcal{A})$ *contains a totally real field of degree* $\dim(\mathcal{A})$.

This means that factors of $\mathcal{J}_0(N)$ have very special and large endomorphism rings. As consequence there is a splitting of Galois representations of $G_\mathbb{Q}$ constructed by the action on Tate modules of $\mathcal{J}_0(N)$ into a sum of two-dimensional representations with real eigenvalues, and these "modular representations" play a most important role in number theory, e.g. for the proof of Fermat's Last Theorem. The narrow relation to arithmetic is reflected by the **Eichler-Shimura** congruence

$$T_\ell = \mathrm{Frob}_\ell + \ell/\mathrm{Frob}_\ell,$$

where Frob is the Frobenius endomorphism on $\mathcal{J}_0(N) \bigotimes \mathbb{F}_\ell$. In particular, Frob_ℓ satisfies the **Eichler-Shimura** equation

$$X^2 - T_\ell \cdot X + \ell = 0.$$

A curve $\mathcal{C}$ whose Jacobian is a factor of $\mathcal{J}_0(N)$ is called **modular of level** N.

Using cusps forms it is possible to determine its period matrix, decide whether it is hyperelliptic, and then compute its Weierstrass equation (see [**73**], [**72**]). The *importance for cryptography* is the fact that Frob satisfies the quadratic Eichler-Shimura equation over a totally real number field, and this can be used for point counting for curves of genus ≥ 2 as in [**26**]. Hence *modular curves of genus* 2 *are potentially usable for DL-systems.*

8.3. Correspondences via monodromy groups. We assume that we have a cover morphism

$$f : \mathcal{C} \to \mathbb{P}^1$$

defined over k of degree n, satisfying some fixed ramification conditions and having a fixed monodromy group $G_f := \mathrm{Mon}\ (f)$. We have morphisms

$$\tilde{f} : \tilde{\mathcal{H}} \xrightarrow{\tilde{h}} \mathcal{C} \xrightarrow{f} \mathbb{P}^1$$

with $\tilde{f}$ a Galois cover of f with Galois group G_f. For simplicity, we assume that the field of constants of $\tilde{\mathcal{H}}$ is k. This setting is motivated by the theory of *Hurwitz spaces* and it is hoped that one can exploit their rich and, over $\mathbb{C}$, well understood theory ([**22**] and [**23**]).

Next we choose subgroups $H_1 \subset G_f$ fixing $\mathcal{C}$ and H_2 containing H_1. Let $\mathcal{H}$ be the curve fixed by H_1 and $\mathcal{D}$ the fixed curve under H_2. So $\mathcal{H}$ covers both $\mathcal{C}$ and $\mathcal{D}$. Let

$$h : \mathcal{H} \to \mathcal{C} \quad \text{and} \quad g : \mathcal{H} \to \mathcal{D}$$

with morphisms induced by the Galois action. Hence the degree of h is equal to $\deg(h) = \frac{|G_f|}{|H_1|\cdot n}$ and the degree of g is equal to $\deg(g) = \frac{|H_2|}{|H_1|}$. We get a correspondence

$$\eta : \mathcal{J}_\mathcal{C} \to \mathcal{J}_\mathcal{D}$$

by applying $g_* \circ h^*$ to the Picard groups. In general, η will be neither injective nor surjective.

LEMMA 15. *Assume that $\mathcal{J}_\mathcal{D}$ is a simple abelian variety, $\dim \mathcal{J}_\mathcal{D} = g_\mathcal{C}$, and that there is a prime divisor $\mathfrak{p}_\infty$ of $\mathcal{C}$ which is totally ramified under h, i.e. there is exactly one prime divisor $\mathfrak{P}_\infty$ of $\mathcal{H}$ with norm $\mathfrak{p}$, and that there is no non-constant morphism of degree $\leq \deg(h)$ from $\mathcal{D}$ to the projective line. Then η is an isogeny.*

PROOF. Since $J_\mathcal{D}$ is simple, it is enough to show that η is not the zero map. Let $\mathfrak{p}'_\infty$ be the norm of $\mathfrak{P}_\infty$ under g. Without loss of generality we can assume that k is algebraically closed. So we find a prime divisor $\mathfrak{P}$ of $\mathcal{H}$ which is different from all prime divisors in $g^{-1}(\mathfrak{p}'_\infty)$.

Let c be the class of $\mathfrak{p} - \mathfrak{p}_\infty$, where $\mathfrak{p} = h_*(\mathfrak{P})$. Then $\eta(c)$ is the class of the divisor

$$D_\mathfrak{p} := \sum_{\mathfrak{P} \in h^{-1}(\mathfrak{p})} g_*(\mathfrak{P}) - \deg(h) \cdot g_*(\mathfrak{P}_\infty).$$

Note that $D_\mathfrak{p} \neq 0$ (as divisor). If the class of $D_\mathfrak{p}$ would be trivial, then there would be a non-constant function on $\mathcal{D}$ with pole order $\leq \deg(h)$ and hence a non-constant map of $\mathcal{D}$ to the projective line of degree $\leq \deg(h)$, which is a contradiction.

□

We shall see in Section 9 that we can realize the situation (over $\bar{k}$) of the lemma for hyperelliptic curves of genus 3 with non-decomposable Jacobian, f a polynomial of degree 6, $G_f = S_4$, H_1 a subgroup of order 2 and H_2 a subgroup of order 6. This leads to isogenies of degree 8 discussed by B. Smith, and generically maps hyperelliptic curves to non-hyperelliptic curves. The *importance for cryptography* is that generic hyperelliptic curves of genus 3 are not usable for DL-systems.

It is an open and challenging problem to find other interesting correspondences of low degree between Jacobian varieties induced by correspondences between curves and (possibly) attached to Hurwitz spaces.

9. Genus 3 curves and cryptography

QUESTION 1. *Can one use curves $\mathcal{C}$ of genus 3 for DL-systems?*

To find equations for random curves of genus 3 is easy: Either take a regular plane quartic (non-hyperelliptic curve) or a curve with equation $Y^2 = f(X)$ with $\deg(f) = 7$. In both cases the addition law is easily implemented and fast. If $\mathcal{C}$ is hyperelliptic, the Cantor algorithm is well-studied and fast, moreover one can transform it into formulas (involving, alas, many special cases), which are sometimes more convenient for implementations near to specialized hardware. The generic cases for addition and doubling are explicitly given by Algorithms 14.52 and 14.53 in [**7**]. The timings are not too far away from additions in groups in elliptic curves (with comparable size order)(see [**7**, Table 14.13]). For non-hyperelliptic curves see [**20**]. But we have already discussed the security problem: One can only use hyperelliptic curves of genus 3 for which the Jacobians do not possess an easily computed isogeny to another principally polarized abelian variety which has a non-hyperelliptic polarization. As we shall see next this will endanger the DL in "generic" hyperelliptic curves of genus 3.

9.1. Isogenies via S_4-covers. As observed by B. Smith [**66**] "many" hyperelliptic curves are isogenous to non-hyperelliptic curves via an isogeny with degree dividing 8. This fact is interpreted in terms of Hurwitz spaces and connected modular spaces in [**22**, **23**]. We refer for details and refinements to these papers.

For our purposes it will be enough to look at the case that the base field k is algebraically closed, which we shall assume from now on. For applications in cryptography one has to study rationality problems; see [**66**] and [**23**]. The construction relies on the so-called trigonal construction of Donagi-Livné [**18**]. We begin with a hyperelliptic curve $\mathcal{C}$ of genus 3 and its uniquely determined hyperelliptic projection $f_1 : \mathcal{C} \to \mathbb{P}^1$ with 8 ramification points $P_1, \cdots, P_8$, which extend to the Weierstrass points of $\mathcal{C}$. By linear algebra we show that there is a map

$$f_2 : \mathbb{P}^1 \to \mathbb{P}^1$$

of degree 3 with the following properties:

- f_2 is unramified in $P_1, \cdots P_8$, its ramification points are denoted by $Q_1, \cdots Q_4$ on the base line $\mathbb{P}^1$. The ramification order in Q_i is 2, and so each Q_i has exactly one unramified extension under f_2 denoted by Q_i'.
- $f_2(\{P_1, \cdots P_8\}) = \{S_1, \cdots S_4\}$ such that, after a suitable numeration, $f_2(P_i) = f_2(P_{4+i})$ for $1 \leq i \leq 4$.

Now we can use Galois theory.

9.1.1. *The monodromy group of f_2.* Obviously, the Galois closure $\tilde{f}_2 = f_2 \circ h_2$ of f_2 has as Galois group the symmetric group S_3 (since f_2 is not Galois because of the ramification type), and h_2 is a degree 2 cover $\mathcal{E}' \overset{h_2}{\to} \mathcal{C}$. From Galois theory we get that $\tilde{f}_2 = \pi \circ \eta$, where

$$\eta : \mathcal{E}' \to \mathcal{E}$$

is a cyclic cover of degree 3 with Galois group equal to the alternating subgroup A_3. Then, $\mathcal{E}$ is a quadratic cover of $\mathbb{P}^1$ ramified exactly at the discriminant

$$\Delta_1 = Q_1 + \cdots + Q_4$$

of f_2. Therefore $\mathcal{E}$ is an elliptic curve with cover map π to $\mathbb{P}^1$. From construction and Abhyankar's lemma it follows that η is unramified. Hence $\mathcal{E}'$ is an elliptic curve, too, and η is an isogeny of degree 3 (after applying a suitable translation).

9.1.2. *The monodromy group of $f = f_2 \circ f_1$.* Since f is a cover of degree 6, its Galois group can be embedded into S_6. But a closer analysis using the specific ramification situation shows; see [**22**, Thm. 3].

LEMMA 16. *The monodromy group of f is isomorphic to S_4.*

Let $\tilde{f} : \tilde{\mathcal{C}} \to \mathbb{P}^1$ be the Galois cover of curves factoring over f with Galois group S_4. Let $\mathcal{C}'$ be the subcover of $\tilde{\mathcal{C}}$ with function field equal to the composite of the function fields of $\mathcal{C}$ and $\mathcal{E}'$, i.e. the normalization of the fiber product of $\mathcal{C}$ with $\mathcal{E}'$. Let

$$\pi_{\mathcal{C}} : \mathcal{C}' \to \mathcal{C}$$

the projection to $\mathcal{C}$, which is a cover of degree 2. The Galois group of $\tilde{\mathcal{C}}/\mathcal{C}$ contains two transpositions. Let σ be one of them, chosen such that with $G_2 = \langle \sigma \rangle$ we get

$\mathcal{C}' := \tilde{\mathcal{C}}/G_2$. Hence, σ is contained in precisely two of the stabilizers $T_1, \ldots, T_4$ of the elements $\{1,2,3,4\}$ on which S_4 acts. Let

$$\pi_T : \tilde{\mathcal{C}} \to \mathcal{D} := \tilde{\mathcal{C}}/T$$

be the quotient map. Then $\tilde{f}$ factors over π_T as $\tilde{f} = g \circ \pi_T$, where $g : \mathcal{D} \to \mathbb{P}^1$ has $\deg(g) = 4$. Note that g is primitive (does not factor over a quadratic subcover). We can use the Hurwitz genus formula to compute the genus of $\mathcal{D}$. For this we have to determine the ramification of $\mathcal{D}/\mathbb{P}^1$ under g.

LEMMA 17. *The genus of $\mathcal{D}$ is equal to 3, and so is equal to the genus of $\mathcal{C}$.*

We are interested in the case that $\mathcal{J}(\mathcal{C})$ is simple. Then we get from Section 8 that:

PROPOSITION 9. *Let $\mathcal{J}_{\mathcal{C}}$ be a simple abelian variety and $\mathcal{D}$ be non-hyperelliptic. The pair of cover maps $(\pi_{\mathcal{C}}, \pi_T)$ from $\mathcal{C}'$ to $(\mathcal{C}, \mathcal{D})$ induces an isogeny*

$$\eta : \mathcal{J}_{\mathcal{C}} \to \mathcal{J}_{\mathcal{D}},$$

whose kernel is elementary-abelian and has degree ≤ 8.

A more detailed analysis due to E. Kani shows that the proposition is true without the assumption that $\mathcal{D}$ is non-hyperelliptic. Thus we have the following:

COROLLARY 9. *The notations are as above. Let k be equal to $\mathbb{F}_q$ and assume that $\mathcal{D}$ is non-hyperelliptic. Then the computation of the Discrete Logarithm in $\mathrm{Pic}^0_{\mathcal{C}}$ has complexity $\mathcal{O}(q)$.*

This result motivates the question whether the assumptions of the Corollary are often satisfied. Empirically, B. Smith has given a positive answer. A rigorous answer is given in [**23**]. We have already explained that by the construction of a $(2,3)$-cover as above we have found a generically finite and dominant morphism from a Hurwitz space $\mathcal{H}_\infty$ to the hyperelliptic locus in the moduli space $\mathcal{M}_3$ of curves of genus 3. Hence $\mathcal{H}_\infty$ is a scheme of dimension 5. Via the trigonal construction we have, to each hyperelliptic curve $\mathcal{C}$, found a curve $\mathcal{D}$ of genus 3 with a cover map

$$g : \mathcal{D} \to \mathbb{P}^1$$

with $\deg(g) = 4$ and the monodromy group of g equal to S_4. Moreover, a detailed study of the construction allows to determine the ramification type of g in the generic case:

There are 8 ramification points of g, exactly 4 points $P_1, \ldots, P_4$ amongst them are of type $(2,2)$ (i.e. $g^*(P_i) = 2(Q_{i,1} + Q_{i,2})$, and the other 4 ramification points are of type $(2,1,1)$. Hence $(\mathcal{D}, g)$ yields a point in a Hurwitz space $\mathcal{H}_2$ of dimension 5.

In [**23**] one discusses the hyperelliptic locus $\mathcal{H}_{hyp}$ in $\mathcal{H}_2$. The computational part of this discussion determines conditions for the coefficients of Weierstrass equations for curves $\mathcal{D}$ lying in $\mathcal{H}_{hyp}$. This is rather complicated, but one sees that generically these coefficients are parametrized by a 4-dimensional space. Rather deep and involved geometric methods have to be used to transfer these computations into scheme-theoretical results and to get

THEOREM 39. *The Hurwitz space $\mathcal{H}_{hyp}$ is a unirational, irreducible variety of dimension 4, provided that char $k > 5$. Moreover, the natural forgetful map*

$$\mu : \mathcal{H}_{hyp} \to \mathcal{M}_3$$

to the moduli space $\mathcal{M}_3$ of genus 3 curves has finite fibers and so its image is also irreducible of dimension 4.

COROLLARY 10. *Assume that k is algebraically closed. There is a 1-codimensional subscheme U of $\mathcal{M}_{3,hyp}$ such that for $\mathcal{C} \notin U$ the isogeny η maps $\mathcal{J}_\mathcal{C}$ to the Jacobian of a non-hyperelliptic curve $\mathcal{D}$.*

Replacing the algebraically closed field k by a finite field $\mathbb{F}_q$ one has to study rationality conditions for η. This is done in [**66**] and [**23**]. As result we get the following:

COROLLARY 11. *There are $\mathcal{O}(q^5)$ isomorphism classes of hyperelliptic curves of genus 3 defined over $\mathbb{F}_q$ for which the discrete logarithm in the divisor class group of degree 0 has complexity $\mathcal{O}(q)$, up to log-factors. Since $|\operatorname{Pic}^0(C)| \sim q^3$, the DL system of these hyperelliptic curves of genus 3 is weak.*

9.2. Point counting. In general, not much is known about fast point counting algorithms on curves of genus 3 (aside of the general fact that for all abelian varieties there is a polynomial time algorithm due to Pila). But as we have seen above, for applications in cryptography we have to restrict ourselves to special hyperelliptic curves (where it is not at all clear what "special" means for a concrete curve), and so we do not lose much by restricting to hyperelliptic curves $\mathcal{C}$ whose Jacobian $\mathcal{J}_\mathcal{C} =: \mathcal{J}$ has a special endomorphism ring $\mathcal{O}_\mathcal{J}$.

9.2.1. *Real multiplication.* A first possibility is to assume that $\mathcal{J}$ has real multiplication. This means that $\mathcal{O}_\mathcal{J}$ contains an order $\mathcal{R}$ of a totally real field of degree 3. An immediate consequence is that there are many isogenies at hand, and in the case of genus 2 this situation has accelerated the point counting dramatically [**30**]. So there is hope that the same could happen for Jacobians of dimension 3. Hence it is interesting to construct hyperelliptic curves $\mathcal{C}$ such that $\mathcal{J}_\mathcal{C}$ has real multiplication. In view of the results about Jacobians of the modular curves $X_0(N)$ in Section 8.2 it is natural to look for curves whose Jacobian is a quotient of $J_0(N)$ for some N. This was successfully done by H. J. Weber ([**72**]). The procedure is as follows:

First, one computes eigenspaces of dimension 3 of the space of cusp forms of level N under the Hecke operators. Using the attached differentials one can compute (over $\mathbb{C}$) the period matrix of the corresponding factor $\mathcal{J}$ of $J_0(N)$ and decides whether it is principally polarized and hence is the Jacobian of a curve $\mathcal{C}$. Using theta-null values one decides whether $\mathcal{C}$ is hyperelliptic.

If so, one can compute invariants of the curve, and (e.g by using Rosenhain models compute a Weierstrass equation (at the end over $\mathbb{Z}[1/N]$) of $\mathcal{C}$ whose Jacobian is a simple factor of $\mathcal{J}_0(N)$. Reduction modulo p gives hyperelliptic curves over $\mathbb{Z}/p$ of genus 3 with (known) real multiplication.

The method works quite well but has one disadvantage: Since there are many non-hyperelliptic curves of genus 3 with real multiplication we are not sure whether the constructed curve is isogenous to a non-hyperelliptic curve under the trigonal construction described above.

9.2.2. *Complex multiplication.* We strengthen the condition on $\operatorname{End}(\mathcal{J})$ and assume that $\mathcal{J}$ has complex multiplication and so is the reduction of a curve defined over a number field. Recall that this means that there is an embedding of $\operatorname{End}(\mathcal{J})$ as order $\mathcal{O}$ into a CM-field K, i.e. K is a totally imaginary quadratic extension of a totally real field K_0 of degree 3 over $\mathbb{Q}$.

The arithmetic of $\mathcal{C}$ and $\mathcal{J}$ is reflected by the arithmetic of orders in K. In particular, one finds the Frobenius endomorphism of reductions of $\mathcal{C}$ modulo prime ideals $\mathfrak{p}$ of K as element in $\mathcal{O}$. This solves the problem of point counting on $\mathcal{C}$ modulo $\mathfrak{p}$ immediately. Moreover, class field theory of K gives both a classification of isomorphism classes of curves $\mathcal{C}$ with CM-field K and methods to find period matrices of $\mathcal{J}$ and so equations of $\mathcal{C}$. Details and more references can be found in [**7**] sections 5.1 and 18.3.

But trying to find examples for hyperelliptic curves attached to CM fields of degree 6 one runs into trouble since these examples seem to be very rare. (Recall that the hyperelliptic locus in $\mathcal{M}_3$ has codimension 1.) This was one of the results of the thesis of A. Weng (Essen 2001). So one has to use some force: If $\mathcal{J}$ has an automorphism of order 4 the curve $\mathcal{C}$ has an automorphism of order at least 2 and if $\mathcal{J}$ is simple the quotient of $\mathcal{C}$ by this automorphism has to be $\mathbb{P}^1$, and so $\mathcal{C}$ is hyperelliptic and has an automorphism of order 4. The existence of J with automorphism φ of order 4 is obtained by a special choice of the CM-field K:

Let K_0 be a totally real field of degree 3 with class number 1 (there are many fields with these properties) and take $K = K_0(\sqrt{-1})$, and for $\mathcal{O}$ take the maximal order of K. In [**73**] one finds in detail how these choices lead to many examples of hyperelliptic curves over finite fields suitable for cryptography.

There is a bit of hope that the following question may have an affirmative answer.

QUESTION 2. *We assume now that $\mathcal{C}$ is a hyperelliptic curve with an automorphism of order* 4. *Is $\mathcal{C}$ resistant against the trigonal attack?*

If the answer would be yes and since automorphisms of degree 4 survive under isogenies of degree prime to 2 one could hope to have a positive answer to the following.

QUESTION 3. *Let $\mathcal{C}$ be a hyperelliptic curve with an automorphism of order* 4 *and with simple Jacobian variety $\mathcal{J}$. Let*

$$\eta : \mathcal{J} \to \mathcal{J}'$$

be an isogeny with $\mathcal{J}'$ principally polarized. Is $\mathcal{J}'$ the Jacobian variety of a hyperelliptic curve?

In the case of a positive answer to the question the CM-curves with CM-field K of degree 4 containing $\sqrt{-1}$ would deliver nice and easy to handle candidates for cryptographically usable DL-systems

10. Genus 2 curves and cryptography

Curves $\mathcal{C}$ of genus 2 with at least one rational Weierstrass point P_∞ are very interesting objects for creating DL-systems and in most aspects they can very well compete with elliptic curves (see Section 11). The research area around these curves is attractive since there is a lot of activity but also a lot of unsolved problems till now.

Security. The hardness of the DL in the Picard groups of randomly chosen curves over prime fields of order q is comparable with the hardness on elliptic curves over prime fields of order q^2, in particular, all known versions of index-calculus attacks have a complexity equal to $\mathcal{O}(q)$ and hence are not more efficient than generic algorithms. (Recall that because of Thm. 12 we can expect that $|\operatorname{Pic}^0_{\mathcal{C}}| \sim q^2$.)

Addition: By our assumption we can assume that $\mathcal{C}$ is given by a Weierstrass equation

$$Y^2Z^3 = f(X,Z)$$

with $f(X,Z)$ homogenous and of degree 5 in X.

Hence we can use Mumford representations of reduced divisors and the Cantor algorithm (see Section 4.3). A detailed analysis including all special cases is done in [**7**, Section 14.3.2], including a determination of complexity (see Table 14.2 and Table 14.13).

Alternatively we use the interpolation formulas given explicitly in Example 5, and we have the choice to chose coordinates taylor-made to soft- and hardware environments. As result we can state that the efficiency of group operations in $\mathrm{Pic}^0_{\mathcal{C}}$ is on the same level as it is for elliptic curves.

If we are only interested in scalar multiplication (e.g. for key exchange) we can use, as in the case of elliptic curves, a "Montgomery ladder" to compute this multiplication. The role of x-coordinates of points on elliptic curves is played by coordinates on the Kummer surface related to the Abelian surface $\mathcal{J}_{\mathcal{C}}$.

Kummer surfaces The following is due to P. Gaudry [**24**] and Gaudry/Lubicz [**31**]. We embed $\mathcal{C}$ into $J_{\mathcal{C}}$ by using P_∞ as base point, and continue the hyperelliptic involution ω of $\mathcal{C}$ to $J_{\mathcal{J}_{\mathcal{C}}}$. Then $J_{\mathcal{C}}/w =: K$ is the Kummer variety of $\mathcal{C}$, and we have an embedding of $\mathbb{P}^1 \cong \mathcal{C}/w$ into K.

On K the action of $\mathbb{Z}$ is induced by the group structure on the Jacobian. One checks that one has a scalar multiplication but no group structure (compare the case of elliptic curves). Hence the usual add-and double algorithm to get a fast scalar multiplication does not work. To repair this one uses the **Montgomery ladder** (see [**7**]) which is well known for elliptic curves.

To make the ladder very fast one uses a remarkable tool: classical modular forms in an abstract setting!

More concretely, P. Gaudry uses in [**24**] classical theory of theta functions, their p-adic interpretation and reduction, exploits "classical" doubling formulas and gets **extremely simple doubling formulas**.

One drawback is that the model used for $\mathcal{C}$ based on Theta functions has bad reduction modulo 2. So in [**24**] Gaudry had to exclude the important case that the ground field has even characteristic. More arithmetic geometry, namely the theory of minimal models enabled him together with D. Lubicz to remove this restriction [**31**].

The third necessary aspect important for the construction of DL-systems is point counting, which has to be so effective that in reasonable time one finds by a random search curves $\mathcal{C}$ and fields $\mathbb{F}_p$ with the property that a large prime number ℓ with size $\sim q^2$ divides $|\mathcal{J}_{\mathcal{C}}(\mathbb{F}_p)|$.

10.1. Point counting on curves of genus 2.

10.1.1. *Point counting on random curves.* A generic method to determine the order of a finite group is given by a variant of Shank's baby-step giant-step method, whose efficiency depends on the size of the interval in which on can place $|\mathcal{J}_{\mathcal{C}}(\mathbb{F}_p)|$. This is used if one knows $|\mathcal{J}_{\mathcal{C}}(\mathbb{F}_p)|$ modulo a rather big number N together with the information given by the Hasse estimate.

To get such a congruence one tries to determine the characteristic polynomial of the Frobenius endomorphism ϕ_p modulo "enough" small numbers by its action on torsion points. So a first step for counting algorithms is to determine polynomial or ideals which vanish on torsion points of a given order. This procedure was already the key part of Schoof's algorithm for elliptic curves (Section 11.2).

But for really fast algorithms for elliptic curves one needs one more ingredient: isogenies and corresponding modular polynomials respectively ideals.

For curves of genus 2 division polynomials and modular polynomials are not so well understood as for elliptic curves but as we have announced already in Example 5 this is an active area of research.

The starting point is the Mumford representation of points on $\mathcal{J}_{\mathcal{C}}(\mathbb{F}_p)$. We assume that $\mathcal{C}$ is given by $y^2 = f(x)$ as before and $D = \langle u, v \rangle$ is a reduced divisor. Most reduced divisors have weight 2, i.e the degree of u is 2. The set of those divisors with strictly lower weight is called Θ. A divisor of weight 1 i.e., with a single point $P = (x_P, y_P)$, is represented by

$$\langle u(x), v(x) \rangle = \langle x - x_P, y_P \rangle. \tag{16}$$

The unique divisor of weight 0, is the identity $\mathcal{O}$ given as $\mathcal{O} = \langle u(x), v(x) \rangle = \langle 1, 0 \rangle$. Any divisor of weight 2 is given as

$$\langle u(x), v(x) \rangle = \langle x^2 + u_1 x + u_2, v_0 x + v_1 \rangle.$$

The following algorithm using **division polynomials** is due to Gaudry and Harley.

For a divisor of weight 1, i.e. given by an ideal $P = \langle x - x_P, y_P \rangle$ in general position we have

$$[l]P = \left\langle x^2 + \frac{d_1^{(l)}(x_p)}{d_0^{(l)}(x_p)} x + \frac{d_2^{(l)}(x_p)}{d_0^{(l)}(x_p)}, y_P \left(\frac{e_1^{(l)}(x_p)}{e_0^{(l)}(x_p)} x + \frac{e_2^{(l)}(x_p)}{e_0^{(l)}(x_p)} \right) \right\rangle$$

where $e_i^{(l)}, d_i^{(l)}$ are polynomials with degrees $\deg d_i^{(l)} = 2l^2 - (i+1)$ and $\deg e_i^{(l)} = 3l^2 - (i+1)$, for $i = 0, 1, 2$; see [**29**] for details.

For a divisor of weight two, we consider it as a sum of two divisors of weight 1, say $D = P_1 + P_2$ where $P_1 = \langle x - x_1, y_1 \rangle$ and $P_2 = \langle x - y_2, y_2 \rangle$, where x_1 and x_2 are roots of $u(x)$ and $y_i = v(x_i)$ and u, v come from the Mumford presentation. Then,

$$[l]D = [l]P_1 + [l]P_2.$$

With these formulas one computes the order of $\operatorname{Jac}_{\mathcal{C}} \mathbb{F}_p$ modulo l; see [**29**, Section 5.4]. The cost of the algorithm is

$$O(l^2)M(l^2) + O(d \log q)M(l^4) + O(l^2 + \log q)M(d),$$

where $M(n)$ is the number of field operations required to multiply two polynomials of degree n, and d is the smallest degree of resultants of $u(x)$ and $v(x)$; see [**29**].

Next one tries to determine "modular equations" for finite subschemes of $\mathcal{J}_{\mathcal{C}}$. Here a paper of P. Gaudry and E. Schost is a remarkable beginning (see [**27**]. Mixing the results and methods and using many tricks Gaudry and Schost succeed in [**32**] to count points on the Jacobian of some hundreds of random curves of genus 2 and finally found one having the security level of AES 128. The development still goes further. The interested reader should have a look at the paper [**1**].

10.2. Modular curves of genus two. In Section 10.1.1 we have seen that in principle one can find cryptographically relevant curves of genus 2 by a search on random curves using an analogue of Schoof's algorithm. But the necessary input of computing capacity (and implementation art) is rather heavy.

So it may be useful to look for classes of special curves for which point counting is easier. Again work of Gaudry, together with David Kohel and Benjamin Smith shows that one can accelerate the algorithm dramatically if the Jacobian of $\mathcal{C}$ has real multiplication (see [**30**].

The new algorithm has, for large p, complexity $\widetilde{\mathcal{O}}(\log^5 q)$. This is used to compute a 256-bit prime-order Jacobian, suitable for cryptographic applications, and also the order of a 1024-bit Jacobian.

Hence it is interesting to construct curves of genus 2 with real multiplication. We describe how this can be done by using factors of Jacobians of modular curves. The basic reference is Wang [**71**].

Let N be a positive integer and $X_0(N)$ the modular curve as described in Section 5.1. Let $S_2(N)$ be the space of cusp forms of weight 2 for $\Gamma_0(N)$ and $f = \sum_{n=1}^{\infty} a_n q^n \in S_2(N)$ be a newform. This newform determines a simple abelian variety $\mathcal{A}_f$ which is a factor of $J_0(N) := \operatorname{Jac} X_0(N)$. The knowledge of the newform is equivalent with the knowledge of holomorphic differentials on the factor and this is used by Wang to compute the period matrix of $\mathcal{A}_f$ by computing the complex integrals of a symplectic basis. Moreover, he determined conditions when this period matrix Ω_f corresponds to a principally polarized abelian variety. He did this for factors of dimension ≥ 2 but we focus on the case when this variety has dimension two.

Having found the period matrix one has to construct the curve. We have mentioned this task already in before and cited [**72**].

10.2.1. *Thetanulls.* Once the period matrix Ω_f is determined, one can compute the theta functions. For any genus 2 curve we have six odd theta characteristics and ten even theta characteristics. The following are the sixteen theta characteristics, where the first ten are even and the last six are odd. For simplicity, we denote them by $\theta_i = \begin{bmatrix} a \\ b \end{bmatrix}$ instead of $\theta_i = \begin{bmatrix} a \\ b \end{bmatrix} (z, \tau)$ where $i = 1, \dots, 10$ for the even theta functions.

$$\theta_1 = \begin{bmatrix} 0 & 0 \\ 0 & 0 \end{bmatrix}, \theta_2 = \begin{bmatrix} 0 & 0 \\ \frac{1}{2} & \frac{1}{2} \end{bmatrix}, \theta_3 = \begin{bmatrix} 0 & 0 \\ \frac{1}{2} & 0 \end{bmatrix}, \theta_4 = \begin{bmatrix} 0 & 0 \\ 0 & \frac{1}{2} \end{bmatrix}, \theta_5 = \begin{bmatrix} \frac{1}{2} & 0 \\ 0 & 0 \end{bmatrix},$$

$$\theta_6 = \begin{bmatrix} \frac{1}{2} & 0 \\ 0 & \frac{1}{2} \end{bmatrix}, \theta_7 = \begin{bmatrix} 0 & \frac{1}{2} \\ 0 & 0 \end{bmatrix}, \theta_8 = \begin{bmatrix} \frac{1}{2} & \frac{1}{2} \\ 0 & 0 \end{bmatrix}, \theta_9 = \begin{bmatrix} 0 & \frac{1}{2} \\ \frac{1}{2} & 0 \end{bmatrix}, \theta_{10} = \begin{bmatrix} \frac{1}{2} & \frac{1}{2} \\ \frac{1}{2} & \frac{1}{2} \end{bmatrix},$$

and the odd theta functions correspond to the following characteristics

$$\begin{bmatrix} 0 & \frac{1}{2} \\ 0 & \frac{1}{2} \end{bmatrix}, \begin{bmatrix} 0 & \frac{1}{2} \\ \frac{1}{2} & \frac{1}{2} \end{bmatrix}, \begin{bmatrix} \frac{1}{2} & 0 \\ \frac{1}{2} & 0 \end{bmatrix}, \begin{bmatrix} \frac{1}{2} & \frac{1}{2} \\ \frac{1}{2} & 0 \end{bmatrix}, \begin{bmatrix} \frac{1}{2} & 0 \\ \frac{1}{2} & \frac{1}{2} \end{bmatrix}, \begin{bmatrix} \frac{1}{2} & \frac{1}{2} \\ 0 & \frac{1}{2} \end{bmatrix}$$

We call fundamental theta constants θ_1, θ_2, θ_3, θ_4. All the other theta constants can be expressed in terms of these four; see [**59**] for details. The classical result of Picard determines the relation between theta characteristics and branch points of a genus two curve.

LEMMA 18 (Picard). *Let a genus 2 curve be given by*

$$Y^2 = X(X-1)(X-\lambda)(X-\mu)(X-\nu). \tag{17}$$

Then, λ, μ, ν can be written as follows:

$$\lambda = \frac{\theta_1^2\theta_3^2}{\theta_2^2\theta_4^2}, \quad \mu = \frac{\theta_3^2\theta_8^2}{\theta_4^2\theta_{10}^2}, \quad \nu = \frac{\theta_1^2\theta_8^2}{\theta_2^2\theta_{10}^2}. \tag{18}$$

Such branch points were expressed in terms of the fundamental theta constants.

LEMMA 19 ([**59**]). *Every genus 2 curve $\mathcal{X}$ can be written in the form:*

$$y^2 = x\,(x-1)\left(x - \frac{\theta_1^2\theta_3^2}{\theta_2^2\theta_4^2}\right)\left(x^2 - \frac{\theta_2^2\,\theta_3^2 + \theta_1^2\,\theta_4^2}{\theta_2^2\,\theta_4^2}\cdot\alpha\,x + \frac{\theta_1^2\theta_3^2}{\theta_2^2\theta_4^2}\,\alpha^2\right),$$

where $\alpha = \frac{\theta_8^2}{\theta_{10}^2}$ and in terms of $\theta_1, \dots, \theta_4$ is given by

$$\alpha^2 + \frac{\theta_1^4 + \theta_2^4 - \theta_3^4 - \theta_4^4}{\theta_1^2\theta_2^2 - \theta_3^2\theta_4^2}\,\alpha + 1 = 0$$

Furthermore, if $\alpha = \pm 1$ then $V_4 \hookrightarrow \mathrm{Aut}(\mathcal{X})$.

From the above we have that $\theta_8^4 = \theta_{10}^4$ implies that $V_4 \hookrightarrow \mathrm{Aut}(\mathcal{X})$. in [**59**, Lemma 15] determines a necessary and equivalent statement when $V_4 \hookrightarrow \mathrm{Aut}(\mathcal{X})$ in terms of thetanulls.

The last part of the lemma above shows that if $\theta_8^4 = \theta_{10}^4$ then all coefficients of the genus 2 curve are given as rational functions of powers of these 4 fundamental theta functions. Such fundamental theta functions determine the field of moduli of the given curve. Hence, the curve is defined over its field of moduli.

Once the fundamental thetanulls are computed, the arithmetic invariants J_2, J_4, J_6, J_{10} can be computed via formulas given in [**36**].

Till now the computations were made over $\mathbb{R}$ with large enough precision. Now we use the $\mathbb{Z}$-structure of $X_0(N)$ to identify the invariants with integers. Reducing modulo primes one finds invariants of curves defined over finite fields. Then by [**51**] we compute an equation of the genus 2 curves over a minimal field of definition for these invariants and so over a finite field.

REMARK 10. *In* [**71**] *there are used absolute invariants instead of the above arithmetic invariants. Moreover, the case of curves with automorphism group of order > 2 doesn't seem to have been considered. Nevertheless, it seems as this was not a problem for $N \leq 200$, which seems to suggest that no such genus two curves appear for such N.*

10.3. CM curves. We further specialize and want to use curves of genus 2 whose Jacobian has complex multiplication. We shall use class field theory and the theory of Taniyama-Shimura of CM-fields to find such curves over number fields.By reduction we find curves with CM over finite fields, and again class field theory of CM-fields reduces point counting modulo p to the computation of the trace of an element in the CM-field with norm p.

Choose a squarefree integer $d \in \mathbb{N}$ such that $K_0 := \mathbb{Q}(\sqrt{d})$ has class number one. Let $\alpha = a + b\sqrt{d}$ be squarefree and $\alpha > 0$. Thus $K = K_0(i\sqrt{\alpha})$ is a CM field of degree 4. We choose d and α such that $K/\mathbb{Q}$ is Galois with group V_4 (i.e. Klein four-group). Since $[K : \mathbb{Q}] = 4$ and K is CM field we have four distinct embeddings φ_i, $i = 1, \dots, 4$ of K into $\mathbb{C}$. A tuple $(K, \Phi) = (K, \{\varphi_1, \varphi_2\})$ is called CM-type. For an ideal $I \subset \mathcal{O}_K$ we define

$$\Phi(I) = \{(\varphi_1(x), \varphi_2(x))^t, x \in I\}.$$

Then $\mathbb{C}^2/\Phi(I)$ is an Abelian variety with complex multiplication by $\mathcal{O}_K$. Conversely every abelian variety $\mathcal{A}$ of CM-type (K, Φ) with complex multiplication by $\mathcal{O}_K$ is isomorphic to an abelian variety $\mathcal{A}_{I,\Phi}$; see Shimura-Taniyama (1961) [**64**].

The period matrix of $\mathcal{A}_{I,\Phi}$ lies in the Siegel upper half plane $\mathbb{H}_2$ and therefore we can equip $\mathcal{A}_{I,\Phi}$ with a principal polarization determined by an element $\gamma \in K$.

10.3.1. *Class polynomials.* For elliptic curves with complex multiplication by $\mathcal{O}_K$ the j-invariant lies in the Hilbert class field of the imaginary quadratic field K. The case of $g = 1$ is simpler due to the fact that the *reflex CM-field* $\hat{K}$ is equal to K (see [**63**]), which is not true for higher genus. The following is mostly due to A. Weng [**74**].

THEOREM 40. *Let K be a CM-field such that $[K : \mathbb{Q}] = 4$.*

i) *For every genus 2 curve $\mathcal{C}$ with CM-type by $\mathcal{O}_K$, the absolute invariants $\mathbf{x}_1, \mathbf{x}_2, \mathbf{x}_3$ are algebraic numbers that lie in a class field over the reflex CM-field $\hat{K}$.*
ii) *For two genus 2 curves $\mathcal{C}$ and $\mathcal{C}'$ with CM with $\mathcal{O}_K$ we have that $\mathbf{x}_i(\mathcal{C})$ and $\mathbf{x}_i(\mathcal{C}')$, for $i = 1, 2, 3$, are Galois conjugates.*
iii) *Let $\{\mathcal{C}_1, \mathcal{C}_2, \ldots, \mathcal{C}_s\}$ be a set of representatives of isomorphism classes of genus 2 curves whose Jacobians have CM with endomorphism ring $\mathcal{O}_K$. Denote by $\mathbf{x}_i^{(j)}$, the i-th absolute invariant of $\mathcal{C}_j$. The polynomials*

$$H_{K,i}(X) := \prod_{j=1}^{s} (X - \mathbf{x}_i^{(j)}),$$

for $i = 1, 2, 3$, have coefficients in $\mathbb{Q}$.

Polynomials $H_{K,1}$, $H_{K,2}$, $H_{K,3}$, are called the **class polynomials**.

THEOREM 41. *Let K be a CM-field of degree 4 and $p \geq 7$ a prime which does not divide the denominators of the class polynomials $H_i(X) := H_{K,i}(X)$, $i = 1, 2, 3$. Then the following hold:*

- *For all $w \in \mathcal{O}_K$ with $w\bar{w} = p$, $H_i(X)$ have a linear factor over $\mathbb{F}_p$ corresponding to w.*
- *For each $\alpha \in \mathbb{F}_p$ there are two $\mathbb{F}_p$-isomorphism classes $\mathcal{A}_{p,1}$ and $\mathcal{A}_{p,2}$ of principally polarized abelian varieties over $\mathbb{F}_p$ with absolute invariants $\mathbf{x}_i = \alpha$, for $i = 1, 2, 3$.*
- *The principally polarized abelian varieties $\mathcal{A}_{p,1}$ and $\mathcal{A}_{p,2}$ have CM by $\mathcal{O}_K$.*
- *The number of $\mathbb{F}_p$-rational points of $\mathcal{A}_{p,j}$, $j = 1, 2$, is given by*

$$\prod_{r=1}^{4} \left(1 + (-1)^j w_r\right)$$

- *The equation $w\bar{w} = p$ for $w \in \mathcal{O}_K$ has (up to conjugacy and sign) at most two different solutions. Hence, for every CM-field of degree 4 there are at most four different possible orders of groups of $\mathbb{F}_p$-rational points of principally polarized abelian varieties defined over $\mathbb{F}_p$ with CM by $\mathcal{O}_K$.*

Once we compute the class polynomials $H_{K,i}$ we can reduce them module p (for large enough p) and get $H_{K,i}(X) \mod p$. The roots of $H_{K,i}(X) \mod p$ are the absolute invariants of genus 2 curves $\mathcal{C}$ modulo p. Now that we know invariants of the curve we can determine its equation as in [**51**]. Then the reduced curve is defined over $\mathbb{F}_p$ or a quadratic extension.

For example, if we are in the first case of the above theorem, say we find elements $w_1, \bar{w}_1 \in \mathcal{O}_K$ such that $w_1\bar{w}_1 = p$ then there exists at most one second solution (up to conjugation) such that $w_2\bar{w}_2 = p$. We set $W := \{\pm w_1, \pm w_2\}$. Then the order of $\mathrm{Jac}(\mathcal{C} \mod p)$, over $\mathbb{F}_p$, is $\{\chi_w(1) \mid w \in W\}$, where $\chi_w(T)$ is the characteristic polynomial of w.

By using the CM-method for curves of genus 2 one gets a very efficient way to construct cryptographically strong DL-systems as extensive tables have shown in the thesis of A. Weng.

One could hope to use these results not only for DL-systems but also for isogeny graphs of Jacobians of dimension 2 and it could be worthwhile to investigate whether they could be used in the way we shall see in Section 11 below.

11. Elliptic curve cryptography

Finally we come to the most interesting and well-understood case. We shall use isogenies between elliptic curves and their computation quite often, and so we begin, for the convenience of the reader, with a fundamental result of J. Vélu; see [**70**].

PROPOSITION 10 (Vélu's formula). *Let E be an elliptic curve, defined over a field k, with equation*

$$E : y^2 = x^3 + ax + b$$

and $G \subset E(\bar{k})$ be a finite subgroup invariant under G_k. The separable isogeny $\phi : E \to E/G$, of kernel G, can be written as follows:

For any $P(x, y)$ we get

$$\phi(P) = \left(x + \sum_{Q \in G \setminus \{\mathcal{O}\}} x(P+Q) - x(Q),\; y + \sum_{Q \in G \setminus \{\mathcal{O}\}} y(P+Q) - y(Q) \right) \tag{19}$$

and the curve E/G has equation $y^2 = x^3 + a'x + b'$, where

$$a' = a - 5 \sum_{Q \in G \setminus \{\mathcal{O}\}} (3x(Q)^2 + a),$$

$$b' = b - 7 \sum_{Q \in G \setminus \{\mathcal{O}\}} (5x(Q)^3 + 3ax(Q) + b).$$

Thus, knowing a finite subgroup G of E we can explicitly construct the corresponding isogeny $E \to E' := E/G$.

11.1. Endomorphism ring of E.

The following results are mostly due to **M. Deuring** and mainly contained in the beautiful paper [**15**].

DEFINITION 42. Let $\mathcal{E}$ be an elliptic curve over k. $\mathcal{E}$ is ordinary if and only if $\mathrm{End}(\mathcal{E})$ is commutative. $\mathcal{E}$ is supersingular if and only if $\mathrm{End}(\mathcal{E})$ is not commutative.

THEOREM 43 (Deuring). *Let $\mathcal{E}$ be an elliptic curve defined over a field k. The following hold:*

i) *If* char(k) = 0, *then $\mathcal{E}$ is ordinary and*
   - $\mathrm{End}_{\bar{k}}(\mathcal{E}) = \mathbb{Z}$ *(generic case) or* $\mathrm{End}_{\bar{k}}(\mathcal{E})$ *is an order* $O_{\mathcal{E}} \subset \mathbb{Q}(\sqrt{-d_{\mathcal{E}}})$, $d_{\mathcal{E}} > 0$ *(CM-case).*

- *Take $\mathcal{E}$ with CM with order $O_{\mathcal{E}}$. Let $S_{\mathcal{E}}$ be the set of k-isomorphism classes of elliptic curves with endomorphism ring $O_{\mathcal{E}}$. Then* $\mathrm{Pic}(O_{\mathcal{E}})$ *acts in a natural and simply transitive way on $S_{\mathcal{E}}$, hence $S_{\mathcal{E}}$ is a principally homogeneous space with translation group* $\mathrm{Pic}(O_{\mathcal{E}})$*: For $c \in \mathrm{Pic}(O_{\mathcal{E}})$, $\mathfrak{A} \in c$ and $\mathbb{C}/O_{\mathcal{E}} = \mathcal{E}_0$ we get $c \cdot [\mathcal{E}_0]$ is the class of $\mathbb{C}/\mathfrak{A}$.*

ii) ***(Deuring's lifting theorem)*** *Let $\mathcal{E}$ be an elliptic curve over $\mathbb{F}_q$ which is ordinary over $\overline{\mathbb{F}_q}$. Then there is, up to $\mathbb{C}$-isomorphisms, exactly one elliptic curve $\tilde{\mathcal{E}}$ with CM over a number field K such that*
- *there is a prime $\mathfrak{p}$ of K with $\tilde{\mathcal{E}}_{\mathfrak{p}} \cong \mathcal{E}$ with $\tilde{\mathcal{E}}_{\mathfrak{p}}$ the reduction of $\tilde{\mathcal{E}}$ modulo $\mathfrak{p}$, and*
- $\mathrm{End}(\tilde{\mathcal{E}}) = \mathrm{End}(\mathcal{E})_{\mathfrak{p}} = O_{\mathcal{E}}$, *with $O_{\mathcal{E}}$ an order in an imaginary quadratic field.*

iii) *If $\mathcal{E}$ is supersingular, then*
- *up to twists, all supersingular elliptic curves in characteristic p are defined over $\mathbb{F}_{p^2}$, i.e. their j-invariant lies in $\mathbb{F}_{p^2}$.*
- $|\mathcal{E}(\mathbb{F}_{p^2})| = (p \pm 1)^2$, *and the sign depends on the twist class of $\mathcal{E}$.*
- $\mathrm{End}_{\overline{\mathbb{F}_p}}(\mathcal{E})$ *is a maximal order in the quaternion algebra $\mathbb{Q}_p$, which is unramified outside of ∞ and p.*

We remark that the endomorphism ring of an elliptic curve over a finite field $\mathbb{F}_q$ is never equal to $\mathbb{Z}$ since there is the Frobenius endomorphism $\phi_{\mathbb{F}_q,\mathcal{E}}$ induced by the Frobenius automorphism of $\mathbb{F}_q$ which has degree q. We give a first application of the lifting theorem.

COROLLARY 12 (Hasse). *Let $\mathcal{E}$ be an ordinary elliptic curve over $\mathbb{F}_q$. Then the Frobenius endomorphism $\phi_{\mathbb{F}_q,\mathcal{E}}$ is an integer in an imaginary quadratic field with norm q, and hence has a minimal polynomial*

$$\chi_{\mathcal{E},q}(T) = T^2 - \mathrm{tr}(\phi_{\mathbb{F}_q,\mathcal{E}}) \cdot T + q$$

with

$$|(\mathrm{tr}(\phi_{\mathbb{F}_q,\mathcal{E}})^2 - 4q| < 0.$$

Recall that the number of $\mathbb{F}_q$-rational points of $\mathcal{E}$ is

$$|\mathcal{E}(\mathbb{F}_q)| =: n_{\mathbb{F}_q,\mathcal{E}} = \chi_{\mathcal{E},q}(1).$$

COROLLARY 13. $|n_{\mathbb{F}_q,\mathcal{E}} - q - 1| < 2\sqrt{q}$.

Using the result iii) in Thm. 43 and the observation that the eigenvalues of $\phi_{\mathbb{F}_{q^d},\mathcal{E}}$ are the d-th power of the eigenvalues of $\phi_{\mathbb{F}_q,\mathcal{E}}$ we get that

$$|n_{\mathbb{F}_q,\mathcal{E}} - q - 1| \leq 2\sqrt{q}$$

for all elliptic curves of $\mathbb{F}_q$. This is the *Hasse bound* for elliptic curves, a special case of the Hasse-Weil bound for points on Jacobian varieties over finite fields (Thm. 12).

11.2. Point counting. Corollary 13 is the key fact for a polynomial time algorithm for computing the order of $\mathcal{E}(\mathbb{F}_q)$ for elliptic curves $\mathcal{E}$ defined over the field $\mathbb{F}_q$, which is called **Schoof's Algorithm** .

The idea is to compute $\chi_{\mathcal{E},q}(T) \mod n$ for small numbers n by computing the action of $\phi_{\mathbb{F}_q,\mathcal{E}}$ on $\mathcal{E}[n]$ (take for instance $n = \ell$ as small prime number or $n = 2^k$ with k small) and then to use CRT and the Hasse bound for the trace of $\phi_{\mathbb{F}_q,\mathcal{E}}$ to determine $\chi_{\mathcal{E},q}(T)$. To do this use the classical n-division polynomials Ψ_n.

The disadvantage is that $\deg(\Psi_n) \sim n^2/2$ and therefore the Schoof algorithm is too slow.

The way out of this problem is to use étale isogenies with cyclic kernel of order n and the fact (see Section 5) that we can interpret these isogenies with the help of points on an explicitly known curve, namely the modular curve $X_0(n)$. An explicit equation for an affine model of $X_0(N)$ is given by the classical modular polynomial $\phi(j, j_N)$. It allows an effective computation of isogenies (as functions including the determination of the image curve) at least if n is of moderate size).

THEOREM 44 (Vélu, Couveignes, Lercier, Elkies, Kohel, and many other contributors:). *The cost for the computation of an isogeny of degree ℓ of an elliptic curve $\mathcal{E}$ over $\mathbb{F}_q$ is*

$$\mathcal{O}(\ell^2 + \ell \log(\ell) \log(q)).$$

The **Idea of Atkin-Elkies** is: Use *étale isogenies* of small degree of $\mathcal{E}$ instead of points, and use the modular polynomial ϕ_n of degree $\sim n$. The resulting *Schoof-Atkin-Elkies algorithm* is very fast, in particular if one assumes as "standard conjecture" the generalized Riemann hypothesis (GRH).

COROLLARY 14 (SAE). *$|\mathcal{E}(\mathbb{F}_q)|$ can be computed (probabilistically, with GRH) with complexity $\mathcal{O}((\log q)^4)$. Therefore we can construct, for primes p sufficiently large, (many) elliptic curves with $|\mathcal{E}(\mathbb{F}_p)| = k \cdot \ell$ with k small (e.g. $k = 1$ if we want) and ℓ a prime so large that (using classical computers and according to our best knowledge) the security level of the discrete logarithm in $\mathcal{E}(\mathbb{F}_p)$ is matching AES 128 (or larger).*

11.3. Looking for post-quantum security. As we have seen in Section 11.2 we can construct elliptic curves over prime fields such that the resulting DL-systems are secure under the known attacks. But the situation changes totally if we allow algorithms based on quantum computers. We shall discuss now how we can use isogenies of elliptic curves (and maybe, of curves of larger genus with convenient endomorphism rings) to find key exchange schemes staying in the frame of Diffie-Hellman type protocols as described in Section 6.

11.3.1. *The isogeny graph.*

DEFINITION 45. Let $\mathcal{E}$ be an elliptic curve over $\mathbb{F}_q$. The **isogeny graph** of $\mathcal{E}$ is a graph where nodes are j-invariants of elliptic curves isogenous to $\mathcal{E}$ and edges are isogenies between the curves attached to the nodes.

For some applications one restricts the degree of the isogenies defining edges. Isogeny graphs of ordinary elliptic curves are discussed by using the lifting theorem of Deuring, identifying these graphs with graphs coming from ideal classes of orders in imaginary quadratic fields , and then using analytic number theory and properties of modular forms. The study of graphs of supersingular elliptic curves uses properties of maximal orders of quaternion algebras. In both cases one gets the following result due to D. Jao, S. D. Miller and R. Vekatesan [**12**, Prop. 2.1].

THEOREM 46. *The isogeny graph of $\mathcal{E}$ is a Ramanujan graph.*

A first application, also to be found in [**12**] is the following:

COROLLARY 15. *Assume the $\mathcal{E}, \mathcal{E}'$ are elliptic curves over $\mathbb{F}_q$ with* $\mathrm{End}(\mathcal{E}) = \mathrm{End}(\mathcal{E}') = \mathcal{O}$. *Then there exists a subexponential algorithm which relates the DLP in $\mathcal{E}(\mathbb{F}_q)$ to the DLP in $\mathcal{E}'(\mathbb{F}_q)$.*

Hence we cannot weaken the DLP by applying isogenies between elliptic curves with the same order.

11.3.2. *The system of Couveignes-Stolbunov.* A second application of isogeny graphs is constructive. We sketch in the following work of Stolbunov [**68**] and Couveignes [**9**]. We use the results of Thm. 43 for an ordinary elliptic curve $\mathcal{E}_0$ over $\mathbb{F}_q$ with ring of endomorphism $\text{End}(\mathcal{E}_0) = O$, which is an order in a quadratic imaginary field.

In analogy to the notation Theorem 43 define $\mathcal{S}_{E_0}$ as set of isomorphism classes of elliptic curves over $\overline{\mathbb{F}}_q$ with ring of endomorphisms O. Then $\mathcal{S}_{E_0}$ is a $\text{Pic}(O)$-set. Hence, we can use it for *Key Exchange protocols* as in Section 6: *The partner P chooses $c \in \text{Pic}(O)$ and publishes the j-invariant of $c \cdot E_0$.*

The exchange is not as fast as DL-systems since we cannot use a *double-and add*-algorithm but it is feasible since one finds enough isogenies that are composites of isogenies of small degree (smoothness); for an example see [**68**]. The **security** depends on the hardness of the following problem:

PROBLEM 1. *Find an isogeny between two given isogenous elliptic curves.*

The following gives an idea of the running time for the solution to this problem.

PROPOSITION 11 (Kohel, Galbraith, Hess, Smart et al.). *The expected number of **bit**-operations for the computation of an isogeny between ordinary elliptic curves over $\mathbb{F}_q$ with endomorphism ring O_{K_E} is*

$$\mathcal{O}(q^{1/4+o(1)} \log^2(q) \log\log(q)).$$

But recall: We are in the situation where an abelian group is acting on a set, and so there is a subexponential algorithm to solve the hidden-shift problem. This means that we can only expect *subexponential* security in the Q-bit world for the key exchange scheme; see results of Childs, Jao, Soukharev in [**6**]. Comparing this with the situation we have nowadays with respect to the widely tolerated RSA-system this may be not so disastrous.

11.3.3. *The key exchange system of De Feo.* The suggestion is to use supersingular elliptic curves over $\mathbb{F}_{p^2}$ and their properties also stated in Thm. 43. Take

$$p = r^a \cdot s^b \cdot f - 1$$

with $p \equiv 1 \mod 4$. Then

$$E_0 : Y^2 Z = X^3 + XZ^2$$

is a supersingular elliptic curve over $\mathbb{F}_{p^2}$. We describe the key exchange scheme invented and implemented by De Feo, Jao and Plût [**12**] in the frame we have introduced in Section 6.

The objects in the categories $\mathcal{C}_i \quad (i = 1, 2)$ have as **objects** the isomorphism classes of supersingular curves E over $\mathbb{F}_{p^2}$ isogenous to $\mathcal{E}_0$ and hence with

$$|E(\mathbb{F}_{p^2})| = (r^a \cdot s^b \cdot f)^2.$$

The **morphisms in $\mathcal{C}_1$** are isogenies φ with $|\ker(\varphi)|$ dividing r^a.

The **morphisms in $\mathcal{C}_2$** are isogenies ψ with $|\ker(\psi)|$ dividing s^b.

For these categories pushouts exist. For additional information choose P_1, P_2 of order r^a and Q_1, Q_2 of order s^b in $\mathcal{E}_0(\mathbb{F}_{p^2})$.

Key exchange:

- The Partner $\mathcal{P}_1$ chooses $n_1, n_2 \in \mathbb{Z}/r^a$ and the isogeny
$$\eta : \mathcal{E}_0 \to \mathcal{E}_0/\langle n_1 P_1 + n_2 P_2 \rangle =: \mathcal{E}_1.$$
- The Partner $\mathcal{P}_2$ chooses $m_1, m_2 \in \mathbb{Z}/s^b$ and computes the isogeny
$$\psi : \mathcal{E}_0 \to \mathcal{E}_0/\langle m_1 Q_1 + m_2 Q_2 \rangle =: \mathcal{E}_2.$$
- $\mathcal{P}_2$ sends $(\mathcal{E}_2, \psi(P_1), \psi(P_2))$.
- $\mathcal{P}_1$ can compute the common secret, the pushout of η and ψ as
$$\mathcal{E}_3 := \mathcal{E}_2/\langle n_1\psi(P_1) + n_2\psi(P_2)\rangle.$$

An analogous procedure enables $\mathcal{P}_2$ to compute the isomorphy class of $\mathcal{E}_3$, which is the common secret of the partners.

Again **security** depends on the hardness to compute an isogeny of two elliptic curves, but now the two elliptic curves are supersingular.

State of the art: The best known algorithms have exponential complexity $p^{1/4}$ (bit-computer) resp. $p^{1/6}$ (quantum computer), and so one can hope that a prime p with 768 bit yields AES128 security level. So we have, compared with other post-quantum suggestions for key exchange schemes, a very small key size.

In contrast to the ordinary case the groups around like the class groups of left ideals in maximal orders **are not abelian**, and so the hidden shift problem is not solved till now in subexponential time.

References

[1] Pierre-Jean Abelard Simon ; Gaudry Pierrick; Spaenlehauer, *Improved complexity bounds for counting points on hyperelliptic curves*, Foundations of Computational Mathematics (2018), to appear.

[2] L. Beshaj and S. Guest, *Weighted projective space of binary sextics*, Algebraic curves and cryptography, 2018.

[3] Reinier Bröker, Everett W. Howe, Kristin E. Lauter, and Peter Stevenhagen, *Genus-2 curves and jacobians with a given number of points* (2014), available at 1403.6911.

[4] Reinier Bröker and Kristin Lauter, *Modular polynomials for genus 2*, LMS J. Comput. Math. **12** (2009), 326–339, DOI 10.1112/S1461157000001546. MR2570930

[5] Gabriel Cardona and Jordi Quer, *Field of moduli and field of definition for curves of genus 2*, Computational aspects of algebraic curves, Lecture Notes Ser. Comput., vol. 13, World Sci. Publ., Hackensack, NJ, 2005, pp. 71–83, DOI 10.1142/9789812701640_0006. MR2181874

[6] Andrew Childs, David Jao, and Vladimir Soukharev, *Constructing elliptic curve isogenies in quantum subexponential time*, J. Math. Cryptol. **8** (2014), no. 1, 1–29, DOI 10.1515/jmc-2012-0016. MR3163097

[7] Henri Cohen, Gerhard Frey, Roberto Avanzi, Christophe Doche, Tanja Lange, Kim Nguyen, and Frederik Vercauteren (eds.), *Handbook of elliptic and hyperelliptic curve cryptography*, Discrete Mathematics and its Applications (Boca Raton), Chapman & Hall/CRC, Boca Raton, FL, 2006. MR2162716

[8] Marc Coppens and Gerriet Martens, *Linear series on 4-gonal curves*, Math. Nachr. **213** (2000), 35–55, DOI 10.1002/(SICI)1522-2616(200005)213:1⟨35::AID-MANA35⟩3.3.CO;2-Q. MR1755245

[9] Jean-Marc Couveignes, *Hard homogeneous spaces*, 2006. Cryptology e-Print Arxive.

[10] Luca De Feo, *Fast algorithms for computing isogenies between ordinary elliptic curves in small characteristic*, J. Number Theory **131** (2011), no. 5, 873–893, DOI 10.1016/j.jnt.2010.07.003. MR2772477

[11] Luca De Feo, Cyril Hugounenq, Jérôme Plût, and Éric Schost, *Explicit isogenies in quadratic time in any characteristic*, LMS J. Comput. Math. **19** (2016), no. suppl. A, 267–282, DOI 10.1112/S146115701600036X. MR3540960

[12] Luca De Feo, David Jao, and Jérôme Plût, *Towards quantum-resistant cryptosystems from supersingular elliptic curve isogenies*, J. Math. Cryptol. **8** (2014), no. 3, 209–247, DOI 10.1515/jmc-2012-0015. MR3259113

[13] P. Deligne and D. Mumford, *The irreducibility of the space of curves of given genus*, Inst. Hautes Études Sci. Publ. Math. **36** (1969), 75–109. MR0262240

[14] P. Deligne and M. Rapoport, *Les schémas de modules de courbes elliptiques* (French), Modular functions of one variable, II (Proc. Internat. Summer School, Univ. Antwerp, Antwerp, 1972), Springer, Berlin, 1973, pp. 143–316. Lecture Notes in Math., Vol. 349. MR0337993

[15] Max Deuring, *Die Typen der Multiplikatorenringe elliptischer Funktionenkörper* (German), Abh. Math. Sem. Hansischen Univ. **14** (1941), 197–272, DOI 10.1007/BF02940746. MR0005125

[16] C. Diem, *On arithmetic and the discrete logarithm problem in class groups of curves*, Ph.D. Thesis, 2008. Habilitationsschrift.

[17] Whitfield Diffie and Martin E. Hellman, *New directions in cryptography*, IEEE Trans. Information Theory **IT-22** (1976), no. 6, 644–654, DOI 10.1109/tit.1976.1055638. MR0437208

[18] Ron Donagi and Ron Livné, *The arithmetic-geometric mean and isogenies for curves of higher genus*, Ann. Scuola Norm. Sup. Pisa Cl. Sci. (4) **28** (1999), no. 2, 323–339. MR1736231

[19] Andreas Enge and Pierrick Gaudry, *A general framework for subexponential discrete logarithm algorithms*, Acta Arith. **102** (2002), no. 1, 83–103, DOI 10.4064/aa102-1-6. MR1884958

[20] Stéphane Flon, Roger Oyono, and Christophe Ritzenthaler, *Fast addition on non-hyperelliptic genus 3 curves*, Algebraic geometry and its applications, 2008, pp. 1–28. MR2484046

[21] Gerhard Frey, *Isogenies in theory and praxis*, Open problems in mathematics and computational science, Springer, Cham, 2014, pp. 37–68. MR3330877

[22] Gerhard Frey and Ernst Kani, *Correspondences on hyperelliptic curves and applications to the discrete logarithm*, 2011, pp. 1–19.

[23] Gerhard Frey and Ernst Kani, *Normal forms of hyperelliptic curves of genus 3*, Des. Codes Cryptogr. **77** (2015), no. 2-3, 677–712, DOI 10.1007/s10623-015-0122-2. MR3403171

[24] P. Gaudry, *Fast genus 2 arithmetic based on theta functions*, J. Math. Cryptol. **1** (2007), no. 3, 243–265, DOI 10.1515/JMC.2007.012. MR2372155

[25] P. Gaudry, F. Hess, and N. P. Smart, *Constructive and destructive facets of Weil descent on elliptic curves*, J. Cryptology **15** (2002), no. 1, 19–46, DOI 10.1007/s00145-001-0011-x. MR1880933

[26] Pierrick Gaudry, David Kohel, and Benjamin Smith, *Counting points on genus 2 curves with real multiplication*, Advances in cryptology—ASIACRYPT 2011, Lecture Notes in Comput. Sci., vol. 7073, Springer, Heidelberg, 2011, pp. 504–519, DOI 10.1007/978-3-642-25385-0_27. MR2935020

[27] P. Gaudry and É. Schost, *Modular equations for hyperelliptic curves*, Math. Comp. **74** (2005), no. 249, 429–454, DOI 10.1090/S0025-5718-04-01682-5. MR2085901

[28] P. Gaudry, E. Thomé, N. Thériault, and C. Diem, *A double large prime variation for small genus hyperelliptic index calculus*, Math. Comp. **76** (2007), no. 257, 475–492, DOI 10.1090/S0025-5718-06-01900-4. MR2261032

[29] Pierrick Gaudry and Robert Harley, *Counting points on hyperelliptic curves over finite fields*, Algorithmic number theory (Leiden, 2000), 2000, pp. 313–332. MR1850614

[30] Pierrick Gaudry, David Kohel, and Benjamin Smith, *Counting points on genus 2 curves with real multiplication*, Advances in cryptology—ASIACRYPT 2011, Lecture Notes in Comput. Sci., vol. 7073, Springer, Heidelberg, 2011, pp. 504–519, DOI 10.1007/978-3-642-25385-0_27. MR2935020

[31] Pierrick Gaudry and David Lubicz, *The arithmetic of characteristic 2 Kummer surfaces and of elliptic Kummer lines*, Finite Fields Appl. **15** (2009), no. 2, 246–260, DOI 10.1016/j.ffa.2008.12.006. MR2494393

[32] Pierrick Gaudry and Éric Schost, *Genus 2 point counting over prime fields*, J. Symbolic Comput. **47** (2012), no. 4, 368–400, DOI 10.1016/j.jsc.2011.09.003. MR2890878

[33] Ki-ichiro Hashimoto and Naoki Murabayashi, *Shimura curves as intersections of Humbert surfaces and defining equations of QM-curves of genus two*, Tohoku Math. J. (2) **47** (1995), no. 2, 271–296, DOI 10.2748/tmj/1178225596. MR1329525

[34] F. Hess, *Computing Riemann-Roch spaces in algebraic function fields and related topics*, J. Symbolic Comput. **33** (2002), no. 4, 425–445, DOI 10.1006/jsco.2001.0513. MR1890579

[35] Jun-ichi Igusa, *Arithmetic variety of moduli for genus two*, Ann. of Math. (2) **72** (1960), 612–649, DOI 10.2307/1970233. MR0114819

[36] Jun-ichi Igusa, *On Siegel modular forms of genus two*, Amer. J. Math. **84** (1962), 175–200, DOI 10.2307/2372812. MR0141643

[37] Antoine Joux, Andrew Odlyzko, and Cécile Pierrot, *The past, evolving present, and future of the discrete logarithm*, Open problems in mathematics and computational science, Springer, Cham, 2014, pp. 5–36. MR3330876

[38] Naoki Kanayama, *Division polynomials and multiplication formulae of Jacobian varieties of dimension 2*, Math. Proc. Cambridge Philos. Soc. **139** (2005), no. 3, 399–409, DOI 10.1017/S0305004105008765. MR2177167

[39] Naoki Kanayama, *Corrections to "Division polynomials and multiplication formulae in dimension 2" [MR2177167]*, Math. Proc. Cambridge Philos. Soc. **149** (2010), no. 1, 189–192, DOI 10.1017/S0305004110000113. MR2651585

[40] Nicholas M. Katz and Barry Mazur, *Arithmetic moduli of elliptic curves*, Annals of Mathematics Studies, vol. 108, Princeton University Press, Princeton, NJ, 1985. MR772569

[41] Chandrashekhar Khare and Jean-Pierre Wintenberger, *Serre's modularity conjecture. II*, Invent. Math. **178** (2009), no. 3, 505–586, DOI 10.1007/s00222-009-0206-6. MR2551764

[42] Mark Kisin, *Modularity of 2-adic Barsotti-Tate representations*, Invent. Math. **178** (2009), no. 3, 587–634, DOI 10.1007/s00222-009-0207-5. MR2551765

[43] M. Kraitchik, *Introduction à la théorie des nombres* (French), Gauthier-Villars, Paris, 1952. MR0051845

[44] Greg Kuperberg, *A subexponential-time quantum algorithm for the dihedral hidden subgroup problem*, SIAM J. Comput. **35** (2005), no. 1, 170–188, DOI 10.1137/S0097539703436345. MR2178804

[45] Serge Lang, *Abelian varieties*, Springer-Verlag, New York-Berlin, 1983. Reprint of the 1959 original. MR713430

[46] Frank Leitenberger, *About the group law for the Jacobi variety of a hyperelliptic curve*, Beiträge Algebra Geom. **46** (2005), no. 1, 125–130. MR2146447

[47] Davide Lombardo, *Computing the geometric endomorphism ring of a genus-2 Jacobian*, Math. Comp. **88** (2019), no. 316, 889–929, DOI 10.1090/mcom/3358. MR3882288

[48] David Lubicz and Damien Robert, *Computing isogenies between abelian varieties*, Compos. Math. **148** (2012), no. 5, 1483–1515, DOI 10.1112/S0010437X12000243. MR2982438

[49] David Lubicz and Damien Robert, *Computing separable isogenies in quasi-optimal time*, LMS J. Comput. Math. **18** (2015), no. 1, 198–216, DOI 10.1112/S146115701400045X. MR3349315

[50] K. Magaard, T. Shaska, S. Shpectorov, and H. Völklein, *The locus of curves with prescribed automorphism group*, Sūrikaisekikenkyūsho Kōkyūroku **1267** (2002), 112–141. Communications in arithmetic fundamental groups (Kyoto, 1999/2001). MR1954371

[51] Andreas Malmendier and Tony Shaska, *A universal genus-two curve from Siegel modular forms*, SIGMA Symmetry Integrability Geom. Methods Appl. **13** (2017), Paper No. 089, 17, DOI 10.3842/SIGMA.2017.089. MR3731039

[52] J. Mandili and T. Shaska, *Computation of heights on weighted projective spaces*, Algebraic curves and cryptography, 2018.

[53] B. Mazur, *Modular curves and the Eisenstein ideal*, Inst. Hautes Études Sci. Publ. Math. **47** (1977), 33–186 (1978). MR488287

[54] Loï c Merel, *Universal Fourier expansions of modular forms*, On Artin's conjecture for odd 2-dimensional representations, 1994, pp. 59–94. MR1322319

[55] Jean-François Mestre, *Construction de courbes de genre* 2 *à partir de leurs modules* (French), Effective methods in algebraic geometry (Castiglioncello, 1990), Progr. Math., vol. 94, Birkhäuser Boston, Boston, MA, 1991, pp. 313–334. MR1106431

[56] David Mumford, *Abelian varieties*, Tata Institute of Fundamental Research Studies in Mathematics, vol. 5, Published for the Tata Institute of Fundamental Research, Bombay; by Hindustan Book Agency, New Delhi, 2008. With appendices by C. P. Ramanujam and Yuri Manin; Corrected reprint of the second (1974) edition. MR2514037

[57] Yoshihiro Ônishi, *Determinant expressions for hyperelliptic functions*, Proc. Edinb. Math. Soc. (2) **48** (2005), no. 3, 705–742, DOI 10.1017/S0013091503000695. With an appendix by Shigeki Matsutani. MR2171194

[58] Frans Oort, *Endomorphism algebras of abelian varieties*, Algebraic geometry and commutative algebra, Vol. II, Kinokuniya, Tokyo, 1988, pp. 469–502. MR977774

[59] E. Previato, T. Shaska, and G. S. Wijesiri, *Thetanulls of cyclic curves of small genus*, Albanian J. Math. **1** (2007), no. 4, 253–270. MR2367218

[60] Oded Regev, *New lattice-based cryptographic constructions*, J. ACM **51** (2004), no. 6, 899–942, DOI 10.1145/1039488.1039490. MR2145258

[61] Tanush Shaska and Jennifer L. Thompson, *On the generic curve of genus 3*, Affine algebraic geometry, Contemp. Math., vol. 369, Amer. Math. Soc., Providence, RI, 2005, pp. 233–243, DOI 10.1090/conm/369/06814. MR2126664

[62] Tony Shaska, *Genus 2 curves with* $(3,3)$*-split Jacobian and large automorphism group*, Algorithmic number theory (Sydney, 2002), 2002, pp. 205–218. MR2041085

[63] Goro Shimura, *Abelian varieties with complex multiplication and modular functions*, Princeton Mathematical Series, vol. 46, Princeton University Press, Princeton, NJ, 1998. MR1492449

[64] Goro Shimura and Yutaka Taniyama, *Complex multiplication of abelian varieties and its applications to number theory*, Publications of the Mathematical Society of Japan, vol. 6, The Mathematical Society of Japan, Tokyo, 1961. MR0125113

[65] Joseph H. Silverman, *The arithmetic of elliptic curves*, Graduate Texts in Mathematics, vol. 106, Springer-Verlag, New York, 1992. Corrected reprint of the 1986 original. MR1329092

[66] Benjamin Smith, *Computing low-degree isogenies in genus 2 with the Dolgachev-Lehavi method*, Arithmetic, geometry, cryptography and coding theory, Contemp. Math., vol. 574, Amer. Math. Soc., Providence, RI, 2012, pp. 159–170, DOI 10.1090/conm/574/11418. MR2961408

[67] Henning Stichtenoth, *Algebraic function fields and codes*, 2nd ed., Graduate Texts in Mathematics, vol. 254, Springer-Verlag, Berlin, 2009. MR2464941

[68] Anton Stolbunov, *Constructing public-key cryptographic schemes based on class group action on a set of isogenous elliptic curves*, Adv. Math. Commun. **4** (2010), no. 2, 215–235, DOI 10.3934/amc.2010.4.215. MR2654134

[69] John Tate, *Endomorphisms of abelian varieties over finite fields*, Invent. Math. **2** (1966), 134–144, DOI 10.1007/BF01404549. MR0206004

[70] Jacques Vélu, *Isogénies entre courbes elliptiques* (French), C. R. Acad. Sci. Paris Sér. A-B **273** (1971), A238–A241. MR0294345

[71] Xiang Dong Wang, *2-dimensional simple factors of* $J_0(N)$, Manuscripta Math. **87** (1995), no. 2, 179–197, DOI 10.1007/BF02570470. MR1334940

[72] Hermann-Josef Weber, *Hyperelliptic simple factors of* $J_0(N)$ *with dimension at least* 3, Experiment. Math. **6** (1997), no. 4, 273–287. MR1606908

[73] Annegret Weng, *A class of hyperelliptic CM-curves of genus three*, J. Ramanujan Math. Soc. **16** (2001), no. 4, 339–372. MR1877806

[74] Annegret Weng, *Constructing hyperelliptic curves of genus 2 suitable for cryptography*, Math. Comp. **72** (2003), no. 241, 435–458, DOI 10.1090/S0025-5718-02-01422-9. MR1933830

[75] Yuri G. Zarhin, *Families of absolutely simple hyperelliptic Jacobians*, Proc. Lond. Math. Soc. (3) **100** (2010), no. 1, 24–54, DOI 10.1112/plms/pdp020. MR2578467

[76] Yuri G. Zarhin, *Endomorphism algebras of superelliptic Jacobians*, Geometric methods in algebra and number theory, Progr. Math., vol. 235, Birkhäuser Boston, Boston, MA, 2005, pp. 339–362, DOI 10.1007/0-8176-4417-2_15. MR2166091

Institut für Experimentelle Mathematik, Universität Duisburg-Essen, 45326 Essen, Germany

Email address: frey@iem.uni-due.de

Department of Mathematics and Statistics, Oakland University, Rochester, Michigan 48309

Email address: shaska@oakland.edu

Selected Published Titles in This Series

724 **Lubjana Beshaj and Tony Shaska, Editors,** Algebraic Curves and Their Applications, 2019

720 **Alexandre Girouard, Editor,** Spectral Theory and Applications, 2018

719 **Florian Sobieczky, Editor,** Unimodularity in Randomly Generated Graphs, 2018

718 **David Ayala, Daniel S. Freed, and Ryan E. Grady, Editors,** Topology and Quantum Theory in Interaction, 2018

717 **Federico Bonetto, David Borthwick, Evans Harrell, and Michael Loss, Editors,** Mathematical Problems in Quantum Physics, 2018

716 **Alex Martsinkovsky, Kiyoshi Igusa, and Gordana Todorov, Editors,** Surveys in Representation Theory of Algebras, 2018

715 **Sergio R. López-Permouth, Jae Keol Park, S. Tariq Rizvi, and Cosmin S. Roman, Editors,** Advances in Rings and Modules, 2018

714 **Jens Gerlach Christensen, Susanna Dann, and Matthew Dawson, Editors,** Representation Theory and Harmonic Analysis on Symmetric Spaces, 2018

713 **Naihuan Jing and Kailash C. Misra, Editors,** Representations of Lie Algebras, Quantum Groups and Related Topics, 2018

712 **Nero Budur, Tommaso de Fernex, Roi Docampo, and Kevin Tucker, Editors,** Local and Global Methods in Algebraic Geometry, 2018

711 **Thomas Creutzig and Andrew R. Linshaw, Editors,** Vertex Algebras and Geometry, 2018

710 **Raphaël Danchin, Reinhard Farwig, Jiří Neustupa, and Patrick Penel, Editors,** Mathematical Analysis in Fluid Mechanics, 2018

709 **Fernando Galaz-García, Juan Carlos Pardo Millán, and Pedro Solórzano, Editors,** Contributions of Mexican Mathematicians Abroad in Pure and Applied Mathematics, 2018

708 **Christian Ausoni, Kathryn Hess, Brenda Johnson, Ieke Moerdijk, and Jérôme Scherer, Editors,** An Alpine Bouquet of Algebraic Topology, 2018

707 **Nitya Kitchloo, Mona Merling, Jack Morava, Emily Riehl, and W. Stephen Wilson, Editors,** New Directions in Homotopy Theory, 2018

706 **Yeonhyang Kim, Sivaram K. Narayan, Gabriel Picioroaga, and Eric S. Weber, Editors,** Frames and Harmonic Analysis, 2018

705 **Graham J. Leuschke, Frauke Bleher, Ralf Schiffler, and Dan Zacharia, Editors,** Representations of Algebras, 2018

704 **Alain Escassut, Cristina Perez-Garcia, and Khodr Shamseddine, Editors,** Advances in Ultrametric Analysis, 2018

703 **Andreas Malmendier and Tony Shaska, Editors,** Higher Genus Curves in Mathematical Physics and Arithmetic Geometry, 2018

702 **Mark Grant, Gregory Lupton, and Lucile Vandembroucq, Editors,** Topological Complexity and Related Topics, 2018

701 **Joan-Carles Lario and V. Kumar Murty, Editors,** Number Theory Related to Modular Curves, 2018

700 **Alexandre Girouard, Dmitry Jakobson, Michael Levitin, Nilima Nigam, Iosif Polterovich, and Frédéric Rochon, Editors,** Geometric and Computational Spectral Theory, 2017

699 **Mark L. Agranovsky, Matania Ben-Artzi, Catherine Bénéteau, Lavi Karp, Dmitry Khavinson, Simeon Reich, David Shoikhet, Gilbert Weinstein, and Lawrence Zalcman, Editors,** Complex Analysis and Dynamical Systems VII, 2017